Probability and Statistics for Computer Scientists

Third Edition

Probability and Statistics for Computer Scientists

Third Edition

Michael Baron

Department of Mathematics and Statistics
College of Arts and Sciences
American University
Washington DC

CRC Press
Taylor & Francis Group
Boca Raton London New York

CRC Press is an imprint of the
Taylor & Francis Group, an **informa** business

A CHAPMAN & HALL BOOK

CRC Press
Taylor & Francis Group
6000 Broken Sound Parkway NW, Suite 300
Boca Raton, FL 33487-2742

© 2019 by Taylor & Francis Group, LLC
CRC Press is an imprint of Taylor & Francis Group, an Informa business

No claim to original U.S. Government works

Printed on acid-free paper

International Standard Book Number-13: 978-1-138-04448-7 (Hardback)

Library of Congress Cataloging-in-Publication Data

Names: Baron, Michael, 1968- author.
Title: Probability and statistics for computer scientists / Michael Baron.
Description: Third edition. | Boca Raton : CRC Press, Taylor & Francis Group, 2019. | Includes index.
Identifiers: LCCN 2019006958 | ISBN 9781138044487 (hardback : alk. paper)
Subjects: LCSH: Probabilities--Textbooks. | Mathematical statistics--Textbooks. | Probabilities--Computer simulation--Textbooks. | Mathematical statistics--Computer simulation--Textbooks.
Classification: LCC QA273 .B2575 2019 | DDC 519.201/13--dc23
LC record available at https://lccn.loc.gov/2019006958

Visit the Taylor & Francis Web site at
http://www.taylorandfrancis.com

and the CRC Press Web site at
http://www.crcpress.com

To my parents – Genrietta and Izrael-Vulf Baron

Contents

9 Statistical Inference I   243

Preface

Starting with the fundamentals of probability, this text leads readers to computer simulations and Monte Carlo methods, stochastic processes and Markov chains, queuing systems, statistical inference, and regression. These areas are heavily used in modern computer science, computer engineering, software engineering, and related fields.

For whom this book is written

The book is primarily intended for junior undergraduate to beginning graduate level students majoring in computer-related fields – computer science, software engineering, information systems, data science, information technology, telecommunications, etc. At the same time, it can be used by electrical engineering, mathematics, statistics, natural science, and other majors for a standard calculus-based introductory statistics course. Standard topics in probability and statistics are covered in Chapters 1–4 and 8–9.

Graduate students can use this book to prepare for probability-based courses such as queuing theory, artificial neural networks, computer performance, etc.

The book can also be used as a standard reference on probability and statistical methods, simulation, and modeling tools.

Recommended courses

The text is recommended for a one-semester course with several open-end options available. At the same time, with the new material added in the second and the third editions, the book can serve as a text for a full two-semester course in Probability and Statistics.

After introducing probability and distributions in Chapters 1–4, instructors may choose the following continuations, see Figure 1.

Probability-oriented course. Proceed to Chapters 6–7 for Stochastic Processes, Markov Chains, and Queuing Theory. Computer science majors will find it attractive to supplement such a course with computer simulations and Monte Carlo methods. Students can learn and practice general simulation techniques in Chapter 5, then advance to the simulation of stochastic processes and rather complex queuing systems in Sections 6.4 and 7.6. Chapter 5 is highly recommended but not required for the rest of the material.

Statistics-focused course. Proceed to Chapters 8–9 directly after the probability core, followed by additional topics in Statistics selected from Chapters 10 and 11. Such a curriculum is more standard, and it is suitable for a wide range of majors. Chapter 5 remains optional but recommended; it discusses statistical methods based on computer simulations. Modern bootstrap techniques in Section 10.3 will attractively continue this discussion.

A course satisfying ABET requirements. Topics covered in this book satisfy ABET (Accreditation Board for Engineering and Technology) requirements for probability and statistics.

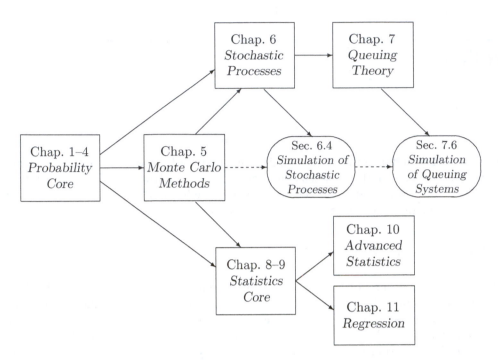

FIGURE 1: *Flow-chart of chapters.*

To meet the requirements, instructors should choose topics from Chapters 1–11. All or some of Chapters 5–7 and 10–11 may be considered optional, depending on the program's ABET objectives.

A *two-semester course* will cover all Chapters 1–11, possibly skipping some sections. The material presented in this book splits evenly between Probability topics for the first semester (Chapters 1–7) and Statistics topics for the second semester (Chapters 8–11).

Prerequisites, and use of the appendix

Working *differentiation and integration skills* are required starting from Chapter 4. They are usually covered in one semester of university calculus.

As a refresher, the appendix has a very brief summary of the minimum calculus techniques required for reading this book (Section A.4). Certainly, this section cannot be used to *learn* calculus "from scratch". It only serves as a reference and student aid.

Next, Chapters 6–7 and Sections 11.3–11.4 rely on very basic matrix computations. Essentially, readers should be able to multiply matrices, solve linear systems (Chapters 6–7), and compute inverse matrices (Section 11.3). A basic refresher of these skills with some examples is in the appendix, Section A.5.

Style and motivation

The book is written in a lively style and reasonably simple language that students find easy to read and understand. Reading this book, students should feel as if an experienced and enthusiastic lecturer is addressing them in person.

Besides computer science applications and multiple motivating examples, the book contains related interesting facts, paradoxical statements, wide applications to other fields, etc. I expect prepared students to enjoy the course, benefit from it, and find it attractive and useful for their careers.

Every chapter contains multiple examples with explicit solutions, many of them motivated by computer science applications. Every chapter is concluded with a short summary and more exercises for homework assignments and self-training. Over 270 problems can be assigned from this book.

Computers, data, demos, illustrations, R, and MATLAB®

Frequent self-explaining figures help readers understand and visualize concepts, formulas, and even some proofs. Moreover, instructors and students are invited to use included short programs for *computer demonstrations*. Randomness, uncertainty, behavior of random variables and stochastic processes, convergence results such as the Central Limit Theorem, and especially Monte Carlo simulations can be nicely visualized by animated graphics.

These short computer codes contain very basic and simple commands, written in **R** and **MATLAB** with detailed commentary. Preliminary knowledge of these languages is not necessary. Readers can also choose another software and reproduce the given commands in it line by line or use them as a *flow-chart*.

Instructors have options of teaching the course with R, MATLAB, both of them, some other software, or with no software at all.

Having understood computerized examples in the text, students will use similar codes to complete the projects and mini-projects proposed in this book.

For educational purposes, data sets used in the book are not large. They are printed in the book, typically at the first place where they are used, and also, placed in our data inventory on the web site `http://fs2.american.edu/baron/www/Book/`. Students can either type them as part of their computer programs, or they can download the data files given in text and comma-separated value formats. All data sets are listed in Section A.1, where we also teach how to read them into R and MATLAB.

Second edition and advanced statistics topics

Broad feedback coming from professors who use this book for their courses in different countries motivated me to work on the second edition. As a result, the Statistical Inference chapter expanded and split into Chapters 9 and 10. The added material is organized in the new sections, according to Table 0.1.

Also, the 2nd edition has about 60 additional exercises. Enjoy practicing, dear students, you will only benefit from extra training!

Third edition and R

The main news in the 3rd edition is the use of **R**, a popular software for statistical computing. MATLAB has fulfilled its mission in earlier editions as a tool for computer demonstrations, simulations, animated graphics, and basic statistical methods with easy-to-read

New sections in the 2nd edition	
Axioms of probability	2.2.1
Standard errors of estimates	8.2.5
Estimation of standard errors	9.1.3

Inference about variances
    Estimation, confidence intervals, and hypothesis testing for
    the population variance and for the ratio of variances.
    Chi-square distribution and F-distribution.
	9.5

Chi-square tests
    Testing distributions and families of distributions;
    goodness of fit tests; contingency tables.
	10.1

Nonparametric statistics
    Sign test; Wilcoxon signed rank test;
    Mann–Whitney–Wilcoxon rank sum test.
	10.2

Bootstrap
    Estimating properties of estimators;
    Bootstrap confidence intervals.
	10.3

TABLE 0.1: *New material in the 2nd edition.*

codes. At the same time, expansion of Statistics chapters, adoption of this book in a number of universities across four continents, and broad feedback from course instructors (thank you, colleagues!) prompted me to add R examples parallel to MATLAB.

R, an unbelievably popular and still developing statistical software, is freely available for a variety of operating systems for anyone to install from the site https://www.r-project.org/. The same site contains links to support, news, and other information about R. Supplementary to basic R, one can invoke numerous additional packages written by different users for various statistical methods. Some of them will be used in this book.

Thanks and acknowledgments

The author is grateful to Taylor & Francis Group for their constant professional help, responsiveness, support, and encouragement. Special thanks are due David Grubbs, Bob Stern, Marcus Fontaine, Jill Jurgensen, Barbara Johnson, Rachael Panthier, and Shashi Kumar. Many thanks go to my colleagues at American University, the University of Texas at Dallas, and other universities for their inspiring support and invaluable feedback, especially to Professors Stephen Casey, Betty Malloy, and Nathalie Japkowicz from American University, Farid Khafizov and Pankaj Choudhary from UT-Dallas, Joan Staniswalis and Amy Wagler from UT-El Paso, Lillian Cassel from Villanova, Alan Sprague from the University of Alabama, Katherine Merrill from the University of Vermont, Alessandro Di Bucchianico from Eindhoven University of Technology, Marc Aerts from Hasselt University, Pratik Shah from the Indian Institute of Information Technology in Vadodara, and Magagula Vusi Mpendulo from the University of Eswatini. I am grateful to Elena Baron for creative illustrations; Kate Pechekhonova for interesting examples; and last but not least, to Eric, Anthony, Masha, and Natasha Baron for their amazing patience and understanding.

MATLAB® is a registered trademark of The MathWorks, Inc. For product information please contact:

The MathWorks, Inc.
3 Apple Hill Drive
Natick, MA, 01760-2098 USA
Tel: 508-647-7000
Fax: 508-647-7001
E-mail: info@mathworks.com
Web: www.mathworks.com

Chapter 1

Introduction and Overview

1.1 Making decisions under uncertainty

This course is about uncertainty, measuring and quantifying uncertainty, and making decisions under uncertainty. Loosely speaking, by *uncertainty* we mean the condition when results, outcomes, the nearest and remote future are not completely determined; their development depends on a number of factors and just on a pure chance.

Simple examples of uncertainty appear when you buy a lottery ticket, turn a wheel of fortune, or toss a coin to make a choice.

Uncertainly appears in virtually all areas of *Computer Science* and *Software Engineering*. Installation of software requires uncertain time and often uncertain disk space. A newly released software contains an uncertain number of defects. When a computer program is executed, the amount of required memory may be uncertain. When a job is sent to a printer, it takes uncertain time to print, and there is always a different number of jobs in a queue ahead of it. Electronic components fail at uncertain times, and the order of their failures cannot be predicted exactly. Viruses attack a system at unpredictable times and affect an unpredictable number of files and directories.

Uncertainty surrounds us in *everyday life*, at home, at work, in business, and in leisure. To take a snapshot, let us listen to the evening news.

Example 1.1. We may find out that the stock market had several ups and downs today which were caused by new contracts being made, financial reports being released, and other events of this sort. Many turns of stock prices remained unexplained. Clearly, nobody would have ever lost a cent in stock trading had the market contained no uncertainty.

We may find out that a launch of a space shuttle was postponed because of weather conditions. Why did not they know it in advance, when the event was scheduled? Forecasting weather precisely, with no error, is not a solvable problem, again, due to uncertainty.

To support these words, a meteorologist predicts, say, a 60% chance of rain. Why cannot she let us know exactly whether it will rain or not, so we'll know whether or not to take our umbrellas? Yes, because of uncertainty. Because she cannot always know the situation with future precipitation for sure.

We may find out that eruption of an active volcano has suddenly started, and it is not clear which regions will have to evacuate.

We may find out that a heavily favored home team unexpectedly lost to an outsider, and a young tennis player won against expectations. Existence and popularity of totalizators, where participants place bets on sports results, show that uncertainty enters sports, results of each game, and even the final standing.

We may also hear reports of traffic accidents, crimes, and convictions. Of course, if that driver knew about the coming accident ahead of time, he would have stayed home.   ◇

Certainly, this list can be continued (at least one thing is certain!). Even when you drive to your college tomorrow, you will see an unpredictable number of green lights when you approach them, you will find an uncertain number of vacant parking slots, you will reach the classroom at an uncertain time, and you cannot be certain now about the number of classmates you will find in the classroom when you enter it.

Realizing that many important phenomena around us bear uncertainty, we have to understand it and deal with it. Most of the time, we are forced *to make decisions under uncertainty*. For example, we have to deal with internet and e-mail knowing that we may not be protected against all kinds of viruses. New software has to be released even if its testing probably did not reveal all the defects. Some memory or disk quota has to be allocated for each customer by servers, internet service providers, etc., without knowing exactly what portion of users will be satisfied with these limitations. And so on.

This book is about measuring and dealing with *uncertainty* and *randomness*. Through basic theory and numerous examples, it teaches

- how to evaluate *probabilities*, or chances of different results (when the exact result is uncertain),

- how to select a suitable *model* for a phenomenon containing uncertainty and use it in subsequent decision making,

- how to evaluate performance characteristics and other important *parameters* for new devices and servers,

- how to make optimal decisions under uncertainty.

Summary and conclusion

Uncertainty is a condition when the situation cannot be predetermined or predicted for sure with no error. Uncertainty exists in computer science, software engineering, in many aspects of science, business, and our everyday life. It is an objective reality, and one has to be able to deal with it. We are forced to make decisions under uncertainty.

1.2 Overview of this book

The next chapter introduces a language that we'll use to describe and quantify uncertainty. It is a language of *Probability*. When outcomes are uncertain, one can identify more likely and less likely ones and assign, respectively, high and low probabilities to them. Probabilities are numbers between 0 and 1, with 0 being assigned to an *impossible event* and 1 being the probability of an event that occurs *for sure*.

Next, using the introduced language, we shall discuss *random variables* as quantities that depend on chance. They assume different values with different probabilities. Due to uncertainty, an exact value of a random variable cannot be computed before this variable is actually observed or measured. Then, the best way to describe its behavior is to list all its *possible values* along with the corresponding probabilities.

Such a collection of probabilities is called a *distribution*. Amazingly, many different phenomena of seemingly unrelated nature can be described by the same distribution or by the same *family of distributions*. This allows a rather general approach to the entire class of situations involving uncertainty. As an application, it will be possible to compute probabilities of interest, once a suitable family of distributions is found. Chapters 3 and 4 introduce families of distributions that are most commonly used in computer science and other fields.

In modern practice, however, one often deals with rather complicated random phenomena where computation of probabilities and other quantities of interest is far from being straightforward. In such situations, we will make use of *Monte Carlo methods* (Chapter 5). Instead of direct computation, we shall learn methods of *simulation* or *generation* of random variables. If we are able to write a computer code for simulation of a certain phenomenon, we can immediately put it in a loop and simulate such a phenomenon thousands or millions of times and simply count how many times our event of interest occurred. This is how we shall distinguish more likely and less likely events. We can then *estimate* probability of an event by computing a proportion of simulations that led to the occurrence of this event.

As a step up to the next level, we shall realize that many random variables depend not only on a chance but also on *time*. That is, they evolve and develop in time while being random at each particular moment. Examples include the number of concurrent users, the number of jobs in a queue, the system's available capacity, intensity of internet traffic, stock prices, air temperature, etc. A random variable that depends on time is called a *stochastic process*. In Chapter 6, we study some commonly used types of stochastic processes and use these models to compute probabilities of events and other quantities of interest.

An important application of virtually all the material acquired so far lies in *queuing systems* (Chapter 7). These are systems of one or several servers performing certain tasks and serving jobs or customers. There is a lot of uncertainty in such systems. Customers arrive at unpredictable times, spend random time waiting in a queue, get assigned to a server, spend random time receiving service, and depart (Figure 1.1). In simple cases, we shall use our methods of computing probabilities and analyzing stochastic processes to compute such important characteristics of a queuing system as the utilization of a server, average waiting time of customers, average response time (from arrival until departure), average number of jobs in the system at any time, or the proportion of time the server is idle. This is extremely important for planning purposes. Performance characteristics can be recalculated for the next year, when, say, the number of customers is anticipated to increase by 5%. As a result, we'll know whether the system will remain satisfactory or will require an upgrade.

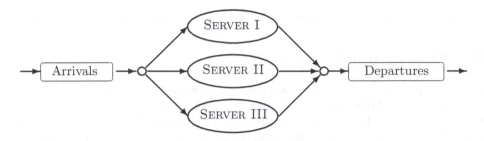

FIGURE 1.1: *A queuing system with 3 servers.*

When direct computation is too complicated, resource consuming, too approximate, or simply not feasible, we shall use Monte Carlo methods. The book contains standard examples of computer codes simulating rather complex queuing systems and evaluating their vital characteristics. The codes are written in R and MATLAB, with detailed explanations of steps, and most of them can be directly translated to other computer languages.

Next, we turn to *Statistical Inference*. While in Probability, we usually deal with more or less clearly described situations (models), in Statistics, all the analysis is based on collected and observed data. Given the data, a suitable model (say, a family of distributions) is fitted, its parameters are estimated, and conclusions are drawn concerning the entire totality of observed and unobserved subjects of interest that should follow the same model.

A typical Probability problem sounds like this.

Example 1.2. A folder contains 50 executable files. When a computer virus or a hacker attacks the system, each file is affected with probability 0.2. Compute the probability that during a virus attack, more than 15 files get affected. ◊

Notice that the situation is rather clearly described, in terms of the total number of files and the chance of affecting each file. The only uncertain quantity is the number of affected files, which cannot be predicted for sure.

A typical Statistics problem sounds like this.

Example 1.3. A folder contains 50 executable files. When a computer virus or a hacker attacks the system, each file is affected with probability p. It has been observed that during a virus attack, 15 files got affected. Estimate p. Is there a strong indication that p is greater than 0.2? ◊

This is a practical situation. A user only knows the objectively observed data: the number of files in the folder and the number of files that got affected. Based on that, he needs to estimate p, the proportion of *all* the files, including the ones in his system and any similar systems. One may provide a point estimator of p, a real number, or may opt to construct a *confidence interval* of "most probable" values of p. Similarly, a meteorologist may predict, say, a temperature of 70°F, which, realistically, does not exclude a possibility of 69 or 72 degrees, or she may give us an interval by promising, say, between 68 and 72 degrees.

Most forecasts are being made from a carefully and suitably chosen model that fits the data. A widely used method is *regression* that utilizes the observed data to find a mathematical form of relationship between two variables (Chapter 11). One variable is called *predictor*, the other is *response*. When the relationship between them is established, one can use the predictor to infer about the response. For example, one can more or less accurately estimate the average installation time of a software given the size of its executable files. An even more accurate inference about the response can be made based on *several predictors* such as the size of executable files, amount of random access memory (RAM), and type of processor and operating system. This type of data analysis will require *multivariate regression*.

Each method will be illustrated by numerous practical examples and exercises. As the ultimate target, by the end of this course, students should be able to read a word problem or a corporate report, realize the uncertainty involved in the described situation, select a suitable probability model, estimate and test its parameters based on real data, compute probabilities of interesting events and other vital characteristics, make meaningful conclusions and forecasts, and explain these results to other people.

Summary and conclusions

In this course, uncertainty is measured and described on a language of Probability. Using this language, we shall study random variables and stochastic processes and learn the most commonly used types of distributions. In particular, we shall be able to find a suitable stochastic model for the described situation and use it to compute probabilities and other quantities of interest. When direct computation is not feasible, Monte Carlo methods will be used based on a random number generator. We shall then learn how to make decisions under uncertainty based on the observed data, how to estimate parameters of interest, test hypotheses, fit regression models, and make forecasts.

Exercises

1.1. List 20 situations involving uncertainty that happened with you yesterday.

1.2. Name 10 random variables that you observed or dealt with yesterday.

1.3. Name 5 stochastic processes that played a role in your actions yesterday.

1.4. In a famous joke, a rather lazy student tosses a coin in order to decide what to do next. If it turns up heads, play a computer game. If tails, watch a video. If it stands on its edge, do the homework. If it hangs in the air, study for an exam.

  (a) Which events should be assigned probability 0, probability 1, and some probability strictly between 0 and 1?

(b) What probability between 0 and 1 would you assign to the event "watch a video", and how does it help you to define "a fair coin"?

1.5. A new software package is being tested by specialists. Every day, a number of defects is found and corrected. It is planned to release this software in 30 days. Is it possible to predict how many defects per day specialists will be finding at the time of the release? What data should be collected for this purpose, what is the predictor, and what is the response?

1.6. Mr. Cheap plans to open a computer store to trade hardware. He would like to stock an optimal number of different hardware products in order to optimize his monthly profit. Data are available on similar computer stores opened in the area. What kind of data should Mr. Cheap collect in order to predict his monthly profit? What should he use as a predictor and as a response in his analysis?

Part I

Probability and Random Variables

Chapter 2

Probability

This chapter introduces the key concept of *probability*, its fundamental rules and properties, and discusses most basic methods of computing probabilities of various events.

2.1 Events and their probabilities

The concept of *probability* perfectly agrees with our intuition. In everyday life, probability of an event is understood as a *chance* that this event will happen.

Example 2.1. If a fair coin is tossed, we say that it has a 50-50 (equal) chance of turning up heads or tails. Hence, the probability of each side equals $1/2$. It does not mean that a coin tossed 10 times will always produce exactly 5 heads and 5 tails. If you don't believe, try it! However, if you toss a coin 1 million times, the proportion of heads is anticipated to be very close to $1/2$.

$\Diamond$

This example suggests that in a long run, probability of an event can be viewed as a *proportion* of times this event happens, or its *relative frequency*. In forecasting, it is common to speak about the probability as a *likelihood* (say, the company's profit is likely to rise during the next quarter). In gambling and lottery, probability is equivalent to *odds*. Having the winning odds of 1 to 99 (1:99) means that the probability to win is 0.01, and the probability to lose is 0.99. It also means, on a relative-frequency language, that if you play long enough, you will win about 1% of the time.

Example 2.2. If there are 5 communication channels in service, and a channel is selected at random when a telephone call is placed, then each channel has a probability $1/5 = 0.2$ of being selected. ◊

Example 2.3. Two competing software companies are after an important contract. Company A is twice as likely to win this competition as company B. Hence, the probability to win the contract equals 2/3 for A and 1/3 for B. ◊

A mathematical definition of probability will be given in Section 2.2.1, after we get acquainted with a few fundamental concepts.

2.1.1 Outcomes, events, and the sample space

Probabilities arise when one considers and weighs possible results of some *experiment*. Some results are more likely than others. An experiment may be as simple as a coin toss, or as complex as starting a new business.

DEFINITION 2.1

> A collection of all elementary results, or **outcomes** of an experiment, is called a **sample space**.

DEFINITION 2.2

> Any set of outcomes is an **event**. Thus, events are subsets of the sample space.

Example 2.4. A tossed die can produce one of 6 possible outcomes: 1 dot through 6 dots. Each outcome is an event. There are other events: observing an even number of dots, an odd number of dots, a number of dots less than 3, etc. ◊

A sample space of N possible outcomes yields 2^N possible events.

PROOF: To count all possible events, we shall see how many ways an event can be constructed. The first outcome can be included into our event or excluded, so there are two possibilities. Then, every next outcome is either included or excluded, so every time the number of possibilities doubles. Overall, we have

$$\overbrace{2 \cdot 2 \cdot \ldots \cdot 2}^{N \text{ times}} = 2^N \tag{2.1}$$

possibilities, leading to a total of 2^N possible events. ☐

Example 2.5. Consider a football game between Washington and Dallas. The sample space consists of 3 outcomes,

$$\Omega = \{ \text{ Washington wins, Dallas wins, they tie } \}$$

Combining these outcomes in all possible ways, we obtain the following $2^3 = 8$ events: Washington wins, loses, ties, gets at least a tie, gets at most a tie, no tie, gets *some* result, and gets *no result*. The event "some result" is the entire sample space Ω, and by common sense, it should have probability 1. The event "no result" is empty, it does not contain any outcomes, so its probability is 0. $\diamond$

$$
\begin{array}{rcl}
\underline{\text{NOTATION}} \quad \Omega & = & \text{sample space} \\
\varnothing & = & \text{empty event} \\
\boldsymbol{P}\{E\} & = & \text{probability of event } E
\end{array}
$$

2.1.2 Set operations

Events are *sets* of outcomes. Therefore, to learn how to compute probabilities of events, we shall discuss some *set operations*. Namely, we shall define unions, intersections, differences, and complements.

DEFINITION 2.3 ————

A **union** of events $A, B, C, \ldots$ is an event consisting of *all* the outcomes in all these events. It occurs if *any* of $A, B, C, \ldots$ occurs, and therefore, corresponds to the word "OR": A or B or C or ... (Figure 2.1a).

Diagrams like Figure 2.1, where events are represented by circles, are called *Venn diagrams*.

DEFINITION 2.4 ————

An **intersection** of events $A, B, C, \ldots$ is an event consisting of outcomes that are *common* in all these events. It occurs if *each* $A, B, C, \ldots$ occurs, and therefore, corresponds to the word "AND": A and B and C and ... (Figure 2.1b).

DEFINITION 2.5 ————

A **complement** of an event A is an event that occurs every time when A does not occur. It consists of outcomes excluded from A, and therefore, corresponds to the word "NOT": not A (Figure 2.1c).

DEFINITION 2.6 ————

A **difference** of events A and B consists of all outcomes included in A but excluded from B. It occurs when A occurs and B does not, and corresponds to "BUT NOT": A but not B (Figure 2.1d).

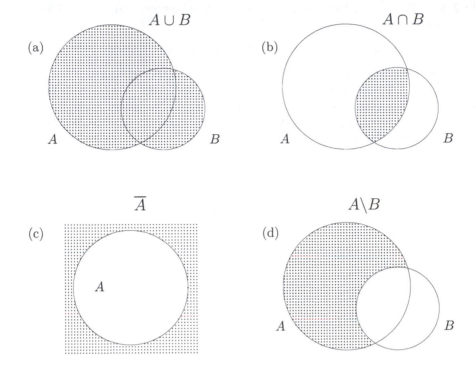

FIGURE 2.1: *Venn diagrams for (a) union, (b) intersection, (c) complement, and (d) difference of events.*

$$
\begin{array}{rcl}
\textsc{Notation} \quad \Big\| \quad A \cup B & = & \text{union} \\
A \cap B & = & \text{intersection} \\
\overline{A} \text{ or } A^c & = & \text{complement} \\
A \backslash B & = & \text{difference}
\end{array} \Big\|
$$

DEFINITION 2.7 ————

Events A and B are **disjoint** if their intersection is empty,

$$A \cap B = \varnothing.$$

Events $A_1, A_2, A_3, \ldots$ are **mutually exclusive** or **pairwise disjoint** if any two of these events are disjoint, i.e.,

$$A_i \cap A_j = \varnothing \quad \text{for any} \quad i \neq j.$$

DEFINITION 2.8 ————

Events $A, B, C, \ldots$ are **exhaustive** if their union equals the whole sample space, i.e.,

$$A \cup B \cup C \cup \ldots = \Omega.$$

Mutually exclusive events will never occur at the same time. Occurrence of any one of them eliminates the possibility for all the others to occur.

Exhaustive events cover the entire Ω, so that "there is nothing left." In other words, among any collection of exhaustive events, at least one occurs for sure.

Example 2.6. When a card is pooled from a deck at random, the four suits are at the same time disjoint and exhaustive. ◇

Example 2.7. Any event A and its complement $\overline{A}$ represent a classical example of disjoint and exhaustive events. ◇

Example 2.8. Receiving a grade of A, B, or C for some course are mutually exclusive events, but unfortunately, they are not exhaustive. ◇

As we see in the next sections, it is often easier to compute probability of an intersection than probability of a union. Taking complements converts unions into intersections, see (2.2).

$$\overline{E_1 \cup \ldots \cup E_n} = \overline{E_1} \cap \ldots \cap \overline{E_n}, \qquad \overline{E_1 \cap \ldots \cap E_n} = \overline{E_1} \cup \ldots \cup \overline{E_n} \qquad (2.2)$$

PROOF OF (2.2): Since the union $E_1 \cup \ldots \cup E_n$ represents the event "at least one event occurs," its complement has the form

$$\begin{aligned} \overline{E_1 \cup \ldots \cup E_n} &= \{ \text{ none of them occurs } \} \\ &= \{ E_1 \text{ does not occur } \cap \ldots \cap E_n \text{ does not occur } \} \\ &= \overline{E_1} \cap \ldots \cap \overline{E_n}. \end{aligned}$$

The other equality in (2.2) is left as Exercise 2.35. □

Example 2.9. Graduating with a GPA of 4.0 is an *intersection* of getting an A in *each* course. Its *complement*, graduating with a GPA below 4.0, is a *union* of receiving a grade below A *at least in one* course. ◇

Rephrasing (2.2), a complement to "nothing" is "something," and "not everything" means "at least one is missing".

2.2 Rules of Probability

Now we are ready for the rigorous definition of probability. All the rules and principles of computing probabilities of events follow from this definition.

Mathematically, *probability* is introduced through several axioms.

2.2.1   Axioms of Probability

First, we choose a *sigma-algebra* $\mathfrak{M}$ of events on a sample space Ω. This is a collection of events whose probabilities we can consider in our problem.

DEFINITION 2.9

A collection $\mathfrak{M}$ of events is a **sigma-algebra** on sample space Ω if

(a) it includes the sample space,

$$\Omega \in \mathfrak{M}$$

(b) every event in $\mathfrak{M}$ is contained along with its complement; that is,

$$E \in \mathfrak{M} \;\Rightarrow\; \overline{E} \in \mathfrak{M}$$

(c) every finite or countable collection of events in $\mathfrak{M}$ is contained along with its union; that is,

$$E_1, E_2, \ldots \in \mathfrak{M} \;\Rightarrow\; E_1 \cup E_2 \cup \ldots \in \mathfrak{M}.$$

Here are a few examples of sigma-algebras.

Example 2.10 (DEGENERATE SIGMA-ALGEBRA). By conditions (a) and (b) in Definition 2.9, every sigma-algebra has to contain the sample space Ω and the empty event $\varnothing$. This minimal collection

$$\mathfrak{M} = \{\Omega, \varnothing\}$$

forms a sigma-algebra that is called *degenerate*. $\diamond$

Example 2.11 (POWER SET). On the other extreme, what is the richest sigma-algebra on a sample space Ω? It is the collection of *all* the events,

$$\mathfrak{M} = 2^\Omega = \{E, \; E \subset \Omega\}.$$

As we know from (2.1), there are 2^N events on a sample space of N outcomes. This explains the notation 2^Ω. This sigma-algebra is called a *power set*. $\diamond$

Example 2.12 (BOREL SIGMA-ALGEBRA). Now consider an experiment that consists of selecting a point on the real line. Then, each outcome is a point $x \in \mathbb{R}$, and the sample space is $\Omega = \mathbb{R}$. Do we want to consider a probability that the point falls in a given interval? Then define a sigma-algebra $\mathfrak{B}$ to be a collection of all the intervals, finite and infinite, open and closed, and all their finite and countable unions and intersections. This sigma-algebra is very rich, but apparently, it is much less than the power set 2^Ω. This is the *Borel sigma-algebra*, after the French mathematician Émile Borel (1871–1956). In fact, it consists of all the real sets that *have length*. $\diamond$

Axioms of Probability are in the following definition.

DEFINITION 2.10 ——————

> Assume a sample space Ω and a sigma-algebra of events $\mathfrak{M}$ on it. **Probability**
>
> $$\boldsymbol{P} \,:\, \mathfrak{M} \to [0,1]$$
>
> is a function of events with the domain $\mathfrak{M}$ and the range $[0,1]$ that satisfies the following two conditions,
>
> *(Unit measure)* The sample space has unit probability, $\boldsymbol{P}(\Omega) = 1$.
>
> *(Sigma-additivity)* For any finite or countable collection of *mutually exclusive* events $E_1, E_2, \ldots \in \mathfrak{M}$,
>
> $$\boldsymbol{P}\{E_1 \cup E_2 \cup \ldots\} = \boldsymbol{P}(E_1) + \boldsymbol{P}(E_2) + \ldots.$$

All the rules of probability are consequences from this definition.

2.2.2 Computing probabilities of events

Armed with the fundamentals of probability theory, we are now able to compute probabilities of many interesting events.

Extreme cases

A sample space Ω consists of all possible outcomes, therefore, it occurs for sure. On the contrary, an empty event $\varnothing$ never occurs. So,

$$\boldsymbol{P}\{\Omega\} = 1 \text{ and } \boldsymbol{P}\{\varnothing\} = 0. \tag{2.3}$$

PROOF: Probability of Ω is given by the definition of probability. By the same definition, $\boldsymbol{P}\{\Omega\} = \boldsymbol{P}\{\Omega \cup \varnothing\} = \boldsymbol{P}\{\Omega\} + \boldsymbol{P}\{\varnothing\}$, because Ω and $\varnothing$ are mutually exclusive. Therefore, $\boldsymbol{P}\{\varnothing\} = 0$. □

Union

Consider an event that consists of some finite or countable collection of mutually exclusive outcomes,

$$E = \{\omega_1, \omega_2, \omega_3, \ldots\}.$$

Summing probabilities of these outcomes, we obtain the probability of the entire event,

$$\boldsymbol{P}\{E\} = \sum_{\omega_k \in E} \boldsymbol{P}\{\omega_k\} = \boldsymbol{P}\{\omega_1\} + \boldsymbol{P}\{\omega_2\} + \boldsymbol{P}\{\omega_3\} \ldots$$

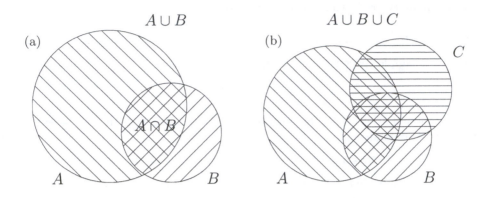

FIGURE 2.2: *(a) Union of two events. (b) Union of three events.*

Example 2.13. If a job sent to a printer appears first in line with probability 60%, and second in line with probability 30%, then with probability 90% it appears either first or second in line.   ◊

It is crucial to notice that only *mutually exclusive* events (those with empty intersections) satisfy the sigma-additivity. If events intersect, their probabilities cannot be simply added. Look at the following example.

Example 2.14. During some construction, a network blackout occurs on Monday with probability 0.7 and on Tuesday with probability 0.5. Then, does it appear on Monday *or* Tuesday with probability $0.7 + 0.5 = 1.2$? Obviously not, because probability should always be between 0 and 1! Probabilities are not additive here because blackouts on Monday and Tuesday are not mutually exclusive. In other words, it is not impossible to see blackouts on both days.   ◊

In Example 2.14, blind application of the rule for the union of mutually exclusive events clearly overestimated the actual probability. The Venn diagram shown in Figure 2.2a explains it. We see that in the sum $P\{A\} + P\{B\}$, all the common outcomes are counted *twice*. Certainly, this caused the overestimation. Each outcome should be counted only once! To correct the formula, subtract probabilities of common outcomes, which is $P\{A \cap B\}$.

$$
\begin{array}{c|l}
\textbf{Probability} & P\{A \cup B\} = P\{A\} + P\{B\} - P\{A \cap B\} \\
\textbf{of a union} & \text{For mutually exclusive events,} \\
& P\{A \cup B\} = P\{A\} + P\{B\}
\end{array}
\qquad (2.4)
$$

Generalization of this formula is not straightforward. For 3 events,

$$
\begin{aligned}
P\{A \cup B \cup C\} =\ & P\{A\} + P\{B\} + P\{C\} - P\{A \cap B\} - P\{A \cap C\} \\
& - P\{B \cap C\} + P\{A \cap B \cap C\}.
\end{aligned}
$$

As seen in Figure 2.2b, when we add probabilities of A, B, and C, each pairwise intersection is counted twice. Therefore, we subtract the probabilities of $P\{A \cap B\}$, etc. Finally, consider the triple intersection $A \cap B \cap C$. Its probability is counted 3 times within each main event, then subtracted 3 times with each pairwise intersection. Thus, it is not counted at all so far! Therefore, we add its probability $P\{A \cap B \cap C\}$ in the end.

For an arbitrary collection of events, see Exercise 2.34.

Example 2.15.   In Example 2.14, suppose there is a probability 0.35 of experiencing network blackouts on both Monday and Tuesday. Then the probability of having a blackout on Monday *or* Tuesday equals

$$0.7 + 0.5 - 0.35 = 0.85.$$

$\diamond$

Complement

Recall that events A and $\overline{A}$ are exhaustive, hence $A \cup \overline{A} = \Omega$. Also, they are disjoint, hence

$$P\{A\} + P\{\overline{A}\} = P\{A \cup \overline{A}\} = P\{\Omega\} = 1.$$

Solving this for $P\{\overline{A}\}$, we obtain a rule that perfectly agrees with the common sense,

Complement rule $\boxed{P\{\overline{A}\} = 1 - P\{A\}}$

Example 2.16.   If a system appears protected against a new computer virus with probability 0.7, then it is exposed to it with probability $1 - 0.7 = 0.3$. $\diamond$

Example 2.17.   Suppose a computer code has no errors with probability 0.45. Then, it has at least one error with probability 0.55. $\diamond$

Intersection of independent events

DEFINITION 2.11 ───

Events $E_1, \ldots, E_n$ are **independent** if they occur independently of each other, i.e., occurrence of one event does not affect the probabilities of others.

The following basic formula can serve as the criterion of independence.

Independent events $\boxed{P\{E_1 \cap \ldots \cap E_n\} = P\{E_1\} \cdot \ldots \cdot P\{E_n\}}$

We shall defer explanation of this formula until Section 2.4 which will also give a rule for intersections of *dependent* events.

2.2.3   Applications in reliability

Formulas of the previous Section are widely used in *reliability*, when one computes the probability for a system of several components to be functional.

Example 2.18 (RELIABILITY OF BACKUPS).   There is a 1% probability for a hard drive to crash. Therefore, it has two backups, each having a 2% probability to crash, and all three components are independent of each other. The stored information is lost only in an unfortunate situation when all three devices crash. What is the probability that the information is saved?

Solution.   Organize the data. Denote the events, say,

$$H = \{ \text{ hard drive crashes } \},$$

$$B_1 = \{ \text{ first backup crashes } \}, \ B_2 = \{ \text{ second backup crashes } \}.$$

It is given that H, B_1, and B_2 are independent,

$$P\{H\} = 0.01, \ \text{and} \ P\{B_1\} = P\{B_2\} = 0.02.$$

Applying rules for the complement and for the intersection of independent events,

$$
\begin{aligned}
P\{ \text{ saved } \} &= 1 - P\{ \text{ lost } \} = 1 - P\{H \cap B_1 \cap B_2\} \\
&= 1 - P\{H\}\, P\{B_1\}\, P\{B_2\} \\
&= 1 - (0.01)(0.02)(0.02) = 0.999996.
\end{aligned}
$$

(This is precisely the reason of having backups, isn't it? Without backups, the probability for information to be saved is only 0.99.)   ◊

When the system's components are connected *in parallel*, it is sufficient for at least one component to work in order for the whole system to function. Reliability of such a system is computed as in Example 2.18. Backups can always be considered as devices connected in parallel.

At the other end, consider a system whose components are connected *in sequel*. Failure of one component inevitably causes the whole system to fail. Such a system is more "vulnerable." In order to function with a high probability, it needs each component to be reliable, as in the next example.

Example 2.19.   Suppose that a shuttle's launch depends on three key devices that operate independently of each other and malfunction with probabilities 0.01, 0.02, and 0.02, respectively. If any of the key devices malfunctions, the launch will be postponed. Compute the probability for the shuttle to be launched on time, according to its schedule.

Solution. In this case,

$$
\begin{aligned}
\boldsymbol{P}\{\text{ on time }\} &= \boldsymbol{P}\{\text{ all devices function }\} \\
&= \boldsymbol{P}\{\overline{H} \cap \overline{B}_1 \cap \overline{B}_2\} \\
&= \boldsymbol{P}\{\overline{H}\}\,\boldsymbol{P}\{\overline{B}_1\}\,\boldsymbol{P}\{\overline{B}_2\} \quad (independence) \\
&= (1 - 0.01)(1 - 0.02)(1 - 0.02) \quad (complement\ rule) \\
&= 0.9508.
\end{aligned}
$$

Notice how with the same probabilities of individual components as in Example 2.18, the system's reliability decreased because the components were connected sequentially. ◇

Many modern systems consist of a great number of devices connected in sequel and in parallel.

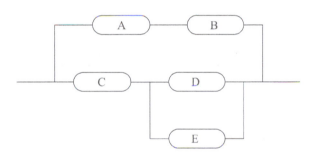

FIGURE 2.3: *Calculate reliability of this system (Example 2.20).*

Example 2.20 (TECHNIQUES FOR SOLVING RELIABILITY PROBLEMS). Calculate reliability of the system in Figure 2.3 if each component is operable with probability 0.92 independently of the other components.

Solution. This problem can be simplified and solved "step by step."

1. The upper link A-B works if both A and B work, which has probability

$$
\boldsymbol{P}\{A \cap B\} = (0.92)^2 = 0.8464.
$$

We can represent this link as one component F that operates with probability 0.8464.

2. By the same token, components D and E, connected in parallel, can be replaced by component G, operable with probability

$$
\boldsymbol{P}\{D \cup E\} = 1 - (1 - 0.92)^2 = 0.9936,
$$

as shown in Figure 2.4a.

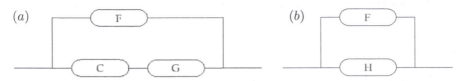

FIGURE 2.4: *Step by step solution of a system reliability problem.*

3. Components C and G, connected sequentially, can be replaced by component H, operable with probability $P\{C \cap G\} = 0.92 \cdot 0.9936 = 0.9141$, as shown in Figure 2.4b.

4. Last step. The system operates with probability

$$P\{F \cup H\} = 1 - (1 - 0.8464)(1 - 0.9141) = \underline{0.9868},$$

   which is the final answer.

In fact, the event "the system is operable" can be represented as $(A \cap B) \cup \{C \cap (D \cup E)\}$, whose probability we found step by step.   ◊

2.3   Combinatorics

2.3.1   Equally likely outcomes

A simple situation for computing probabilities is the case of *equally likely outcomes*. That is, when the sample space Ω consists of n possible outcomes, $\omega_1, \ldots, \omega_n$, each having the same probability. Since

$$\sum_1^n P\{\omega_k\} = P\{\Omega\} = 1,$$

we have in this case $P\{\omega_k\} = 1/n$ for all k. Further, a probability of any event E consisting of t outcomes, equals

$$P\{E\} = \sum_{\omega_k \in E} \left(\frac{1}{n}\right) = t\left(\frac{1}{n}\right) = \frac{\text{number of outcomes in } E}{\text{number of outcomes in } \Omega}.$$

The outcomes forming event E are often called "favorable." Thus we have a formula

Equally likely outcomes	$P\{E\} = \dfrac{\text{number of favorable outcomes}}{\text{total number of outcomes}} = \dfrac{N_F}{N_T}$	(2.5)

where index "F" means "favorable" and "T" means "total."

Example 2.21. Tossing a die results in 6 equally likely possible outcomes, identified by the number of dots from 1 to 6. Applying (2.5), we obtain,

$$P\{1\} = 1/6, \quad P\{\text{ odd number of dots }\} = 3/6, \quad P\{\text{ less than 5 }\} = 4/6.$$

$\Diamond$

The solution and even the answer to such problems may depend on our choice of outcomes and a sample space. Outcomes should be defined in such a way that they appear equally likely, otherwise formula (2.5) does not apply.

Example 2.22. A card is drawn from a bridge 52-card deck at random. Compute the probability that the selected card is a spade.

<u>First solution</u>. The sample space consists of 52 equally likely outcomes—cards. Among them, there are 13 favorable outcomes—spades. Hence, $P\{\text{spade}\} = 13/52 = 1/4$.

<u>Second solution</u>. The sample space consists of 4 equally likely outcomes—suits: clubs, diamonds, hearts, and spades. Among them, one outcome is favorable—spades. Hence, $P\{\text{spade}\} = 1/4$. $\Diamond$

These two solutions relied on different sample spaces. However, in both cases, the defined outcomes were equally likely, therefore (2.5) was applicable, and we obtained the same result.

However, the situation may be different.

Example 2.23. A young family plans to have two children. What is the probability of two girls?

<u>Solution 1 (wrong)</u>. There are 3 possible families with 2 children: two girls, two boys, and one of each gender. Therefore, the probability of two girls is $1/3$.

<u>Solution 2 (right)</u>. Each child is (supposedly) equally likely to be a boy or a girl. Genders of the two children are (supposedly) independent. Therefore,

$$P\{\text{two girls}\} = \left(\frac{1}{2}\right)\left(\frac{1}{2}\right) = 1/4.$$

$\Diamond$

The second solution implies that the sample space consists of four, not three, equally likely outcomes: two boys, two girls, a boy and a girl, a girl and a boy. Each outcome in this sample has probability $1/4$. Notice that the last two outcomes are counted separately, with the meaning, say, "the first child is a boy, the second one is a girl" and "the first child is a girl, the second one is a boy."

It is all right to define a sample space as in Solution 1. However, one must know in this case that the defined outcomes are *not equally likely*. Indeed, from Solution 2, we see that having one child of each gender is the most likely outcome, with probability of $1/4 + 1/4 = 1/2$. It was a mistake to apply (2.5) for such a sample space in Solution 1.

Example 2.24 (PARADOX). There is a simple but controversial "situation" about a family with two children. Even some graduate students often fail to resolve it.

A family has two children. You met one of them, Lev, and he is a boy. What is the probability that the other child is also a boy?

On one hand, why would the other child's gender be affected by Lev? Lev should have a brother or a sister with probabilities 1/2 and 1/2.

On the other hand, see Example 2.23. The sample space consists of 4 equally likely outcomes, $\{GG, BB, BG, GB\}$. You have already met one boy, thus the first outcome is automatically eliminated: $\{BB, BG, GB\}$. Among the remaining three outcomes, Lev has a brother in one case and a sister in two cases. Thus, isn't the probability of a boy equal 1/3?

Where is the catch? Apparently, the sample space Ω has not been clearly defined in this example. The experiment is more complex than in Example 2.23 because we are now concerned not only about the gender of children but also about meeting one of them. What is the mechanism, what are the probabilities for you to meet one or the other child? And once you met Lev, do the outcomes $\{BB, BG, GB\}$ remain equally likely?

A complete solution to this paradox is broken into steps in Exercise 2.31. ◇

In reality, business-related, sports-related, and political events are typically not equally likely. One outcome is usually more likely than another. For example, one team is always stronger than the other. Equally likely outcomes are usually associated with conditions of "a fair game" and "selected at random." In fair gambling, all cards, all dots on a die, all numbers in a roulette are equally likely. Also, when a survey is conducted, or a sample is selected "at random," the outcomes are "as close as possible" to being equally likely. This means all the subjects have the same chance to be selected into a sample (otherwise, it is not a fair sample, and it can produce "biased" results).

2.3.2 Permutations and combinations

Formula (2.5) is simple, as long as its numerator and denominator can be easily evaluated. This is rarely the case; often the sample space consists of a multitude of outcomes. *Combinatorics* provides special techniques for the computation of $\mathcal{N}_T$ and $\mathcal{N}_F$, the total number and the number of favorable outcomes.

We shall consider a generic situation when objects are selected *at random* from a set of n. This general model has a number of useful applications.

The objects may be selected with replacement or without replacement. They may also be *distinguishable* or *indistinguishable*.

DEFINITION 2.12 ───────

> Sampling **with replacement** means that every sampled item is replaced into the initial set, so that any of the objects can be selected with probability $1/n$ at any time. In particular, the same object may be sampled more than once.

Sampling **without replacement** means that every sampled item is removed from further sampling, so the set of possibilities reduces by 1 after each selection.

Objects are **distinguishable** if sampling of exactly the same objects *in a different order* yields a different outcome, that is, a different element of the sample space. For **indistinguishable** objects, the order is not important, it only matters which objects are sampled and which ones aren't. Indistinguishable objects arranged in a different order do not generate a new outcome.

Example 2.25 (COMPUTER-GENERATED PASSWORDS). When random passwords are generated, the order of characters is important because a different order yields a different password. Characters are distinguishable in this case. Further, if a password has to consist of different characters, they are sampled from the alphabet without replacement. ◊

Example 2.26 (POLLS). When a sample of people is selected to conduct a poll, the same participants produce the same responses regardless of their order. They can be considered indistinguishable. ◊

Permutations with replacement

Possible selections of k *distinguishable* objects from a set of n are called *permutations*. When we sample with replacement, each time there are n possible selections, hence the total number of permutations is

Permutations with replacement

$$P_r(n,k) = \overbrace{n \cdot n \cdot \ldots \cdot n}^{k \text{ terms}} = n^k$$

Example 2.27 (BREAKING PASSWORDS). From an alphabet consisting of 10 digits, 26 lower-case and 26 capital letters, one can create $P_r(62,8) = 218{,}340{,}105{,}584{,}896$ (over 218 trillion) different 8-character passwords. At a speed of 1 million passwords per second, it will take a spy program almost 7 years to try all of them. Thus, on the average, it will guess your password in about 3.5 years.

At this speed, the spy program can test 604,800,000,000 passwords within 1 week. The probability that it guesses your password in 1 week is

$$\frac{N_F}{N_T} = \frac{\text{number of favorable outcomes}}{\text{total number of outcomes}} = \frac{604{,}800{,}000{,}000}{218{,}340{,}105{,}584{,}896} = 0.00277.$$

However, if capital letters are not used, the number of possible passwords is reduced to $P_r(36, 8) = 2{,}821{,}109{,}907{,}456$. On the average, it takes the spy only 16 days to guess such a password! The probability that it will happen in 1 week is 0.214. A wise recommendation to include all three types of characters in our passwords and to change them once a year is perfectly clear to us now...   ◊

Permutations without replacement

During sampling without replacement, the number of possible selections reduces by 1 each time an object is sampled. Therefore, the number of permutations is

Permutations without replacement	$$P(n,k) = \overbrace{n(n-1)(n-2) \cdot \ldots \cdot (n-k+1)}^{k \text{ terms}} = \frac{n!}{(n-k)!}$$

where $n! = 1 \cdot 2 \cdot \ldots n$ (*n-factorial*) denotes the product of all integers from 1 to n.

The number of permutations without replacement also equals the number of possible allocations of k distinguishable objects among n available slots.

Example 2.28. In how many ways can 10 students be seated in a classroom with 15 chairs?

Solution. Students are distinguishable, and each student needs a separate seat. Thus, the number of possible allocations is the number of permutations without replacement, $P(15, 10) = 15 \cdot 14 \cdot \ldots \cdot 6 = 1.09 \cdot 10^{10}$. Notice that if students enter the classroom one by one, the first student has 15 choices of seats, then one seat is occupied, and the second student has only 14 choices, etc., and the last student takes one of 6 chairs available at that time.   ◊

Combinations without replacement

Possible selections of k *indistinguishable* objects from a set of n are called *combinations*. The number of combinations without replacement is also called "n choose k" and is denoted by $C(n,k)$ or $\binom{n}{k}$.

The only difference from $P(n,k)$ is disregarding the order. Now the same objects sampled in a different order produce the same outcome. Thus, $P(k,k) = k!$ different permutations (rearrangements) of the same objects yield only 1 combination. The total number of combinations is then

$$
\begin{array}{c}
\textbf{Combinations} \\
\textbf{without} \\
\textbf{replacement}
\end{array}
\qquad
\boxed{C(n,k) = \binom{n}{k} = \frac{P(n,k)}{P(k,k)} = \frac{n!}{k!(n-k)!}}
\qquad (2.6)
$$

Example 2.29. An antivirus software reports that 3 folders out of 10 are infected. How many possibilities are there?

<u>Solution</u>. Folders A, B, C and folders C, B, A represent the same outcome, thus, the order is not important. A software clearly detected 3 different folders, thus it is sampling without replacement. The number of possibilities is

$$
\binom{10}{3} = \frac{10!}{3!\,7!} = \frac{10 \cdot 9 \cdot \ldots \cdot 1}{(3 \cdot 2 \cdot 1)(7 \cdot \ldots \cdot 1)} = \frac{10 \cdot 9 \cdot 8}{3 \cdot 2 \cdot 1} = 120.
$$

$\Diamond$

Computational shortcuts

Instead of computing $C(n,k)$ directly by the formula, we can simplify the fraction. At least, the numerator and denominator can both be divided by either $k!$ or $(n-k)!$ (choose the larger of these for greater reduction). As a result,

$$
C(n,k) = \binom{n}{k} = \frac{n \cdot (n-1) \cdot \ldots \cdot (n-k+1)}{k \cdot (k-1) \cdot \ldots \cdot 1},
$$

the top and the bottom of this fraction being products of k terms. It is also handy to notice that

$$
\begin{aligned}
C(n,k) &= C(n,n-k) \text{ for any } k \text{ and } n \\
C(n,0) &= 1 \\
C(n,1) &= n
\end{aligned}
$$

Example 2.30. There are 20 computers in a store. Among them, 15 are brand new and 5 are refurbished. Six computers are purchased for a student lab. From the first look, they are indistinguishable, so the six computers are selected at random. Compute the probability that among the chosen computers, two are refurbished.

<u>Solution</u>. Compute the total number and the number of favorable outcomes. The *total* number of ways in which 6 computers are selected from 20 is

$$
\mathcal{N}_T = \binom{20}{6} = \frac{20 \cdot 19 \cdot 18 \cdot 17 \cdot 16 \cdot 15}{6 \cdot 5 \cdot 4 \cdot 3 \cdot 2 \cdot 1}.
$$

FIGURE 2.5: *Counting combinations with replacement. Vertical bars separate different classes of items.*

We applied the mentioned computational shortcut. Next, for the number of *favorable* outcomes, 2 refurbished computers are selected from a total of 5, and the remaining 4 new ones are selected from a total of 15. There are

$$\mathcal{N}_F = \binom{5}{2}\binom{15}{4} = \left(\frac{5 \cdot 4}{2 \cdot 1}\right)\left(\frac{15 \cdot 14 \cdot 13 \cdot 12}{4 \cdot 3 \cdot 2 \cdot 1}\right)$$

favorable outcomes. With further reduction of fractions, the probability equals

$$\boldsymbol{P}\{\text{ two refurbished computers }\} = \frac{\mathcal{N}_F}{\mathcal{N}_T} = \frac{7 \cdot 13 \cdot 5}{19 \cdot 17 \cdot 4} = 0.3522.$$

Combinations with replacement

For combinations with replacement, the order is not important, and each object may be sampled more than once. Then each outcome consists of counts, how many times each of n objects appears in the sample. In Figure 2.5, we draw a circle for each time object #1 is sampled, then draw a separating bar, then a circle for each time object #2 is sampled, etc. Two bars next to each other mean that the corresponding object has never been sampled.

The resulting picture has to have k circles for a sample of size k and $(n-1)$ bars separating n objects. Each picture with these conditions represents an outcome. How many outcomes are there? It is the number of allocations of k circles and $(n-1)$ bars among $(k+n-1)$ slots available for them. Hence,

Combinations with replacement	$C_r(n,k) = \binom{k+n-1}{k} = \dfrac{(k+n-1)!}{k!(n-1)!}$

<u>NOTATION</u>

$$
\begin{aligned}
P_r(n,k) &= \text{number of permutations with replacement} \\
P(n,k) &= \text{number of permutations without replacement} \\
C_r(n,k) &= \text{number of combinations with replacement} \\
\left.\begin{array}{l} C(n,k) \\ \binom{n}{k} \end{array}\right\} &= \text{number of combinations without replacement}
\end{aligned}
$$

2.4 Conditional probability and independence

Conditional probability

Suppose you are meeting someone at an airport. The flight is likely to arrive on time; the probability of that is 0.8. Suddenly it is announced that the flight departed one hour behind the schedule. Now it has the probability of only 0.05 to arrive on time. New information affected the probability of meeting this flight on time. The new probability is called *conditional probability*, where the new information, that the flight departed late, is a *condition*.

DEFINITION 2.15 ——————

> **Conditional probability** of event A given event B is the probability that A occurs when B is *known to occur*.

<u>NOTATION</u> ‖ $P\{A \mid B\}$ = conditional probability of A given B ‖

How does one compute the conditional probability? First, consider the case of equally likely outcomes. In view of the new information, occurrence of the condition B, only the outcomes contained in B still have a non-zero chance to occur. Counting only such outcomes, the *unconditional* probability of A,

$$P\{A\} = \frac{\text{number of outcomes in } A}{\text{number of outcomes in } \Omega},$$

is now replaced by the *conditional probability* of A given B,

$$P\{A \mid B\} = \frac{\text{number of outcomes in } A \cap B}{\text{number of outcomes in } B} = \frac{P\{A \cap B\}}{P\{B\}}.$$

This appears to be the general formula.

$$\boxed{\begin{array}{l}\textbf{Conditional} \\ \textbf{probability}\end{array} \quad P\{A \mid B\} = \frac{P\{A \cap B\}}{P\{B\}}} \qquad (2.7)$$

Rewriting (2.7) in a different way, we obtain the general formula for the probability of intersection.

$$\boxed{\begin{array}{l}\textbf{Intersection,} \\ \textbf{general case}\end{array} \quad P\{A \cap B\} = P\{B\}\, P\{A \mid B\}} \qquad (2.8)$$

Independence

Now we can give an intuitively very clear definition of *independence*.

DEFINITION 2.16 ─────────────

Events A and B are **independent** if occurrence of B does not affect the probability of A, i.e.,

$$P\{A \mid B\} = P\{A\}.$$

According to this definition, *conditional* probability equals *unconditional* probability in case of independent events. Substituting this into (2.8) yields

$$P\{A \cap B\} = P\{A\}\,P\{B\}.$$

This is our old formula for independent events.

Example 2.31. Ninety percent of flights depart on time. Eighty percent of flights arrive on time. Seventy-five percent of flights depart on time and arrive on time.

(a) Eric is meeting Alyssa's flight, which departed on time. What is the probability that Alyssa will arrive on time?

(b) Eric has met Alyssa, and she arrived on time. What is the probability that her flight departed on time?

(c) Are the events, departing on time and arriving on time, independent?

<u>Solution.</u> Denote the events,

$$A = \{\text{arriving on time}\},$$
$$D = \{\text{departing on time}\}.$$

We have:
$$P\{A\} = 0.8, \quad P\{D\} = 0.9, \quad P\{A \cap D\} = 0.75.$$

(a) $P\{A \mid D\} = \dfrac{P\{A \cap D\}}{P\{D\}} = \dfrac{0.75}{0.9} = \underline{0.8333}.$

(b) $P\{D \mid A\} = \dfrac{P\{A \cap D\}}{P\{A\}} = \dfrac{0.75}{0.8} = \underline{0.9375}.$

(c) Events are not independent because

$$P\{A \mid D\} \neq P\{A\}, \quad P\{D \mid A\} \neq P\{D\}, \quad P\{A \cap D\} \neq P\{A\}\,P\{D\}.$$

Actually, any one of these inequalities is sufficient to prove that A and D are dependent. Further, we see that $P\{A \mid D\} > P\{A\}$ and $P\{D \mid A\} > P\{D\}$. In other words, departing on time increases the probability of arriving on time, and vice versa. This perfectly agrees with our intuition.   ◇

Bayes Rule

The last example shows that two conditional probabilities, $P\{A \mid B\}$ and $P\{B \mid A\}$, are not the same, in general. Consider another example.

Example 2.32 (RELIABILITY OF A TEST). There exists a test for a certain viral infection (including a virus attack on a computer network). It is 95% reliable for infected patients and 99% reliable for the healthy ones. That is, if a patient has the virus (event V), the test shows that (event S) with probability $P\{S \mid V\} = 0.95$, and if the patient does not have the virus, the test shows that with probability $P\{\overline{S} \mid \overline{V}\} = 0.99$.

Consider a patient whose test result is positive (i.e., the test shows that the patient has the virus). Knowing that sometimes the test is wrong, naturally, the patient is eager to know the probability that he or she indeed has the virus. However, this conditional probability, $P\{V \mid S\}$, is not stated among the given characteristics of this test. $\Diamond$

This example is applicable to any testing procedure including software and hardware tests, pregnancy tests, paternity tests, alcohol tests, academic exams, etc. The problem is to connect the given $P\{S \mid V\}$ and the quantity in question, $P\{V \mid S\}$. This was done in the eighteenth century by English minister *Thomas Bayes* (1702–1761) in the following way.

Notice that $A \cap B = B \cap A$. Therefore, using (2.8), $P\{B\} P\{A \mid B\} = P\{A\} P\{B \mid A\}$.

Solve for $P\{B \mid A\}$ to obtain

$$\text{Bayes Rule} \qquad \boxed{P\{B \mid A\} = \frac{P\{A \mid B\} P\{B\}}{P\{A\}}} \qquad (2.9)$$

Example 2.33 (SITUATION ON A MIDTERM EXAM). On a midterm exam, students X, Y, and Z forgot to sign their papers. Professor knows that they can write a good exam with probabilities 0.8, 0.7, and 0.5, respectively. After the grading, he notices that two unsigned exams are good and one is bad. Given this information, and assuming that students worked independently of each other, what is the probability that the bad exam belongs to student Z?

Solution. Denote good and bad exams by G and B. Also, let GGB denote two good and one bad exams, XG denote the event "student X wrote a good exam," etc. We need to find $P\{ZB \mid GGB\}$ given that $P\{G \mid X\} = 0.8$, $P\{G \mid Y\} = 0.7$, and $P\{G \mid Z\} = 0.5$.

By the *Bayes Rule*,

$$P\{ZB \mid GGB\} = \frac{P\{GGB \mid ZB\} P\{ZB\}}{P\{GGB\}}.$$

Given ZB, event GGB occurs only when both X and Y write good exams. Thus, $P\{GGB \mid ZB\} = (0.8)(0.7)$.

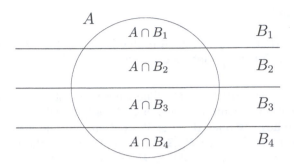

FIGURE 2.6: *Partition of the sample space Ω and the event A.*

Event GGB consists of three outcomes depending on the student who wrote the bad exam. Adding their probabilities, we get

$$
\begin{aligned}
&\boldsymbol{P}\{GGB\} \\
&= \boldsymbol{P}\{XG \cap YG \cap ZB\} + \boldsymbol{P}\{XG \cap YB \cap ZG\} + \boldsymbol{P}\{XB \cap YG \cap ZG\} \\
&= (0.8)(0.7)(0.5) + (0.8)(0.3)(0.5) + (0.2)(0.7)(0.5) = 0.47.
\end{aligned}
$$

Then

$$
\boldsymbol{P}\{ZB \mid GGB\} = \frac{(0.8)(0.7)(0.5)}{0.47} = \underline{0.5957}.
$$

$\Diamond$

In the Bayes Rule (2.9), the denominator is often computed by the Law of Total Probability.

Law of Total Probability

This law relates the unconditional probability of an event A with its conditional probabilities. It is used every time when it is easier to compute conditional probabilities of A given additional information.

Consider some partition of the sample space Ω with mutually exclusive and exhaustive events $B_1, \ldots, B_k$. It means that

$$
B_i \cap B_j = \varnothing \text{ for any } i \neq j \text{ and } B_1 \cup \ldots \cup B_k = \Omega.
$$

These events also partition the event A,

$$
A = (A \cap B_1) \cup \ldots \cup (A \cap B_k),
$$

and this is also a union of mutually exclusive events (Figure 2.6). Hence,

$$
\boldsymbol{P}\{A\} = \sum_{j=1}^{k} \boldsymbol{P}\{A \cap B_j\},
$$

and we arrive to the following rule.

Law of Total Probability	$$P\{A\} = \sum_{j=1}^{k} P\{A \mid B_j\} P\{B_j\}$$ In case of two events $(k = 2)$, $$P\{A\} = P\{A \mid B\} P\{B\} + P\{A \mid \overline{B}\} P\{\overline{B}\}$$

(2.10)

Together with the Bayes Rule, it makes the following popular formula

Bayes Rule for two events	$$P\{B \mid A\} = \frac{P\{A \mid B\} P\{B\}}{P\{A \mid B\} P\{B\} + P\{A \mid \overline{B}\} P\{\overline{B}\}}$$

Example 2.34 (RELIABILITY OF A TEST, CONTINUED). Continue Example 2.32. Suppose that 4% of all the patients are infected with the virus, $P\{V\} = 0.04$. Recall that $P\{S \mid V\} = 0.95$ and $P\{\overline{S} \mid \overline{V}\} = 0.99$. If the test shows positive results, the (conditional) probability that a patient has the virus equals

$$
\begin{aligned}
P\{V \mid S\} &= \frac{P\{S \mid V\} P\{V\}}{P\{S \mid V\} P\{V\} + P\{S \mid \overline{V}\} P\{\overline{V}\}} \\
&= \frac{(0.95)(0.04)}{(0.95)(0.04) + (1 - 0.99)(1 - 0.04)} = \underline{0.7983}.
\end{aligned}
$$

$\Diamond$

Example 2.35 (DIAGNOSTICS OF COMPUTER CODES). A new computer program consists of two modules. The first module contains an error with probability 0.2. The second module is more complex; it has a probability of 0.4 to contain an error, independently of the first module. An error in the first module alone causes the program to crash with probability 0.5. For the second module, this probability is 0.8. If there are errors in both modules, the program crashes with probability 0.9. Suppose the program crashed. What is the probability of errors in both modules?

Solution. Denote the events,

$$A = \{\text{errors in module I}\}, \quad B = \{\text{errors in module II}\}, \quad C = \{\text{crash}\}.$$

Further,

$$
\begin{aligned}
\{\text{errors in module I alone}\} &= A \backslash B = A \backslash (A \cap B) \\
\{\text{errors in module II alone}\} &= B \backslash A = B \backslash (A \cap B).
\end{aligned}
$$

It is given that $P\{A\} = 0.2$, $P\{B\} = 0.4$, $P\{A \cap B\} = (0.2)(0.4) = 0.08$, by independence, $P\{C \mid A \backslash B\} = 0.5$, $P\{C \mid B \backslash A\} = 0.8$, and $P\{C \mid A \cap B\} = 0.9$.

We need to compute $P\{A \cap B \mid C\}$. Since A is a union of disjoint events $A \backslash B$ and $A \cap B$, we compute

$$P\{A \backslash B\} = P\{A\} - P\{A \cap B\} = 0.2 - 0.08 = 0.12.$$

Similarly,

$$P\{B \backslash A\} = 0.4 - 0.08 = 0.32.$$

Events $(A \backslash B)$, $(B \backslash A)$, $A \cap B$, and $\overline{(A \cup B)}$ form a partition of Ω, because they are mutually exclusive and exhaustive. The last of them is the event of no errors in the entire program. Given this event, the probability of a crash is 0. Notice that A, B, and $(A \cap B)$ are neither mutually exclusive nor exhaustive, so they cannot be used for the Bayes Rule. Now organize the data.

Location of errors			Probability of a crash		
$P\{A \backslash B\}$	=	0.12	$P\{C \mid A \backslash B\}$	=	0.5
$P\{B \backslash A\}$	=	0.32	$P\{C \mid B \backslash A\}$	=	0.8
$P\{A \cap B\}$	=	0.08	$P\{C \mid A \cap B\}$	=	0.9
$P\{\overline{A \cup B}\}$	=	0.48	$P\{C \mid \overline{A \cup B}\}$	=	0

Combining the Bayes Rule and the Law of Total Probability,

$$P\{A \cap B \mid C\} = \frac{P\{C \mid A \cap B\} P\{A \cap B\}}{P\{C\}},$$

where

$$\begin{aligned} P\{C\} = \ & P\{C \mid A \backslash B\} P\{A \backslash B\} + P\{C \mid B \backslash A\} P\{B \backslash A\} \\ & + P\{C \mid A \cap B\} P\{A \cap B\} + P\{C \mid \overline{A \cup B}\} P\{\overline{A \cup B}\}. \end{aligned}$$

Then

$$P\{A \cap B \mid C\} = \frac{(0.9)(0.08)}{(0.5)(0.12) + (0.8)(0.32) + (0.9)(0.08) + 0} = \underline{0.1856}.$$

$\Diamond$

Summary and conclusions

Probability of any event is a number between 0 and 1. The empty event has probability 0, and the sample space has probability 1. There are rules for computing probabilities of unions, intersections, and complements. For a union of disjoint events, probabilities are added. For an intersection of independent events, probabilities are multiplied. Combining these rules, one evaluates reliability of a system given reliabilities of its components.

In the case of equally likely outcomes, probability is a ratio of the number of favorable outcomes to the total number of outcomes. Combinatorics provides tools for computing these numbers in frequent situations involving permutations and combinations, with or without replacement.

Given occurrence of event B, one can compute conditional probability of event A. Unconditional probability of A can be computed from its conditional probabilities by the Law of Total Probability. The Bayes Rule, often used in testing and diagnostics, relates conditional probabilities of A given B and of B given A.

Exercises

2.1. Out of six computer chips, two are defective. If two chips are randomly chosen for testing (without replacement), compute the probability that both of them are defective. List all the outcomes in the sample space.

2.2. Suppose that after 10 years of service, 40% of computers have problems with motherboards (MB), 30% have problems with hard drives (HD), and 15% have problems with both MB and HD. What is the probability that a 10-year old computer still has fully functioning MB and HD?

2.3. A new computer virus can enter the system through e-mail or through the internet. There is a 30% chance of receiving this virus through e-mail. There is a 40% chance of receiving it through the internet. Also, the virus enters the system simultaneously through e-mail and the internet with probability 0.15. What is the probability that the virus does not enter the system at all?

2.4. Among employees of a certain firm, 70% know Java, 60% know Python, and 50% know both languages. What portion of programmers

(a) does not know Python?

(b) does not know Python and does not know Java?

(c) knows Java but not Python?

(d) knows Python but not Java?

(e) If someone knows Python, what is the probability that he/she knows Java too?

(f) If someone knows Java, what is the probability that he/she knows Python too?

2.5. A computer program is tested by 3 *independent* tests. When there is an error, these tests will discover it with probabilities 0.2, 0.3, and 0.5, respectively. Suppose that the program contains an error. What is the probability that it will be found by at least one test?

2.6. Under good weather conditions, 80% of flights arrive on time. During bad weather, only 30% of flights arrive on time. Tomorrow, the chance of good weather is 60%. What is the probability that your flight will arrive on time?

2.7. A system may become infected by some spyware through the internet or e-mail. Seventy percent of the time the spyware arrives via the internet, thirty percent of the time via e-mail. If it enters via the internet, the system detects it immediately with probability 0.6. If via e-mail, it is detected with probability 0.8. What percentage of times is this spyware detected?

2.8. A shuttle's launch depends on three key devices that may fail independently of each other with probabilities 0.01, 0.02, and 0.02, respectively. If any of the key devices fails, the launch will be postponed. Compute the probability for the shuttle to be launched on time, according to its schedule.

2.9. Successful implementation of a new system is based on three independent modules. Module 1 works properly with probability 0.96. For modules 2 and 3, these probabilities equal 0.95 and 0.90. Compute the probability that at least one of these three modules fails to work properly.

2.10. Three computer viruses arrived as an e-mail attachment. Virus A damages the system with probability 0.4. Independently of it, virus B damages the system with probability 0.5. Independently of A and B, virus C damages the system with probability 0.2. What is the probability that the system gets damaged?

2.11. A computer program is tested by 5 independent tests. If there is an error, these tests will discover it with probabilities 0.1, 0.2, 0.3, 0.4, and 0.5, respectively. Suppose that the program contains an error. What is the probability that it will be found

   (a) by at least one test?

   (b) by at least two tests?

   (c) by all five tests?

2.12. A building is examined by policemen with four dogs that are trained to detect the scent of explosives. If there are explosives in a certain building, and each dog detects them with probability 0.6, independently of other dogs, what is the probability that the explosives will be detected by at least one dog?

2.13. An important module is tested by three independent teams of inspectors. Each team detects a problem in a defective module with probability 0.8. What is the probability that at least one team of inspectors detects a problem in a defective module?

2.14. A spyware is trying to break into a system by guessing its password. It does not give up until it tries 1 million different passwords. What is the probability that it will guess the password and break in if by rules, the password must consist of

   (a) 6 different lower-case letters?

   (b) 6 different letters, some may be upper-case, and it is case-sensitive?

   (c) any 6 letters, upper- or lower-case, and it is case-sensitive?

   (d) any 6 characters including letters and digits?

2.15. A computer program consists of two blocks written independently by two different programmers. The first block has an error with probability 0.2. The second block has an error with probability 0.3. If the program returns an error, what is the probability that there is an error in both blocks?

2.16. A computer maker receives parts from three suppliers, S1, S2, and S3. Fifty percent come from S1, twenty percent from S2, and thirty percent from S3. Among all the parts supplied by S1, 5% are defective. For S2 and S3, the portion of defective parts is 3% and 6%, respectively.

  (a) What portion of all the parts is defective?

  (b) A customer complains that a certain part in her recently purchased computer is defective. What is the probability that it was supplied by S1?

2.17. A computer assembling company receives 24% of parts from supplier X, 36% of parts from supplier Y, and the remaining 40% of parts from supplier Z. Five percent of parts supplied by X, ten percent of parts supplied by Y, and six percent of parts supplied by Z are defective. If an assembled computer has a defective part in it, what is the probability that this part was received from supplier Z?

2.18. A problem on a multiple-choice quiz is answered correctly with probability 0.9 if a student is prepared. An unprepared student guesses between 4 possible answers, so the probability of choosing the right answer is 1/4. Seventy-five percent of students prepare for the quiz. If Mr. X gives a correct answer to this problem, what is the chance that he did not prepare for the quiz?

2.19. At a plant, 20% of all the produced parts are subject to a special electronic inspection. It is known that any produced part which was inspected electronically has no defects with probability 0.95. For a part that was not inspected electronically this probability is only 0.7. A customer receives a part and find defects in it. What is the probability that this part went through an electronic inspection?

2.20. All athletes at the Olympic games are tested for performance-enhancing steroid drug use. The imperfect test gives positive results (indicating drug use) for 90% of all steroid-users but also (and incorrectly) for 2% of those who do not use steroids. Suppose that 5% of all registered athletes use steroids. If an athlete is tested negative, what is the probability that he/she uses steroids?

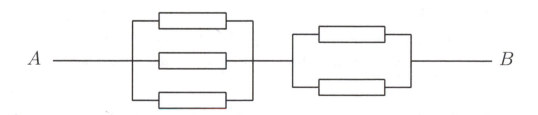

FIGURE 2.7: *Calculate reliability of this system (Exercise 2.21).*

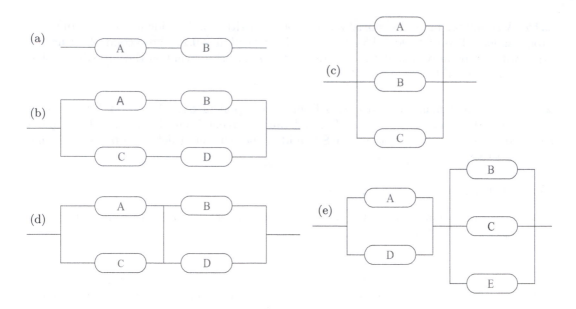

FIGURE 2.8: *Calculate reliability of each system (Exercise 2.23).*

2.21. In the system in Figure 2.7, each component fails with probability 0.3 independently of other components. Compute the system's reliability.

2.22. Three highways connect city A with city B. Two highways connect city B with city C. During a rush hour, each highway is blocked by a traffic accident with probability 0.2, independently of other highways.

 (a) Compute the probability that there is at least one open route from A to C.

 (b) How will a new highway, also blocked with probability 0.2 independently of other highways, change the probability in (a) if it is built

   (α) between A and B?
   (β) between B and C?
   (γ) between A and C?

2.23. Calculate the reliability of each system shown in Figure 2.8, if components A, B, C, D, and E function properly with probabilities 0.9, 0.8, 0.7, 0.6, and 0.5, respectively.

2.24. Among 10 laptop computers, five are good and five have defects. Unaware of this, a customer buys 6 laptops.

 (a) What is the probability of exactly 2 defective laptops among them?

 (b) Given that *at least* 2 purchased laptops are defective, what is the probability that *exactly* 2 are defective?

2.25. Two out of six computers in a lab have problems with hard drives. If three computers are selected at random for inspection, what is the probability that none of them has hard drive problems?

2.26. For the University Cheese Club, Danielle and Anthony buy four brands of cheese, three brands of crackers, and two types of grapes. Each club member has to taste two different cheeses, two types of crackers, and one bunch of grapes. If 40 club members show up, can they all have different meals?

2.27. This is known as *the Birthday Problem*.

(a) Consider a class with 30 students. Compute the probability that at least two of them have their birthdays on the same day. (For simplicity, ignore the leap year.)

(b) How many students should be in class in order to have this probability above 0.5?

2.28. Among eighteen computers in some store, six have defects. Five randomly selected computers are bought for the university lab. Compute the probability that all five computers have no defects.

2.29. A quiz consists of 6 multiple-choice questions. Each question has 4 possible answers. A student is unprepared, and he has no choice but guessing answers completely at random. He passes the quiz if he gets at least 3 questions correctly. What is the probability that he will pass?

2.30. An internet search engine looks for a keyword in 9 databases, searching them in a random order. Only 5 of these databases contain the given keyword. Find the probability that it will be found in at least 2 of the first 4 searched databases.

2.31. Consider the situation described in Example 2.24 on p. 22, but this time let us define the sample space clearly. Suppose that one child is older, and the other is younger, their gender is independent of their age, and the child you met is one or the other with probabilities 1/2 and 1/2.

(a) List all the outcomes in this sample space. Each outcome should tell the children's gender, which child is older, and which child you have met.

(b) Show that *unconditional* probabilities of outcomes BB, BG, and GB are equal.

(c) Show that *conditional* probabilities of BB, BG, and GB, after you met Lev, are not equal.

(d) Show that Lev has a brother with *conditional* probability 1/2.

2.32. Show that events $A, B, C, \ldots$ are disjoint if and only if $\overline{A}, \overline{B}, \overline{C}, \ldots$ are exhaustive.

2.33. Events A and B are independent. Show, intuitively and mathematically, that:

(a) Their complements are also independent.

(b) If they are disjoint, then $P\{A\} = 0$ or $P\{B\} = 0$.

(c) If they are exhaustive, then $P\{A\} = 1$ or $P\{B\} = 1$.

2.34. Derive a computational formula for the probability of a union of N arbitrary events. Assume that probabilities of all individual events and their intersections are given.

2.35. Prove that
$$\overline{E_1 \cap \ldots \cap E_n} = \overline{E_1} \cup \ldots \cup \overline{E_n}$$
for arbitrary events $E_1, \ldots, E_n$.

2.36. From the "addition" rule of probability, derive the "subtraction" rule:

$$\text{if } B \subset A, \text{ then } \boldsymbol{P}\{A \backslash B\} = \boldsymbol{P}(A) - \boldsymbol{P}(B).$$

2.37. Prove "subadditivity": $\boldsymbol{P}\{E_1 \cup E_2 \cup \ldots\} \leq \sum \boldsymbol{P}\{E_i\}$ for any events $E_1, E_2, \ldots \in \mathfrak{M}$.

Chapter 3

Discrete Random Variables and Their Distributions

This chapter introduces the concept of a random variable and studies discrete distributions in detail. Continuous distributions are discussed in Chapter 4.

3.1 Distribution of a random variable

3.1.1 Main concepts

> A **random variable** is a function of an outcome,
>
> $$X = f(\omega).$$
>
> In other words, it is a quantity that depends on chance.

The domain of a random variable is the sample space Ω. Its range can be the set of all real numbers $\mathbb{R}$, or only the positive numbers $(0, +\infty)$, or the integers $\mathbb{Z}$, or the interval $(0, 1)$, etc., depending on what possible values the random variable can potentially take.

Once an experiment is completed, and the outcome ω is known, the value of random variable $X(\omega)$ becomes determined.

Example 3.1. Consider an experiment of tossing 3 fair coins and counting the number of heads. Certainly, the same model suits the number of girls in a family with 3 children, the number of 1's in a random binary string of 3 characters, etc.

Let X be the number of heads (girls, 1's). Prior to an experiment, its value is not known. All we can say is that X has to be an integer between 0 and 3. Since assuming each value is an event, we can compute probabilities,

$$\boldsymbol{P}\{X = 0\} = \boldsymbol{P}\{\text{three tails}\} = \boldsymbol{P}\{TTT\} = \left(\frac{1}{2}\right)\left(\frac{1}{2}\right)\left(\frac{1}{2}\right) = \frac{1}{8}$$

$$\boldsymbol{P}\{X = 1\} = \boldsymbol{P}\{HTT\} + \boldsymbol{P}\{THT\} + \boldsymbol{P}\{TTH\} = \frac{3}{8}$$

$$\boldsymbol{P}\{X = 2\} = \boldsymbol{P}\{HHT\} + \boldsymbol{P}\{HTH\} + \boldsymbol{P}\{THH\} = \frac{3}{8}$$

$$\boldsymbol{P}\{X = 3\} = \boldsymbol{P}\{HHH\} = \frac{1}{8}$$

Summarizing,

x	$\boldsymbol{P}\{X = x\}$
0	1/8
1	3/8
2	3/8
3	1/8
Total	1

$\Diamond$

This table contains everything that is known about random variable X prior to the experiment. Before we know the outcome ω, we cannot tell what X equals to. However, we can list all the possible values of X and determine the corresponding probabilities.

DEFINITION 3.2 ─────────

> Collection of all the probabilities related to X is the **distribution** of X. The function
>
> $$P(x) = \boldsymbol{P}\{X = x\}$$
>
> is the **probability mass function**, or **pmf**. The **cumulative distribution function**, or **cdf** is defined as
>
> $$F(x) = \boldsymbol{P}\{X \le x\} = \sum_{y \le x} \boldsymbol{P}(y). \qquad (3.1)$$
>
> The set of possible values of X is called the **support** of the distribution F.

For every outcome ω, the variable X takes one and only one value x. This makes events $\{X = x\}$ disjoint and exhaustive, and therefore,

$$\sum_x P(x) = \sum_x \boldsymbol{P}\{X = x\} = 1.$$

Looking at (3.1), we can conclude that the cdf $F(x)$ is a non-decreasing function of x, always between 0 and 1, with

$$\lim_{x \downarrow -\infty} F(x) = 0 \quad \text{and} \quad \lim_{x \uparrow +\infty} F(x) = 1.$$

Between any two subsequent values of X, $F(x)$ is constant. It jumps by $P(x)$ at each possible value x of X (see Figure 3.1, right).

Recall that one way to compute the probability of an event is to add probabilities of all the outcomes in it. Hence, for any set A,

$$\boldsymbol{P}\{X \in A\} = \sum_{x \in A} P(x).$$

When A is an interval, its probability can be computed directly from the cdf $F(x)$,

$$\boldsymbol{P}\{a < X \le b\} = F(b) - F(a).$$

Example 3.2. The pmf and cdf of X in Example 3.1 are shown in Figure 3.1.

◇

COMPUTER DEMO. To illustrate Examples 3.1 and 3.2, the following computer codes simulate 3 coin tosses 10,000 times and produce a *histogram* of the obtained values of X on Figure 3.2.

─── R ─────────

```
N <- 10000                    # Number of simulations
U <- matrix(runif(3*N),3,N)   # A 3-by-N matrix of random numbers from [0,1]
Y <- (U < 0.5)                # Y=1 (heads) if U < 0.5, otherwise Y=0 (tails)
X <- colSums(Y)               # Sums across columns. X = number of heads
hist(X)                       # Histogram of X
```

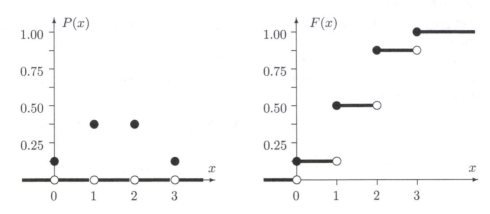

FIGURE 3.1: *The probability mass function $P(x)$ and the cumulative distribution function $F(x)$ for Example 3.1. White circles denote excluded points.*

—— MATLAB ————

```
N = 10000;      % Number of simulations
U = rand(3,N);  % A 3-by-N matrix of random numbers from [0,1]
Y = (U < 0.5);  % Y=1 (heads) if U < 0.5, otherwise Y=0 (tails)
X = sum(Y);     % Sums across columns. X = number of heads
hist(X);        % Histogram of X
```

On the obtained histogram, the two middle columns for $X = 1$ and $X = 2$ are about 3 times higher than the columns on each side, for $X = 0$ and $X = 3$. That is, in a run of 10,000 simulations, values 1 and 2 are attained three times more often than 0 and 3. This agrees with the pmf $P(0) = P(3) = 1/8$, $P(1) = P(2) = 3/8$.

Did you notice that the two middle columns are not exactly equal? That is because we used a random number generator. Every time when we run one of these computer programs, we'll get a slightly different histogram. Is your picture also a little different? This always happens when we use *generated* random variables to draw conclusions – every time results may differ a little. However, the differences are smaller when the number of simulations is large (like $N = 10,000$ in our case).

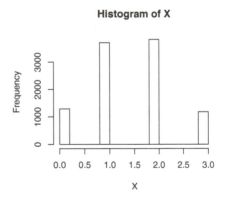

FIGURE 3.2: *A histogram of values of X for Examples 3.1, 3.2, obtained from $N = 10,000$ random generations.*

Finally, we note that both R and MATLAB have special commands to generate X in one step. Here, we derived X from generated random variables *uniformly distributed* between 0 and 1, a tool that is available in most software languages.

Characters # and % mark comments. They are included for us, the users, and R and MATLAB will simply ignore the text that follows them.

We'll discuss Monte Carlo simulations in detail in Chapter 5 and histograms in Section 8.3.1.

Example 3.3 (ERRORS IN INDEPENDENT MODULES). A program consists of two modules. The number of errors X_1 in the first module has the pmf $P_1(x)$, and the number of errors X_2 in the second module has the pmf $P_2(x)$, independently of X_1, where

x	$P_1(x)$	$P_2(x)$
0	0.5	0.7
1	0.3	0.2
2	0.1	0.1
3	0.1	0

Find the pmf and cdf of $Y = X_1 + X_2$, the total number of errors.

<u>Solution.</u> We break the problem into steps. First, determine all possible values of Y, then compute the probability of each value. Clearly, the number of errors Y is an integer that can be as low as $0 + 0 = 0$ and as high as $3 + 2 = 5$. Since $P_2(3) = 0$, the second module has at most 2 errors. Next,

$$
\begin{aligned}
P_Y(0) &= P\{Y = 0\} = \boldsymbol{P}\{X_1 = X_2 = 0\} = P_1(0)P_2(0) \\
&= (0.5)(0.7) = 0.35 \\
P_Y(1) &= P\{Y = 1\} = P_1(0)P_2(1) + P_1(1)P_2(0) \\
&= (0.5)(0.2) + (0.3)(0.7) = 0.31 \\
P_Y(2) &= P\{Y = 2\} = P_1(0)P_2(2) + P_1(1)P_2(1) + P_1(2)P_2(0) \\
&= (0.5)(0.1) + (0.3)(0.2) + (0.1)(0.7) = 0.18 \\
P_Y(3) &= P\{Y = 3\} = P_1(1)P_2(2) + P_1(2)P_2(1) + P_1(3)P_2(0) \\
&= (0.3)(0.1) + (0.1)(0.2) + (0.1)(0.7) = 0.12 \\
P_Y(4) &= P\{Y = 4\} = P_1(2)P_2(2) + P_1(3)P_2(1) \\
&= (0.1)(0.1) + (0.1)(0.2) = 0.03 \\
P_Y(5) &= P\{Y = 5\} = P_1(3)P_2(2) = (0.1)(0.1) = 0.01
\end{aligned}
$$

Now check:

$$
\sum_{y=0}^{5} P_Y(y) = 0.35 + 0.31 + 0.18 + 0.12 + 0.03 + 0.01 = 1,
$$

thus we *probably* counted all the possibilities and did not miss any. (We just wanted to emphasize that simply getting $\sum P(x) = 1$ does not guarantee that we made no mistake in our solution. However, if this equality is not satisfied, we have a mistake for sure.)

The cumulative function can be computed as

$$
\begin{aligned}
F_Y(0) &= P_Y(0) = 0.35 \\
F_Y(1) &= F_Y(0) + P_Y(1) = 0.35 + 0.31 = 0.66 \\
F_Y(2) &= F_Y(1) + P_Y(2) = 0.66 + 0.18 = 0.84 \\
F_Y(3) &= F_Y(2) + P_Y(3) = 0.84 + 0.12 = 0.96 \\
F_Y(4) &= F_Y(3) + P_Y(4) = 0.96 + 0.03 = 0.99 \\
F_Y(5) &= F_Y(4) + P_Y(5) = 0.99 + 0.01 = 1.00
\end{aligned}
$$

Between the values of Y, $F(x)$ is constant. $\diamond$

3.1.2 Types of random variables

So far, we are dealing with *discrete random variables*. These are variables whose range is finite or countable. In particular, it means that their values can be listed, or arranged in a sequence. Examples include the number of jobs submitted to a printer, the number of errors, the number of error-free modules, the number of failed components, and so on. Discrete variables don't have to be integers. For example, the *proportion* of defective components in a lot of 100 can be 0, 1/100, 2/100, ..., 99/100, or 1. This variable assumes 101 different values, so it is discrete, although not an integer.

On the contrary, *continuous random variables* assume a whole interval of values. This could be a bounded interval (a, b), or an unbounded interval such as $(a, +\infty)$, $(-\infty, b)$, or $(-\infty, +\infty)$. Sometimes, it may be a union of several such intervals. Intervals are uncountable, therefore, all values of a random variable cannot be listed in this case. Examples of continuous variables include various times (software installation time, code execution time, connection time, waiting time, lifetime), also physical variables like weight, height, voltage, temperature, distance, the number of miles per gallon, etc. We shall discuss continuous random variables in detail in Chapter 4.

Example 3.4. For comparison, observe that a long jump is formally a continuous random variable because an athlete can jump any distance within some range. Results of a high jump, however, are discrete because the bar can only be placed on a finite number of heights. $\Diamond$

Notice that rounding a continuous random variable, say, to the nearest integer makes it discrete.

Sometimes we can see *mixed random variables* that are discrete on some range of values and continuous elsewhere. For example, see a remark about waiting times on page 186.

Example 3.5. A job is sent to a printer. Let X be the waiting time before the job starts printing. With some probability, this job appears first in line and starts printing immediately, $X = 0$. It is also possible that the job is first in line but it takes 20 seconds for the printer to warm up, in which case $X = 20$. So far, the variable has a discrete behavior with a positive pmf $P(x)$ at $x = 0$ and $x = 20$. However, if there are other jobs in a queue, then X depends on the time it takes to print them, which is a continuous random variable. Using a popular jargon, besides *"point masses"* at $x = 0$ and $x = 20$, the variable is continuous, taking values in $(0, +\infty)$. Thus, X is neither discrete nor continuous. It is mixed. $\Diamond$

3.2 Distribution of a random vector

Often we deal with several random variables simultaneously. We may look at the size of a RAM and the speed of a CPU, the price of a computer and its capacity, temperature and humidity, technical and artistic performance, etc.

3.2.1 Joint distribution and marginal distributions

DEFINITION 3.3 ———

> If X and Y are random variables, then the pair (X, Y) is a **random vector**. Its distribution is called the **joint distribution** of X and Y. Individual distributions of X and Y are then called the **marginal distributions**.

Although we talk about two random variables in this section, all the concepts extend to a vector $(X_1, X_2, \ldots, X_n)$ of n components and its joint distribution.

Similarly to a single variable, the *joint distribution* of a vector is a collection of probabilities for a vector (X, Y) to take a value (x, y). Recall that two vectors are equal,

$$(X, Y) = (x, y),$$

if $X = x$ <u>and</u> $Y = y$. This "and" means the intersection, therefore, the *joint probability mass function* of X and Y is

$$P(x, y) = \boldsymbol{P}\{(X, Y) = (x, y)\} = \boldsymbol{P}\{X = x \cap Y = y\}.$$

Again, $\{(X, Y) = (x, y)\}$ are exhaustive and mutually exclusive events for different pairs (x, y), therefore,

$$\sum_x \sum_y P(x, y) = 1.$$

The joint distribution of (X, Y) carries the complete information about the behavior of this random vector. In particular, the marginal probability mass functions of X and Y can be obtained from the joint pmf by the Addition Rule.

$$
\textbf{Addition Rule} \quad
\boxed{
\begin{aligned}
P_X(x) &= \boldsymbol{P}\{X = x\} &= \sum_y P_{(X,Y)}(x, y) \\[2mm]
P_Y(y) &= \boldsymbol{P}\{Y = y\} &= \sum_x P_{(X,Y)}(x, y)
\end{aligned}
}
\quad (3.2)
$$

That is, to get the marginal pmf of one variable, we add the joint probabilities over all values of the other variable.

The Addition Rule is illustrated in Figure 3.3. Events $\{Y = y\}$ for different values of y partition the sample space Ω. Hence, their intersections with $\{X = x\}$ partition the event $\{X = x\}$ into mutually exclusive parts. By the rule for the union of mutually exclusive events, formula (2.4) on p. 16, their probabilities should be added. These probabilities are precisely $P_{(X,Y)}(x, y)$.

In general, the joint distribution cannot be computed from marginal distributions because they carry no information about interrelations between random variables. For example, marginal distributions cannot tell whether variables X and Y are independent or dependent.

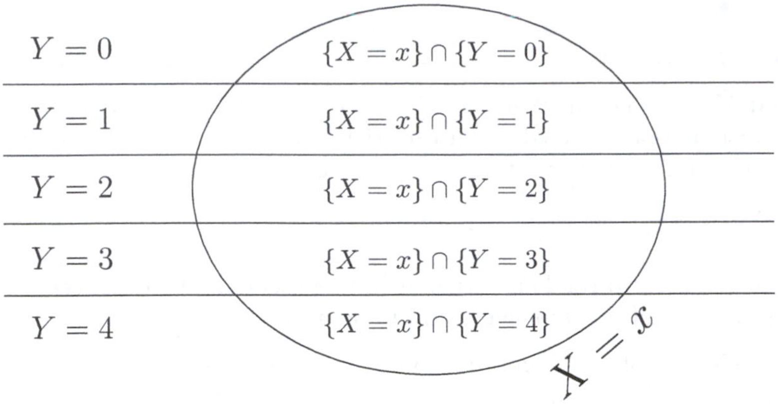

FIGURE 3.3: *Addition Rule: computing marginal probabilities from the joint distribution.*

3.2.2   Independence of random variables

DEFINITION 3.4 ——————————————————————————————

Random variables X and Y are **independent** if

$$P_{(X,Y)}(x,y) = P_X(x)P_Y(y)$$

for *all* values of x and y. This means, events $\{X = x\}$ and $\{Y = y\}$ are independent for all x and y; in other words, variables X and Y take their values independently of each other.

In problems, to show independence of X and Y, we have to check whether the joint pmf factors into the product of marginal pmfs for *all* pairs x and y. To prove dependence, we only need to present one counterexample, a pair (x, y) with $P(x, y) \neq P_X(x)P_Y(y)$.

Example 3.6. A program consists of two modules. The number of errors, X, in the first module and the number of errors, Y, in the second module have the joint distribution, $P(0,0) = P(0,1) = P(1,0) = 0.2$, $P(1,1) = P(1,2) = P(1,3) = 0.1$, $P(0,2) = P(0,3) = 0.05$. Find (a) the marginal distributions of X and Y, (b) the probability of no errors in the first module, and (c) the distribution of the total number of errors in the program. Also, (d) find out if errors in the two modules occur independently.

Solution. It is convenient to organize the joint pmf of X and Y in a table. Adding rowwise and columnwise, we get the marginal pmfs,

$P_{(X,Y)}(x,y)$		y				$P_X(x)$
		0	1	2	3	
x	0	0.20	0.20	0.05	0.05	0.50
	1	0.20	0.10	0.10	0.10	0.50
$P_Y(y)$		0.40	0.30	0.15	0.15	1.00

This solves (a).

(b) $P_X(0) = 0.50$.

(c) Let $Z = X + Y$ be the total number of errors. To find the distribution of Z, we first identify its possible values, then find the probability of each value. We see that Z can be as small as 0 and as large as 4. Then,

$$
\begin{aligned}
P_Z(0) &= \boldsymbol{P}\{X+Y=0\} = \boldsymbol{P}\{X=0 \cap Y=0\} = P(0,0) = 0.20, \\
P_Z(1) &= \boldsymbol{P}\{X=0 \cap Y=1\} + \boldsymbol{P}\{X=1 \cap Y=0\} \\
&= P(0,1) + P(1,0) = 0.20 + 0.20 = 0.40, \\
P_Z(2) &= P(0,2) + P(1,1) = 0.05 + 0.10 = 0.15, \\
P_Z(3) &= P(0,3) + P(1,2) = 0.05 + 0.10 = 0.15, \\
P_Z(4) &= P(1,3) = 0.10.
\end{aligned}
$$

It is a good check to verify that $\sum_z P_Z(z) = 1$.

(d) To decide on the independence of X and Y, check if their joint pmf factors into a product of marginal pmfs. We see that $P_{(X,Y)}(0,0) = 0.2$ indeed equals $P_X(0)P_Y(0) = (0.5)(0.4)$. Keep checking... Next, $P_{(X,Y)}(0,1) = 0.2$ whereas $P_X(0)P_Y(1) = (0.5)(0.3) = 0.15$. There is no need to check further. We found a pair of x and y that violates the formula for independent random variables. Therefore, the numbers of errors in two modules are dependent. $\Diamond$

3.3 Expectation and variance

The distribution of a random variable or a random vector, the full collection of related probabilities, contains the entire information about its behavior. This detailed information can be summarized in a few vital characteristics describing the average value, the most likely value of a random variable, its spread, variability, etc. The most commonly used are the *expectation, variance, standard deviation, covariance,* and *correlation,* introduced in this section. Also rather popular and useful are the *mode, moments, quantiles,* and *interquartile range* that we discuss in Sections 8.1 and 9.1.1.

3.3.1 Expectation

DEFINITION 3.5 ───────

> **Expectation** or **expected value** of a random variable X is its mean, the average value.

We know that X can take different values with different probabilities. For this reason, its average value is *not* just the average of all its values. Rather, it is a *weighted average.*

(a) $\mathbf{E}(X) = 0.5$ (b) $\mathbf{E}(X) = 0.25$

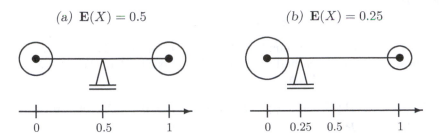

FIGURE 3.4: *Expectation as a center of gravity.*

Example 3.7. Consider a variable that takes values 0 and 1 with probabilities $P(0) = P(1) = 0.5$. That is,

$$X = \begin{cases} 0 & \text{with probability } 1/2 \\ 1 & \text{with probability } 1/2 \end{cases}$$

Observing this variable many times, we shall see $X = 0$ about 50% of times and $X = 1$ about 50% of times. The average value of X will then be close to 0.5, so it is reasonable to have $\mathbf{E}(X) = 0.5$. ◊

Now let's see how unequal probabilities of $X = 0$ and $X = 1$ affect the expectation $\mathbf{E}(X)$.

Example 3.8. Suppose that $P(0) = 0.75$ and $P(1) = 0.25$. Then, in a long run, X is equal 1 only 1/4 of times, otherwise it equals 0. Suppose we earn \$1 every time we see $X = 1$. On the average, we earn \$1 every four times, or \$0.25 per each observation. Therefore, in this case $\mathbf{E}(X) = 0.25$. ◊

A physical model for these two examples is shown in Figure 3.4. In Figure 3.4a, we put two equal masses, 0.5 units each, at points 0 and 1 and connect them with a firm but weightless rod. The masses represent probabilities $P(0)$ and $P(1)$. Now we look for a point at which the system will be balanced. It is symmetric, hence its balance point, *the center of gravity*, is in the middle, 0.5.

Figure 3.4b represents a model for the second example. Here the masses at 0 and 1 equal 0.75 and 0.25 units, respectively, according to $P(0)$ and $P(1)$. This system is balanced at 0.25, which is also its center of gravity.

Similar arguments can be used to derive the general formula for the expectation.

Expectation, discrete case	$\mu = \mathbf{E}(X) = \sum_x xP(x)$	(3.3)

This formula returns the center of gravity for a system with masses $P(x)$ allocated at points x. Expected value is often denoted by a Greek letter μ.

In a certain sense, expectation is the best forecast of X. The variable itself is random. It takes different values with different probabilities $P(x)$. At the same time, it has just one expectation $\mathbf{E}(X)$ which is non-random.

3.3.2 Expectation of a function

Often we are interested in another variable, Y, that is a function of X. For example, downloading time depends on the connection speed, profit of a computer store depends on the number of computers sold, and bonus of its manager depends on this profit. Expectation of $Y = g(X)$ is computed by a similar formula,

$$\mathbf{E}\{g(X)\} = \sum_x g(x)P(x). \tag{3.4}$$

Remark: Indeed, if g is a one-to-one function, then Y takes each value $y = g(x)$ with probability $P(x)$, and the formula for $\mathbf{E}(Y)$ can be applied directly. If g is not one-to-one, then some values of $g(x)$ will be repeated in (3.4). However, they are still multiplied by the corresponding probabilities. When we add in (3.4), these probabilities are also added, thus each value of $g(x)$ is still multiplied by the probability $P_Y(g(x))$.

3.3.3 Properties

The following *linear* properties of expectations follow directly from (3.3) and (3.4). For *any* random variables X and Y and any non-random numbers a, b, and c, we have

$$
\begin{array}{c}
\textbf{Properties} \\
\textbf{of} \\
\textbf{expectations}
\end{array}
\left|
\begin{array}{rcl}
\mathbf{E}(aX + bY + c) & = & a\,\mathbf{E}(X) + b\,\mathbf{E}(Y) + c \\[4pt]
\text{In particular,} & & \\
\mathbf{E}(X + Y) & = & \mathbf{E}(X) + \mathbf{E}(Y) \\
\mathbf{E}(aX) & = & a\,\mathbf{E}(X) \\
\mathbf{E}(c) & = & c \\[4pt]
\text{For } \textbf{independent } X \text{ and } Y, & & \\
\mathbf{E}(XY) & = & \mathbf{E}(X)\,\mathbf{E}(Y)
\end{array}
\right.
\tag{3.5}
$$

PROOF: The first property follows from the Addition Rule (3.2). For any X and Y,

$$\mathbf{E}(aX + bY + c) = \sum_x \sum_y (ax + by + c)P_{(X,Y)}(x,y)$$

$$= \sum_x ax \sum_y P_{(X,Y)}(x,y) + \sum_y by \sum_x P_{(X,Y)}(x,y) + c \sum_x \sum_y P_{(X,Y)}(x,y)$$

$$= a \sum_x xP_X(x) + b \sum_y yP_Y(y) + c.$$

The next three equalities are special cases. To prove the last property, we recall that $P_{(X,Y)}(x,y) = P_X(x)P_Y(y)$ for independent X and Y, and therefore,

$$\mathbf{E}(XY) = \sum_x \sum_y (xy)P_X(x)P_Y(y) = \sum_x xP_X(x) \sum_y yP_Y(y) = \mathbf{E}(X)\,\mathbf{E}(Y). \qquad \square$$

Remark: The last property in (3.5) holds for some dependent variables too, hence it cannot be used to verify independence of X and Y.

Example 3.9. In Example 3.6 on p. 46,

$$
\begin{aligned}
\mathbf{E}(X) &= (0)(0.5) + (1)(0.5) = 0.5 \text{ and} \\
\mathbf{E}(Y) &= (0)(0.4) + (1)(0.3) + (2)(0.15) + (3)(0.15) = 1.05,
\end{aligned}
$$

therefore, the expected total number of errors is

$$\mathbf{E}(X + Y) = 0.5 + 1.05 = 1.65.$$

$\Diamond$

Remark: Clearly, the program will never have 1.65 errors, because the number of errors is always integer. Then, should we round 1.65 to 2 errors? Absolutely not, it would be a mistake. Although both X and Y are integers, their expectations, or average values, do not have to be integers at all.

3.3.4 Variance and standard deviation

Expectation shows where the average value of a random variable is located, or where the variable is *expected* to be, plus or minus some error. How large could this "error" be, and how much can a variable *vary* around its expectation? Let us introduce some measures of variability.

Example 3.10. Here is a rather artificial but illustrative scenario. Consider two users. One receives either 48 or 52 e-mail messages per day, with a 50-50% chance of each. The other receives either 0 or 100 e-mails, also with a 50-50% chance. What is a common feature of these two distributions, and how are they different?

We see that both users receive the same average number of e-mails:

$$\mathbf{E}(X) = \mathbf{E}(Y) = 50.$$

However, in the first case, the actual number of e-mails is always close to 50, whereas it always differs from it by 50 in the second case. The first random variable, X, is more stable; it has *low variability*. The second variable, Y, has *high variability*. $\Diamond$

This example shows that variability of a random variable is measured by its distance from the mean $\mu = \mathbf{E}(X)$. In its turn, this distance is random too, and therefore, cannot serve as a characteristic of a distribution. It remains to square it and take the expectation of the result.

DEFINITION 3.6 ────────────

> **Variance** of a random variable is defined as the expected squared deviation from the mean. For discrete random variables, variance is
>
> $$\sigma^2 = \text{Var}(X) = \mathbf{E}\left(X - \mathbf{E}X\right)^2 = \sum_x (x - \mu)^2 P(x)$$

Remark: Notice that if the distance to the mean is not squared, then the result is always $\mu - \mu = 0$ bearing no information about the distribution of X.

According to this definition, variance is always non-negative. Further, it equals 0 only if $x = \mu$ for all values of x, i.e., when X is constantly equal to μ. Certainly, a constant (non-random) variable has zero variability.

Variance can also be computed as

$$\text{Var}(X) = \mathbf{E}(X^2) - \mu^2. \tag{3.6}$$

A proof of this is left as Exercise 3.39a.

DEFINITION 3.7

Standard deviation is a square root of variance,

$$\sigma = \text{Std}(X) = \sqrt{\text{Var}(X)}$$

Continuing the Greek-letter tradition, variance is often denoted by σ^2. Then, standard deviation is σ.

If X is measured in some units, then its mean μ has the same measurement unit as X. Variance σ^2 is measured in *squared units*, and therefore, it cannot be compared with X or μ. No matter how funny it sounds, it is rather normal to measure variance of profit in *squared dollars*, variance of class enrollment in *squared students*, and variance of available disk space in *squared* gigabytes. When a squared root is taken, the resulting standard deviation σ is again measured in the same units as X. This is the main reason of introducing yet another measure of variability, σ.

3.3.5 Covariance and correlation

Expectation, variance, and standard deviation characterize the distribution of a single random variable. Now we introduce measures of *association* of two random variables.

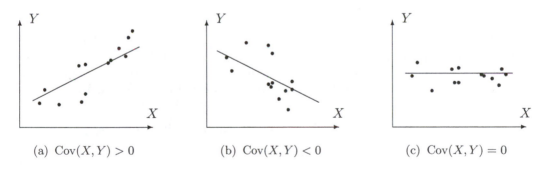

(a) $\text{Cov}(X, Y) > 0$ (b) $\text{Cov}(X, Y) < 0$ (c) $\text{Cov}(X, Y) = 0$

FIGURE 3.5: *Positive, negative, and zero covariance.*

DEFINITION 3.8

Covariance $\sigma_{XY} = \text{Cov}(X, Y)$ is defined as

$$\begin{aligned} \text{Cov}(X, Y) &= \mathbf{E}\{(X - \mathbf{E}X)(Y - \mathbf{E}Y)\} \\ &= \mathbf{E}(XY) - \mathbf{E}(X)\mathbf{E}(Y) \end{aligned}$$

It summarizes interrelation of two random variables.

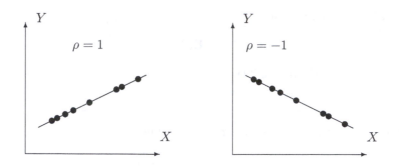

FIGURE 3.6: *Perfect correlation: $\rho = \pm 1$.*

Covariance is the expected product of deviations of X and Y from their respective expectations. If $\text{Cov}(X, Y) > 0$, then positive deviations $(X - \mathbf{E}X)$ are more likely to be multiplied by positive $(Y - \mathbf{E}Y)$, and negative $(X - \mathbf{E}X)$ are more likely to be multiplied by negative $(Y - \mathbf{E}Y)$. In short, large X imply large Y, and small X imply small Y. These variables are *positively correlated*, Figure 3.5a.

Conversely, $\text{Cov}(X, Y) < 0$ means that large X generally correspond to small Y and small X correspond to large Y. These variables are *negatively correlated*, Figure 3.5b.

If $\text{Cov}(X, Y) = 0$, we say that X and Y are *uncorrelated*, Figure 3.5c.

DEFINITION 3.9

Correlation coefficient between variables X and Y is defined as

$$\rho = \frac{\text{Cov}(X, Y)}{(\text{Std}X)(\text{Std}Y)}$$

Correlation coefficient is a rescaled, normalized covariance. Notice that covariance $\text{Cov}(X, Y)$ has a measurement unit. It is measured in units of X multiplied by units of Y. As a result, it is not clear from its value whether X and Y are strongly or weakly correlated. Really, one has to compare $\text{Cov}(X, Y)$ with the magnitude of X and Y. Correlation coefficient performs such a comparison, and as a result, it is dimensionless.

How do we interpret the value of ρ? What possible values can it take?

As a special case of famous *Cauchy-Schwarz inequality*,

$$-1 \leq \rho \leq 1,$$

where $|\rho| = 1$ is possible only when all values of X and Y lie on a straight line, as in Figure 3.6. Further, values of ρ near 1 indicate strong positive correlation, values near (-1) show strong negative correlation, and values near 0 show weak correlation or no correlation.

3.3.6   Properties

The following properties of variances, covariances, and correlation coefficients hold for any random variables X, Y, Z, and W and any non-random numbers a, b, c and d.

Properties of variances and covariances

$$\text{Var}(aX + bY + c) = a^2\,\text{Var}(X) + b^2\,\text{Var}(Y) + 2ab\,\text{Cov}(X, Y)$$

$$\text{Cov}(aX + bY, cZ + dW)$$
$$= ac\,\text{Cov}(X, Z) + ad\,\text{Cov}(X, W) + bc\,\text{Cov}(Y, Z) + bd\,\text{Cov}(Y, W)$$

$$\text{Cov}(X, Y) \quad = \quad \text{Cov}(Y, X)$$

$$\rho(X, Y) \quad = \quad \rho(Y, X)$$

In particular,

$$\text{Var}(aX + b) \quad = \quad a^2\,\text{Var}(X)$$
$$\text{Cov}(aX + b, cY + d) \quad = \quad ac\,\text{Cov}(X, Y)$$
$$\rho(aX + b, cY + d) \quad = \quad \rho(X, Y)$$

For independent X and Y,

$$\text{Cov}(X, Y) \quad = \quad 0$$
$$\text{Var}(X + Y) \quad = \quad \text{Var}(X) + \text{Var}(Y)$$

(3.7)

PROOF: To prove the first two formulas, we only need to multiply parentheses in the definitions of variance and covariance and apply (3.5):

$$\text{Var}(aX + bY + c) = \mathbf{E}\left\{aX + bY + c - \mathbf{E}(aX + bY + c)\right\}^2$$
$$= \mathbf{E}\left\{(aX - a\,\mathbf{E}X) + (bY - b\,\mathbf{E}Y) + (c - c)\right\}^2$$
$$= \mathbf{E}\left\{a(X - \mathbf{E}X)\right\}^2 + \mathbf{E}\left\{b(Y - \mathbf{E}Y)\right\}^2 + \mathbf{E}\left\{a(X - \mathbf{E}X)b(Y - \mathbf{E}Y)\right\}$$
$$\quad + \mathbf{E}\left\{b(Y - \mathbf{E}Y)a(X - \mathbf{E}X)\right\}$$
$$= a^2\,\text{Var}(X) + b^2\,\text{Var}(Y) + 2ab\,\text{Cov}(X, Y).$$

A formula for $\text{Cov}(aX + bY, cZ + dW)$ is proved similarly and is left as an exercise.

For *independent* X and Y, we have $\mathbf{E}(XY) = \mathbf{E}(X)\,\mathbf{E}(Y)$ from (3.5). Then, according to the definition of covariance, $\text{Cov}(X, Y) = 0$.

The other formulas follow directly from general cases, and their proofs are omitted.

$\square$

We see that *independent* variables are always *uncorrelated*. The reverse is not always true. There exist some variables that are uncorrelated but not independent.

Notice that adding a constant does not affect the variables' variance or covariance. It *shifts* the whole distribution of X without changing its variability or degree of dependence of another variable. The correlation coefficient does not change even when multiplied by a constant because it is recomputed to the unit scale, due to $\text{Std}(X)$ and $\text{Std}(Y)$ in the denominator in Definition 3.9.

Example 3.11. Continuing Example 3.6, we compute

x	$P_X(x)$	$xP_X(x)$	$x - \mathbf{E}X$	$(x - \mathbf{E}X)^2 P_X(x)$
0	0.5	0	-0.5	0.125
1	0.5	0.5	0.5	0.125
	$\mu_X = 0.5$			$\sigma_X^2 = 0.25$

and (using the second method of computing variances)

y	$P_Y(y)$	$yP_Y(y)$	y^2	$y^2P_Y(y)$
0	0.4	0	0	0
1	0.3	0.3	1	0.3
2	0.15	0.3	4	0.6
3	0.15	0.45	9	1.35
	$\mu_Y = 1.05$		$\mathbf{E}(Y^2) = 2.25$	

<u>Result</u>: $\text{Var}(X) = 0.25$, $\text{Var}(Y) = 2.25 - 1.05^2 = 1.1475$, $\text{Std}(X) = \sqrt{0.25} = 0.5$, and $\text{Std}(Y) = \sqrt{1.1475} = 1.0712$.

Also,

$$\mathbf{E}(XY) = \sum_x \sum_y xy P(x,y) = (1)(1)(0.1) + (1)(2)(0.1) + (1)(3)(0.1) = 0.6$$

(the other five terms in this sum are 0). Therefore,

$$\text{Cov}(X,Y) = \mathbf{E}(XY) - \mathbf{E}(X)\mathbf{E}(Y) = 0.6 - (0.5)(1.05) = 0.075$$

and

$$\rho = \frac{\text{Cov}(X,Y)}{(\text{Std}X)(\text{Std}Y)} = \frac{0.075}{(0.5)(1.0712)} = 0.1400.$$

Thus, the numbers of errors in two modules are *positively and not very strongly correlated*. $\Diamond$

NOTATION

μ or $\mathbf{E}(X)$	=	expectation
σ_X^2 or $\text{Var}(X)$	=	variance
σ_X or $\text{Std}(X)$	=	standard deviation
σ_{XY} or $\text{Cov}(X,Y)$	=	covariance
ρ_{XY}	=	correlation coefficient

3.3.7   Chebyshev's inequality

Knowing just the expectation and variance, one can find the range of values most likely taken by this variable. Russian mathematician *Pafnuty Chebyshev* (1821–1894) showed that any random variable X with expectation $\mu = \mathbf{E}(X)$ and variance $\sigma^2 = \text{Var}(X)$ belongs to the interval $\mu \pm \varepsilon = [\mu - \varepsilon, \mu + \varepsilon]$ with probability of at least $1 - (\sigma/\varepsilon)^2$. That is,

Chebyshev's inequality

$$\boldsymbol{P}\{|X - \mu| > \varepsilon\} \le \left(\frac{\sigma}{\varepsilon}\right)^2$$

for any distribution with expectation μ and variance σ^2 and for any positive ε.

(3.8)

PROOF: Here we consider discrete random variables. For other types, the proof is similar. According to Definition 3.6,

$$\sigma^2 = \sum_{\text{all } x} (x - \mu)^2 P(x) \geq \sum_{\text{only } x:|x-\mu|>\varepsilon} (x - \mu)^2 P(x)$$

$$\geq \sum_{x:|x-\mu|>\varepsilon} \varepsilon^2 P(x) = \varepsilon^2 \sum_{x:|x-\mu|>\varepsilon} P(x) = \varepsilon^2 \boldsymbol{P}\{|x - \mu| > \varepsilon\}.$$

Hence, $\boldsymbol{P}\{|x - \mu| > \varepsilon\} \leq \sigma^2/\varepsilon^2$. ☐

Chebyshev's inequality shows that only a large variance may allow a variable X to differ significantly from its expectation μ. In this case, the *risk* of seeing an extremely low or extremely high value of X increases. For this reason, risk is often measured in terms of a variance or standard deviation.

Example 3.12. Suppose the number of errors in a new software has expectation $\mu = 20$ and a standard deviation of 2. According to (3.8), there are more than 30 errors with probability

$$\boldsymbol{P}\{X > 30\} \leq \boldsymbol{P}\{|X - 20| > 10\} \leq \left(\frac{2}{10}\right)^2 = 0.04.$$

However, if the standard deviation is 5 instead of 2, then the probability of more than 30 errors can only be bounded by $\left(\frac{5}{10}\right)^2 = 0.25$. ◇

Chebyshev's inequality is universal because it works for *any* distribution. Often it gives a rather loose bound for the probability of $|X - \mu| > \varepsilon$. With more information about the distribution, this bound may be improved.

3.3.8 Application to finance

Chebyshev's inequality shows that in general, higher variance implies higher probabilities of *large deviations*, and this increases the *risk* for a random variable to take values far from its expectation.

This finds a number of immediate applications. Here we focus on evaluating risks of financial deals, allocating funds, and constructing optimal portfolios. This application is intuitively simple. The same methods can be used for the optimal allocation of computer memory, CPU time, customer support, or other resources.

Example 3.13 (CONSTRUCTION OF AN OPTIMAL PORTFOLIO). We would like to invest $10,000 into shares of companies XX and YY. Shares of XX cost $20 per share. The market analysis shows that their expected return is $1 per share with a standard deviation of $0.5. Shares of YY cost $50 per share, with an expected return of $2.50 and a standard deviation of $1 per share, and returns from the two companies are independent. In order to maximize the expected return and minimize the risk (standard deviation or variance), is it better to invest (A) all $10,000 into XX, (B) all $10,000 into YY, or (C) $5,000 in each company?

Solution. Let X be the actual (random) return from each share of XX, and Y be the actual return from each share of YY. Compute the expectation and variance of the return for each of the proposed portfolios (A, B, and C).

(a) At \$20 a piece, we can use \$10,000 to buy 500 shares of XX collecting a profit of $A = 500X$. Using (3.5) and (3.7),

$$
\begin{aligned}
\mathbf{E}(A) &= 500\,\mathbf{E}(X) = (500)(1) = 500; \\
\mathrm{Var}(A) &= 500^2\,\mathrm{Var}(X) = 500^2(0.5)^2 = 62,500.
\end{aligned}
$$

(b) Investing all \$10,000 into YY, we buy 10,000/50=200 shares of it and collect a profit of $B = 200Y$,

$$
\begin{aligned}
\mathbf{E}(B) &= 200\,\mathbf{E}(Y) = (200)(2.50) = 500; \\
\mathrm{Var}(A) &= 200^2\,\mathrm{Var}(Y) = 200^2(1)^2 = 40,000.
\end{aligned}
$$

(c) Investing \$5,000 into each company makes a portfolio consisting of 250 shares of XX and 100 shares of YY; the profit in this case will be $C = 250X + 100Y$. Following (3.7) for independent X and Y,

$$
\begin{aligned}
\mathbf{E}(C) &= 250\,\mathbf{E}(X) + 100\,\mathbf{E}(Y) = 250 + 250 = 500; \\
\mathrm{Var}(C) &= 250^2\,\mathrm{Var}(X) + 100^2\,\mathrm{Var}(Y) = 250^2(0.5)^2 + 100^2(1)^2 = 25,625.
\end{aligned}
$$

Result: The expected return is the same for each of the proposed three portfolios because each share of each company is expected to return 1/20 or 2.50/50, which is 5%. In terms of the expected return, all three portfolios are *equivalent*. Portfolio C, where investment is split between two companies, has the lowest variance; therefore, it is the least risky. This supports one of the basic principles in finance: *to minimize the risk, diversify the portfolio.*

$\Diamond$

Example 3.14 (OPTIMAL PORTFOLIO, CORRELATED RETURNS). Suppose now that the individual stock returns X and Y are no longer independent. If the correlation coefficient is $\rho = 0.4$, how will it change the results of the previous example? What if they are negatively correlated with $\rho = -0.2$?

Solution. Only the volatility of the diversified portfolio C changes due to the correlation coefficient. Now $\mathrm{Cov}(X, Y) = \rho\,\mathrm{Std}(X)\,\mathrm{Std}(Y) = (0.4)(0.5)(1) = 0.2$, thus the variance of C increases by $2(250)(100)(0.2) = 10,000$,

$$
\begin{aligned}
\mathrm{Var}(C) &= \mathrm{Var}(250X + 100Y) \\
&= 250^2\,\mathrm{Var}(X) + 100^2\,\mathrm{Var}(Y) + 2(250)(100)\,\mathrm{Cov}(X, Y) \\
&= 25,625 + 10,000 = 35,625.
\end{aligned}
$$

Nevertheless, the diversified portfolio C is still optimal.

Why did the risk of portfolio C increase due to positive correlation of the two stocks? When X and Y are positively correlated, low values of X are likely to accompany low values of Y; therefore, the probability of the overall low return is higher, increasing the risk of the portfolio.

Conversely, negative correlation means that low values of X are likely to be compensated by high values of Y, and vice versa. Thus, the risk is reduced. Say, with the given $\rho = -0.2$, we compute $\mathrm{Cov}(X, Y) = \rho\,\mathrm{Std}(X)\,\mathrm{Std}(Y) = (-0.2)(0.5)(1) = -0.1$, and

$$
\mathrm{Var}(C) = 25,625 + 2(250)(100)\,\mathrm{Cov}(X, Y) = 25,625 - 5,000 = 20,625.
$$

Diversified portfolios consisting of negatively correlated components are the least risky. $\Diamond$

Example 3.15 (OPTIMIZING EVEN FURTHER). So, after all, with $10,000 to invest, what is the most optimal portfolio consisting of shares of XX and YY, given their correlation coefficient of $\rho = -0.2$?

This is an *optimization* problem. Suppose t dollars are invested into XX and $(10,000 - t)$ dollars into YY, with the resulting profit is C_t. This amounts for $t/20$ shares of X and $(10,000 - t)/50 = 200 - t/50$ shares of YY. Plans A and B correspond to $t = 10,000$ and $t = 0$.

The expected return remains constant at $500. If $\text{Cov}(X, Y) = -0.1$, as in Example 3.14 above, then

$$
\begin{aligned}
\text{Var}(C_t) &= \text{Var}\left\{(t/20)X + (200 - t/50)Y\right\} \\
&= (t/20)^2 \text{Var}(X) + (200 - t/50)^2 \text{Var}(Y) + 2(t/20)(200 - t/50)\text{Cov}(X, Y) \\
&= (t/20)^2 (0.5)^2 + (200 - t/50)^2 (1)^2 + 2(t/20)(200 - t/50)(-0.1) \\
&= \frac{49t^2}{40,000} - 10t + 40,000.
\end{aligned}
$$

See the graph of this function in Figure 3.7. Minimum of this variance is found at $t^* =$

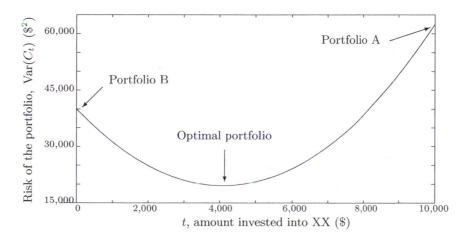

FIGURE 3.7: *Variance of a diversified portfolio.*

$10/(\frac{49t^2}{40,000}) = 4081.63$. Thus, for the most optimal portfolio, we should invest $4081.63 into XX and the remaining $5919.37 into YY. Then we achieve the smallest possible risk (variance) of (^{2})19,592, measured, as we know, in squared dollars. ◊

3.4 Families of discrete distributions

Next, we introduce the most commonly used families of discrete distributions. Amazingly, absolutely different phenomena can be adequately described by the same mathematical model, or a family of distributions. For example, as we shall see below, the number of virus attacks, received e-mails, error messages, network blackouts, telephone calls, traffic accidents, earthquakes, and so on can all be modeled by the same Poisson family of distributions.

3.4.1   Bernoulli distribution

The simplest random variable (excluding non-random ones!) takes just two possible values. Call them 0 and 1.

DEFINITION 3.10 ————————————————————————————————

> A random variable with two possible values, 0 and 1, is called a **Bernoulli variable**, its distribution is **Bernoulli distribution**, and any experiment with a *binary outcome* is called a **Bernoulli trial**.

This distribution is named after a Swiss mathematician *Jacob Bernoulli* (1654-1705) who discovered not only Bernoulli but also Binomial distribution.

Good or defective components, parts that pass or fail tests, transmitted or lost signals, working or malfunctioning hardware, benign or malicious attachments, sites that contain or do not contain a keyword, girls and boys, heads and tails, and so on, are examples of Bernoulli trials. All these experiments fit the same Bernoulli model, where we shall use generic names for the two outcomes: *"successes"* and *"failures."* These are nothing but commonly used generic names; in fact, successes do not have to be good, and failures do not have to be bad.

If $P(1) = p$ is the probability of a *success*, then $P(0) = q = 1 - p$ is the probability of a *failure*. We can then compute the expectation and variance as

$$
\begin{aligned}
\mathbf{E}(X) &= \sum_x P(x) = (0)(1-p) + (1)(p) = p, \\
\mathrm{Var}(X) &= \sum_x (x-p)^2 P(x) = (0-p)^2(1-p) + (1-p)^2 p \\
&= p(1-p)(p+1-p) = p(1-p).
\end{aligned}
$$

$$
\begin{array}{c|rcl}
& p & = & \text{probability of success} \\
\textbf{Bernoulli} & P(x) & = & \begin{cases} q = 1-p & \text{if} \quad x = 0 \\ p & \text{if} \quad x = 1 \end{cases} \\
\textbf{distribution} & \mathbf{E}(X) & = & p \\
& \mathrm{Var}(X) & = & pq
\end{array}
$$

In fact, we see that there is a whole *family of Bernoulli distributions*, indexed by a *parameter* p. Every p between 0 and 1 defines another Bernoulli distribution. The distribution with $p = 0.5$ carries the highest level of uncertainty because $\mathrm{Var}(X) = pq$ is maximized by $p = q = 0.5$. Distributions with lower or higher p have lower variances. Extreme parameters $p = 0$ and $p = 1$ define non-random variables 0 and 1, respectively; their variance is 0.

3.4.2 Binomial distribution

Now consider a sequence of independent Bernoulli trials and count the number of successes in it. This may be the number of defective computers in a shipment, the number of updated files in a folder, the number of girls in a family, the number of e-mails with attachments, etc.

DEFINITION 3.11 ——————

A variable described as the number of successes in a sequence of independent Bernoulli trials has **Binomial distribution**. Its parameters are n, the number of trials, and p, the probability of success.

Remark: "Binomial" can be translated as "two numbers," *bi* meaning "two" and *nom* meaning "a number," thus reflecting the concept of binary outcomes.

Binomial probability mass function is

$$P(x) = \boldsymbol{P}\{X = x\} = \binom{n}{x} p^x q^{n-x}, \qquad x = 0, 1, \ldots, n, \tag{3.9}$$

which is the probability of exactly x successes in n trials. In this formula, p^x is the probability of x successes, probabilities being multiplied due to independence of trials. Also, q^{n-x} is the probability of the remaining $(n-x)$ trials being failures. Finally, $\binom{n}{x} = \frac{n!}{x!(n-x)!}$ is the number of elements of the sample space Ω that form the event $\{X = x\}$. This is the number of possible orderings of x successes and $(n-x)$ failures among n trials, and it is computed as $C(n, x)$ in (2.6).

Due to a somewhat complicated form of (3.9), practitioners use a *table of Binomial distribution*, Table A2. Its entries are values of the Binomial cdf $F(x)$. If we need a pmf instead of a cdf, we can always compute it from the table as

$$P(x) = F(x) - F(x - 1).$$

Example 3.16. As part of a business strategy, randomly selected 20% of new internet service subscribers receive a special promotion from the provider. A group of 10 neighbors signs for the service. What is the probability that at least 4 of them get a special promotion?

<u>Solution.</u> We need to find the probability $\boldsymbol{P}\{X \geq 4\}$, where X is the number of people, out of 10, who receive a special promotion. This is the number of successes in 10 Bernoulli trials, therefore, X has Binomial distribution with parameters $n = 10$ and $p = 0.2$. From Table A2,

$$\boldsymbol{P}\{X \geq 4\} = 1 - F(3) = 1 - 0.8791 = \underline{0.1209}.$$

$\diamond$

Table A2 of Binomial distribution does not go beyond $n = 20$. For large n, we shall learn to use some good approximations.

Let's turn to the Binomial expectation and variance. Computing the expected value directly by (3.3) results in a complicated formula,

$$\mathbf{E}(X) = \sum_{x=0}^{n} x \binom{n}{x} p^x q^{n-x} = \dots ?$$

A shortcut can be obtained from the following important property.

Each Bernoulli trial is associated with a Bernoulli variable that equals 1 if the trial results in a success and 0 in case of a failure. Then, a sum of these variables is the overall number of successes. Thus, *any Binomial variable X can be represented as a sum of independent Bernoulli variables,*

$$X = X_1 + \dots + X_n.$$

We can use this to compute (referring to (3.5) and (3.7))

$$\mathbf{E}(X) = \mathbf{E}(X_1 + \dots + X_n) = \mathbf{E}(X_1) + \dots + \mathbf{E}(X_n) = p + \dots + p = np$$

and

$$\text{Var}(X) = \text{Var}(X_1 + \dots + X_n) = \text{Var}(X_1) + \dots + \text{Var}(X_n) = npq.$$

Binomial distribution

n	$=$	number of trials
p	$=$	probability of success
$P(x)$	$=$	$\binom{n}{x} p^x q^{n-x}$
$\mathbf{E}(X)$	$=$	np
$\text{Var}(X)$	$=$	npq

Example 3.17. An exciting computer game is released. Sixty percent of players complete all the levels. Thirty percent of them will then buy an advanced version of the game. Among 15 users, what is the expected number of people who will buy the advanced version? What is the probability that at least two people will buy it?

Solution. Let X be the number of people (successes), among the mentioned 15 users (trials), who will buy the advanced version of the game. It has Binomial distribution with $n = 15$ trials and the probability of success

$$
\begin{aligned}
p &= \mathbf{P}\{\text{buy advanced}\} \\
&= \mathbf{P}\{\text{buy advanced} \mid \text{complete all levels}\} \, \mathbf{P}\{\text{complete all levels}\} \\
&= (0.30)(0.60) = 0.18.
\end{aligned}
$$

Then we have

$$\mathbf{E}(X) = np = (15)(0.18) = \underline{2.7}$$

and

$$\mathbf{P}\{X \geq 2\} = 1 - P(0) - P(1) = 1 - (1-p)^n - np(1-p)^{n-1} = \underline{0.7813}.$$

The last probability was computed directly by formula (3.9) because the probability of success, 0.18, is not in Table A2.   ◊

Computer notes

All the distribution families discussed in this book are included in R and in MATLAB Statistics Toolbox.

In R, Binomial cdf can be calculated by the command `pbinom(x,n,p)`, where n and p are parameters. Also, `dbinom(x,n,p)` computes the probability mass function, and `rbinom(k,n,p)` generates k Binomial random variable with the given parameters.

In MATLAB, `binocdf(x,n,p)` is used for the Binomial cdf, `binopdf(x,n,p)` for the pdf and `binornd(n,p)` for generating a random variable. Alternatively, we can write `cdf('binomial',x,n,p)` for the cdf and `pdf('binomial',x,n,p)` for the pdf.

R and MATLAB commands for other distributions are constructed in the same way; see our inventory in Section A.2.

3.4.3   Geometric distribution

Again, consider a sequence of independent Bernoulli trials. Each trial results in a "success" or a "failure."

DEFINITION 3.12 ─────────────────

> The number of Bernoulli trials needed to get the first success has **Geometric distribution**.

Example 3.18. A search engine goes through a list of sites looking for a given key phrase. Suppose the search terminates as soon as the key phrase is found. The number of sites visited is Geometric. ◇

Example 3.19. A hiring manager interviews candidates, one by one, to fill a vacancy. The number of candidates interviewed until one candidate receives an offer has Geometric distribution. ◇

Geometric random variables can take any integer value from 1 to infinity, because one needs at least 1 trial to have the first success, and the number of trials needed is not limited by any specific number. (For example, there is no guarantee that among the first 10 coin tosses there will be at least one head.) The only parameter is p, the probability of a "success."

Geometric probability mass function has the form

$$P(x) = \boldsymbol{P}\{ \text{ the 1st success occurs on the } x\text{-th trial} \} = (1-p)^{x-1}p, \quad x = 1, 2, \ldots,$$

which is the probability of $(x-1)$ failures followed by one success. Comparing with (3.9), there is no number of combinations in this formula because only one outcome has the first success coming on the x-th trial.

This is the first time we see an *unbounded* random variable, that is, with no upper bound. Variable X can take any positive integer value, from 1 to ∞. It is insightful to check whether $\sum_x P(x) = 1$, as it should hold for all pmf. Indeed,

$$\sum_x P(x) = \sum_{x=1}^{\infty}(1-p)^{x-1}p = \frac{(1-p)^0}{1-(1-p)}p = 1,$$

where we noticed that the sum on the left is a *geometric series* that gave its name to Geometric distribution.

Finally, Geometric distribution has expectation $\mu = 1/p$ and variance $\sigma^2 = (1-p)/p^2$.

PROOF: The geometric series $s(q) = \sum_0^{\infty} q^x$ equals $(1-q)^{-1}$. Taking derivatives with respect to q,

$$\frac{1}{(1-q)^2} = \left(\frac{1}{1-q}\right)' = s'(q) = \left(\sum_0^{\infty} q^x\right)' = \sum_0^{\infty} xq^{x-1} = \sum_1^{\infty} xq^{x-1}.$$

It follows that for a Geometric variable X with parameter $p = 1 - q$,

$$\mathbf{E}(X) = \sum_{x=1}^{\infty} xq^{x-1}p = \frac{1}{(1-q)^2}p = \frac{1}{p}.$$

Derivation of the variance is similar. One has to take the second derivative of $s(q)$ and get, after a few manipulations, an expression for $\sum x^2 q^{x-1}$. $\qquad\square$

Geometric distribution	p	$=$ probability of success
	$P(x)$	$= (1-p)^{x-1}p, \quad x = 1, 2, \ldots$
	$\mathbf{E}(X)$	$= \dfrac{1}{p}$
	$\text{Var}(X)$	$= \dfrac{1-p}{p^2}$

(3.10)

Example 3.20 (ST. PETERSBURG PARADOX). This paradox was noticed by a Swiss mathematician *Daniel Bernoulli* (1700–1782), a nephew of Jacob. It describes *a gambling strategy that enables one to win any desired amount of money with probability one.*

Isn't it a very attractive strategy? It's real, there is no cheating!

Consider a game that can be played any number of times. Rounds are independent, and each time your winning probability is p. The game does not have to be favorable to you or even fair; this p can be any positive probability. For each round, you bet some amount x. In case of a success, you win x. If you lose a round, you lose x.

The strategy is simple. Your initial bet is the amount that you desire to win eventually. Then, if you win a round, stop. If you lose a round, double your bet and continue.

Let the desired profit be $100. The game will progress as follows.

Round	Bet	Balance... ... if lose	... if win
1	100	−100	+100 and stop
2	200	−300	+100 and stop
3	400	−700	+100 and stop
...	...	...	...

Sooner or later, the game will stop, and at this moment, your balance will be $100. Guaranteed! However, this is not what D. Bernoulli called a paradox.

How many rounds should be played? Since each round is a Bernoulli trial, the number of them, X, until the first win is a Geometric random variable with parameter p.

Is the game endless? No, on the average, it will last $\mathbf{E}(X) = 1/p$ rounds. In a fair game with $p = 1/2$, one will need 2 rounds, on the average, to win the desired amount. In an "unfair" game, with $p < 1/2$, it will take longer to win, but still a finite number of rounds. For example, if $p = 0.2$, i.e., one win in five rounds, then on the average, one stops after $1/p = 5$ rounds. This is not a paradox yet.

Finally, how much money does one need to have in order to be able to follow this strategy? Let Y be the amount of the last bet. According to the strategy, $Y = 100 \cdot 2^{X-1}$. It is a discrete random variable whose expectation equals

$$
\begin{aligned}
\mathbf{E}(Y) &= \sum_x \left(100 \cdot 2^{x-1}\right) P_X(x) = 100 \sum_{x=1}^{\infty} 2^{x-1}(1-p)^{x-1} p \\
&= 100p \sum_{x=1}^{\infty} (2(1-p))^{x-1} = \begin{cases} \dfrac{100p}{2(1-p)} & \text{if} \quad p > 1/2 \\ +\infty & \text{if} \quad p \le 1/2. \end{cases}
\end{aligned}
$$

This is the St. Petersburg Paradox! A random variable that is always finite has an infinite expectation! Even when the game is fair offering a 50-50 chance to win, one has to be (on the average!) infinitely wealthy to follow this strategy.

To the best of our knowledge, every casino has a limit on the maximum bet, making sure that gamblers cannot fully execute this St. Petersburg strategy. When such a limit is enforced, it can be proved theoretically that a winning strategy does not exist. ◇

3.4.4 Negative Binomial distribution

When we studied Geometric distribution and St. Petersburg paradox in Section 3.4.3, we played a game until the first win. Now keep playing until we reach k wins. The number of games played is then *Negative Binomial*.

DEFINITION 3.13 ─────────────

> In a sequence of independent Bernoulli trials, the number of trials needed to obtain k successes has **Negative Binomial distribution**.

In some sense, Negative Binomial distribution is opposite to Binomial distribution. Binomial variables count the number of successes in a fixed number of trials whereas Negative Binomial variables count the number of trials needed to see a fixed number of successes. Other than this, there is nothing "negative" about this distribution.

Negative Binomial probability mass function is

$$P(x) = \boldsymbol{P}\{ \text{ the } x\text{-th trial results in the } k\text{-th success } \}$$

$$= \boldsymbol{P}\left\{ \begin{array}{c} (k-1) \text{ successes in the first } (x-1) \text{ trials,} \\ \text{and the last trial is a success} \end{array} \right\}$$

$$= \binom{x-1}{k-1}(1-p)^{x-k}p^k.$$

This formula accounts for the probability of k successes, the remaining $(x-k)$ failures, and the number of outcomes–sequences with the k-th success coming on the x-th trial.

Negative Binomial distribution has two parameters, k and p. With $k = 1$, it becomes Geometric. Also, each Negative Binomial variable can be represented as a sum of independent Geometric variables,

$$X = X_1 + \ldots + X_k, \tag{3.11}$$

with the same probability of success p. Indeed, the number of trials until the k-th success consists of a Geometric number of trials X_1 until the first success, an additional Geometric number of trials X_2 until the second success, etc.

Because of (3.11), we have

$$\mathbf{E}(X) = \mathbf{E}(X_1 + \ldots + X_k) = \frac{k}{p};$$

$$\mathrm{Var}(X) = \mathrm{Var}(X_1 + \ldots + X_k) = \frac{k(1-p)}{p^2}.$$

$$
\begin{array}{l|ll}
 & k & = \text{ number of successes} \\
 & p & = \text{ probability of success} \\
\textbf{Negative} & P(x) & = \dbinom{x-1}{k-1}(1-p)^{x-k}p^k, \quad x = k, k+1, \ldots \\
\textbf{Binomial} & & \\
\textbf{distribution} & \mathbf{E}(X) & = \dfrac{k}{p} \\
 & & \\
 & \mathrm{Var}(X) & = \dfrac{k(1-p)}{p^2}
\end{array}
\tag{3.12}
$$

Example 3.21 (SEQUENTIAL TESTING). In a recent production, 5% of certain electronic components are defective. We need to find 12 non-defective components for our 12 new computers. Components are tested until 12 non-defective ones are found. What is the probability that more than 15 components will have to be tested?

Solution. Let X be the number of components tested until 12 non-defective ones are found. It is a number of trials needed to see 12 successes, hence X has Negative Binomial distribution with $k = 12$ and $p = 0.05$.

We need $\boldsymbol{P}\{X > 15\} = \sum_{16}^{\infty} P(x)$ or $1 - F(15)$; however, there is no table of Negative Binomial distribution in the Appendix, and applying the formula for $P(x)$ directly is rather cumbersome. What would be a quick solution?

Virtually any Negative Binomial problem can be solved by a Binomial distribution. Although X is not Binomial at all, the probability $P\{X > 15\}$ can be related to some Binomial variable. In our example,

$$
\begin{aligned}
P\{X > 15\} &= P\{\text{ more than 15 trials needed to get 12 successes }\} \\
&= P\{\text{ 15 trials are not sufficient }\} \\
&= P\{\text{ there are fewer than 12 successes in 15 trials }\} \\
&= P\{Y < 12\},
\end{aligned}
$$

where Y is the number of successes (non-defective components) in 15 trials, which is a Binomial variable with parameters $n = 15$ and $p = 0.95$. From Table A2 on p. 427,

$$
P\{X > 15\} = P\{Y < 12\} = P\{Y \le 11\} = F(11) = \underline{0.0055}.
$$

This technique, expressing a probability about one random variable in terms of another random variable, is rather useful. Soon it will help us relate Gamma and Poisson distributions and simplify computations significantly. ◇

3.4.5   Poisson distribution

The next distribution is related to a concept of *rare events*, or Poissonian events. Essentially it means that two such events are extremely unlikely to occur simultaneously or within a very short period of time. Arrivals of jobs, telephone calls, e-mail messages, traffic accidents, network blackouts, virus attacks, errors in software, floods, and earthquakes are examples of rare events. The rigorous definition of rare events is given in Section 6.3.2.

DEFINITION 3.14 ————————————————————————

> The number of rare events occurring within a fixed period of time has **Poisson distribution**.

This distribution bears the name of a famous French mathematician *Siméon-Denis Poisson* (1781–1840).

Poisson distribution

$$
\begin{aligned}
\lambda &= \text{frequency, average number of events} \\
P(x) &= e^{-\lambda}\frac{\lambda^x}{x!},\ x = 0, 1, 2, \ldots \\
\mathbf{E}(X) &= \lambda \\
\mathrm{Var}(X) &= \lambda
\end{aligned}
$$

PROOF: This probability mass function satisfies $\sum_0^\infty P(x) = 1$ because the Taylor series for $f(\lambda) = e^\lambda$ at $\lambda = 0$ is

$$
e^\lambda = \sum_{x=0}^\infty \frac{\lambda^x}{x!}, \tag{3.13}
$$

and this series converges for all λ.

The expectation $\mathbf{E}(X)$ can be derived from (3.13) as

$$\mathbf{E}(X) = \sum_{x=0}^{\infty} x e^{-\lambda} \frac{\lambda^x}{x!} = 0 + e^{-\lambda} \sum_{x=1}^{\infty} \frac{\lambda^{x-1} \cdot \lambda}{(x-1)!} = e^{-\lambda} \cdot e^{\lambda} \cdot \lambda = \lambda.$$

Similarly,

$$\mathbf{E}(X^2) - \mathbf{E}(X) = \sum_{x=0}^{\infty} (x^2 - x) P(x) = \sum_{x=2}^{\infty} x(x-1) e^{-\lambda} \frac{\lambda^x}{x!} = e^{-\lambda} \sum_{x=2}^{\infty} \frac{\lambda^{x-2} \cdot \lambda^2}{(x-2)!} = \lambda^2,$$

and therefore,

$$\text{Var}(X) = \mathbf{E}(X^2) - \mathbf{E}^2(X) = (\lambda^2 + \mathbf{E}(X)) - \mathbf{E}^2(X) = \lambda^2 + \lambda - \lambda^2 = \lambda.$$

$\square$

A Poisson variable can take any nonnegative integer value because there may be no rare events within the chosen period, on one end, and the possible number of events is not limited, on the other end. Poisson distribution has one parameter, $\lambda > 0$, which is the average number of the considered rare events. Values of its cdf are given in Table A3 on p. 430.

Example 3.22 (NEW ACCOUNTS). Customers of an internet service provider initiate new accounts at the average rate of 10 accounts per day.

(a) What is the probability that more than 8 new accounts will be initiated today?
(b) What is the probability that more than 16 accounts will be initiated within 2 days?

Solution.   (a) New account initiations qualify as rare events because no two customers open accounts simultaneously. Then the number X of today's new accounts has Poisson distribution with parameter $\lambda = 10$. From Table A3,

$$P\{X > 8\} = 1 - F_X(8) = 1 - 0.333 = \underline{0.667}.$$

(b) The number of accounts, Y, opened within 2 days does *not* equal $2X$. Rather, Y is another Poisson random variable whose parameter equals 20. Indeed, the parameter is the average number of rare events, which, over the period of two days, doubles the one-day average. Using Table A3 with $\lambda = 20$,

$$P\{Y > 16\} = 1 - F_Y(16) = 1 - 0.221 = \underline{0.779}.$$

$\Diamond$

3.4.6   Poisson approximation of Binomial distribution

Poisson distribution can be effectively used to approximate Binomial probabilities when the number of trials n is large, and the probability of success p is small. Such an approximation is adequate, say, for $n \geq 30$ and $p \leq 0.05$, and it becomes more accurate for larger n.

Example 3.23 (NEW ACCOUNTS, CONTINUED). Indeed, the situation in Example 3.22 can be viewed as a sequence of Bernoulli trials. Suppose there are $n = 400,000$ potential internet users in the area, and on any specific day, each of them opens a new account with probability $p = 0.000025$. We see that the number of new accounts is the number of successes, hence a Binomial model with expectation $\mathbf{E}(X) = np = 10$ is possible. However, a distribution with such extreme n and p is unlikely to be found in any table, and computing its pmf by hand is tedious. Instead, one can use Poisson distribution with the same expectation $\lambda = 10$. $\Diamond$

Poisson approximation to Binomial	Binomial$(n, p) \approx$ Poisson(λ) where $n \geq 30$, $p \leq 0.05$, $np = \lambda$	(3.14)

Remark: Mathematically, it means closeness of Binomial and Poisson pmf,

$$\lim_{\substack{n \to \infty \\ p \to 0 \\ np \to \lambda}} \binom{n}{x} p^x (1-p)^{n-x} = e^{-\lambda} \frac{\lambda^x}{x!}$$

and this is what S. D. Poisson has shown.

When p is large ($p \geq 0.95$), the Poisson approximation is applicable too. The probability of a failure $q = 1 - p$ is small in this case. Then, we can approximate the number of failures, which is also Binomial, by a Poisson distribution.

Example 3.24. Ninety-seven percent of electronic messages are transmitted with no error. What is the probability that out of 200 messages, at least 195 will be transmitted correctly?

<u>Solution</u>. Let X be the number of correctly transmitted messages. It is the number of successes in 200 Bernoulli trials, thus X is Binomial with $n = 200$ and $p = 0.97$. Poisson approximation cannot be applied to X because p is too large. However, the number of failures Y is also Binomial, with parameters $n = 200$ and $q = 0.03$, and it is approximately Poisson with $\lambda = nq = 6$. From Table A3,

$$P\{X \geq 195\} = P\{Y \leq 5\} = F_Y(5) \approx \underline{0.446}.$$

$\Diamond$

There is a great variety of applications involving a large number of trials with a small probability of success. If the trials are not independent, the number of successes is not Binomial, in general. However, if dependence is weak, the use of Poisson approximation in such problems can still produce remarkably accurate results.

Example 3.25 (BIRTHDAY PROBLEM). This continues Exercise 2.27 on p. 37. Consider a class with $N \geq 10$ students. Compute the probability that at least two of them have their birthdays on the same day. How many students should be in class in order to have this probability above 0.5?

<u>Solution</u>. Poisson approximation will be used for the number of shared birthdays among all

$$n = \binom{N}{2} = \frac{N(N-1)}{2}$$

pairs of students in this class. In each pair, both students are born on the same day with probability $p = 1/365$. Each pair is a Bernoulli trial because the two birthdays either match or don't match. Besides, matches in two different pairs are "nearly" independent. Therefore, X, the number of pairs sharing birthdays, is "almost" Binomial. For $N \geq 10$, $n \geq 45$ is large, and p is small, thus, we shall use Poisson approximation with $\lambda = np = N(N-1)/730$,

$$P\{\text{there are two students sharing birthday}\} = 1 - P\{\text{no matches}\}$$

$$= 1 - P\{X = 0\} \approx 1 - e^{-\lambda} \approx 1 - e^{-N^2/730}.$$

Solving the inequality $1 - e^{-N^2/730} > 0.5$, we obtain $N > \sqrt{730 \ln 2} \approx 22.5$. That is, in a class of at least $N = 23$ students, there is a more than 50% chance that at least two students were born on the same day of the year! ◊

The introduced method can only be applied to very small and very large values of p. For moderate p $(0.05 \leq p \leq 0.95)$, the Poisson approximation may not be accurate. These cases are covered by the Central Limit Theorem on p. 93.

Summary and conclusions

Discrete random variables can take a finite or countable number of isolated values with different probabilities. Collection of all such probabilities is a distribution, which describes the behavior of a random variable. Random vectors are sets of random variables; their behavior is described by the joint distribution. Marginal probabilities can be computed from the joint distribution by the Addition Rule.

The average value of a random variable is its expectation. Variability around the expectation is measured by the variance and the standard deviation. Covariance and correlation measure association of two random variables. For any distribution, probabilities of large deviations can be bounded by Chebyshev's inequality, using only the expectation and variance.

Different phenomena can often be described by the same probabilistic model, or a family of distributions. The most commonly used discrete families are Binomial, including Bernoulli, Negative Binomial, including Geometric, and Poisson. Each family of distributions is used for a certain general type of situations; it has its parameters and a clear formula and/or a table for computing probabilities. These families are summarized in Section A.2.1.

Exercises

3.1. A computer virus is trying to corrupt two files. The first file will be corrupted with probability 0.4. Independently of it, the second file will be corrupted with probability 0.3.

   (a) Compute the probability mass function (pmf) of X, the number of corrupted files.

   (b) Draw a graph of its cumulative distribution function (cdf).

3.2. Every day, the number of network blackouts has a distribution (probability mass function)

x	0	1	2
$P(x)$	0.7	0.2	0.1

A small internet trading company estimates that each network blackout results in a $500 loss. Compute expectation and variance of this company's daily loss due to blackouts.

3.3. There is one error in one of five blocks of a program. To find the error, we test three randomly selected blocks. Let X be the number of errors in these three blocks. Compute $\mathbf{E}(X)$ and $\mathrm{Var}(X)$.

3.4. Tossing a fair die is an experiment that can result in any integer number from 1 to 6 with equal probabilities. Let X be the number of dots on the top face of a die. Compute $\mathbf{E}(X)$ and $\mathrm{Var}(X)$.

3.5. A software package consists of 12 programs, five of which must be upgraded. If 4 programs are randomly chosen for testing,

   (a) What is the probability that at least two of them must be upgraded?

   (b) What is the expected number of programs, out of the chosen four, that must be upgraded?

3.6. A computer program contains one error. In order to find the error, we split the program into 6 blocks and test two of them, selected at random. Let X be the number of errors in these blocks. Compute $\mathbf{E}(X)$.

3.7. The number of home runs scored by a certain team in one baseball game is a random variable with the distribution

x	0	1	2
$P(x)$	0.4	0.4	0.2

The team plays 2 games. The number of home runs scored in one game is independent of the number of home runs in the other game. Let Y be the *total* number of home runs. Find $\mathbf{E}(Y)$ and $\mathrm{Var}(Y)$.

3.8. A computer user tries to recall her password. She knows it can be one of 4 possible passwords. She tries her passwords until she finds the right one. Let X be the number of wrong passwords she uses before she finds the right one. Find $\mathbf{E}(X)$ and $\mathrm{Var}(X)$.

3.9. It takes an average of 40 seconds to download a certain file, with a standard deviation of 5 seconds. The actual distribution of the download time is unknown. Using Chebyshev's inequality, what can be said about the probability of spending more than 1 minute for this download?

3.10. Every day, the number of traffic accidents has the probability mass function

x	0	1	2	more than 2
$P(x)$	0.6	0.2	0.2	0

independently of other days. What is the probability that there are more accidents on Friday than on Thursday?

3.11. Two dice are tossed. Let X be *the smaller* number of points. Let Y be *the larger* number of points. If both dice show the same number, say, z points, then $X = Y = z$.

(a) Find the joint probability mass function of (X, Y).

(b) Are X and Y independent? Explain.

(c) Find the probability mass function of X.

(d) If $X = 2$, what is the probability that $Y = 5$?

3.12. Two random variables, X and Y, have the joint distribution $P(x, y)$,

$P(x,y)$		x	
		0	1
y	0	0.5	0.2
	1	0.2	0.1

(a) Are X and Y independent? Explain.

(b) Are $(X + Y)$ and $(X - Y)$ independent? Explain.

3.13. Two random variables X and Y have the joint distribution, $P(0,0) = 0.2$, $P(0,2) = 0.3$, $P(1,1) = 0.1$, $P(2,0) = 0.3$, $P(2,2) = 0.1$, and $P(x, y) = 0$ for all other pairs (x, y).

(a) Find the probability mass function of $Z = X + Y$.

(b) Find the probability mass function of $U = X - Y$.

(c) Find the probability mass function of $V = XY$.

3.14. An internet service provider charges its customers for the time of the internet use rounding it up to the nearest hour. The joint distribution of the used time (X, hours) and the charge per hour (Y, cents) is given in the table below.

$P(x,y)$		x			
		1	2	3	4
	1	0	0.06	0.06	0.10
y	2	0.10	0.10	0.04	0.04
	3	0.40	0.10	0	0

Each customer is charged $Z = X \cdot Y$ cents, which is the number of hours multiplied by the price of each hour. Find the distribution of Z.

3.15. Let X and Y be the number of hardware failures in two computer labs in a given month. The joint distribution of X and Y is given in the table below.

$P(x,y)$		x		
		0	1	2
	0	0.52	0.20	0.04
y	1	0.14	0.02	0.01
	2	0.06	0.01	0

(a) Compute the probability of at least one hardware failure.

(b) From the given distribution, are X and Y independent? Why or why not?

3.16. The number of hardware failures, X, and the number of software failures, Y, on any day in a small computer lab have the joint distribution $P(x, y)$, where $P(0, 0) = 0.6$, $P(0, 1) = 0.1$, $P(1, 0) = 0.1$, $P(1, 1) = 0.2$. Based on this information,

(a) Are X and Y (hardware and software failures) independent?

(b) Compute $\mathbf{E}(X + Y)$, i.e., the expected total number of failures during 1 day.

3.17. Shares of company A are sold at \$10 per share. Shares of company B are sold at \$50 per share. According to a market analyst, 1 share of each company can either gain \$1, with probability 0.5, or lose \$1, with probability 0.5, independently of the other company. Which of the following portfolios has the lowest risk:

(a) 100 shares of A

(b) 50 shares of A + 10 shares of B

(c) 40 shares of A + 12 shares of B

3.18. Shares of company A cost \$10 per share and give a profit of $X\%$. Independently of A, shares of company B cost \$50 per share and give a profit of $Y\%$. Deciding how to invest \$1,000, Mr. X chooses between 3 portfolios:

(a) 100 shares of A,

(b) 50 shares of A and 10 shares of B,

(c) 20 shares of B.

The distribution of X is given by probabilities:

$$P\{X = -3\} = 0.3, P\{X = 0\} = 0.2, P\{X = 3\} = 0.5.$$

The distribution of Y is given by probabilities:

$$P\{Y = -3\} = 0.4, P\{Y = 3\} = 0.6.$$

Compute expectations and variances of the total dollar profit generated by portfolios (a), (b), and (c). What is the least risky portfolio? What is the most risky portfolio?

3.19. A and B are two competing companies. An investor decides whether to buy

(a) 100 shares of A, or

(b) 100 shares of B, or

(c) 50 shares of A and 50 shares of B.

A profit made on 1 share of A is a random variable X with the distribution $P(X = 2) = P(X = -2) = 0.5$.

A profit made on 1 share of B is a random variable Y with the distribution $P(Y = 4) = 0.2, P(Y = -1) = 0.8$.

If X and Y are independent, compute the expected value and variance of the total profit for strategies (a), (b), and (c).

3.20. A quality control engineer tests the quality of produced computers. Suppose that 5% of computers have defects, and defects occur independently of each other.

(a) Find the probability of exactly 3 defective computers in a shipment of twenty.

(b) Find the probability that the engineer has to test at least 5 computers in order to find 2 defective ones.

3.21. A lab network consisting of 20 computers was attacked by a computer virus. This virus enters each computer with probability 0.4, independently of other computers. Find the probability that it entered at least 10 computers.

3.22. Five percent of computer parts produced by a certain supplier are defective. What is the probability that a sample of 16 parts contains more than 3 defective ones?

3.23. Every day, a lecture may be canceled due to inclement weather with probability 0.05. Class cancelations on different days are independent.

(a) There are 15 classes left this semester. Compute the probability that at least 4 of them get canceled.

(b) Compute the probability that the tenth class this semester is the third class that gets canceled.

3.24. An internet search engine looks for a certain keyword in a sequence of independent web sites. It is believed that 20% of the sites contain this keyword.

(a) Compute the probability that at least 5 of the first 10 sites contain the given keyword.

(b) Compute the probability that the search engine had to visit at least 5 sites in order to find the first occurrence of a keyword.

3.25. About ten percent of users do not close Windows properly. Suppose that Windows is installed in a public library that is used by random people in a random order.

(a) On the average, how many users of this computer *do not* close Windows properly before someone *does* close it properly?

(b) What is the probability that exactly 8 of the next 10 users will close Windows properly?

3.26. After a computer virus entered the system, a computer manager checks the condition of all important files. She knows that each file has probability 0.2 to be damaged by the virus, independently of other files.

(a) Compute the probability that at least 5 of the first 20 files are damaged.

(b) Compute the probability that the manager has to check at least 6 files in order to find 3 undamaged files.

3.27. Natasha is learning a poem for the New Year show. She recites it until she can finally read the whole verse without a single mistake. However, each time there is only a 20% probability of that, and it seems independent of previous recitations.

(a) Compute the probability that it will take Natasha more than 10 recitations.

(b) What is the expected number of times Natasha will recite the poem?

(c) Masha suggests that Natasha should keep working on this until she reads the whole verse without a single mistake three times. Repeat (a) and (b) assuming that Natasha agrees with this recommendation.

3.28. Messages arrive at an electronic message center at random times, with an average of 9 messages per hour.

(a) What is the probability of receiving *at least* five messages during the next hour?

(b) What is the probability of receiving *exactly* five messages during the next hour?

3.29. The number of received electronic messages has Poisson distribution with some parameter λ. Using Chebyshev inequality, show that the probability of receiving more than 4λ messages does not exceed $1/(9\lambda)$.

3.30. An insurance company divides its customers into 2 groups. Twenty percent of customers are in the high-risk group, and eighty percent are in the low-risk group. The high-risk customers make an average of 1 accident per year while the low-risk customers make an average of 0.1 accidents per year. Eric had no accidents last year. What is the probability that he is a high-risk driver?

3.31. Eric from Exercise 3.30 continues driving. After three years, he still has no traffic accidents. Now, what is the conditional probability that he is a high-risk driver?

3.32. Before the computer is assembled, its vital component (motherboard) goes through a special inspection. Only 80% of components pass this inspection.

(a) What is the probability that at least 18 of the next 20 components pass inspection?

(b) On the average, how many components should be inspected until a component that passes inspection is found?

3.33. On the average, 1 computer in 800 crashes during a severe thunderstorm. A certain company had 4,000 working computers when the area was hit by a severe thunderstorm.

(a) Compute the probability that less than 10 computers crashed.

(b) Compute the probability that exactly 10 computers crashed.

You may want to use a suitable approximation.

3.34. The number of computer shutdowns during any month has a Poisson distribution, averaging 0.25 shutdowns per month.

(a) What is the probability of at least 3 computer shutdowns during the next year?

(b) During the next year, what is the probability of at least 3 months (out of 12) with exactly 1 computer shutdown in each?

3.35. A dangerous computer virus attacks a folder consisting of 250 files. Files are affected by the virus independently of one another. Each file is affected with the probability 0.032. What is the probability that more than 7 files are affected by this virus?

3.36. In some city, the probability of a thunderstorm on any day is 0.6. During a thunderstorm, the number of traffic accidents has Poisson distribution with parameter 10. Otherwise, the number of traffic accidents has Poisson distribution with parameter 4. If there were 7 accidents yesterday, what is the probability that there was a thunderstorm?

3.37. An interactive system consists of ten terminals that are connected to the central computer. At any time, each terminal is ready to transmit a message with probability 0.7, independently of other terminals. Find the probability that exactly 6 terminals are ready to transmit at 8 o'clock.

3.38. Network breakdowns are unexpected rare events that occur every 3 weeks, on the average. Compute the probability of more than 4 breakdowns during a 21-week period.

3.39. Simplifying expressions, derive from the definitions of variance and covariance that

(a) $\mathrm{Var}(X) = \mathbf{E}(X^2) - \mathbf{E}^2(X)$;

(b) $\mathrm{Cov}(X, Y) = \mathbf{E}(XY) - \mathbf{E}(X)\mathbf{E}(Y)$.

3.40. Show that

$$\mathrm{Cov}(aX + bY + c, dZ + eW + f)$$
$$= ad\,\mathrm{Cov}(X, Z) + ae\,\mathrm{Cov}(X, W) + bd\,\mathrm{Cov}(Y, Z) + be\,\mathrm{Cov}(Y, W)$$

for any random variables X, Y, Z, W, and any non-random numbers a, b, c, d, e, f.

Chapter 4

Continuous Distributions

Recall that any discrete distribution is concentrated on a finite or countable number of isolated values. Conversely, *continuous variables can take any value of an interval,* (a, b), $(a, +\infty)$, $(-\infty, +\infty)$, etc. Various times like service time, installation time, download time, failure time, and also physical measurements like weight, height, distance, velocity, temperature, and connection speed are examples of continuous random variables.

4.1   Probability density

For all continuous variables, the probability mass function (pmf) is always equal to zero,[1]

$$P(x) = 0 \quad \text{for all } x.$$

As a result, the pmf does not carry any information about a random variable. Rather, we can use the *cumulative distribution function* (cdf) $F(x)$. In the continuous case, it equals

$$F(x) = P\{X \leq x\} = P\{X < x\}.$$

These two expressions for $F(x)$ differ by $P\{X = x\} = P(x) = 0$.

In both continuous and discrete cases, the cdf $F(x)$ is a non-decreasing function that ranges from 0 to 1. Recall from Chapter 3 that in the discrete case, the graph of $F(x)$ has *jumps* of magnitude $P(x)$. For continuous distributions, $P(x) = 0$, which means no jumps. The cdf in this case is a continuous function (see, for example, Figure 4.2 on p. 77).

---

[1]**Remark:** In fact, any probability mass function $P(x)$ can give positive probabilities to a finite or countable set only. They key is that we must have $\sum_x P(x) = 1$. So, there can be at most 2 values of x with $P(x) \geq 1/2$, at most 4 values with $P(x) \geq 1/4$, etc. Continuing this way, we can list all x where $P(x) > 0$. And since all such values with positive probabilities can be listed, their set is at most countable. It cannot be an interval because any interval is uncountable. This agrees with $P(x) = 0$ for continuous random variables that take entire intervals of values.

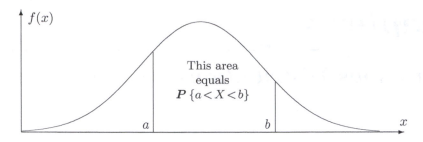

FIGURE 4.1: *Probabilities are areas under the density curve.*

Assume, additionally, that $F(x)$ has a derivative. This is the case for all commonly used continuous distributions, but in general, it is not guaranteed by continuity and monotonicity (the famous Cantor function is a counterexample).

DEFINITION 4.1 ─────────────────────────────────

Probability density function (pdf, density) is the derivative of the cdf, $f(x) = F'(x)$. The distribution is called **continuous** if it has a density.

Then, $F(x)$ is an *antiderivative* of a density. By the Fundamental Theorem of Calculus, the integral of a density from a to b equals to the difference of antiderivatives, i.e.,

$$\int_a^b f(x)dx = F(b) - F(a) = \boldsymbol{P}\left\{a < X < b\right\},$$

where we notice again that the probability in the right-hand side also equals $\boldsymbol{P}\left\{a \le X < b\right\}$, $\boldsymbol{P}\left\{a < X \le b\right\}$, and $\boldsymbol{P}\left\{a \le X \le b\right\}$.

Probability density function

$$f(x) = F'(x)$$
$$\boldsymbol{P}\left\{a < X < b\right\} = \int_a^b f(x)dx$$

Thus, probabilities can be calculated by integrating a density over the given sets. Furthermore, the integral $\int_a^b f(x)dx$ equals the area below the density curve between the points a and b. Therefore, geometrically, probabilities are represented by *areas* (Figure 4.1). Substituting $a = -\infty$ and $b = +\infty$, we obtain

$$\int_{-\infty}^b f(x)dx = \boldsymbol{P}\left\{-\infty < X < b\right\} = F(b) \text{ and } \int_{-\infty}^{+\infty} f(x)dx = \boldsymbol{P}\left\{-\infty < X < +\infty\right\} = 1.$$

That is, the total area below the density curve equals 1.

Looking at Figure 4.1, we can see why $P(x) = 0$ for all continuous random variables. That is because

$$P(x) = \boldsymbol{P}\left\{x \le X \le x\right\} = \int_x^x f = 0.$$

Geometrically, it is the area below the density curve, where two sides of the region collapse into one.

Example 4.1. The lifetime, in years, of some electronic component is a continuous random variable with the density

$$f(x) = \begin{cases} \dfrac{k}{x^3} & \text{for} \quad x \geq 1 \\ 0 & \text{for} \quad x < 1. \end{cases}$$

Find k, draw a graph of the cdf $F(x)$, and compute the probability for the lifetime to exceed 5 years.

<u>Solution.</u> Find k from the condition $\int f(x)dx = 1$:

$$\int_{-\infty}^{+\infty} f(x)dx = \int_1^{+\infty} \frac{k}{x^3}dx = -\left.\frac{k}{2x^2}\right|_{x=1}^{+\infty} = \frac{k}{2} = 1.$$

Hence, $k = 2$. Integrating the density, we get the cdf,

$$F(x) = \int_{-\infty}^{x} f(y)dy = \int_1^{x} \frac{2}{y^3}dy = -\left.\frac{1}{y^2}\right|_{y=1}^{x} = 1 - \frac{1}{x^2}$$

for $x > 1$. Its graph is shown in Figure 4.2.

Next, compute the probability for the lifetime to exceed 5 years,

$$\boldsymbol{P}\{X > 5\} = 1 - F(5) = 1 - \left(1 - \frac{1}{5^2}\right) = 0.04.$$

We can also obtain this probability by integrating the density,

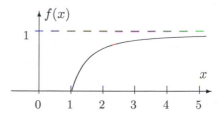

FIGURE 4.2: *Cdf for Example 4.1.*

$$\boldsymbol{P}\{X > 5\} = \int_5^{+\infty} f(x)dx = \int_5^{+\infty} \frac{2}{x^3}dx = -\left.\frac{1}{x^2}\right|_{x=5}^{+\infty} = \frac{1}{25} = 0.04.$$

$\diamond$

Distribution	Discrete	Continuous
Definition	$P(x) = \boldsymbol{P}\{X = x\}$ (pmf)	$f(x) = F'(x)$ (pdf)
Computing probabilities	$\boldsymbol{P}\{X \in A\} = \sum_{x \in A} P(x)$	$\boldsymbol{P}\{X \in A\} = \int_A f(x)dx$
Cumulative distribution function	$F(x) = \boldsymbol{P}\{X \leq x\} = \sum_{y \leq x} P(y)$	$F(x) = \boldsymbol{P}\{X \leq x\} = \int_{-\infty}^{x} f(y)dy$
Total probability	$\sum_x P(x) = 1$	$\int_{-\infty}^{\infty} f(x)dx = 1$

TABLE 4.1: *Pmf $P(x)$ versus pdf $f(x)$.*

Analogy: pmf versus pdf

The role of a density for continuous distributions is very similar to the role of the probability mass function for discrete distributions. Most vital concepts can be translated from the discrete case to the continuous case by replacing pmf $P(x)$ with pdf $f(x)$ and integrating instead of summing, as in Table 4.1.

Joint and marginal densities

DEFINITION 4.2

For a vector of random variables, the **joint cumulative distribution function** is defined as

$$F_{(X,Y)}(x,y) = \boldsymbol{P}\{X \le x \ \cap Y \le y\}.$$

The **joint density** is the *mixed derivative* of the joint cdf,

$$f_{(X,Y)}(x,y) = \frac{\partial^2}{\partial x \partial y} F_{(X,Y)}(x,y).$$

Similarly to the discrete case, a marginal density of X or Y can be obtained by integrating out the other variable. Variables X and Y are *independent* if their joint density factors into the product of marginal densities. Probabilities about X and Y can be computed by integrating the joint density over the corresponding set of vector values $(x,y) \in \mathbb{R}^2$. This is also analogous to the discrete case; see Table 4.2.

These concepts are directly extended to three or more variables.

Expectation and variance

Continuing our analogy with the discrete case, *expectation* of a continuous variable is also defined as a center of gravity,

Distribution	Discrete	Continuous
Marginal distributions	$P(x) = \sum_y P(x,y)$ $P(y) = \sum_x P(x,y)$	$f(x) = \int f(x,y)dy$ $f(y) = \int f(x,y)dx$
Independence	$P(x,y) = P(x)P(y)$	$f(x,y) = f(x)f(y)$
Computing probabilities	$\boldsymbol{P}\{(X,Y) \in A\}$ $= \sum \sum_{(x,y) \in A} P(x,y)$	$\boldsymbol{P}\{(X,Y) \in A\}$ $= \iint_{(x,y) \in A} f(x,y)\,dx\,dy$

TABLE 4.2: *Joint and marginal distributions in discrete and continuous cases.*

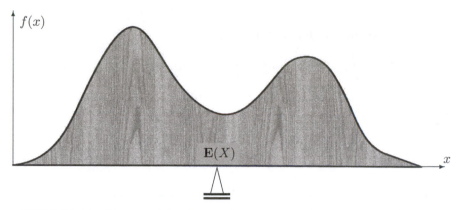

FIGURE 4.3: *Expectation of a continuous variable as a center of gravity.*

$$\mu = \mathbf{E}(X) = \int x f(x) dx$$

(compare with the discrete case on p. 48, Figure 3.4). This time, if the entire region below the density curve is cut from a piece of wood, then it will be balanced at a point with coordinate $\mathbf{E}(X)$, as shown in Figure 4.3.

Variance, standard deviation, covariance, and correlation of continuous variables are defined similarly to the discrete case, see Table 4.3. All the properties in (3.5), (3.7), and (3.8) extend to the continuous distributions. In calculations, don't forget to replace a pmf with a pdf, and a summation with an integral.

Example 4.2. A random variable X in Example 4.1 has density

$$f(x) = 2x^{-3} \text{ for } x \geq 1.$$

Its expectation equals

$$\mu = \mathbf{E}(X) = \int x f(x) dx = \int_1^\infty 2x^{-2} dx = -2x^{-1}\Big|_1^\infty = 2.$$

Discrete	Continuous
$\mathbf{E}(X) = \sum_x x P(x)$	$\mathbf{E}(X) = \int x f(x) dx$
$\begin{aligned} \text{Var}(X) &= \mathbf{E}(X - \mu)^2 \\ &= \sum_x (x - \mu)^2 P(x) \\ &= \sum_x x^2 P(x) - \mu^2 \end{aligned}$	$\begin{aligned} \text{Var}(X) &= \mathbf{E}(X - \mu)^2 \\ &= \int (x - \mu)^2 f(x) dx \\ &= \int x^2 f(x) dx - \mu^2 \end{aligned}$
$\begin{aligned} \text{Cov}(X, Y) &= \mathbf{E}(X - \mu_X)(Y - \mu_Y) \\ &= \sum_x \sum_y (x - \mu_X)(y - \mu_Y) P(x, y) \\ &= \sum_x \sum_y (xy) P(x, y) - \mu_x \mu_y \end{aligned}$	$\begin{aligned} \text{Cov}(X, Y) &= \mathbf{E}(X - \mu_X)(Y - \mu_Y) \\ &= \iint (x - \mu_X)(y - \mu_Y) f(x, y)\, dx\, dy \\ &= \iint (xy) f(x, y)\, dx\, dy - \mu_x \mu_y \end{aligned}$

TABLE 4.3: *Moments for discrete and continuous distributions.*

Computing its variance, we run into a "surprise,"

$$\sigma^2 = \text{Var}(X) = \int x^2 f(x)dx - \mu^2 = \int_1^\infty 2x^{-1}dx - 4 = 2\ln x\big|_1^\infty - 4 = +\infty.$$

This variable does not have a finite variance! Its values are always finite, but its variance is infinity. In this regard, also see Example 3.20 and St. Petersburg paradox on p. 62.   ◊

4.2   Families of continuous distributions

As in the discrete case, varieties of phenomena can be described by relatively few families of continuous distributions. Here, we shall discuss Uniform, Exponential, Gamma, and Normal families, adding Student's t, Pearson's χ^2, and Fisher's F distributions in later chapters.

4.2.1   Uniform distribution

Uniform distribution plays a unique role in stochastic modeling. As we shall see in Chapter 5, a random variable with any thinkable distribution can be generated from a Uniform random variable. Many computer languages and software are equipped with a random number generator that produces Uniform random variables. Users can convert them into variables with desired distributions and use for computer simulation of various events and processes.

Also, Uniform distribution is used in any situation when a value is picked "at random" from a given interval; that is, without any preference to lower, higher, or medium values. For example, locations of errors in a program, birthdays throughout a year, and many continuous random variables modulo 1, modulo 0.1, 0.01, etc., are uniformly distributed over their corresponding intervals.

To give equal preference to all values, the Uniform distribution has a *constant* density (Figure 4.4). On the interval (a, b), its density equals

$$f(x) = \frac{1}{b-a}, \quad a < x < b,$$

because the rectangular area below the density graph must equal 1.

For the same reason, $|b - a|$ has to be a finite number. There does not exist a Uniform distribution on the entire real line. In other words, if you are asked to choose a random number from $(-\infty, +\infty)$, you cannot do it uniformly.

The Uniform property

For any $h > 0$ and $t \in [a, b - h]$, the probability

$$\boldsymbol{P}\{\, t < X < t + h \,\} = \int_t^{t+h} \frac{1}{b-a}\, dx = \frac{h}{b-a}$$

is *independent* of t. This is the *Uniform property*: the probability is only determined by the length of the interval, but not by its location.

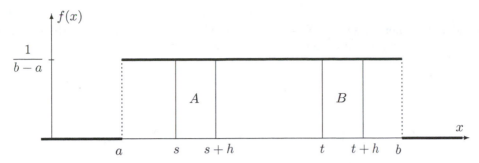

FIGURE 4.4: *The Uniform density and the Uniform property.*

Example 4.3. In Figure 4.4, rectangles A and B have the same area, showing that
$P\{s < X < s + h\} = P\{t < X < t + h\}$. ◇

Example 4.4. If a flight scheduled to arrive at 5 pm actually arrives at a Uniformly distributed time between 4:50 and 5:10, then it is equally likely to arrive before 5 pm and after 5 pm, equally likely before 4:55 and after 5:05, etc. ◇

Standard Uniform distribution

The Uniform distribution with $a = 0$ and $b = 1$ is called *Standard Uniform distribution*. The Standard Uniform density is $f(x) = 1$ for $0 < x < 1$. Most random number generators return a Standard Uniform random variable.

All the Uniform distributions are related in the following way. If X is a Uniform(a, b) random variable, then

$$Y = \frac{X - a}{b - a}$$

is Standard Uniform. Likewise, if Y is Standard Uniform, then

$$X = a + (b - a)Y$$

is Uniform(a, b). Check that $X \in (a, b)$ if and only if $Y \in (0, 1)$.

A number of other families of distributions have a "standard" member. Typically, a simple transformation converts a standard random variable into a non-standard one, and vice versa.

Expectation and variance

For a Standard Uniform variable Y,

$$\mathbf{E}(Y) = \int yf(y)dy = \int_0^1 ydy = \frac{1}{2}$$

and

$$\text{Var}(Y) = \mathbf{E}(Y^2) - \mathbf{E}^2(Y) = \int_0^1 y^2 dy - \left(\frac{1}{2}\right)^2 = \frac{1}{3} - \frac{1}{4} = \frac{1}{12}.$$

Now, consider the general case. Let $X = a + (b-a)Y$ which has a Uniform(a, b) distribution. By the properties of expectations and variances in (3.5) and (3.7),

$$\mathbf{E}(X) = \mathbf{E}\{a + (b-a)Y\} = a + (b-a)\,\mathbf{E}(Y) = a + \frac{b-a}{2} = \frac{a+b}{2}$$

and

$$\mathrm{Var}(X) = \mathrm{Var}\{a + (b-a)Y\} = (b-a)^2\,\mathrm{Var}(Y) = \frac{(b-a)^2}{12}.$$

The expectation is precisely the middle of the interval $[a, b]$. Giving no preference to left or right sides, this agrees with the Uniform property and with the physical meaning of $\mathbf{E}(X)$ as a center of gravity.

$$
\boxed{
\begin{array}{lcl}
(a, b) & = & \text{range of values} \\[2mm]
f(x) & = & \dfrac{1}{b-a}, \quad a < x < b \\[3mm]
\mathbf{E}(X) & = & \dfrac{a+b}{2} \\[3mm]
\mathrm{Var}(X) & = & \dfrac{(b-a)^2}{12}
\end{array}
}
$$

Uniform distribution

4.2.2 Exponential distribution

Exponential distribution is often used to model *time*: waiting time, interarrival time, hardware lifetime, failure time, time between telephone calls, etc. As we shall see below, in a sequence of rare events, when the number of events is Poisson, the time between events is Exponential.

Exponential distribution has density

$$f(x) = \lambda e^{-\lambda x} \text{ for } x > 0. \tag{4.1}$$

With this density, we compute the Exponential cdf, mean, and variance as

$$F(x) = \int_0^x f(t)dt = \int_0^x \lambda e^{-\lambda t} dt = 1 - e^{-\lambda x} \quad (x > 0), \tag{4.2}$$

$$\mathbf{E}(X) = \int t f(t)dt = \int_0^\infty t\lambda e^{-\lambda t} dt = \frac{1}{\lambda} \quad \left(\begin{array}{c}\text{integrating}\\\text{by parts}\end{array}\right), \tag{4.3}$$

$$\mathrm{Var}(X) = \int t^2 f(t)dt - \mathbf{E}^2(X)$$

$$= \int_0^\infty t^2 \lambda e^{-\lambda t} dt - \left(\frac{1}{\lambda}\right)^2 \quad (\text{by parts twice})$$

$$= \frac{2}{\lambda^2} - \frac{1}{\lambda^2} = \frac{1}{\lambda^2}. \tag{4.4}$$

The quantity λ is a parameter of Exponential distribution, and its meaning is clear from $\mathbf{E}(X) = 1/\lambda$. If X is time, measured in minutes, then λ is a frequency, measured in min^{-1}. For example, if arrivals occur every half a minute, on the average, then $\mathbf{E}(X) = 0.5$ and $\lambda = 2$, saying that they occur with a frequency (arrival rate) of 2 arrivals per minute. This λ has the same meaning as the parameter of Poisson distribution in Section 3.4.5.

Times between rare events are Exponential

What makes Exponential distribution a good model for interarrival times? Apparently, this is not only experimental, but also a mathematical fact.

As in Section 3.4.5, consider the sequence of rare events, where the number of occurrences during time t has Poisson distribution with a parameter proportional to t. This process is rigorously defined in Section 6.3.2 where we call it *Poisson process*.

Event "the time T until the next event is greater than t" can be rephrased as "zero events occur by the time t," and further, as "$X = 0$," where X is the number of events during the time interval $[0, t]$. This X has Poisson distribution with parameter λt. It equals 0 with probability

$$P_X(0) = e^{-\lambda t}\frac{(\lambda t)^0}{0!} = e^{-\lambda t}.$$

Then we can compute the cdf of T as

$$F_T(t) = 1 - \mathbf{P}\{T > t\} = 1 - \mathbf{P}\{X = 0\} = 1 - e^{-\lambda t}, \tag{4.5}$$

and here we recognize the Exponential cdf. Therefore, the time until the next arrival has Exponential distribution.

Example 4.5. Jobs are sent to a printer at an average rate of 3 jobs per hour.
(a) What is the expected time between jobs?
(b) What is the probability that the next job is sent within 5 minutes?

<u>Solution.</u> Job arrivals represent rare events, thus the time T between them is Exponential with the given parameter $\lambda = 3$ hrs^{-1} (jobs per hour).
(a) $\mathbf{E}(T) = 1/\lambda = 1/3$ hours or 20 minutes between jobs;
(b) Convert to the same measurement unit: 5 min = $(1/12)$ hrs. Then,

$$\mathbf{P}\{T < 1/12 \text{ hrs}\} = F(1/12) = 1 - e^{-\lambda(1/12)} = 1 - e^{-1/4} = \underline{0.2212}.$$

$\Diamond$

Memoryless property

It is said that "Exponential variables lose memory." What does it mean?

Suppose that an Exponential variable T represents waiting time. Memoryless property means that the fact of having waited for t minutes gets "forgotten," and it does not affect the future waiting time. Regardless of the event $T > t$, when the total waiting time exceeds

t, the remaining waiting time still has Exponential distribution with the same parameter. Mathematically,

$$P\{T > t + x \mid T > t\} = P\{T > x\} \qquad \text{for } t, x > 0. \tag{4.6}$$

In this formula, t is the already elapsed portion of waiting time, and x is the additional, remaining time.

PROOF: From (4.2), $P\{T > x\} = e^{-\lambda x}$. Also, by the formula (2.7) for conditional probability,

$$P\{T > t + x \mid T > t\} = \frac{P\{T > t + x \cap T > t\}}{P\{T > t\}} = \frac{P\{T > t + x\}}{P\{T > t\}} = \frac{e^{-\lambda(t + x)}}{e^{-\lambda t}} = e^{-\lambda x}.$$

$\square$

This property is unique for Exponential distribution. No other continuous variable $X \in (0, \infty)$ is memoryless. Among discrete variables, such a property belongs to *Geometric distribution* (Exercise 4.36).

In a sense, Geometric distribution is a discrete analogue of Exponential. Connection between these two families of distributions will become clear in Section 5.2.3.

Exponential distribution	λ =	frequency parameter, the number of events per time unit
	$f(x)$ =	$\lambda e^{-\lambda x}, \quad x > 0$
	$\mathbf{E}(X)$ =	$\dfrac{1}{\lambda}$
	$\mathrm{Var}(X)$ =	$\dfrac{1}{\lambda^2}$

4.2.3   Gamma distribution

When a certain procedure consists of α independent steps, and each step takes Exponential(λ) amount of time, then the total time has *Gamma distribution* with parameters α and λ.

Thus, Gamma distribution can be widely used for the total time of a multistage scheme, for example, related to downloading or installing a number of files. In a process of rare events, with Exponential times between any two consecutive events, the time of the α-th event has Gamma distribution because it consists of α independent Exponential times.

Example 4.6 (INTERNET PROMOTIONS). Users visit a certain internet site at the average rate of 12 hits per minute. Every sixth visitor receives some promotion that comes in a form of a flashing banner. Then the time between consecutive promotions has Gamma distribution with parameters $\alpha = 6$ and $\lambda = 12$. $\Diamond$

Having two parameters, the Gamma distribution family offers a variety of models for positive random variables. Besides the case when a Gamma variable represents a sum of independent

Exponential variables, Gamma distribution is often used for the amount of money being paid, amount of a commodity being used (gas, electricity, etc.), a loss incurred by some accident, etc.

Gamma distribution has a density

$$f(x) = \frac{\lambda^\alpha}{\Gamma(\alpha)} x^{\alpha-1} e^{-\lambda x}, \quad x > 0. \tag{4.7}$$

The denominator contains a *Gamma function*, Section A.4.5. With certain techniques, this density can be mathematically derived for integer α by representing a Gamma variable X as a sum of Exponential variables each having a density (4.1).

In fact, α can take any positive value, not necessarily integer. With different α, the Gamma density takes different shapes (Figure 4.5). For this reason, α is called a *shape parameter*.

Notice two important special cases of a Gamma distribution. When $\alpha = 1$, the Gamma distribution becomes *Exponential*. This can be seen comparing (4.7) and (4.1) for $\alpha = 1$. Another special case with $\lambda = 1/2$ and any $\alpha > 0$ results in a so-called *Chi-square distribution* with (2α) degrees of freedom, which we discuss in Section 9.5.1 and use a lot in Chapter 10.

$$
\begin{array}{rcl}
\text{Gamma}(1, \lambda) & = & \text{Exponential}(\lambda) \\
\text{Gamma}(\alpha, 1/2) & = & \text{Chi-square}(2\alpha)
\end{array}
$$

Expectation, variance, and some useful integration remarks

Gamma cdf has the form

$$F(t) = \int_0^t f(x)dx = \frac{\lambda^\alpha}{\Gamma(\alpha)} \int_0^t x^{\alpha-1} e^{-\lambda x} dx. \tag{4.8}$$

This expression, related to a so-called *incomplete Gamma function*, does not simplify, and thus, computing probabilities is not always trivial. Let us offer several computational shortcuts.

First, let us notice that $\int_0^\infty f(x)dx = 1$ for Gamma and all the other densities. Then, integrating (4.7) from 0 to ∞, we obtain that

$$\int_0^\infty x^{\alpha-1} e^{-\lambda x} dx = \frac{\Gamma(\alpha)}{\lambda^\alpha} \quad \text{for any } \alpha > 0 \text{ and } \lambda > 0 \tag{4.9}$$

Substituting $\alpha + 1$ and $\alpha + 2$ in place of α, we get for a Gamma variable X,

$$\mathbf{E}(X) = \int x f(x)dx = \frac{\lambda^\alpha}{\Gamma(\alpha)} \int_0^\infty x^\alpha e^{-\lambda x} dx = \frac{\lambda^\alpha}{\Gamma(\alpha)} \cdot \frac{\Gamma(\alpha+1)}{\lambda^{\alpha+1}} = \frac{\alpha}{\lambda} \tag{4.10}$$

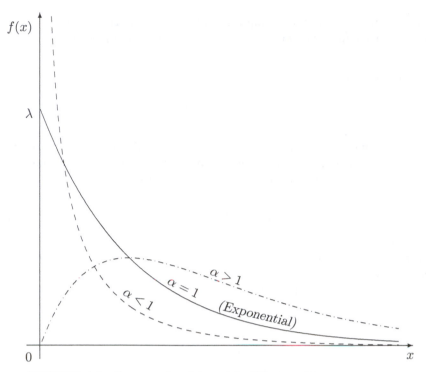

FIGURE 4.5: *Gamma densities with different shape parameters α.*

(using the equality $\Gamma(t+1) = t\Gamma(t)$ that holds for all $t > 0$),

$$\mathbf{E}(X^2) = \int x^2 f(x) dx = \frac{\lambda^\alpha}{\Gamma(\alpha)} \int_0^\infty x^{\alpha+1} e^{-\lambda x} dx = \frac{\lambda^\alpha}{\Gamma(\alpha)} \cdot \frac{\Gamma(\alpha+2)}{\lambda^{\alpha+2}} = \frac{(\alpha+1)\alpha}{\lambda^2},$$

and therefore,

$$\text{Var}(X) = \mathbf{E}(X^2) - \mathbf{E}^2(X) = \frac{(\alpha+1)\alpha - \alpha^2}{\lambda^2} = \frac{\alpha}{\lambda^2}. \tag{4.11}$$

For $\alpha = 1$, this agrees with (4.3) and (4.4). Moreover, for any integer α, (4.10) and (4.11) can be obtained directly from (4.3) and (4.4) by representing a Gamma variable X as a sum of independent Exponential(λ) variables $X_1, \ldots, X_\alpha$,

$$\mathbf{E}(X) = \mathbf{E}(X_1 + \ldots + X_\alpha) = \mathbf{E}(X_1) + \ldots + \mathbf{E}(X_\alpha) = \alpha\left(\frac{1}{\lambda}\right),$$

$$\text{Var}(X) = \text{Var}(X_1 + \ldots + X_\alpha) = \text{Var}(X_1) + \ldots + \text{Var}(X_\alpha) = \alpha\left(\frac{1}{\lambda^2}\right).$$

Gamma distribution	α = shape parameter	
	λ = frequency parameter	
	$f(x) = \dfrac{\lambda^\alpha}{\Gamma(\alpha)} x^{\alpha-1} e^{-\lambda x}, \quad x > 0$	
	$\mathbf{E}(X) = \dfrac{\alpha}{\lambda}$	
	$\text{Var}(X) = \dfrac{\alpha}{\lambda^2}$	

(4.12)

Example 4.7 (TOTAL COMPILATION TIME). Compilation of a computer program consists of 3 blocks that are processed sequentially, one after another. Each block takes Exponential time with the mean of 5 minutes, independently of other blocks.

(a) Compute the expectation and variance of the total compilation time.

(b) Compute the probability for the entire program to be compiled in less than 12 minutes.

<u>Solution</u>. The total time T is a sum of three independent Exponential times, therefore, it has Gamma distribution with $\alpha = 3$. The frequency parameter λ equals $(1/5)$ min^{-1} because the Exponential compilation time of each block has expectation $1/\lambda = 5$ min.

(a) For a Gamma random variable T with $\alpha = 3$ and $\lambda = 1/5$,

$$\mathbf{E}(T) = \frac{3}{1/5} = 15 \ (\text{min}) \quad \text{and} \quad \text{Var}(T) = \frac{3}{(1/5)^2} = 75 \ (\text{min}^2).$$

(b) A direct solution involves two rounds of integration by parts,

$$
\begin{aligned}
\mathbf{P}\{T < 12\} &= \int_0^{12} f(t)dt = \frac{(1/5)^3}{\Gamma(3)} \int_0^{12} t^2 e^{-t/5} dt \\
&= \frac{(1/5)^3}{2!} \left(-5t^2 e^{-t/5} \Big|_{t=0}^{t=12} + \int_0^{12} 10t e^{-t/5} dt \right) \\
&= \frac{1/125}{2} \left(-5t^2 e^{-t/5} - 50t e^{-t/5} \Big|_{t=0}^{t=12} + \int_0^{12} 50 e^{-t/5} dt \right) \\
&= \frac{1}{250} \left(-5t^2 e^{-t/5} - 50t e^{-t/5} - 250 e^{-t/5} \right) \Big|_{t=0}^{t=12} \\
&= 1 - e^{-2.4} - 2.4 e^{-2.4} - 2.88 e^{-2.4} = \underline{0.4303}.
\end{aligned}
\tag{4.13}
$$

A much shorter way is to apply the Gamma-Poisson formula below (Example 4.8). ◇

Gamma-Poisson formula

Computation of Gamma probabilities can be significantly simplified by thinking of a Gamma variable as the time between some rare events. In particular, one can avoid lengthy integration by parts, as in Example 4.7, and use Poisson distribution instead.

Indeed, let T be a Gamma variable with an integer parameter α and some positive λ. This is a distribution of the time of the α-th rare event. Then, the event $\{T > t\}$ means that the α-th rare event occurs after the moment t, and therefore, *fewer than* α *rare events occur before the time* t. We see that

$$\{T > t\} = \{X < \alpha\},$$

where X is the number of events that occur before the time t. This number of rare events X has Poisson distribution with parameter (λt); therefore, the probability

$$\mathbf{P}\{T > t\} = \mathbf{P}\{X < \alpha\}$$

and the probability of a complement

$$\mathbf{P}\{T \le t\} = \mathbf{P}\{X \ge \alpha\}$$

can both be computed using the Poisson distribution of X.

<table>
<tr><td>Gamma-Poisson formula</td><td>For a Gamma(α, λ) variable T and a Poisson(λt) variable X,

$P\{T > t\} = P\{X < \alpha\}$

$P\{T \leq t\} = P\{X \geq \alpha\}$</td><td>(4.14)</td></tr>
</table>

Remark: Recall that $P\{T > t\} = P\{T \geq t\}$ and $P\{T < t\} = P\{T \leq t\}$ for a Gamma variable T, because it is continuous. Hence, (4.14) can also be used for the computation of $P\{T \geq t\}$ and $P\{T < t\}$. Conversely, the probability of $\{X = \alpha\}$ cannot be neglected for the Poisson (discrete!) variable X, thus the signs in the right-hand sides of (4.14) cannot be altered.

Example 4.8 (TOTAL COMPILATION TIME, CONTINUED). Here is an alternative solution to Example 4.7(b). According to the Gamma-Poisson formula with $\alpha = 3$, $\lambda = 1/5$, and $t = 12$,
$$P\{T < 12\} = P\{X \geq 3\} = 1 - F(2) = 1 - 0.5697 = \underline{0.430}$$
from Table A3 for the Poisson distribution of X with parameter $\lambda t = 2.4$.

Furthermore, we notice that the four-term mathematical expression that we obtained in (4.13) after integrating by parts represents precisely
$$P\{X \geq 3\} = 1 - P(0) - P(1) - P(2).$$

$\Diamond$

Example 4.9. Lifetimes of computer memory chips have Gamma distribution with expectation $\mu = 12$ years and standard deviation $\sigma = 4$ years. What is the probability that such a chip has a lifetime between 8 and 10 years?

<u>Solution.</u>

STEP 1, PARAMETERS. From the given data, compute parameters of this Gamma distribution. Using (4.12), obtain a system of two equations and solve them for α and λ,
$$\begin{cases} \mu &= \alpha/\lambda \\ \sigma^2 &= \alpha/\lambda^2 \end{cases} \Rightarrow \begin{cases} \alpha &= \mu^2/\sigma^2 &= (12/4)^2 &= 9, \\ \lambda &= \mu/\sigma^2 &= 12/4^2 &= 0.75. \end{cases}$$

STEP 2, PROBABILITY. We can now compute the probability,
$$P\{8 < T < 10\} = F_T(10) - F_T(8). \tag{4.15}$$
For each term in (4.15), we use the Gamma-Poisson formula with $\alpha = 9$, $\lambda = 0.75$, and $t = 8, 10$,
$$F_T(10) = P\{T \leq 10\} = P\{X \geq 9\} = 1 - F_X(8) = 1 - 0.662 = 0.338$$
from Table A3 for a Poisson variable X with parameter $\lambda t = (0.75)(10) = 7.5$;
$$F_T(8) = P\{T \leq 8\} = P\{X \geq 9\} = 1 - F_X(8) = 1 - 0.847 = 0.153$$
from the same table, this time with parameter $\lambda t = (0.75)(8) = 6$. Then,
$$P\{8 < T < 10\} = 0.338 - 0.153 = \underline{0.185}.$$

$\Diamond$

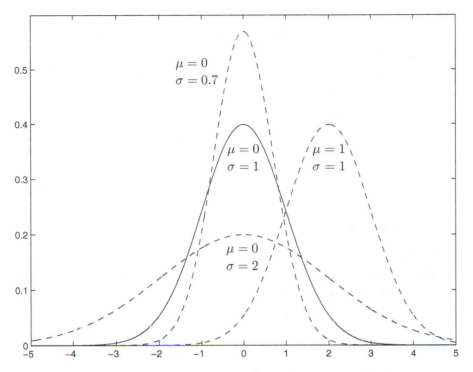

FIGURE 4.6: *Normal densities with different location and scale parameters.*

4.2.4 Normal distribution

Normal distribution plays a vital role in Probability and Statistics, mostly because of the Central Limit Theorem, according to which sums and averages often have approximately Normal distribution. Due to this fact, various fluctuations and measurement errors that consist of accumulated number of small terms appear normally distributed.

Remark: As said by a French mathematician *Jules Henri Poincaré*, "Everyone believes in the Normal law of errors, the experimenters because they think it is a mathematical theorem, the mathematicians because they think it is an experimental fact."

Besides sums, averages, and errors, Normal distribution is often found to be a good model for physical variables like weight, height, temperature, voltage, pollution level, and for instance, household incomes or student grades.

Normal distribution has a density

$$f(x) = \frac{1}{\sigma\sqrt{2\pi}} \exp\left\{ \frac{-(x-\mu)^2}{2\sigma^2} \right\}, \quad -\infty < x < +\infty,$$

where parameters μ and σ have a simple meaning of the expectation $\mathbf{E}(X)$ and the standard deviation $\text{Std}(X)$. This density is known as the bell-shaped curve, symmetric and centered at μ, its spread being controlled by σ. As seen in Figure 4.6, changing μ shifts the curve to the left or to the right without affecting its shape, while changing σ makes it more concentrated or more flat. Often μ and σ are called *location* and *scale* parameters.

$$
\begin{array}{rcl}
\mu & = & \text{expectation, location parameter} \\
\sigma & = & \text{standard deviation, scale parameter} \\
f(x) & = & \dfrac{1}{\sigma\sqrt{2\pi}}\exp\left\{\dfrac{-(x-\mu)^2}{2\sigma^2}\right\}, \quad -\infty < x < \infty \\
\mathbf{E}(X) & = & \mu \\
\text{Var}(X) & = & \sigma^2
\end{array}
$$

Normal distribution

Standard Normal distribution

DEFINITION 4.3

Normal distribution with "standard parameters" $\mu = 0$ and $\sigma = 1$ is called **Standard Normal distribution**.

NOTATION

$$
\begin{array}{rcl}
Z & = & \text{Standard Normal random variable} \\
\phi(x) & = & \dfrac{1}{\sqrt{2\pi}}e^{-x^2/2}, \text{ Standard Normal pdf} \\
\Phi(x) & = & \displaystyle\int_{-\infty}^{x}\dfrac{1}{\sqrt{2\pi}}e^{-z^2/2}dz, \text{ Standard Normal cdf}
\end{array}
$$

A Standard Normal variable, usually denoted by Z, can be obtained from a non-standard Normal(μ,σ) random variable X by *standardizing*, that is, subtracting the mean and dividing by the standard deviation,

$$
Z = \frac{X-\mu}{\sigma}. \tag{4.16}
$$

Unstandardizing Z, we can reconstruct the initial variable X,

$$
X = \mu + \sigma Z. \tag{4.17}
$$

Using these transformations, any Normal random variable can be obtained from a Standard Normal variable Z; therefore, we need a table of Standard Normal Distribution only (Table A4).

To find $\Phi(z)$ from Table A4, we locate a row with the first two digits of z and a column with the third digit of z and read the probability $\Phi(z)$ at their intersection. Notice that $\Phi(z) \approx 0$ (is "practically" zero) for all $z < -3.9$, and $\Phi(z) \approx 1$ (is "practically" one) for all $z > 3.9$.

Example 4.10 (COMPUTING STANDARD NORMAL PROBABILITIES). For a Standard Normal random variable Z,

$$
\begin{array}{rcl}
\boldsymbol{P}\{Z < 1.35\} & = & \Phi(1.35) = 0.9115 \\
\boldsymbol{P}\{Z > 1.35\} & = & 1 - \Phi(1.35) = 0.0885 \\
\boldsymbol{P}\{-0.77 < Z < 1.35\} & = & \Phi(1.35) - \Phi(-0.77) = 0.9115 - 0.2206 = 0.6909.
\end{array}
$$

according to Table A4. Notice that $P\{Z < -1.35\} = 0.0885 = P\{Z > 1.35\}$, which is explained by the symmetry of the Standard Normal density in Figure 4.6. Due to this symmetry, "the left tail," or the area to the left of (-1.35) equals "the right tail," or the area to the right of 1.35. ◇

In fact, the symmetry of the Normal density, mentioned in this example, allows to obtain the first part of Table A4 on p. 432 directly from the second part,

$$\Phi(-z) = 1 - \Phi(z) \qquad \text{for} \quad -\infty < z < +\infty.$$

To compute probabilities about an arbitrary Normal random variable X, we have to standardize it first, as in (4.16), then use Table A4.

Example 4.11 (COMPUTING NON-STANDARD NORMAL PROBABILITIES). Suppose that the average household income in some country is 900 coins, and the standard deviation is 200 coins. Assuming the Normal distribution of incomes, compute the proportion of "the middle class," whose income is between 600 and 1200 coins.

<u>Solution</u>. Standardize and use Table A4. For a Normal($\mu = 900$, $\sigma = 200$) variable X,

$$P\{600 < X < 1200\} = P\left\{\frac{600 - \mu}{\sigma} < \frac{X - \mu}{\sigma} < \frac{1200 - \mu}{\sigma}\right\}$$

$$= P\left\{\frac{600 - 900}{200} < Z < \frac{1200 - 900}{200}\right\} = P\{-1.5 < Z < 1.5\}$$

$$= \Phi(1.5) - \Phi(-1.5) = 0.9332 - 0.0668 = \underline{0.8664}.$$

◇

So far, we were computing probabilities of clearly defined events. These are *direct* problems. A number of applications require solution of an *inverse problem*, that is, finding a value of x given the corresponding probability.

Example 4.12 (INVERSE PROBLEM). The government of the country in Example 4.11 decides to issue food stamps to the poorest 3% of households. Below what income will families receive food stamps?

<u>Solution</u>. We need to find such income x that $P\{X < x\} = 3\% = 0.03$. This is an equation that can be solved in terms of x. Again, we standardize first, then use the table:

$$P\{X < x\} = P\left\{Z < \frac{x - \mu}{\sigma}\right\} = \Phi\left(\frac{x - \mu}{\sigma}\right) = 0.03,$$

from where

$$x = \mu + \sigma\Phi^{-1}(0.03).$$

In Table A4, we have to find the probability, the *table entry* of 0.03. We see that $\Phi(-1.88) \approx 0.03$. Therefore, $\Phi^{-1}(0.03) = -1.88$, and

$$x = \mu + \sigma(-1.88) = 900 + (200)(-1.88) = \underline{524} \text{ (coins)}$$

is the answer. In the literature, the value $\Phi^{-1}(\alpha)$ is often denoted by $z_{1-\alpha}$. ◇

As seen in this example, in order to solve an inverse problem, we use the table first, then *unstandardize*, as in (4.17), and find the required value of x.

4.3   Central Limit Theorem

We now turn our attention to *sums* of random variables,

$$S_n = X_1 + \ldots + X_n,$$

that appear in many applications. Let $\mu = \mathbf{E}(X_i)$ and $\sigma = \mathrm{Std}(X_i)$ for all $i = 1, \ldots, n$. How does S_n behave for large n?

COMPUTER DEMO. The following R and MATLAB codes produce an animated illustration to the behavior of partial sums S_n.

— R —

```
N <- 1000                        # Number of steps of Sn
S <- rep(0, N)                   # Initialize Sn = 0 for all n
Z <- rnorm(N)                    # Standard Normal increments Zn
for (n in 2:N){S[n] <- S[n-1]+Z[n]}  # Random walk Sn defined in a for-loop
    # Three line plots with ranges xlim and ylim on x-axes and y-axes
for (n in 1:N){plot(S[1:n], type='l', xlim=c(0,N), ylim=max(abs(S))*c(-1,1))}
for (n in 1:N){plot(S[1:n]/(1:n), type='l', xlim=c(0,N), ylim=c(-2,2)) }
for (n in 1:N){plot(S[1:n]/sqrt(1:n), type='l', xlim=c(0,N), ylim=c(-2,2)) }
```

— MATLAB —

```
N = 1000;                        % Number of steps of Sn
S = zeros(N,1);                  % Initialize Sn = 0 for all n
Z = randn(N,1);                  % Standard Normal increments Zn
for n=2:N; S(n)=S(n-1)+Z(n); end;   % Random walk Sn defined in a for-loop
n=1:N; comet(n,S); pause(3);     % Behavior of S(n)
comet(n,S./n); pause(3);         % Behavior of S(n)/n
comet(n,S./sqrt(n)); pause(3);   % Behavior of S(n)/sqrt(n)
```

Apparently (R and MATLAB users can run these codes and see it "in real time"),

- The *pure sum* S_n diverges on Figure 4.7a. In fact, this should be anticipated because

$$\mathrm{Var}(S_n) = n\sigma^2 \to \infty,$$

so that variability of S_n grows unboundedly as n goes to infinity.

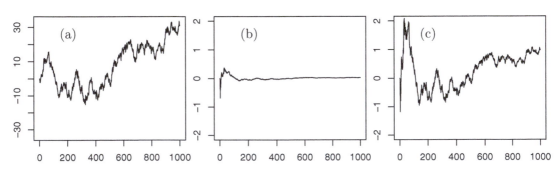

FIGURE 4.7: *Behavior of (a) random walk S_n, (b) S_n/n, and (c) $S_n/\sqrt{n}$ for increasing values of n.*

- The *average* S_n/n *converges* on Figure 4.7b. Indeed, in this case, we have

$$\text{Var}(S_n/n) = \text{Var}(S_n)/n^2 = n\sigma^2/n^2 = \sigma^2/n \to 0,$$

so that variability of (S_n/n) vanishes as $n \to \infty$.

- An interesting normalization factor is $1/\sqrt{n}$. We can see from computer simulations that $S_n/\sqrt{n}$ *neither diverges nor converges* on Figure 4.7c. It does not tend to leave 0, but it does not converge to 0 either. Rather, it behaves like some random variable. The following theorem states that this variable approaches a Standard Normal distribution for large n.

Theorem 1 (CENTRAL LIMIT THEOREM) *Let* $X_1, X_2, \ldots$ *be independent random variables with the same expectation* $\mu = \mathbf{E}(X_i)$ *and the same standard deviation* $\sigma = \text{Std}(X_i)$, *and let*

$$S_n = \sum_{i=1}^{n} X_i = X_1 + \ldots + X_n.$$

As $n \to \infty$, *the standardized sum*

$$Z_n = \frac{S_n - \mathbf{E}(S_n)}{\text{Std}(S_n)} = \frac{S_n - n\mu}{\sigma\sqrt{n}}$$

converges in distribution to a Standard Normal random variable, that is,

$$F_{Z_n}(z) = \mathbf{P}\left\{ \frac{S_n - n\mu}{\sigma\sqrt{n}} \leq z \right\} \to \Phi(z) \tag{4.18}$$

for all z.

This theorem is very powerful because it can be applied to random variables $X_1, X_2, \ldots$ having virtually any thinkable distribution with finite expectation and variance. As long as n is large (the rule of thumb is $n > 30$), one can use Normal distribution to compute probabilities about S_n.

Theorem 1 is only one basic version of the Central Limit Theorem. Over the last two centuries, it has been extended to large classes of dependent variables and vectors, stochastic processes, and so on.

Example 4.13 (ALLOCATION OF DISK SPACE). A disk has free space of 330 megabytes. Is it likely to be sufficient for 300 independent images, if each image has expected size of 1 megabyte with a standard deviation of 0.5 megabytes?

Solution. We have $n = 300$, $\mu = 1$, $\sigma = 0.5$. The number of images n is large, so the Central Limit Theorem applies to their total size S_n. Then,

$$
\begin{aligned}
\mathbf{P}\left\{\text{sufficient space}\right\} &= \mathbf{P}\left\{S_n \leq 330\right\} = \mathbf{P}\left\{ \frac{S_n - n\mu}{\sigma\sqrt{n}} \leq \frac{330 - (300)(1)}{0.5\sqrt{300}} \right\} \\
&\approx \Phi(3.46) = 0.9997.
\end{aligned}
$$

This probability is very high, hence, the available disk space is very likely to be sufficient.

$\diamond$

In the special case of *Normal* variables $X_1, X_2, \ldots$, the distribution of S_n is always Normal, and (4.18) becomes exact equality for arbitrary, even small n.

Example 4.14 (ELEVATOR). You wait for an elevator, whose capacity is 2000 pounds. The elevator comes with ten adult passengers. Suppose your own weight is 150 lbs, and you heard that human weights are normally distributed with the mean of 165 lbs and the standard deviation of 20 lbs. Would you board this elevator or wait for the next one?

Solution. In other words, is overload likely? The probability of an overload equals

$$P\left\{S_{10} + 150 > 2000\right\} = P\left\{\frac{S_{10} - (10)(165)}{20\sqrt{10}} > \frac{2000 - 150 - (10)(165)}{20\sqrt{10}}\right\}$$
$$= 1 - \Phi(3.16) = 0.0008.$$

So, with probability 0.9992 it is safe to take this elevator. It is now for you to decide.

$\Diamond$

Among the random variables discussed in Chapters 3 and 4, at least three have a form of S_n:

Binomial variable	=	sum of independent Bernoulli variables
Negative Binomial variable	=	sum of independent Geometric variables
Gamma variable	=	sum of independent Exponential variables

Hence, the Central Limit Theorem applies to all these distributions with sufficiently large n in the case of Binomial, k for Negative Binomial, and α for Gamma variables.

In fact, *Abraham de Moivre* (1667–1754) obtained the first version of the Central Limit Theorem as the approximation of Binomial distribution.

Normal approximation to Binomial distribution

Binomial variables represent a special case of $S_n = X_1 + \ldots + X_n$, where all X_i have Bernoulli distribution with some parameter p. We know from Section 3.4.5 that small p allows to approximate Binomial distribution with Poisson, and large p allows such an approximation for the number of failures. For the moderate values of p (say, $0.05 \leq p \leq 0.95$) and for large n, we can use Theorem 1:

$$\text{Binomial}(n, p) \approx Normal\left(\mu = np, \sigma = \sqrt{np(1-p)}\right) \qquad (4.19)$$

COMPUTER DEMO. To visualize how Binomial distribution gradually takes the shape of Normal distribution when $n \to \infty$, you may execute the following computer codes that graph Binomial(n, p) pmf for increasing values of n and $p = 0.4$. Commands `Sys.sleep` in R and `pause` in MATLAB make a 0.5-second pause, so we can enjoy the computer animation.

— R ——————

```
N <- 50; p <- 0.4; x <- 0:n; for (n in 1:N){
plot(dbinom(x,n,p),type='b',main='Binomial PMF for increasing values of n');
Sys.sleep(0.5);   }
```

—— MATLAB ————

```
N=50; p=0.4; for n=1:N; x=0:n;
plot(x,binopdf(x,n,p)); title('Binomial PMF for increasing values of n');
pause(0.5); end;
```

If you are using a software that does not have a built-in Binomial pmf like **dbinom** in R or **binopdf** in MATLAB, it is not a problem. You will simply have to enter the formula, for example, `pmf = gamma(n+1)./gamma(x+1)./gamma(n-x+1).*p.^x.*(1-p).^(n-x);`

Continuity correction

This correction is needed when we approximate a discrete distribution (Binomial in this case) by a continuous distribution (Normal). Recall that the probability $P\{X = x\}$ may be positive if X is discrete, whereas it is always 0 for continuous X. Thus, a direct use of (4.19) will always approximate this probability by 0. It is obviously a poor approximation.

This is resolved by introducing a *continuity correction*. Expand the interval by 0.5 units in each direction, then use the Normal approximation. Notice that

$$P_X(x) = P\{X = x\} = P\{x - 0.5 < X < x + 0.5\}$$

is true for a Binomial variable X; therefore, the continuity correction does not change the event and preserves its probability. It makes a difference for the Normal distribution, so every time when we approximate some discrete distribution with some continuous distribution, we should be using a continuity correction. Now it is the probability of an interval instead of one number, and it is not zero.

Example 4.15. A new computer virus attacks a folder consisting of 200 files. Each file gets damaged with probability 0.2 independently of other files. What is the probability that fewer than 50 files get damaged?

Solution. The number X of damaged files has Binomial distribution with $n = 200$, $p = 0.2$, $\mu = np = 40$, and $\sigma = \sqrt{np(1-p)} = 5.657$. Applying the Central Limit Theorem with the continuity correction,

$$P\{X < 50\} = P\{X < 49.5\} = P\left\{\frac{X - 40}{5.657} < \frac{49.5 - 40}{5.657}\right\}$$
$$= \Phi(1.68) = \underline{0.9535}.$$

Notice that the properly applied continuity correction replaces 50 with 49.5, not 50.5. Indeed, we are interested in the event that X is *strictly* less than 50. This includes all values up to 49 and corresponds to the interval $[0, 49]$ that we *expand* to $[0, 49.5]$. In other words, events $\{X < 50\}$ and $\{X < 49.5\}$ are the same; they include the same possible values of X. Events $\{X < 50\}$ and $\{X < 50.5\}$ are different because the former includes $X = 50$, and the latter does not. Replacing $\{X < 50\}$ with $\{X < 50.5\}$ would have changed its probability and would have given a wrong answer. $\diamond$

When a continuous distribution (say, Gamma) is approximated by another continuous distribution (Normal), the continuity correction is not needed. In fact, it would be an error to use it in this case because it would no longer preserve the probability.

Summary and conclusions

Continuous distributions are used to model various times, sizes, measurements, and all other random variables that assume an entire interval of possible values.

Continuous distributions are described by their densities that play a role analogous to probability mass functions of discrete variables. Computing probabilities essentially reduces to integrating a density over the given set. Expectations and variances are defined similarly to the discrete case, replacing a probability mass function by a density and summation by integration.

In different situations, one uses Uniform, Exponential, Gamma, or Normal distributions. A few other families are studied in later chapters.

The Central Limit Theorem states that a standardized sum of a large number of independent random variables is approximately Normal, thus Table A4 can be used to compute related probabilities. A continuity correction should be used when a discrete distribution is approximated by a continuous distribution.

Characteristics of continuous families are summarized in Section A.2.2.

Exercises

4.1. The lifetime, in years, of some electronic component is a continuous random variable with the density

$$f(x) = \begin{cases} \dfrac{k}{x^4} & \text{for} \quad x \geq 1 \\ 0 & \text{for} \quad x < 1. \end{cases}$$

Find k, the cumulative distribution function, and the probability for the lifetime to exceed 2 years.

4.2. The time, in minutes, it takes to reboot a certain system is a continuous variable with the density

$$f(x) = \begin{cases} C(10 - x)^2, & \text{if } 0 < x < 10 \\ 0, & \text{otherwise} \end{cases}$$

(a) Compute C.

(b) Compute the probability that it takes between 1 and 2 minutes to reboot.

4.3. The installation time, in hours, for a certain software module has a probability density function $f(x) = k(1 - x^3)$ for $0 < x < 1$. Find k and compute the probability that it takes less than 1/2 hour to install this module.

4.4. Lifetime of a certain hardware is a continuous random variable with density

$$f(x) = \begin{cases} K - x/50 & \text{for } 0 < x < 10 \text{ years} \\ 0 & \text{for all other } x \end{cases}$$

(a) Find K.

(b) What is the probability of a failure within the first 5 years?

(c) What is the expectation of the lifetime?

4.5. Two continuous random variables X and Y have the joint density

$$f(x, y) = C(x^2 + y), \quad -1 \le x \le 1, \quad 0 \le y \le 1.$$

(a) Compute the constant C.

(b) Find the marginal densities of X and Y. Are these two variables independent?

(c) Compute probabilities $P\{Y < 0.6\}$ and $P\{Y < 0.6 \mid X < 0.5\}$.

4.6. A program is divided into 3 blocks that are being compiled on 3 parallel computers. Each block takes an Exponential amount of time, 5 minutes on the average, independently of other blocks. The program is completed when *all* the blocks are compiled. Compute the expected time it takes the program to be compiled.

4.7. The time it takes a printer to print a job is an Exponential random variable with the expectation of 12 seconds. You send a job to the printer at 10:00 am, and it appears to be third in line. What is the probability that your job will be ready before 10:01?

4.8. For some electronic component, the time until failure has Gamma distribution with parameters $\alpha = 2$ and $\lambda = 2$ (years^{-1}). Compute the probability that the component fails within the first 6 months.

4.9. On the average, a computer experiences breakdowns every 5 months. The time until the first breakdown and the times between any two consecutive breakdowns are independent Exponential random variables. After the third breakdown, a computer requires a special maintenance.

(a) Compute the probability that a special maintenance is required within the next 9 months.

(b) Given that a special maintenance was not required during the first 12 months, what is the probability that it will not be required within the next 4 months?

4.10. Two computer specialists are completing work orders. The first specialist receives 60% of all orders. Each order takes her Exponential amount of time with parameter $\lambda_1 = 3$ hrs^{-1}. The second specialist receives the remaining 40% of orders. Each order takes him Exponential amount of time with parameter $\lambda_2 = 2$ hrs^{-1}.

A certain order was submitted 30 minutes ago, and it is still not ready. What is the probability that the first specialist is working on it?

4.11. Consider a satellite whose work is based on a certain block A. This block has an independent backup B. The satellite performs its task until both A and B fail. The lifetimes of A and B are exponentially distributed with the mean lifetime of 10 years.

(a) What is the probability that the satellite will work for more than 10 years?

(b) Compute the expected lifetime of the satellite.

4.12. A computer processes tasks in the order they are received. Each task takes an Exponential amount of time with the average of 2 minutes. Compute the probability that a package of 5 tasks is processed in less than 8 minutes.

4.13. On the average, it takes 25 seconds to download a file from the internet. If it takes an Exponential amount of time to download one file, then what is the probability that it will take more than 70 seconds to download 3 independent files?

4.14. The time X it takes to reboot a certain system has Gamma distribution with $\mathbf{E}(X) = 20$ min and $\mathrm{Std}(X) = 10$ min.

(a) Compute parameters of this distribution.

(b) What is the probability that it takes less than 15 minutes to reboot this system?

4.15. A certain system is based on two independent modules, A and B. A failure of any module causes a failure of the whole system. The lifetime of each module has a Gamma distribution, with parameters α and λ given in the table,

Component	α	λ (years^{-1})
A	3	1
B	2	2

(a) What is the probability that the system works at least 2 years without a failure?

(b) Given that the system failed during the first 2 years, what is the probability that it failed due to the failure of component B (but not component A)?

4.16. Let Z be a Standard Normal random variable. Compute

(a) $P(Z < 1.25)$ (b) $P(Z \leq 1.25)$ (c) $P(Z > 1.25)$
(d) $P(|Z| \leq 1.25)$ (e) $P(Z < 6.0)$ (f) $P(Z > 6.0)$
(g) With probability 0.8, variable Z does not exceed what value?

4.17. For a Standard Normal random variable Z, compute

(a) $P(Z \geq 0.99)$ (b) $P(Z \leq -0.99)$ (c) $P(Z < 0.99)$
(d) $P(|Z| > 0.99)$ (e) $P(Z < 10.0)$ (f) $P(Z > 10.0)$
(g) With probability 0.9, variable Z is less than what?

4.18. For a Normal random variable X with $\mathbf{E}(X) = -3$ and $\text{Var}(X) = 4$, compute

(a) $P(X \leq 2.39)$ (b) $P(Z \geq -2.39)$ (c) $P(|X| \geq 2.39)$
(d) $P(|X + 3| \geq 2.39)$ (e) $P(X < 5)$ (f) $P(|X| < 5)$
(g) With probability 0.33, variable X exceeds what value?

4.19. According to one of the Western Electric rules for quality control, a produced item is considered conforming if its measurement falls within three standard deviations from the target value. Suppose that the process is in control so that the expected value of each measurement equals the target value. What percent of items will be considered conforming, if the distribution of measurements is

(a) Normal(μ, σ)?
(b) Uniform(a, b)?

4.20. Refer to Exercise 4.19. What percent of items falls *beyond* 1.5 standard deviations from the mean, if the distribution of measurements is

(a) Normal(μ, σ)?
(b) Uniform(a, b)?

4.21. The average height of professional basketball players is around 6 feet 7 inches, and the standard deviation is 3.89 inches. Assuming Normal distribution of heights within this group,

(a) What percent of professional basketball players are taller than 7 feet?

(b) If your favorite player is within the tallest 20% of all players, what can his height be?

4.22. Refer to the country in Example 4.11 on p. 91, where household incomes follow Normal distribution with $\mu = 900$ coins and $\sigma = 200$ coins.

(a) A recent economic reform made households with the income below 640 coins qualify for a free bottle of milk at every breakfast. What portion of the population qualifies for a free bottle of milk?

(b) Moreover, households with an income within the lowest 5% of the population are entitled to a free sandwich. What income qualifies a household to receive free sandwiches?

4.23. The lifetime of a certain electronic component is a random variable with the expectation of 5000 hours and a standard deviation of 100 hours. What is the probability that the average lifetime of 400 components is less than 5012 hours?

4.24. Installation of some software package requires downloading 82 files. On the average, it takes 15 sec to download one file, with a variance of 16 sec^2. What is the probability that the software is installed in less than 20 minutes?

4.25. Among all the computer chips produced by a certain factory, 6 percent are defective. A sample of 400 chips is selected for inspection.

(a) What is the probability that this sample contains between 20 and 25 defective chips (including 20 and 25)?

(b) Suppose that each of 40 inspectors collects a sample of 400 chips. What is the probability that at least 8 inspectors will find between 20 and 25 defective chips in their samples?

4.26. An average scanned image occupies 0.6 megabytes of memory with a standard deviation of 0.4 megabytes. If you plan to publish 80 images on your web site, what is the probability that their total size is between 47 megabytes and 50 megabytes?

4.27. A certain computer virus can damage any file with probability 35%, independently of other files. Suppose this virus enters a folder containing 2400 files. Compute the probability that between 800 and 850 files get damaged (including 800 and 850).

4.28. For the inaugural meeting of the University Cheese Club, Danielle and Anthony plan to have at least 1/4 pounds of cheese for each attending club member to taste. They anticipate at least 100 club members at the meeting, although the exact number is unknown. Members are recommended to bring some cheese to the meeting. It is estimated that each attending club member, independently of others, will bring 1 pound of cheese with probability 0.1, 1/2 pounds with probability 0.2, 1/4 pounds with probability 0.4, and will bring no cheese with probability 0.3.

(a) Can the club leaders conclude that with a probability of 0.95 or higher, the amount of cheese brought to the meeting is at least 1/4 pounds per each attending club member?

(b) How many club members must attend the meeting to make sure that there is at least 1/4 pounds of cheese per each attending club member with a probability of 0.95 or higher?

4.29. Natasha has recently learned to write. Now, she can write each letter in a readable way with probability 0.8. Compute the probability that

(a) You can read your name, when Natasha writes it.

(b) You can read at least 90% of Natasha's 20-letter phrase.

(c) You can read at least 90% of Natasha's 100-letter verse.

(d) You can read at least 90% of Natasha's 500-letter story.

4.30. Seventy independent messages are sent from an electronic transmission center. Messages are processed sequentially, one after another. Transmission time of each message is Exponential with parameter $\lambda = 5$ min^{-1}. Find the probability that all 70 messages are transmitted in less than 12 minutes. Use the Central Limit Theorem.

4.31. A computer lab has two printers. Printer I handles 40% of all the jobs. Its printing time is Exponential with the mean of 2 minutes. Printer II handles the remaining 60% of jobs. Its printing time is Uniform between 0 minutes and 5 minutes. A job was printed in less than 1 minute. What is the probability that it was printed by Printer I?

4.32. An internet service provider has two connection lines for its customers. Eighty percent of customers are connected through Line I, and twenty percent are connected through Line II. Line I has a Gamma connection time with parameters $\alpha = 3$ and $\lambda = 2$ min^{-1}. Line II has a Uniform(a, b) connection time with parameters $a = 20$ sec and $b = 50$ sec. Compute the probability that it takes a randomly selected customer more than 30 seconds to connect to the internet.

4.33. Upgrading a certain software package requires installation of 68 new files. Files are installed consecutively. The installation time is random, but on the average, it takes 15 sec to install one file, with a variance of 11 sec^2.

(a) What is the probability that the whole package is upgraded in less than 12 minutes?

(b) A new version of the package is released. It requires only N new files to be installed, and it is promised that 95% of the time upgrading takes less than 10 minutes. Given this information, compute N.

4.34. Two independent customers are scheduled to arrive in the afternoon. Their arrival times are uniformly distributed between 2 pm and 8 pm. Compute

(a) the expected time of the first (earlier) arrival;

(b) the expected time of the last (later) arrival.

4.35. Let X and Y be independent Standard Uniform random variables.

(a) Find the probability of an event $\{0.5 < (X + Y) < 1.5\}$.

(b) Find the conditional probability $P\{0.3 < X < 0.7 \mid Y > 0.5\}$.

This problem can be solved analytically as well as geometrically.

4.36. Prove the memoryless property of Geometric distribution. That is, if X has Geometric distribution with parameter p, show that

$$P\{X > x + y \mid X > y\} = P\{X > x\}$$

for any integer $x, y \geq 0$.

Chapter 5

Computer Simulations and Monte Carlo Methods

5.1 Introduction

Computer simulations refer to a regeneration of a process by writing a suitable computer program and observing its results. *Monte Carlo methods* are those based on computer simulations involving random numbers.

The main purpose of simulations is estimating such quantities whose direct computation is complicated, risky, consuming, expensive, or impossible. For example, suppose a complex device or machine is to be built and launched. Before it happens, its performance is simulated, and this allows experts to evaluate its adequacy and associated risks carefully and safely. For example, one surely prefers to evaluate reliability and safety of a new module of a space station by means of computer simulations rather than during the actual mission.

Monte Carlo methods are mostly used for the computation of probabilities, expected values, and other distribution characteristics. Recall that probability can be defined as a *long-run proportion*. With the help of random number generators, computers can actually simulate a *long run*. Then, probability can be estimated by a mere computation of the associated observed frequency. The longer run is simulated, the more accurate result is obtained. Similarly, one can estimate expectations, variances, and other distribution characteristics from a long run of simulated random variables.

Monte Carlo methods and Monte Carlo studies inherit their name from Europe's most famous *Monte Carlo casino* (Figure 5.1) located in Principality of Monaco on the Mediterranean coast since the 1850s. Probability distributions involved in gambling are often complicated, but they can be assessed via simulations. In early times, mathematicians generated extra income by estimating vital probabilities, devising optimal gambling strategies, and selling them to professional gamblers.

FIGURE 5.1: *Casino Monte Carlo in Principality of Monaco.*

5.1.1   Applications and examples

Let us briefly look at a few examples, where distributions are rather complicated, and thus, Monte Carlo simulation appears simpler than any direct computation. Notice that in these examples, instead of generating the *actual* devices, computer systems, networks, viruses, and so on, we only simulate the associated *random variables*. For the study of probabilities and distributions, this is entirely sufficient.

Example 5.1 (FORECASTING).   Given just a basic distribution model, it is often very difficult to make reasonably *remote predictions*. Often a one-day development depends on the results obtained during all the previous days. Then prediction for tomorrow may be straightforward whereas computation of a one-month forecast is already problematic.

On the other hand, *simulation* of such a process can be easily performed day by day (or even minute by minute). Based on present results, we simulate the next day. Now we "know it," and thus, we can simulate the day after that, etc. For every time n, we simulate X_{n+1} based on already known $X_1, X_2, ..., X_n$. Controlling the length of this do-loop, we obtain forecasts for the next days, weeks, or months. Such simulations result, for example, in beautiful animated weather maps that meteorologists often show to us on TV news. They help predict future paths of storms and hurricanes as well as possibilities of flash floods.

Simulation of future failures reflects reliability of devices and systems. Simulation of future stock and commodity prices plays a crucial role in finance, as it allows valuations of options and other financial deals.   ◊

Example 5.2 (PERCOLATION).   Consider a network of nodes. Some nodes are connected, say, with transmission lines, others are not (mathematicians would call such a network a *graph*). A signal is sent from a certain node. Once a node k receives a signal, it sends it along each of its output lines with some probability p_k. After a certain period of time, one desires to estimate the proportion of nodes that received a signal, the probability for a certain node to receive it, etc.

This general *percolation* model describes the way many phenomena may *spread*. The role of a signal may be played by a computer virus spreading from one computer to another, or by rumors spreading among people, or by fire spreading through a forest, or by a disease spreading between residents.

Technically, simulation of such a network reduces to generating Bernoulli random variables with parameters p_i. Line i transmits if the corresponding generated variable $X_i = 1$. In the end, we simply count the number of nodes that got the signal, or verify whether the given node received it.   ◊

Example 5.3 (Queuing). A queuing system is described by a number of random variables. It involves spontaneous arrivals of jobs, their random waiting time, assignment to servers, and finally, their random service time and departure. In addition, some jobs may exit prematurely, others may not enter the system if it appears full, and also, intensity of the incoming traffic and the number of servers on duty may change during the day.

When designing a queuing system or a server facility, it is important to evaluate its vital performance characteristics. This will include the job's average waiting time, the average length of a queue, the proportion of customers who had to wait, the proportion of "unsatisfied customers" (that exit prematurely or cannot enter), the proportion of jobs spending, say, more than one hour in the system, the expected usage of each server, the average number of available (idle) servers at the time when a job arrives, and so on. Queuing systems are discussed in detail in Chapter 7; methods of their simulation are in Section 7.6.   ◊

Example 5.4 (Markov chain Monte Carlo). There is a modern technique of generating random variables from rather complex, often intractable distributions, as long as *conditional distributions* have a reasonably simple form. In semiconductor industry, for example, the joint distribution of good and defective chips on a produced wafer has a rather complicated correlation structure. As a result, it can only be written explicitly for rather simplified artificial models. On the other hand, the quality of each chip is predictable based on the quality of the surrounding, neighboring chips. Given its neighborhood, conditional probability for a chip to fail can be written, and thus, its quality can be simulated by generating a corresponding Bernoulli random variable with $X_i = 1$ indicating a failure.

According to the *Markov chain Monte Carlo* (MCMC) methodology, a long sequence of random variables is generated from conditional distributions. A wisely designed MCMC will then produce random variables that have the desired *unconditional* distribution, no matter how complex it is.   ◊

In all the examples, we saw how different types of phenomena can be computer-simulated. However, one simulation is not enough for estimating probabilities and expectations. After we understand how to program the given phenomenon once, we can embed it in a do-loop and repeat similar simulations a large number of times, generating a *long run*. Since the simulated variables are random, we will generally obtain a number of different *realizations*, from which we calculate probabilities and expectations as long-run frequencies and averages.

5.2   Simulation of random variables

As we see, implementation of Monte Carlo methods reduces to generation of random variables from given distributions. Hence, it remains to design algorithms for generating random variables and vectors that will have the desired distributions.

Statistics software packages like R, SAS, Splus, SPSS, Minitab, and others have built-in procedures for the generation of random variables from the most common discrete and continuous distributions. Similar tools are found in recent versions of MATLAB, Microsoft Excel, and some libraries.

Majority of computer languages have a *random number generator* that returns only *Uniformly distributed* independent random variables. This section discusses general methods of transforming Uniform random variables, obtained by a standard random number generator, into variables with desired distributions.

5.2.1 Random number generators

Obtaining a good random variable is not a simple task. How do we know that it is "truly random" and does not have any undesired patterns? For example, quality random number generation is so important in coding and password creation that people design special tests to verify the "randomness" of generated numbers.

In the fields sensitive to good random numbers, their generation is typically related to an accurate measurement of some physical variable, for example, the computer time or noise. Certain transformations of this variable will generally give the desired result.

More often than not, a *pseudo-random number generator* is utilized. This is nothing but a very long list of numbers. A user specifies a *random number seed* that points to the location from which this list will be read. It should be noted that if the same computer code is executed with the same seed, then the same random numbers get generated, leading to identical results. Often each seed is generated within the system, which of course improves the quality of random numbers.

Instead of a computer, a **table of random numbers** is often used for small-size studies. For example, we can use Table A1 in Appendix.

Can a table adequately replace a computer random number generator? There are at least two issues that one should be aware of.

1. Generating a random number seed. A rule of thumb suggests to close your eyes and put your finger "somewhere" on the table. Start reading random numbers from this point, either horizontally, or vertically, or even diagonally, etc. Either way, this method is not pattern-free, and it does not guarantee "perfect" randomness.

2. Using the same table more than once. Most likely, we cannot find a new table of random numbers for each Monte Carlo study. As soon as we use the same table again, we may no longer consider our results independent of each other. Should we be concerned about it? It depends on the situation. For totally unrelated projects, this should not be a problem.

One way or another, a random number generator or a table of random numbers delivers to us Uniform random variables $U_1, U_2, \ldots \in (0, 1)$. The next three subsections show how to transform them into a random variable (vector) X with the given desired distribution $F(x)$.

$$\text{NOTATION} \; \Big\| \; U; U_1, U_2, \ldots \; = \; \begin{array}{c} \text{generated Uniform(0,1)} \\ \text{random variables} \end{array} \; \Big\|$$

5.2.2 Discrete methods

At this point, we have obtained one or several independent Uniform(0,1) random variables by means of a random number generator or a table of random numbers. Variables from certain simple distributions can be immediately generated from this.

Example 5.5 (BERNOULLI). First, simulate a Bernoulli trial with probability of success p. For a Standard Uniform variable U, define

$$X = \begin{cases} 1 & \text{if} \quad U < p \\ 0 & \text{if} \quad U \geq p \end{cases}$$

We call it "a success" if $X = 1$ and "a failure" if $X = 0$. Using the Uniform distribution of U, we find that

$$\boldsymbol{P}\{\text{ success }\} = \boldsymbol{P}\{U < p\} = p.$$

Thus, we have generated a Bernoulli trial, and X has Bernoulli distribution with the desired probability p, see Figure 5.2a.

R and MATLAB codes for this scheme are

```
— R ————————        — MATLAB ————
U <- runif(1);            U = rand;       % Standard Uniform variable U
X <- 1*(U < p);           X = (U < p);    % converted into Bernoulli X
```

The value of p should be defined prior to this program.   ◇

Example 5.6 (BINOMIAL). Once we know how to generate Bernoulli variables, we can obtain a Binomial variable as a sum of n independent Bernoulli. For this purpose, we start with n Uniform random numbers, for example:

```
— R ————————        — MATLAB ————
n <- 20; p <- 0.68;       n = 20; p = 0.68;
U <- runif(n);            U = rand(n,1);
X <- sum(U < p);          X = sum(U < p)
```
                                                                                           ◇

Example 5.7 (GEOMETRIC). A while-loop of Bernoulli trials will generate a Geometric random variable, literally according to its definition on p. 61. We run the loop of trials until the first success occurs. Variable X counts the number of failures:

```
— R ————————             — MATLAB ————
X <- 1; U <- runif(1);         X = 1; U = rand;       % Need at least one trial
while ( U > p ){               while U > p;           % Continue while there
 X <- X+1; U <- runif(1);       X = X+1; U = rand;    %      are failures
} X                            end; X                 % Stop at the 1st success
```
                                                                                           ◇

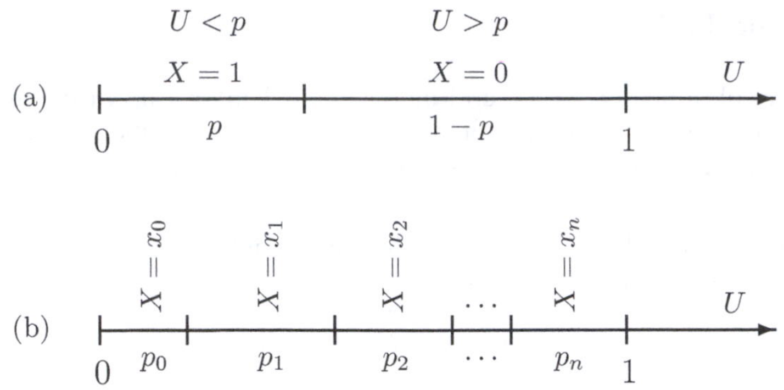

FIGURE 5.2: *Generating discrete random variables. The value of X is determined by the region where the generated value of U belongs.*

Example 5.8 (NEGATIVE BINOMIAL).   Just as in Example 5.6, once we know how to generate a Geometric variable, we can generate a number of them and obtain a Negative Binomial(k, p) variable as a sum of k independent Geometric(p) variables.   ◊

Arbitrary discrete distribution

Examples 5.5-5.8 show how to generate Bernoulli, Binomial, Geometric, and Negative Binomial variables using a random number generator that produces Standard Uniform variables. However, these four are rather standard and simple distributions. Many software languages, including R and MATLAB, even contain built-in functions that can do this work (Section A.2.1 in the Appendix). But what should we do if our distribution is not one of these?

General methods are discussed next.

Apparently, Example 5.5 can be extended to any arbitrary discrete distribution. In this Example, knowing that the random number generator returns a number between 0 and 1, we divided the interval $[0, 1]$ into two parts, p and $(1 - p)$ in length. Then we determined the value of X according to the part where the generated value of U fell, as in Figure 5.2a.

Now consider an arbitrary discrete random variable X that takes values x_0, x_1, ... with probabilities p_0, p_1, ...,

$$p_i = P\{X = x_i\}, \qquad \sum_i p_i = 1.$$

The scheme similar to Example 5.5 can be applied as follows.

Algorithm 5.1 *(Generating discrete variables)*

 1. Divide the interval $[0, 1]$ into subintervals as shown in Figure 5.2b,

$$\begin{aligned}
A_0 &= [0, \ p_0) \\
A_1 &= [p_0, \ p_0 + p_1) \\
A_2 &= [p_0 + p_1, \ p_0 + p_1 + p_2) \\
&\text{etc.}
\end{aligned}$$

Subinterval A_i will have length p_i; there may be a finite or infinite number of them, according to possible values of X.

2. Obtain a Standard Uniform random variable from a random number generator or a table of random numbers.

3. If U belongs to A_i, let $X = x_i$.

From the Uniform distribution, it follows again that

$$\boldsymbol{P}\{X = x_i\} = \boldsymbol{P}\{U \in A_i\} = p_i.$$

Hence, the generated variable X has the desired distribution.

Notice that contrary to Examples 5.6 and 5.8, this algorithm is economic as it requires only one Uniform random number for each generated variable.

Values x_i can be written in any order, but they have to correspond to their probabilities p_i.

Example 5.9 (POISSON). Let us use Algorithm 5.1 to generate a Poisson variable with parameter $\lambda = 5$.

Recall from Section 3.4.5 that a Poisson variable takes values $x_0 = 0$, $x_1 = 1$, $x_2 = 2$, ... with probabilities

$$p_i = \boldsymbol{P}\{X = x_i\} = e^{-\lambda}\frac{\lambda^i}{i!} \ \text{ for } \ i = 0, 1, 2, \ldots$$

Following the algorithm, we generate a Uniform random number U and find the set A_i containing U, so that

$$p_0 + \ldots + p_{i-1} \le U < p_0 + \ldots + p_{i-1} + p_i,$$

or, in terms of the cumulative distribution function,

$$F(i-1) \le U < F(i).$$

This can be done in R or MATLAB by a simple while-loop:

— R ——————

```
lambda <- 5;                  # Parameter: choose any positive lambda
U <- runif(1);                # Generated Uniform variable
i <- 0; F <- exp(-lambda);    # Initial value, F(0)
while (U >= F){               # The loop ends when U < F(i)
  F <- F + exp(-lambda) * lambda^i / factorial(i);
  i <- i + 1;
};  X <- i; X                 # Result: we generated a Poisson variable X
```

— MATLAB ———

```
lambda = 5;                   % Parameter: choose any positive lambda
U = rand;                     % Generated Uniform variable
i = 0; F = exp(-lambda);      % Initial value, F(0)
while (U >= F);               % The loop ends when U < F(i)
  F = F + exp(-lambda) * lambda^i / factorial(i);
  i = i + 1;
end;   X = i                  % Result: we generated a Poisson variable X
```
◇

Remark: Statistical toolbox of MATLAB has built-in tools for generating random variables from certain given distributions. Generating Binomial, Geometric, Poisson, and some other variables can be done in one line `random(name,parameters)`, where "name" indicates the desired family of distributions, or by commands `binornd`, `poissrnd`, etc., listed in our distribution inventory in Section A.2. Similarly, R has built-in functions `rbino`, `rpois`, etc. This chapter discusses general methods of generating random variables from arbitrary distributions, including those that are not found in software languages or published tables. The codes given here are general too – they can be used as flow-charts and be directly translated into other languages.

5.2.3   Inverse transform method

As Algorithm 5.1 applies to all discrete distributions, we now turn attention to the *generation of continuous random variables*. The method will be based on the following simple yet surprising fact.

Theorem 2 *Let X be a continuous random variable with cdf $F_X(x)$. Define a random variable $U = F_X(X)$. The distribution of U is Uniform(0,1).*

PROOF: First, we notice that $0 \le F(x) \le 1$ for all x, therefore, values of U lie in $[0, 1]$. Second, for any $u \in [0, 1]$, find the cdf of U,

$$
\begin{aligned}
F_U(u) &= \mathbf{P}\{U \le u\} \\
&= \mathbf{P}\{F_X(X) \le u\} \\
&= \mathbf{P}\{X \le F_X^{-1}(u)\} \quad \text{(solve the inequality for X)} \\
&= F_X(F_X^{-1}(u)) \quad \text{(by definition of cdf)} \\
&= u \quad \text{(F_X and F_X^{-1} cancel)}
\end{aligned}
$$

In the case if F_X is not invertible, let $F_X^{-1}(u)$ denote the smallest x such that $F_X(x) = u$.

We see that U has cdf $F_U(u) = u$ and density $f_U(u) = F_U'(u) = 1$ for $0 \le u \le 1$. This is the Uniform(0,1) density, hence, U has Uniform(0,1) distribution. $\square$

Regardless of the initial distribution of X, it becomes Uniform(0,1), once X is substituted into its own cumulative distribution function!

Arbitrary continuous distribution

In order to generate variable X with the given continuous cdf F, let us revert the formula $U = F(X)$. Then X can be obtained from a generated Standard Uniform variable U as $X = F^{-1}(U)$.

Algorithm 5.2 *(Generating continuous variables)*

1. Obtain a Standard Uniform random variable from a random number generator.

2. Compute $X = F^{-1}(U)$. In other words, solve the equation $F(X) = U$ for X.

Does it follow directly from Theorem 2 that X has the desired distribution? Details are left in Exercise 5.13.

Example 5.10 (EXPONENTIAL). How shall we generate an Exponential variable with parameter λ? According to Algorithm 5.2, we start by generating a Uniform random variable U. Then we recall that the Exponential cdf is $F(x) = 1 - e^{-\lambda x}$ and solve the equation

$$1 - e^{-\lambda X} = U.$$

The result is

$$X = -\frac{1}{\lambda}\ln(1 - U). \tag{5.1}$$

Can this formula be simplified? By any rule of algebra, the answer is "no." On the other hand, $(1 - U)$ has the same distribution as U, Standard Uniform. Therefore, we can replace U by $(1 - U)$, and variable

$$X_1 = -\frac{1}{\lambda}\ln(U), \tag{5.2}$$

although different from X, will also have the desired Exponential(λ) distribution.

Let us check if our generated variables, X and X_1, have a suitable range of values. We know that $0 < U < 1$ with probability 1, hence $\ln(U)$ and $\ln(1 - U)$ are negative numbers, so that both X and X_1 are positive, as Exponential variables should be. Just a check. $\Diamond$

Example 5.11 (GAMMA). Gamma cdf has a complicated integral form (4.8), not mentioning F^{-1}, needed for Algorithm 5.2. There are numerical methods of solving the equation $F(X) = U$, but there is a simpler option of generating a Gamma variable for any integer α as a sum of α independent Exponential variables:

— R ————————
```
X <- sum( -1/lambda * log(runif(alpha)) )
```

— MATLAB ————
```
X = sum( -1/lambda * log(rand(alpha,1)) )
```
$\Diamond$

Discrete distributions revisited

Algorithm 5.2 is not directly applicable to discrete distributions because the inverse function F^{-1} does not exist in the discrete case. In other words, the key equation $F(X) = U$ may have either infinitely many roots or no roots at all (see Figure 3.1 on p. 42).

Moreover, a discrete variable X has a finite or countable range of possible values, and so does $F(X)$. The probability that U coincidentally equals one of these values is 0. We conclude that the equation $F(X) = U$ has no roots with probability 1.

Let us modify the scheme in order to accommodate discrete variables. Instead of solving $F(x) = U$ exactly, which is impossible, we solve it approximately by finding x, the smallest possible value of X such that $F(x) > U$.

The resulting algorithm is described below.

Algorithm 5.3 *(Generating discrete variables, revisited)*

1. Obtain a Standard Uniform random variable from a random number generator or a table of random numbers.

2. Compute $X = \min \{x \in S \text{ such that } F(x) > U\}$, where S is a set of possible values of X.

Algorithms 5.1 and 5.3 are equivalent if values x_i are arranged in their increasing order in Algorithm 5.1. Details are left in Exercise 5.14.

Example 5.12 (GEOMETRIC REVISITED). Applying Algorithm 5.3 to Geometric cdf

$$F(x) = 1 - (1-p)^x,$$

we need to find the smallest integer x satisfying the inequality (solve it for x)

$$1 - (1-p)^x > U, \qquad (1-p)^x < 1 - U, \qquad x\ln(1-p) < \ln(1-U),$$

$$x > \frac{\ln(1-U)}{\ln(1-p)} \qquad \text{(changing the sign because } \ln(1-p) < 0\text{)}$$

The smallest such integer is the ceiling[1] of $\ln(1-U)/\ln(1-p)$, so that

$$X = \left\lceil \frac{\ln(1-U)}{\ln(1-p)} \right\rceil. \tag{5.3}$$

$\Diamond$

Exponential-Geometric relation

The formula (5.3) appears very similar to (5.1). With $\lambda = -\ln(1-p)$, our generated Geometric variable is just the ceiling of an Exponential variable! In other words, the ceiling of an Exponential variable has Geometric distribution.

This is not a coincidence. In a sense, Exponential distribution is a continuous analogue of a Geometric distribution. Recall that an Exponential variable describes the time until the next "rare event" whereas a Geometric variable is the time (the number of Bernoulli trials) until the next success. Also, both distributions have a memoryless property distinguishing them from other distributions (see (4.6) and Exercise 4.36).

5.2.4   Rejection method

Besides a good computer, one thing is needed to generate continuous variables using the inverse transform method of Section 5.2.3. It is a reasonably *simple form of the cdf* $F(x)$ that allows direct computation of $X = F^{-1}(U)$.

When $F(x)$ has a complicated form but a density $f(x)$ is available, random variables with this density can be generated by *rejection method*.

Consider a point (X, Y) chosen at random from under the graph of density $f(x)$, as shown in Figure 5.3. What is the distribution of X, its first coordinate?

---

[1]Remark: Ceiling function $\lceil x \rceil$ is defined as the smallest integer that is no less than x.

Theorem 3 *Let a pair* (X, Y) *have Uniform distribution over the region*

$$A = \{(x, y) \mid 0 \le y \le f(x)\}$$

for some density function f. *Then* f *is the density of* X.

PROOF: Uniform distribution has a constant density. In case of a pair (X, Y), this density equals 1 because

$$\iint_A 1 \, dy \, dx = \int_x \left(\int_{y=0}^{f(x)} 1 \, dy \right) dx = \int_x f(x) \, dx = 1$$

because $f(x)$ is a density.

The marginal density of X obtains from the joint density by integration,

$$f_X(x) = \int f_{(X,Y)}(x, y) \, dy = \int_{y=0}^{f(x)} 1 \, dy = f(x).$$

$\square$

It remains to generate Uniform points in the region A. For this purpose, we select a *bounding box* around the graph of $f(x)$ as shown in Figure 5.3, generate random points in this rectangle, and reject all the points not belonging to A. The remaining points are Uniformly distributed in A.

Algorithm 5.4 *(Rejection method)*

1. Find such numbers a, b, and c that $0 \le f(x) \le c$ for $a \le x \le b$. The bounding box stretches along the x-axis from a to b and along the y-axis from 0 to c.

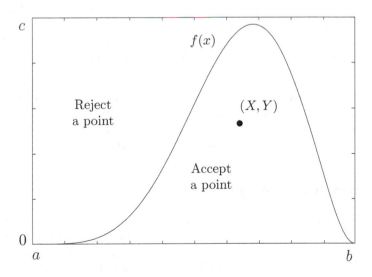

FIGURE 5.3: *Rejection method. A pair* (X, Y) *should have a Uniform distribution in the region under the graph of* $f(x)$.

2. Obtain Standard Uniform random variables U and V from a random number generator or a table of random numbers.

3. Define $X = a + (b - a)U$ and $Y = cV$. Then X has Uniform(a, b) distribution, Y is Uniform$(0, c)$, and the point (X, Y) is Uniformly distributed in the bounding box.

4. If $Y > f(X)$, reject the point and return to step 2. If $Y \leq f(X)$, then X is the desired random variable having the density $f(x)$.

Example 5.13 (Rejection method for Beta distribution). Beta distribution has density

$$f(x) = \frac{\Gamma(\alpha + \beta)}{\Gamma(\alpha)\Gamma(\beta)} x^{\alpha-1}(1 - x)^{\beta-1} \text{ for } 0 \leq x \leq 1.$$

For $\alpha = 5.5$ and $\beta = 3.1$, this density is graphed in Figure 5.3. It never exceeds 2.5, therefore we choose the bounding box with $a = 0$, $b = 1$, and $c = 2.5$. MATLAB commands for generating a Beta(α, β) random variable by the rejection method are:

— R ——————
```
alpha<-5.5; beta<-3.1; a<-0; b<-1; c<-2.5; X<-0; Y<-c;   # Initialization
while (Y > gamma(alpha+beta)/gamma(alpha)/gamma(beta)    # To be continued
       * X^(alpha-1) * (1-X)^(beta-1))  {                # Reject Y > f(X)
   U<-runif(1); V<-runif(1); X<-a+(b-a)*U; Y<-c*V;       # New pair (X,Y)
};   X                                                   # Result
```

— Matlab ————
```
alpha=5.5; beta=3.1; a=0; b=1; c=2.5; X=0; Y=c;          % Initialization
while Y > gamma(alpha+beta)/gamma(alpha)/gamma(beta)...  % To be continued
       * X^(alpha-1) * (1-X)^(beta-1);                   % Reject Y > f(X)
   U=rand; V=rand; X=a+(b-a)*U; Y=c*V;                   % New pair (X,Y)
end; X                                                   % Result
```

It took us two lines to write a formula for the Beta density in these codes. Dots "..." let MATLAB know that the command will continue on the next line. R expected this continuation because the "while" condition was not finished and since the number of "(" exceeded the number of ")". When R knows that the code has to continue, it returns a plus "+" on the next line instead of the usual prompt ">".

A histogram of 10,000 random variables generated (in a loop) by this algorithm is shown in Figure 5.4. Let's compare its shape with the graph of the density $f(x)$ in Figure 5.3. It is not as smooth as $f(x)$ because of randomness and a finite sample size, but if the simulation algorithm is designed properly, the shapes should be similar. ◇

5.2.5 Generation of random vectors

Along the same lines, we can use rejection method to generate *random vectors* having desired joint densities. A bounding box now becomes a multidimensional cube, where we generate a Uniformly distributed random point $(X_1, X_2, \ldots, X_n, Y)$, accepted only if $Y \leq f(X_1, \ldots, X_n)$. Then, the generated vector $(X_1, \ldots, X_n)$ has the desired joint density

$$f_{X_1,\ldots,X_n}(x_1, \ldots, x_n) = \int f_{X_1,\ldots,X_n,Y}(x_1, \ldots, x_n, y)dy = \int_0^{f(x_1,\ldots,x_n)} 1 \, dy = f(x_1, \ldots, x_n).$$

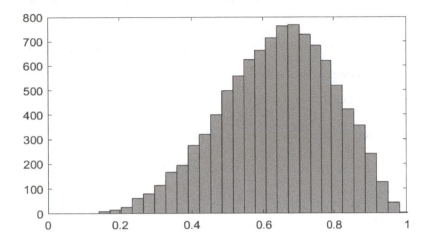

FIGURE 5.4: *A histogram of Beta random variables generated by rejection method. Compare with Figure 5.3.*

The inverse transform method can also be used to generate random vectors. Let us generate X_1 with the marginal cdf F_{X_1}. Observe its value, x_1, and generate X_2 from the *conditional cdf*

$$F_{X_2|X_1}(x_2|x_1) = \begin{cases} \dfrac{\sum_{x \leq x_2} P_{X_1, X_2}(x_1, x)}{P_{X_1}(x_1)} & \text{in the discrete case} \\[2em] \dfrac{\int_{-\infty}^{x_2} f_{X_1, X_2}(x_1, x)dx}{f_{X_1}(x_1)} & \text{in the continuous case} \end{cases}$$

Then generate X_3 from the cdf $F_{X_3|X_1, X_2}(x_3|x_1, x_2)$, etc.

The inverse transform method requires tractable expressions for the cdf F and conditional cdf's. Rejection method requires only the joint density; however, it needs more random number generations due to rejections. This is not a significant disadvantage for relatively small Monte Carlo studies or relatively tight bounding boxes.

5.2.6 Special methods

Because of a complex form of their cdf $F(x)$, some random variables are generated by methods other than inverse transforms.

Poisson distribution

An alternative way of generating a Poisson variable is to count the number of "rare events" occurring during one unit of time.

Recall from Sections 3.4.5 and 4.2.2 that the number of "rare events" has Poisson distribution whereas the time between any two events is Exponential. So, let's keep generating Exponential times between events according to (5.2) while their sum does not exceed 1. The number of generated times equals the number of events, and this is our generated Poisson variable. In short,

1. Obtain Uniform variables $U_1, U_2, \ldots$ from a random number generator.

2. Compute Exponential variables $T_i = -\frac{1}{\lambda} \ln(U_i)$.

3. Let $X = \max \{k : T_1 + \ldots + T_k \leq 1\}$.

This algorithm can be simplified if we notice that

$$T_1 + \ldots + T_k = -\frac{1}{\lambda} \left(\ln(U_1) + \ldots + \ln(U_k) \right) = -\frac{1}{\lambda} \ln(U_1 \cdot \ldots \cdot U_k)$$

and therefore, X can be computed as

$$
\begin{aligned}
X &= \max \left\{ k : -\frac{1}{\lambda} \ln(U_1 \cdot \ldots \cdot U_k) \leq 1 \right\} \\
&= \max \left\{ k : U_1 \cdot \ldots \cdot U_k \geq e^{-\lambda} \right\}.
\end{aligned}
\tag{5.4}
$$

This formula for generating Poisson variables is rather popular.

Normal distribution

Box-Muller transformation

$$
\begin{cases}
Z_1 &= \sqrt{-2 \ln(U_1)} \cos(2\pi U_2) \\
Z_2 &= \sqrt{-2 \ln(U_2)} \sin(2\pi U_2)
\end{cases}
$$

converts a pair of generated Standard Uniform variables (U_1, U_2) into a pair of independent Standard Normal variables (Z_1, Z_2). This is a rather economic algorithm. To see why it works, solve Exercise 5.15.

5.3   Solving problems by Monte Carlo methods

We have learned how to generate random variables from any given distribution. Once we know how to generate one variable, we can put the algorithm in a loop and generate many variables, a "long run." Then, we shall estimate probabilities by the long-run proportions, expectations by the long-run averages, etc.

5.3.1   Estimating probabilities

This section discusses the most basic and most typical application of Monte Carlo methods. Keeping in mind that probabilities are long-run proportions, we generate a long run of experiments and compute the proportion of times when our event occurred.

For a random variable X, the probability $p = \boldsymbol{P}\{X \in A\}$ is estimated by

$$\widehat{p} = \widehat{\boldsymbol{P}}\{X \in A\} = \frac{\text{number of } X_1, \ldots, X_N \in A}{N},$$

where N is the size of a Monte Carlo experiment, $X_1, \ldots, X_N$ are generated random variables with the same distribution as X, and a "hat" means the estimator. The latter is a very common and standard notation:

$$\underline{\text{NOTATION}} \; \left\| \; \widehat{\theta} \;\; = \;\; \text{estimator of an unknown quantity } \theta \; \right\|$$

How accurate is this method? To answer this question, compute $\boldsymbol{E}(\widehat{p})$ and $\text{Std}(\widehat{p})$. Since the number of $X_1, \ldots, X_N$ that fall within set A has Binomial(N, p) distribution with expectation (Np) and variance $Np(1 - p)$, we obtain

$$\begin{aligned} \boldsymbol{E}(\widehat{p}) &= \frac{1}{N}(Np) = p, \text{ and} \\ \text{Std}(\widehat{p}) &= \frac{1}{N}\sqrt{Np(1-p)} = \sqrt{\frac{p(1-p)}{N}}. \end{aligned}$$

The first result, $\boldsymbol{E}(\widehat{p}) = p$, shows that our Monte Carlo estimator of p is *unbiased*, so that over a long run, it will on the average return the desired quantity p.

The second result, $\text{Std}(\widehat{p}) = \sqrt{p(1-p)/N}$, indicates that the standard deviation of our estimator $\widehat{p}$ decreases with N at the rate of $1/\sqrt{N}$. Larger Monte Carlo experiments produce more accurate results. A 100-fold increase in the number of generated variables reduces the standard deviation (therefore, enhancing accuracy) by a factor of 10.

Accuracy of a Monte Carlo study

In practice, how does it help to know the standard deviation of $\widehat{p}$?

First, we can assess the accuracy of our results. For large N, we use Normal approximation of the Binomial distribution of $N\widehat{p}$, as in (4.19) on p. 94. According to it,

$$\frac{N\widehat{p} - Np}{\sqrt{Np(1-p)}} = \frac{\widehat{p} - p}{\sqrt{\frac{p(1-p)}{N}}} \approx \text{Normal}(0, 1),$$

therefore,

$$\boldsymbol{P}\{|\widehat{p} - p| > \varepsilon\} = \boldsymbol{P}\left\{ \frac{|\widehat{p} - p|}{\sqrt{\frac{p(1-p)}{N}}} > \frac{\varepsilon}{\sqrt{\frac{p(1-p)}{N}}} \right\} \approx 2\Phi\left(-\frac{\varepsilon\sqrt{N}}{\sqrt{p(1-p)}} \right). \quad (5.5)$$

We have computed probabilities of this type in Section 4.2.4.

Second, we can design a Monte Carlo study that attains desired accuracy. That is, we can choose some small ε and α and conduct a Monte Carlo study of such a size N that will guarantee an error not exceeding ε with high probability $(1 - \alpha)$. In other words, we can find such N that

$$\boldsymbol{P}\{|\widehat{p} - p| > \varepsilon\} \leq \alpha. \quad (5.6)$$

If we knew the value of p, we could have equated the right-hand side of (5.5) to α and could have solved the resulting equation for N. This would have shown how many Monte Carlo simulations are needed in order to achieve the desired accuracy with the desired high probability. However, p is unknown (if p is known, why do we need a Monte Carlo study to estimate it?). Then, we have two possibilities:

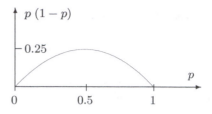

FIGURE 5.5: *Function $p(1-p)$ attains its maximum at $p = 0.5$.*

1. Use an "intelligent guess" (preliminary estimate) of p, if it is available.

2. Bound $p\,(1-p)$ by its largest possible value (see Figure 5.5),

$$p\,(1-p) \le 0.25 \quad \text{for} \quad 0 \le p \le 1.$$

In the first case, if p^* is an "intelligent guess" of p, then we solve the inequality

$$2\Phi\left(-\frac{\varepsilon\sqrt{N}}{\sqrt{p^*(1-p^*)}}\right) \le \alpha$$

in terms of N. In the second case, we solve

$$2\Phi\left(-2\varepsilon\sqrt{N}\right) \le \alpha$$

to find the conservatively sufficient size of the Monte Carlo study.

Solutions of these inequalities give us the following rule.

Size of a Monte Carlo study

In order to guarantee that $\boldsymbol{P}\left\{|\widehat{p} - p| > \varepsilon\right\} \le \alpha$, one needs to simulate

$$N \ge p^*(1 - p^*)\left(\frac{z_{\alpha/2}}{\varepsilon}\right)^2$$

random variables, where p^* is a preliminary estimator of p, or

$$N \ge 0.25\left(\frac{z_{\alpha/2}}{\varepsilon}\right)^2$$

random variables, if no such estimator is available.

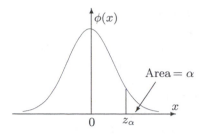

FIGURE 5.6: *Critical value z_α.*

Recall (from Example 4.12) that $z_\alpha = \Phi^{-1}(1-\alpha)$ is such a value of a Standard Normal variable Z that can be exceeded with probability α. It can be obtained from Table A4. The area under the Normal curve to the right of z_α equals α (Figure 5.6).

If this formula returns N that is too small for the Normal approximation, we can use Chebyshev's inequality (3.8) on p. 54. We then obtain that

$$N \geq \frac{p^*(1-p^*)}{\alpha \varepsilon^2}, \qquad (5.7)$$

if an "intelligent guess" p^* is available for p, and

$$N \geq \frac{1}{4\alpha\varepsilon^2}, \qquad (5.8)$$

otherwise, satisfying the desired condition (5.6) (Exercise 5.16).

Example 5.14 (SHARED COMPUTER). The following problem does not have a simple analytic solution (by hand), therefore we use the Monte Carlo method.

A supercomputer is shared by $N = 250$ independent subscribers. Each day, each subscriber uses the facility with probability $p = 0.3$. The number of tasks sent by each active user has Geometric distribution with parameter $q = 0.15$, and each task takes a Gamma($\alpha = 10, \lambda = 3$) distributed computer time (in minutes). Tasks are processed consecutively. What is the probability that all the tasks will be processed, that is, the total requested computer time is less than 24 hours? Estimate this probability, attaining the margin of error ± 0.01 with probability 0.99.

Solution. Every day, a Binomial(C, p) number X of active users send $Y = Y_1 + \ldots + Y_X$ tasks, where each Y_i is Geometric(q). Given X, the number of tasks each day is Negative Binomial(X, q). Their total time T consists of Y Gamma(α, λ) distributed times. Thus, given X and Y, T has Gamma distribution with parameters αY and λ. To summarize:

$$
\begin{aligned}
X &\sim \text{Binomial}(C, p) \\
Y &\sim \text{Negative Binomial}(X, q), \quad \text{given } X \\
T &\sim \text{Gamma}(\alpha Y, \lambda), \quad \text{given } Y
\end{aligned}
$$

The overall distribution of T is too complex; let's use the Monte Carlo approach.

It is hard to come up with an "intelligent guess" of the probability of interest $P\{T < 24 \text{ hrs}\}$. To attain the required accuracy ($\alpha = 0.01$, $\varepsilon = 0.01$), we use

$$N \geq 0.25 \left(\frac{z_{0.01/2}}{\varepsilon}\right)^2 = 0.25 \left(\frac{2.575}{0.01}\right)^2 = 16,577$$

simulations, where $z_{0.01/2} = z_{0.005} = 2.575$ is found from Table A4. The obtained number N is large enough to justify the Normal approximation.

Next, generate the number of active users X and the number of tasks Y. Given Y, generate the time T required by all these tasks. Repeat this algorithm for N days, saving the obtained times $T_1, \ldots, T_N$. The probability of interest is estimated by the proportion of days when the total time is less than 24 hrs, or 1440 min. The following programs can be used:

```
— R ————————              — MATLAB ————————
N <- 16577;                     N = 16577;
C <- 250; p <- 0.3; q <- 0.15;  C = 250; p = 0.3; q = 0.15;
alpha <- 10; lambda <- 3;       alpha = 10; lambda = 3;
T <- rep(0,N);                  T = zeros(N,1);
for (k in 1:N){                 for k = 1:N;
   X <- rbinom(1,C,p);             X = binornd(C,p);
   Y <- rnbinom(1,X,q) + X;        Y = nbinrnd(X,q) + X;
   T[k] <- rgamma(1,alpha*Y, lambda);  T(k) = gamrnd(alpha*Y, 1/lambda);
}                               end;
P.est <- mean(T < 1440);        P_est = mean(T < 1440);
```

The resulting estimated probability should be close to 0.17. We can conclude from this Monte Carlo study that the computer is not likely to process all the tasks. ◊

Remark: Generating random variables, we need to watch for different forms of our distributions. In the last Example, notice that we added X to each generated Negative Binomial variable Y. This is because both R and MATLAB use a *zero-based* version of the Negative Binomial distribution, where $x \geq 0$ instead of $x \geq k$. We fix this by adding the parameter back to the generated variable. Also, a *scale* parameter $\beta = 1/\lambda$ of Gamma distribution is used instead of a *frequency* parameter λ. We handled this by replacing λ with $1/\lambda$. No such substitution is needed in R which treats the Gamma parameter as *frequency*, just as we do in this book.

5.3.2 Estimating means and standard deviations

Estimation of means, standard deviations, and other distribution characteristics is based on the same principle. We generate a Monte Carlo sequence of random variables $X_1, \ldots, X_N$ and compute the necessary long-run averages. The mean $\mathbf{E}(X)$ is estimated by the average (denoted by $\overline{X}$ and pronounced "X-bar")

$$\overline{X} = \frac{1}{N}\left(X_1 + \ldots + X_N\right).$$

If the distribution of $X_1, \ldots, X_N$ has mean μ and standard deviation σ,

$$\mathbf{E}(\overline{X}) = \frac{1}{N}\left(\mathbf{E}X_1 + \ldots + \mathbf{E}X_N\right) = \frac{1}{N}(N\mu) = \mu, \text{ and} \tag{5.9}$$

$$\text{Var}(\overline{X}) = \frac{1}{N^2}\left(\text{Var}X_1 + \ldots + \text{Var}X_N\right) = \frac{1}{N^2}(N\sigma^2) = \frac{\sigma^2}{N}. \tag{5.10}$$

From (5.9), we conclude that the estimator $\overline{X}$ is *unbiased* for estimating μ. From (5.10), its standard deviation is $\text{Std}(\overline{X}) = \sigma/\sqrt{N}$, and it decreases like $1/\sqrt{N}$. For large N, we can use the Central Limit Theorem again to assess accuracy of our results and to design a Monte Carlo study that attains the desired accuracy (the latter is possible if we have a "guess" about σ).

Variance of a random variable is defined as the expectation of $(X - \mu)^2$. Similarly, we estimate it by a long-run average, replacing the unknown value of μ by its Monte Carlo estimator. The resulting estimator is usually denoted by s^2,

$$s^2 = \frac{1}{N-1}\sum_{i=1}^{N}\left(X_i - \overline{X}\right)^2.$$

Remark: Does the coefficient $\frac{1}{N-1}$ seem surprising? Proper averaging should divide the sum of N numbers by N; however, only $(N-1)$ in the denominator guarantees that $\mathbf{E}(s^2) = \sigma^2$, that is, s^2 is unbiased for σ^2. This fact will be proved in Section 8.2.4. Sometimes, coefficients $\frac{1}{N}$ and even $\frac{1}{N+1}$ are also used when estimating σ^2. The resulting estimates are not unbiased, but they have other attractive properties. For large N, the differences between all three estimators are negligible.

Example 5.15 (SHARED COMPUTER, CONTINUED). In Example 5.14, we generated the total requested computer time T for each of N days and computed the proportion of days when this time was less than 24 hours. Further, we can estimate *the expectation of requested time* and the *standard deviation* as follows.

— R ———————————   — MATLAB ————

```
ExpectedTime = mean(T);
StandardDeviation = sd(T);
```

```
ExpectedTime = mean(T);
StandardDeviation = std(T);
```

The mean requested time appears to be about 1667 minutes, exceeding the 24-hour day by 227 minutes. The standard deviation is approximately 243 minutes. Now it is clear why all the tasks get processed with such a low probability as estimated in Example 5.14. $\diamond$

5.3.3 Forecasting

Forecasting the future is often an attractive but uneasy task. In order to predict what happens in t days, one usually has to study events occurring every day between now and t days from now. It is often found that tomorrow's events depend on today, yesterday, etc. In such common situations, exact computation of probabilities may be long and difficult (although we'll learn some methods in Chapters 6 and 7). However, Monte Carlo forecasts are feasible.

The currently known information can be used to generate random variables for tomorrow. Then, they can be considered "known" information, and we can generate random variables for the day after tomorrow, and so on, until day t.

The rest is similar to the previous sections. We generate N Monte Carlo runs for the next t days and estimate probabilities, means, and standard deviations of day t variables by long-run averaging.

In addition to forecasts for a specific day t, we can predict *how long* a certain process will last, *when* a certain event will occur, and *how many* events will occur.

Example 5.16 (NEW SOFTWARE RELEASE). Here is a stochastic model for the number of errors found in a new software release. Every day, software developers find a random number of errors and correct them. The number of errors X_t found on day t is modeled by a Poisson(λ_t) distribution whose parameter is the smallest number of errors found during the previous 3 days,

$$\lambda_t = \min\{X_{t-1}, X_{t-2}, X_{t-3}\}.$$

Suppose that during the first three days, software developers find 28, 22, and 18 errors.

(a) Predict the time it will take to find all the errors.
(b) Estimate the probability that some errors will remain undetected after 21 days.
(c) Predict the total number of errors in this new release.

Solution. Let us generate $N = 1000$ Monte Carlo runs. In each run, we generate the number of errors found on each day until one day this number equals 0. According to our model, no more errors can be found after that, and we conclude that all errors have been detected.

The following computer codes solve all these questions. They use method (5.4) of generating Poisson random variables, although R and MATLAB users can certainly take advantage of built-in functions rpois and poissrnd.

—— R ——————

```
N <- 1000;                          # number of Monte Carlo runs
Time <- rep(0,N);                   # the day the last error is found
Nerrors <- rep(0,N);                # total number of errors
for (k in 1:N) {                    # do-loop of N Monte Carlo runs
    Last3 <- c(28,22,18);           # errors found during last 3 days
    DE <- sum(Last3);               # detected errors so far
    T <- 3; X <- Last3[3];          # T = # days, X = # errors on day T
    while (X > 0) {                 # while-loop until no errors are found
        lambda <- min(Last3);       # parameter λ for day T
        U <- runif(1); X <- 0;      # initial values
        while (U >= exp(-lambda)) {
            X <- X+1; U <- U*runif(1);   # according to (5.4), X is Poisson(λ)
        }
        T <- T+1; DE <- DE+X;       # update after day T
        Last3 <- cbind(Last3[2:3],X);
    }                               # the loop ends when X=0 on day T
    Nerrors[k] <- DE;               # Save the total number of errors
    Time[k] <- T-1;                 # Save the number of days it took
}
```

—— MATLAB ——————

```
N = 1000;                           % number of Monte Carlo runs
Time = zeros(N,1);                  % the day the last error is found
Nerrors = zeros(N,1);               % total number of errors
for k = 1:N;                        % do-loop of N Monte Carlo runs
    Last3 = [28,22,18];             % errors found during last 3 days
    DE = sum(Last3);                % detected errors so far
    T = 3; X = Last3(3);            % T = # days, X = # errors on day T
    while X > 0;                    % while-loop until no errors are found
        lambda = min(Last3);        % parameter λ for day T
        U = rand; X = 0;            % initial values
        while U >= exp(-lambda);
            X = X+1; U = U*rand;    % according to (5.4), X is Poisson(λ)
        end;
        T = T+1; DE = DE+X;         % update after day T
        Last3 = [Last3(2:3), X];
    end;                            % the loop ends when X = 0 on day T
    Nerrors(k) = DE;                % Save the total number of errors
    Time(k) = T-1;                  % Save the number of days it took
end;
```

Now we estimate the expected time it takes to detect all the errors by `mean(Time)`, the probability of errors remaining after 21 days by `mean(Time>21)`, and the expected total number of errors by `mean(Nerrors)`. This Monte Carlo study should predict the expected time of about 19.7 days to detect all the errors, the probability 0.34 that errors remain after 21 days, and about 222 errors overall. ◇

5.3.4 Estimating lengths, areas, and volumes

Lengths

A Standard Uniform variable U has density $f_U(u) = 1$ for $0 \leq u \leq 1$. Hence, U belongs to a set $A \subset [0,1]$ with probability

$$\boldsymbol{P}\{U \in A\} = \int_A 1 \, du = \text{ length of } A. \tag{5.11}$$

Monte Carlo methods can be used to estimate the probability in the left-hand side. At the same time, we estimate the right-hand side of (5.11), the length of A. Generate a long run of Standard Uniform random variables $U_1, U_2, \ldots, U_n$ and estimate the length of A by the proportion of U_i that fall into A.

What if set A does not lie within a unit interval? Well, we can always choose a suitable system of coordinates, with a suitable origin and scale, to make the interval $[0,1]$ cover the given bounded set as long as the latter is bounded. Alternatively, we can cover A with some interval $[a, b]$ and generate *non-standard* Uniform variables on $[a, b]$, in which case the estimated probability $\boldsymbol{P}\{U \in A\}$ should be multiplied by $(b - a)$.

Areas and volumes

Computing lengths rarely represents a serious problem; therefore, one would rarely use Monte Carlo methods for this purpose. However, the situation is different with estimating *areas* and *volumes*.

The method described for estimating lengths is directly translated into higher dimensions. Two independent Standard Uniform variables U and V have the *joint* density $f_{U,V}(u, v) = 1$ for $0 \leq u, v \leq 1$, hence,

$$\boldsymbol{P}\{(U,V) \in B\} = \iint_B 1 \, du \, dv = \text{ area of } B$$

for any two-dimensional set B that lies within a unit square $[0, 1] \times [0, 1]$. Thus, the area of B can be estimated as a long-run frequency of vectors (U_i, V_i) that belong to set B.

Algorithm 5.5 *(Estimating areas)*

1. Obtain a large *even* number of independent Standard Uniform variables from a random number generator, call them $U_1, \ldots, U_n; V_1, \ldots, V_n$.

2. Count the number of pairs (U_i, V_i) such that the point with coordinates (U_i, V_i) belongs to set B. Call this number N_B.

3. Estimate the area of B by N_B/n.

Similarly, a long run of Standard Uniform triples (U_i, V_i, W_i) allows to estimate the volume of any three-dimensional set.

Areas of arbitrary regions with unknown boundaries

Notice that in order to estimate lengths, areas, and volumes by Monte Carlo methods, *knowing exact boundaries is not necessary*. To apply Algorithm 5.5, it is sufficient to determine which points belong to the given set.

Also, the sampling region does not have to be a square. With different scales along the axes, random points may be generated on a rectangle or even a more complicated figure. One way to generate a random point in a region of arbitrary shape is to draw a larger square or rectangle around it and generate uniformly distributed coordinates until the corresponding point belongs to the region. In fact, by estimating the probability for a random point to fall into the area of interest, we estimate the proportion this area makes of the entire sampling region.

Example 5.17 (SIZE OF THE EXPOSED REGION). Consider the following situation. An emergency is reported at a nuclear power plant. It is necessary to assess the size of the region exposed to radioactivity. Boundaries of the region cannot be determined; however, the level of radioactivity can be measured at any given location.

Algorithm 5.5 can be applied as follows. A rectangle of 10 by 8 miles is chosen that is likely to cover the exposed area. Pairs of Uniform random numbers (U_i, V_i) are generated, and the level of radioactivity is measured at all the obtained random locations. The area of dangerous exposure is then estimated as the proportion of measurements above the normal level, multiplied by the area of the sampling rectangle.

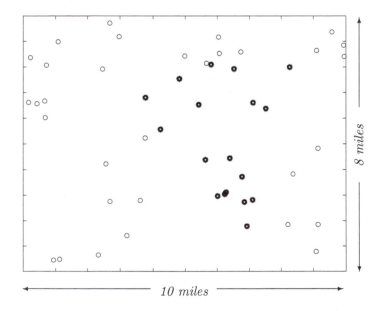

FIGURE 5.7: *Monte Carlo area estimation. Fifty sites are randomly selected; the marked sites belong to the exposed region, Example 5.17.*

In Figure 5.7, radioactivity is measured at 50 random sites, and it is found above the normal level at 18 locations. The exposed area is then estimated as

$$\frac{18}{50}(80 \text{ sq. miles}) = \underline{28.8 \text{ sq. miles}}.$$

$\diamond$

Notice that different scales on different axes in Figure 5.7 allowed to represent a rectangle as a unit square. Alternatively, we could have generated points with Uniform(0,10) x-coordinate and Uniform(0,8) y-coordinate.

5.3.5 Monte Carlo integration

We have seen how Monte Carlo methods can be used to estimate lengths, areas, and volumes. We can extend the method to *definite integrals* estimating areas below or above the graphs of corresponding functions. Computer codes for estimating an integral

$$\mathcal{I} = \int_0^1 g(x)dx$$

are

— R ———————————   — MATLAB ———————

```
N <- 1000             N = 1000;              % Number of simulations
U <- runif(N)         U = rand(N,1);         % (U,V) is a random point
V <- runif(N)         V = rand(N,1);         % in the bounding box
I <- mean( V < g(U) ) I = mean( V < g(U) )   % Estimator of integral I
```

The logical expression `V < g(U)` returns an $N \times 1$ vector. Each component equals "true" or 1 if the inequality holds for the given pair (U_i, V_i), hence this point appears below the graph of $g(x)$, and 0 otherwise. The average of these 0s and 1s, calculated by `mean`, is the proportion of 1s, and this is the Monte Carlo estimator of integral I.

Remark: We assumed here that $0 \leq x \leq 1$ and $0 \leq g(x) \leq 1$. If this is not the case, we transform U and V into non-standard Uniform variables $X = a + (b - a)U$ and $Y = cV$, as in the rejection method, then the obtained estimator I should be multiplied by the box area $c(b - a)$.

Introducing functions in R and MATLAB

The last computer codes for the Monte Carlo integration assume that function g is already defined by the user. For example, one can write `g = function(x){return(sin(sqrt(x)))}` in R to define a function $g(x) = \sin\sqrt{x}$. Such a function can be used in the program, say, for computing `g(3)` or drawing its graph such as `curve(g,0,100)`.

In MATLAB, one should create a file "g.m" with the following text in it,

— MATLAB ———————

```
function y = g(x)
y = sin(sqrt(x));
end
```

If the function has not been defined like this, its full expression should be written in place of g(U).

Accuracy of results

So far, we estimated lengths, areas, volumes, and integrals by long-run proportions, the method described in Section 5.3.1. As we noted there, our estimates are unbiased, and their standard deviation is

$$\text{Std}\left(\widehat{\mathcal{I}}\right) = \sqrt{\frac{\mathcal{I}(1-\mathcal{I})}{N}}, \tag{5.12}$$

where $\mathcal{I}$ is the actual quantity of interest.

Turns out, there are Monte Carlo integration methods that can beat this rate. Next, we derive an unbiased area estimator with a lower standard deviation. Also, it will not be restricted to an interval $[0, 1]$ or even $[a, b]$.

Improved Monte Carlo integration method

First, we notice that a definite integral

$$\mathcal{I} = \int_a^b g(x)dx = \frac{1}{b-a}\int_a^b (b-a)g(x)dx = \mathbf{E}\left\{(b-a)g(X)\right\}$$

equals the expectation of $(b-a)g(X)$ for a Uniform(a, b) variable X. Hence, instead of using proportions, we can estimate $\mathcal{I}$ by averaging $(b-a)g(X_i)$ for some large number of Uniform(a, b) variables $X_1, \ldots, X_N$.

Furthermore, with a proper adjustment, we can use *any continuous distribution* in place of Uniform(a, b). To do this, we choose some density $f(x)$ and write $\mathcal{I}$ as

$$\mathcal{I} = \int_a^b g(x)dx = \int_a^b \frac{g(x)}{f(x)}f(x)\, dx = \mathbf{E}\left(\frac{g(X)}{f(X)}\right),$$

where X has density $f(x)$. It remains to generate a long run of variables $X_1, \ldots, X_N$ with this density and compute the average of $g(X_i)/f(X_i)$.

In particular, we are no longer limited to a finite interval $[a, b]$. For example, by choosing $f(x)$ to be a Standard Normal density, we can perform Monte Carlo integration from $a = -\infty$ to $b = +\infty$:

```
—— R ——————        —— Matlab ——————
N <- 1000;          N = 1000;              % Number of simulations
Z <- rnorm(N)       Z = randn(N,1);        % Normal variables
f <- dnorm(Z)       f = normpdf(Z);        % Normal density
Iest<-mean(g(Z)/f(Z))  Iest=mean(g(Z)./f(Z))  % Estimator of ∫_{-∞}^{∞} g(x) dx
```

Remark: recall that R understands a plain expression such as A*B as a *pointwise* operation, where each element of A is multiplied by the corresponding element of B. Conversely, MATLAB understands such expressions as *matrix* operations. Pointwise operations are marked in MATLAB by "dot" expressions such as g(Z)./f(Z) (also, see matrix operations in the Appendix Section A.5).

Accuracy of the improved method

As we already know from (5.9), using long-run averaging returns an *unbiased* estimator $\widehat{\mathcal{I}}$, hence $\mathbf{E}(\widehat{\mathcal{I}}) = \mathcal{I}$. We also know from (5.10) that

$$\text{Std}(\widehat{\mathcal{I}}) = \frac{\sigma}{\sqrt{N}},$$

where σ is the standard deviation of a random variable $R = g(X)/f(X)$. Hence, the estimator $\widehat{\mathcal{I}}$ is more reliable if σ is small and N is large. Small σ can be obtained by choosing a density $f(x)$ that is approximately proportional to $g(x)$ making R nearly a constant with $\text{Std}(R) \approx 0$. However, generating variables from such a density will usually be just as difficult as computing the integral $\mathcal{I}$.

In fact, using a simple Standard Uniform distribution of X, we already obtain a lower standard deviation than in (5.12). Indeed, suppose that $0 \le g(x) \le 1$ for $0 \le x \le 1$. For a Uniform$(0,1)$ variable X, we have $f(X) = 1$ so that

$$\sigma^2 = \text{Var } R = \text{Var } g(X) = \mathbf{E}g^2(X) - \mathbf{E}^2 g(X) = \int_0^1 g^2(x)dx - \mathcal{I}^2 \le \mathcal{I} - \mathcal{I}^2,$$

because $g^2 \le g$ for $0 \le g \le 1$. We conclude that for this method,

$$\text{Std}(\widehat{\mathcal{I}}) \le \sqrt{\frac{\mathcal{I} - \mathcal{I}^2}{N}} = \sqrt{\frac{\mathcal{I}(1 - \mathcal{I})}{N}}.$$

Comparing with (5.12), we see that with the same number of simulations N, the latter method gives more accurate results. We can also say that to attain the same desired accuracy, the second method requires fewer simulations.

This can be extended to an arbitrary interval $[a, b]$ and any function $g \in [0, c]$ (Exercise 5.18).

Summary and conclusions

Monte Carlo methods are effectively used for estimating probabilities, expectations, and other distribution characteristics in complex situations when computing these quantities by hand is difficult. According to Monte Carlo methodology, we generate a long sequence of random variables $X_1, \ldots, X_N$ from the distribution of interest and estimate probabilities by long-run proportions, expectations by long-run averages, etc. We extend these methods to the estimation of lengths, areas, volumes, and integrals. Similar techniques are used for forecasting.

All the discussed methods produce unbiased results. Standard deviations of the proposed estimators decrease at the rate of $1/\sqrt{N}$. Knowing the standard deviation enables us to assess the accuracy of obtained estimates, and also, to design a Monte Carlo study that attains the desired accuracy with the desired high probability.

In this chapter, we learned the inverse transform method of generating random variables, rejection method, discrete method, and some special methods. Monte Carlo simulation of more advanced models will be considered in Chapters 6 and 7, where we simulate stochastic processes, Markov chains, and queuing systems.

Exercises

5.1. Derive a formula and explain how to generate a random variable with the density

$$f(x) = (1.5)\sqrt{x} \quad \text{for } 0 < x < 1$$

if your random number generator produces a Standard Uniform random variable U. Use the inverse transform method. Compute this variable if $U = 0.001$.

5.2. Let U be a Standard Uniform random variable. Show all the steps required to generate

(a) an Exponential random variable with the parameter $\lambda = 2.5$;

(b) a Bernoulli random variable with the probability of success 0.77;

(c) a Binomial random variable with parameters $n = 15$ and $p = 0.4$;

(d) a discrete random variable with the distribution $P(x)$, where $P(0) = 0.2$, $P(2) = 0.4$, $P(7) = 0.3$, $P(11) = 0.1$;

(e) a continuous random variable with the density $f(x) = 3x^2$, $0 < x < 1$;

(f) a continuous random variable with the density $f(x) = 1.5x^2$, $-1 < x < 1$;

(g) a continuous random variable with the density $f(x) = \frac{1}{12}\sqrt[3]{x}$, $0 \leq x \leq 8$.

If a computer generates U and the result is $U = 0.3972$, compute the variables generated in (a)–(g).

5.3. Explain how one can generate a random variable X that has a pdf

$$f(x) = \begin{cases} \frac{1}{2}(1+x) & \text{if } -1 \leq x \leq 1 \\ 0 & \text{otherwise} \end{cases},$$

given a computer-generated Standard Uniform variable U. Generate X using Table A1.

5.4. To evaluate the system parameters, one uses Monte Carlo methodology and simulates one of its vital characteristics X, a continuous random variable with the density

$$f(x) = \begin{cases} \frac{2}{9}(1+x) & \text{if } -1 \leq x \leq 2 \\ 0 & \text{otherwise} \end{cases}$$

Explain how one can generate X, given a computer-generated Standard Uniform variable U. If $U = 0.2396$, compute X.

5.5. Give an expression that transforms a Standard Uniform variable U into a variable X with the following density,

$$f(x) = \frac{1}{3}x^2, \quad -1 < x < 2.$$

Compute X if a computer returns a value of $U = 0.8$.

5.6. Two mechanics are changing oil filters for the arrived customers. The service time has an Exponential distribution with the parameter $\lambda = 5$ hrs^{-1} for the first mechanic, and $\lambda = 20$ hrs^{-1} for the second mechanic. Since the second mechanic works faster, he is serving 4 times more customers than his partner. Therefore, when you arrive to have your oil filter changed, your probability of being served by the faster mechanic is 4/5. Let X be your service time. Explain how to generate the random variable X.

5.7. Explain how to estimate the following probabilities.

(a) $P\{X > Y\}$, where X and Y are independent Poisson random variables with parameters 3 and 5, respectively.

(b) The probability of a royal-flush (a ten, a jack, a queen, a king, and an ace of the same suit) in poker, if 5 cards are selected at random from a deck of 52 cards.

(c) The probability that it will take more than 35 minutes to have your oil filter changed in Exercise 5.6.

(d) With probability 0.95, we need to estimate each of the probabilities listed in (a-c) with a margin of error not exceeding 0.005. What should be the size of our Monte Carlo study in each case?

(e) (COMPUTER MINI-PROJECT) Conduct a Monte Carlo study and estimate probabilities (a-c) with an error not exceeding 0.005 with probability 0.95.

5.8. (COMPUTER MINI-PROJECT) Area of a unit circle equals π. Cover a circle with a 2 by 2 square and follow Algorithm 5.5 to estimate number π based on 100, 1,000, and 10,000 random numbers. Compare results with the exact value $\pi = 3.14159265358...$ and comment on precision.

5.9. (COMPUTER MINI-PROJECTS) Use Monte Carlo integration to estimate the following integrals.

(a) $\int_0^1 \left| \sin\left(\frac{1}{x}\right) \right| dx$ (b) $\int_0^5 \left| \sin\left(\frac{1}{x}\right) \right| dx$ (c) $\int_0^1 \sin\left(\frac{1}{x}\right) dx$

(d) $\int_{-2}^2 e^{-x^2} dx$ (e) $\int_{-\infty}^\infty e^{-x^2} dx$ (f) $\int_0^\infty e^{-\sqrt{x}} dx$

5.10. (COMPUTER PROJECT) Twenty computers are connected in a network. One computer becomes infected with a virus. Every day, this virus spreads from any infected computer to any uninfected computer with probability 0.1. Also, every day, a computer technician takes 5 infected computers at random (or all infected computers, if their number is less than 5) and removes the virus from them. Estimate:

(a) the expected time it takes to remove the virus from the whole network;

(b) the probability that each computer gets infected at least once;

(c) the expected number of computers that get infected.

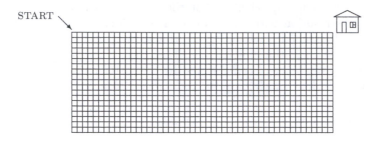

FIGURE 5.8: *The northwestern corner of the forest catches fire (Exercise 5.11).*

5.11. (Computer project) A forest consists of 1,000 trees forming a perfect 50×20 rectangle, Figure 5.8. The northwestern (top-left) corner tree catches fire. Wind blows from the west, therefore, the probability that any tree catches fire from its burning left neighbor is 0.8. The probabilities to catch fire from trees immediately to the right, above, or below are all equal 0.3.

(a) Conduct a Monte Carlo study to estimate the probability that more than 30% of the forest will eventually be burning. With probability 0.95, your answer should differ from the true value by no more than 0.005.

(b) Based on the same study, predict the total number of affected trees X.

(c) Estimate $\text{Std}(X)$ and comment on the accuracy of your estimator of X.

(d) What is the probability that the actual number of affected trees differs from your estimator by more than 25 trees?

(e) A wooden house is located in the northeastern corner of the forest. Would you advise the owner that her house is in real danger?

5.12. (Computer mini-project) Michael has four children, and two of them celebrate their birthdays just two days apart. Is this unusually close?

Let X be the shortest distance between birthdays of four children. Assume that children are born on random, uniformly distributed days of the year, and that birthdays of siblings are independent. Use Monte Carlo simulations to estimate the distribution of X under the stated assumptions and ignoring the leap years. That is, estimate all the probabilities $P(x)$ for $x = 0, 1, \ldots, 91$. Then verify that $\sum P(x) = 1$ and comment on how unusual $X = 2$ is, to answer the question above. What is the mode of the distribution, which is the most likely value of X?

Keep in mind that December 31 and January 1 are only one day apart.

5.13. Let F be a continuous cdf, and U be a Standard Uniform random variable. Show that random variable X obtained from U via a formula $X = F^{-1}(U)$ has cdf F.

5.14. Show that Algorithms 5.1 and 5.3 produce the same discrete variable X if they are based on the same value of a Uniform variable U and values x_i in Algorithm 5.1 are arranged in the increasing order, $x_0 < x_1 < x_2 < \ldots$.

5.15*. Prove that the Box-Muller transformation

$$
\begin{aligned}
Z_1 &= \sqrt{-2\ln(U_1)}\cos(2\pi U_2) \\
Z_2 &= \sqrt{-2\ln(U_1)}\sin(2\pi U_2)
\end{aligned}
$$

returns a pair of independent Standard Normal random variables by showing equality
$\boldsymbol{P}\{Z_1 \le a \;\cap\; Z_2 \le b\} = \Phi(a)\Phi(b)$ for all a and b.

(This requires a substitution of variables in a double integral.)

5.16. A Monte Carlo study is being designed to estimate some probability p by a long-run proportion $\widehat{p}$. Suppose that condition (5.6) can be satisfied by N that is too small to allow the Normal approximation of the distribution of $\widehat{p}$. Use Chebyshev's inequality instead. Show that the size N computed according to (5.7) or (5.8) satisfies the required condition (5.6).

5.17. A random variable is generated from a density $f(x)$ by rejection method. For this purpose, a bounding box with $a \le x \le b$ and $0 \le y \le c$ is selected. Show that this method requires a Geometric(p) number of generated pairs of Standard Uniform random variables and find p.

5.18. We estimate an integral

$$
\mathcal{I} = \int_a^b g(x)\,dx
$$

for arbitrary a and b and a function $0 \le g(x) \le c$ using both Monte Carlo integration methods introduced in this chapter. First, we generate N pairs of Uniform variables (U_i, V_i), $a \le U_i \le b$, $0 \le V_i \le c$, and estimate $\mathcal{I}$ by the properly rescaled proportion of pairs with $V_i \le g(U_i)$. Second, we generate only U_i and estimate $\mathcal{I}$ by the average value of $(b-a)g(U_i)$. Show that the second method produces more accurate results.

Part II

Stochastic Processes

Chapter 6

Stochastic Processes

Let us summarize what we have already accomplished. Our ultimate goal was to learn to make decisions under uncertainty. We introduced a language of *probability* in Chapter 2 and learned how to measure uncertainty. Then, through Chapters 3–5, we studied *random variables*, *random vectors*, and their *distributions*. Have we learned enough to describe a situation involving uncertainty and be able to make good decisions?

Let us look around. If you say "Freeze!" and everything freezes for a moment, the situation will be completely described by random variables surrounding us at a particular moment of time. However, the real world is dynamic. Many variables develop and change in real time: air temperatures, stock prices, interest rates, football scores, popularity of politicians, and also, the CPU usage, the speed of internet connection, the number of concurrent users, the number of running processes, available memory, and so on.

We now start the discussion of *stochastic processes*, which are random variables that evolve and change in time.

6.1 Definitions and classifications

DEFINITION 6.1 ————

> A **stochastic process** is a random variable that also depends on time. It is
> therefore a function of two arguments, $X(t, \omega)$, where:
>
> - $t \in \mathcal{T}$ is time, with $\mathcal{T}$ being a set of possible times, usually $[0, \infty)$, $(-\infty, \infty)$,
> $\{0, 1, 2, \ldots\}$, or $\{\ldots, -2, -1, 0, 1, 2, \ldots\}$;
>
> - $\omega \in \Omega$, as before, is an outcome of an experiment, with Ω being the whole
>   sample space.
>
> Values of $X(t, \omega)$ are called *states*.

At any fixed time t, we see a random variable $X_t(\omega)$, a function of a random outcome. On
the other hand, if we fix ω, we obtain a function of time $X_\omega(t)$. This function is called a
realization, a *sample path*, or a *trajectory* of a process $X(t, \omega)$.

Example 6.1 (CPU USAGE). Looking at the past usage of the central processing unit
(CPU), we see a realization of this process until the current time (Figure 6.1a). However,
the future behavior of the process is unclear. Depending on which outcome ω will actually
take place, the process can develop differently. For example, see two different trajectories
for $\omega = \omega_1$ and $\omega = \omega_2$, two elements of the sample space Ω, on Figure 6.1b.

Remark: You can observe a similar stochastic process on your personal computer. In the latest versions of Windows, Ctrl-Alt-Del, pressed simultaneously and followed by "Windows Task Manager,"
will show the real-time sample path of CPU usage under the tab "Performance."   ◊

Depending on possible values of $\mathcal{T}$ and X, stochastic processes are classified as follows.

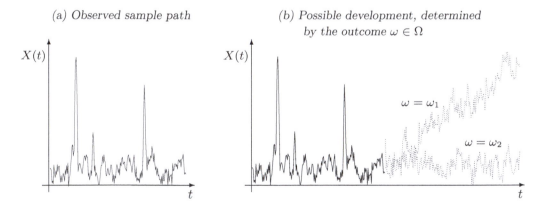

FIGURE 6.1: *Sample paths of CPU usage stochastic process.*

DEFINITION 6.2

> Stochastic process $X(t, \omega)$ is **discrete-state** if variable $X_t(\omega)$ is discrete for each time t, and it is a **continuous-state** if $X_t(\omega)$ is continuous.

DEFINITION 6.3

> Stochastic process $X(t, \omega)$ is a **discrete-time process** if the set of times $\mathcal{T}$ is discrete, that is, it consists of separate, isolated points. It is a **continuous-time process** if $\mathcal{T}$ is a connected, possibly unbounded interval.

Example 6.2. The CPU usage process, in percents, is continuous-state and continuous-time, as we can see in Figure 6.1a. $\Diamond$

Example 6.3. The *actual* air temperature $X(t, \omega)$ at time t is a continuous-time, continuous-state stochastic process. Indeed, it changes smoothly and never jumps from one value to another. However, the temperature $Y(t, \omega)$ reported on a radio every 10 minutes is a discrete-time process. Moreover, since the reported temperature is usually rounded to the nearest degree, it is also a discrete-state process. $\Diamond$

Example 6.4. In a printer shop, let $X(n, \omega)$ be the amount of time required to print the n-th job. This is a discrete-time, continuous-state stochastic process, because $n = 1, 2, 3, \ldots$, and $X \in (0, \infty)$.

Let $Y(n, \omega)$ be the number of pages of the n-th printing job. Now, $Y = 1, 2, 3, \ldots$ is discrete; therefore, this process is discrete-time and discrete-state. $\Diamond$

From now on, we shall not write ω as an argument of $X(t, \omega)$. Just keep in mind that behavior of a stochastic process depends on chance, just as we did with random variables and random vectors.

Another important class of stochastic processes is defined in the next section.

6.2 Markov processes and Markov chains

DEFINITION 6.4

> Stochastic process $X(t)$ is **Markov** if for any $t_1 < \ldots < t_n < t$ and any sets $A; A_1, \ldots, A_n$
>
> $$\boldsymbol{P}\{X(t) \in A \mid X(t_1) \in A_1, \ldots, X(t_n) \in A_n\}$$
> $$= \boldsymbol{P}\{X(t) \in A \mid X(t_n) \in A_n\}. \tag{6.1}$$

Let us look at the equation (6.1). It means that the conditional distribution of $X(t)$ is the same under two different conditions,

(1) given observations of the process X at several moments in the past;

(2) given only *the latest* observation of X.

If a process is Markov, then its future behavior is the same under conditions (1) and (2). In other words, knowing the present, we get no information from the past that can be used to predict the future,

$$\boldsymbol{P}\{\text{ future } \mid \text{ past, present }\} = \boldsymbol{P}\{\text{ future } \mid \text{ present }\}$$

Then, for the future development of a Markov process, only its present state is important, and it does not matter *how* the process arrived to this state.

Some processes satisfy the Markov property, and some don't.

Example 6.5 (INTERNET CONNECTIONS). Let $X(t)$ be the total number of internet connections registered by some internet service provider by the time t. Typically, people connect to the internet at random times, regardless of how many connections have already been made. Therefore, the number of connections in a minute will only depend on the current number. For example, if 999 connections have been registered by 10 o'clock, then their total number will exceed 1000 during the next minute regardless of *when* and *how* these 999 connections were made in the past. This process is *Markov*.   ◊

Example 6.6 (STOCK PRICES). Let $Y(t)$ be the value of some stock or some market index at time t. If we know $Y(t)$, do we also want to know $Y(t-1)$ in order to predict $Y(t+1)$? One may argue that if $Y(t-1) < Y(t)$, then the market is rising, therefore, $Y(t+1)$ is likely (but not certain) to exceed $Y(t)$. On the other hand, if $Y(t-1) > Y(t)$, we may conclude that the market is falling and may expect $Y(t+1) < Y(t)$. It looks like knowing the past *in addition* to the present did help us to predict the future. Then, this process is *not Markov*.
◊

Due to a well-developed theory and a number of simple techniques available for Markov processes, it is important to know whether the process is Markov or not. The idea of Markov dependence was proposed and developed by *Andrei Markov* (1856–1922) who was a student of P. Chebyshev (p. 54) at St. Petersburg University in Russia.

6.2.1   Markov chains

DEFINITION 6.5

> A **Markov chain** is a discrete-time, discrete-state Markov stochastic process.

Introduce a few convenient simplifications. The time is discrete, so let us define the time set as $\mathcal{T} = \{0, 1, 2, \ldots\}$. We can then look at a Markov chain as a random sequence

$$\{X(0), X(1), X(2), \ldots\}.$$

The state set is also discrete, so let us enumerate the states as $1, 2, \ldots, n$. Sometimes we'll start enumeration from state 0, and sometimes we'll deal with a Markov chain with infinitely many (discrete) states, then we'll have $n = \infty$.

The *Markov* property means that only the value of $X(t)$ matters for predicting $X(t+1)$, so the conditional probability

$$p_{ij}(t) = \boldsymbol{P}\left\{X(t+1) = j \mid X(t) = i\right\} \tag{6.2}$$
$$= \boldsymbol{P}\left\{X(t+1) = j \mid X(t) = i, \ X(t-1) = h, \ X(t-2) = g, \ldots\right\}$$

depends on i, j, and t only and equals the probability for the Markov chain X to make a *transition* from state i to state j at time t.

DEFINITION 6.6

Probability $p_{ij}(t)$ in (6.2) is called a **transition probability**. Probability

$$p_{ij}^{(h)}(t) = \boldsymbol{P}\left\{X(t+h) = j \mid X(t) = i\right\}$$

of moving from state i to state j by means of h transitions is an h-**step transition probability**.

DEFINITION 6.7

A Markov chain is **homogeneous** if all its transition probabilities are independent of t. Being homogeneous means that transition from i to j has the same probability at any time. Then $p_{ij}(t) = p_{ij}$ and $p_{ij}^{(h)}(t) = p_{ij}^{(h)}$.

Characteristics of a Markov chain

What do we need to know to describe a Markov chain?

By the Markov property, each next state should be predicted from the previous state only. Therefore, it is sufficient to know the distribution of its initial state $X(0)$ and the mechanism of transitions from one state to another.

The distribution of a Markov chain is completely determined by the initial distribution P_0 and one-step transition probabilities p_{ij}. Here P_0 is the probability mass function of X_0,

$$P_0(x) = \boldsymbol{P}\left\{X(0) = x\right\} \text{ for } x \in \{1, 2, \ldots, n\}$$

Based on this data, we would like to compute:

- *h-step transition probabilities $p_{ij}^{(h)}$;*

- *P_h, the distribution of states at time h, which is our forecast for $X(h)$;*

- *the limit of $p_{ij}^{(h)}$ and P_h as $h \to \infty$, which is our long-term forecast.*

Indeed, when making forecasts for many transitions ahead, computations will become rather lengthy, and thus, it will be more efficient to take the limit.

NOTATION	p_{ij}	$=$	$P\{X(t+1)=j \mid X(t)=i\}$, transition probability
	$p_{ij}^{(h)}$	$=$	$P\{X(t+h)=j \mid X(t)=i\}$, h-step transition probability
	$P_t(x)$	$=$	$P\{X(t)=x\}$, distribution of $X(t)$, distribution of states at time t
	$P_0(x)$	$=$	$P\{X(0)=x\}$, initial distribution

Example 6.7 (WEATHER FORECASTS). In some town, each day is either sunny or rainy. A sunny day is followed by another sunny day with probability 0.7, whereas a rainy day is followed by a sunny day with probability 0.4.

It rains on Monday. Make forecasts for Tuesday, Wednesday, and Thursday.

<u>Solution</u>. Weather conditions in this problem represent a homogeneous Markov chain with 2 states: state 1 = "sunny" and state 2 = "rainy." Transition probabilities are:

$$p_{11} = 0.7, \ p_{12} = 0.3, \ p_{21} = 0.4, \ p_{22} = 0.6,$$

where p_{12} and p_{22} were computed by the complement rule.

If it rains on Monday, then Tuesday is sunny with probability $p_{21} = 0.4$ (making a transition from a rainy to a sunny day), and Tuesday is rainy with probability $p_{22} = 0.6$. We can predict a 60% chance of rain.

Wednesday forecast requires 2-step transition probabilities, making one transition from Monday to Tuesday, $X(0)$ to $X(1)$, and another one from Tuesday to Wednesday, $X(1)$ to $X(2)$. We'll have to condition on the weather situation on Tuesday and use the Law of Total Probability from p. 31,

$$
\begin{aligned}
p_{21}^{(2)} &= \boldsymbol{P}\{\text{ Wednesday is sunny } \mid \text{ Monday is rainy }\} \\
&= \sum_{i=1}^{2} \boldsymbol{P}\{X(1)=i \mid X(0)=2\}\, \boldsymbol{P}\{X(2)=1 \mid X(1)=i\} \\
&= \boldsymbol{P}\{X(1)=1 \mid X(0)=2\}\, \boldsymbol{P}\{X(2)=1 \mid X(1)=1\} \\
&\quad + \boldsymbol{P}\{X(1)=2 \mid X(0)=2\}\, \boldsymbol{P}\{X(2)=1 \mid X(1)=2\} \\
&= p_{21}p_{11} + p_{22}p_{21} = (0.4)(0.7) + (0.6)(0.4) = 0.52.
\end{aligned}
$$

By the Complement Rule, $p_{22}^{(2)} = 0.48$, and thus, we predict a 52% chance of sun and a 48% chance of rain on Wednesday.

For the Thursday forecast, we need to compute 3-step transition probabilities $p_{ij}^{(3)}$ because it takes 3 transitions to move from Monday to Thursday. We have to use the Law of Total Probability conditioning on *both* Tuesday and Wednesday. For example, going from rainy Monday to sunny Thursday means going from rainy Monday to either rainy or sunny Tuesday, then to either rainy or sunny Wednesday, and finally, to sunny Thursday,

$$p_{21}^{(3)} = \sum_{i=1}^{2}\sum_{j=1}^{2} p_{2i}p_{ij}p_{j1}.$$

This corresponds to a sequence of states $2 \to i \to j \to 1$. However, we have already computed 2-step transition probabilities $p_{21}^{(2)}$ and $p_{22}^{(2)}$, describing transition from Monday to Wednesday. It remains to add one transition to Thursday, hence,

$$p_{21}^{(3)} = p_{21}^{(2)}p_{11} + p_{22}^{(2)}p_{21} = (0.52)(0.7) + (0.48)(0.4) = 0.556.$$

So, we predict a 55.6% chance of sun on Thursday and a 44.4% chance of rain.

$\diamond$

The following *transition diagram* (Figure 6.2) reflects the behavior of this Markov chain. Arrows represent all possible one-step transitions, along with the corresponding probabilities. Check this diagram against the transition probabilities stated in Example 6.7. To obtain, say, a 3-step transition probability $p_{21}^{(3)}$, find all 3-arrow paths from state 2 "rainy" to state 1 "sunny." Multiply probabilities along each path and add over all 3-step paths.

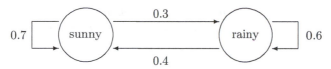

FIGURE 6.2: *Transition diagram for the Markov chain in Example 6.7.*

Example 6.8 (WEATHER, CONTINUED). Suppose now that it does not rain yet, but meteorologists predict an 80% chance of rain on Monday. How does this affect our forecasts?

In Example 6.7, we have computed forecasts under the condition of rain on Monday. Now, a sunny Monday (state 1) is also possible. Therefore, in addition to probabilities $p_{2j}^{(h)}$ we also need to compute $p_{1j}^{(h)}$ (say, using the transition diagram, see Figure 6.2),

$$\begin{aligned} p_{11}^{(2)} &= (0.7)(0.7) + (0.3)(0.4) = 0.61, \\ p_{11}^{(3)} &= (0.7)^3 + (0.7)(0.3)(0.4) + (0.3)(0.4)(0.7) + (0.3)(0.6)(0.4) = 0.583. \end{aligned}$$

The initial distribution $P_0(x)$ is given as

$$P_0(1) = \boldsymbol{P}\{\text{ sunny Monday }\} = 0.2, \qquad P_0(2) = \boldsymbol{P}\{\text{ rainy Monday }\} = 0.8.$$

Then, for each forecast, we use the Law of Total Probability, conditioning on the weather on Monday,

$$P_1(1) = P\{X(1) = 1\} = P_0(1)p_{11} + P_0(2)p_{21} = 0.46 \qquad \text{for Tuesday}$$

$$P_2(1) = P\{X(2) = 1\} = P_0(1)p_{11}^{(2)} + P_0(2)p_{21}^{(2)} = 0.538 \qquad \text{for Wednesday}$$

$$P_3(1) = P\{X(3) = 1\} = P_0(1)p_{11}^{(3)} + P_0(2)p_{21}^{(3)} = 0.5614 \qquad \text{for Thursday}$$

These are probabilities of a sunny day (state 1), respectively, on Tuesday, Wednesday, and Thursday. Then, the chance of rain (state 2) on these days is $P_1(2) = 0.54$, $P_2(2) = 0.462$, and $P_3(2) = 0.4386$.

$\diamond$

Noticeably, more remote forecasts require more lengthy computations. For a t-day ahead forecast, we have to account for all t-step paths on diagram Figure 6.2. Or, we use the Law of Total Probability, conditioning on *all* the intermediate states $X(1), X(2), \ldots, X(t-1)$.

To simplify the task, we shall employ *matrices*. If you are not closely familiar with basic matrix operations, refer to Section A.5 in the Appendix.

6.2.2 Matrix approach

All one-step transition probabilities p_{ij} can be conveniently written in an $n \times n$ *transition probability matrix*

$$
P = \begin{pmatrix}
p_{11} & p_{12} & \cdots & p_{1n} \\
p_{21} & p_{22} & \cdots & p_{2n} \\
\vdots & \vdots & \vdots & \vdots \\
p_{n1} & p_{n2} & \cdots & p_{nn}
\end{pmatrix}
\quad
\begin{array}{l}
\text{From} \\
\text{state:} \\
1 \\
2 \\
\vdots \\
n
\end{array}
$$

$$\text{To state:} \quad 1 \quad 2 \quad \cdots \quad n$$

The entry on the intersection of the i-th row and the j-th column is p_{ij}, the transition probability from state i to state j.

From each state, a Markov chain makes a transition to one and only one state. States-destinations are disjoint and exhaustive events, therefore, *each row total equals 1*,

$$p_{i1} + p_{i2} + \ldots + p_{in} = 1. \tag{6.3}$$

We can also say that probabilities $p_{i1}, p_{i2}, \ldots, p_{in}$ form the *conditional distribution* of $X(1)$, given $X(0)$, so they have to add to 1.

In general, this does not hold for column totals. Some states may be "more favorable" than others, then they are visited more often the others, thus their column total will be larger. Matrices with property (6.3) are called *stochastic*.

Similarly, h-step transition probabilities can be written in an *h-step transition probability matrix*

$$
P^{(h)} = \begin{pmatrix}
p_{11}^{(h)} & p_{12}^{(h)} & \cdots & p_{1n}^{(h)} \\
p_{21}^{(h)} & p_{22}^{(h)} & \cdots & p_{2n}^{(h)} \\
\vdots & \vdots & \vdots & \vdots \\
p_{n1}^{(h)} & p_{n2}^{(h)} & \cdots & p_{nn}^{(h)}
\end{pmatrix}
$$

This matrix is also stochastic because each row represents the conditional distribution of $X(h)$, given $X(0)$ (which is a good way to check our results if we compute $P^{(h)}$ by hand).

Computing h-step transition probabilities

A simple matrix formula connects matrices $P^{(h)}$ and P. Let's start with the 2-step transition probabilities.

By the Law of Total Probability, conditioning and adding over all values of $k = X(1)$,

$$
\begin{aligned}
p_{ij}^{(2)} &= \mathbf{P}\{X(2) = j \mid X(0) = i\} \\[2mm]
&= \sum_{k=1}^{n} \mathbf{P}\{X(1) = k \mid X(0) = i\}\, \mathbf{P}\{X(2) = j \mid X(1) = k\} \\[2mm]
&= \sum_{k=1}^{n} p_{ik} p_{kj} = (p_{i1}, \ldots, p_{in}) \begin{pmatrix} p_{1j} \\ \vdots \\ p_{nj} \end{pmatrix}.
\end{aligned}
$$

Each probability $p_{ij}^{(2)}$ is computed as a sum of $\boldsymbol{P}\{i \to k \to j\}$, over all 2-step paths leading from state i to state j (also, see Example 6.7). As a result, each $p_{ij}^{(2)}$ is a product of the i-th row and the j-th column of matrix P. Hence, the entire 2-step transition probability matrix is

$$
\begin{array}{c|c}
\textbf{2-step transition} & \\
\textbf{probability matrix} & P^{(2)} = P \cdot P = P^2
\end{array}
$$

Further, h-step transition probabilities can be obtained from $(h-1)$-step transition probabilities by conditioning on $X(h-1) = k$,

$$
\begin{aligned}
p_{ij}^{(h)} &= \boldsymbol{P}\{X(h) = j \mid X(0) = i\} \\
&= \sum_{k=1}^{n} \boldsymbol{P}\{X(h-1) = k \mid X(0) = i\}\,\boldsymbol{P}\{X(h) = j \mid X(h-1) = k\} \\
&= \sum_{k=1}^{n} p_{ik}^{(h-1)} p_{kj} = \left(p_{i1}^{(h-1)}, \ldots, p_{in}^{(h-1)} \right)
\begin{pmatrix} p_{1j} \\ \vdots \\ p_{nj} \end{pmatrix}.
\end{aligned}
$$

This time, we consider all h-step paths from i to j. They consist of all $(h-1)$-step paths from i to some state k, followed by one step from k to j. Now, the result is the i-th row of matrix $P^{(h-1)}$ multiplied by the j-th column of P. Hence, $P^{(h)} = P^{(h-1)} \cdot P$, and we have a general formula

$$
\begin{array}{c|c}
\textbf{h-step transition} & \\
\textbf{probability matrix} & P^{(h)} = \underbrace{P \cdot P \cdot \ldots \cdot P}_{h \text{ times}} = P^h
\end{array}
$$

Example 6.9 (SHARED DEVICE). A computer is shared by 2 users who send tasks to a computer remotely and work independently. At any minute, any connected user may disconnect with probability 0.5, and any disconnected user may connect with a new task with probability 0.2. Let $X(t)$ be the number of concurrent users at time t (minutes). This is a Markov chain with 3 states: 0, 1, and 2.

Compute transition probabilities. Suppose $X(0) = 0$, i.e., there are no users at time $t = 0$. Then $X(1)$ is the number of new connections within the next minute. It has Binomial(2,0.2) distribution, therefore,

$$
p_{00} = (.8)^2 = .64, \quad p_{01} = 2(.2)(.8) = .32, \quad p_{02} = (.2)^2 = .04.
$$

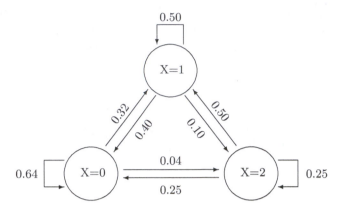

FIGURE 6.3: *Transition diagram for the Markov chain in Example 6.9.*

Next, suppose $X(0) = 1$, i.e., one user is connected, and the other is not. The number of new connections is Binomial(1,0.2), and the number of disconnections is Binomial(1,0.5). Considering all the possibilities, we obtain (verify)

$$p_{10} = (.8)(.5) = .40, \quad p_{11} = (.2)(.5) + (.8)(.5) = .50, \quad p_{12} = (.2)(.5) = .10.$$

Finally, when $X(0) = 2$, no new users can connect, and the number of disconnections is Binomial(2,0.5), so that

$$p_{20} = .25, \quad p_{21} = .50, \quad p_{22} = .25.$$

We obtain the following transition probability matrix,

$$P = \begin{pmatrix} .64 & .32 & .04 \\ .40 & .50 & .10 \\ .25 & .50 & .25 \end{pmatrix}.$$

The transition diagram corresponding to this matrix is shown in Figure 6.3.

The 2-step transition probability matrix is then computed as

$$P^2 = \begin{pmatrix} .64 & .32 & .04 \\ .40 & .50 & .10 \\ .25 & .50 & .25 \end{pmatrix} \begin{pmatrix} .64 & .32 & .04 \\ .40 & .50 & .10 \\ .25 & .50 & .25 \end{pmatrix} = \begin{pmatrix} .5476 & .3848 & .0676 \\ .4810 & .4280 & .0910 \\ .4225 & .4550 & .1225 \end{pmatrix}.$$

For example, if both users are connected at 10:00, then at 10:02 there will be no users with probability $p_{20}^{(2)} = 0.4225$, one user with probability $p_{21}^{(2)} = 0.4550$, and two users with probability $p_{22}^{(2)} = 0.1225$.

Check that both matrices P and P^2 are stochastic. ◊

Computing the distribution of $X(h)$

The distribution of states after h transitions, or the probability mass function of $X(h)$, can be written in a $1 \times n$ matrix,

$$P_h = (P_h(1), \cdots, P_h(n)).$$

By the Law of Total Probability, this time conditioning on $X(0) = k$, we compute

$$
P_h(j) \;=\; \boldsymbol{P}\{X(h) = j\} = \sum_k \boldsymbol{P}\{X(0) = k\}\,\boldsymbol{P}\{X(h) = j \mid X(0) = k\}
$$

$$
=\; \sum_k P_0(k)p_{kj}^{(h)} = \left(\begin{array}{ccc} P_0(1) & \cdots & P_0(n) \end{array} \right) \begin{pmatrix} p_{1j}^{(h)} \\ \vdots \\ p_{1n}^{(h)} \end{pmatrix}.
$$

Each probability $P_h(j)$ is obtained when the entire row P_0 (initial distribution of X) is multiplied by the entire j-th column of matrix P. Hence,

$$
\boxed{\; P_h = P_0 P^h \;} \qquad \text{(6.4)}
$$

Distribution of $X(h)$

Example 6.10 (SHARED DEVICE, CONTINUED). If we know that there are 2 users connected at 10:00, we can write the initial distribution as

$$
P_0 = (0, 0, 1).
$$

Then the distribution of the number of users at 10:02, after $h = 2$ transitions, is computed as

$$
P_2 = P_0 P^2 = (0,0,1) \begin{pmatrix} .5476 & .3848 & .0676 \\ .4810 & .4280 & .0910 \\ .4225 & .4550 & .1225 \end{pmatrix} = (.4225, .4550, .1225),
$$

just as we concluded in the end of Example 6.9.

Suppose that all states are equally likely at 10:00. How can we compute the probability of 1 connected user at 10:02? Here, the initial distribution is

$$
P_0 = (1/3, 1/3, 1/3).
$$

The distribution of $X(2)$ is

$$
P_2 = P_0 P^2 = (1/3, 1/3, 1/3) \begin{pmatrix} .5476 & .3848 & .0676 \\ .4810 & .4280 & .0910 \\ .4225 & .4550 & .1225 \end{pmatrix} = (\ldots, .4226, \ldots).
$$

Thus, $\boldsymbol{P}\{X(2) = 1\} = 0.4226$. We only computed $P_2(1)$. The question did not require computation of the entire distribution P_2. $\diamond$

In practice, knowing the distribution of the number of users at 10:00, one would rarely be interested in the distribution at 10:02. How should one figure the distribution at 11:00, the next day, or the next month?

In other words, how does one compute the distribution of $X(h)$ for large h? A direct solution is $P_h = P_0 \cdot P^h$, which can be computed for moderate n and h. For example, the distribution of the number of users at 11:00, $h = 60$ transitions after 10:00, can be computed in R and MATLAB as

─── R ───────

```
P = matrix( c(.64,.40,.25,.32,.50,
        .50,.04,.10,.25), 3, 3 )
P0 = c(1/3, 1/3, 1/3)
h = 60
# Power of a matrix is in R package "expm"
install.packages("expm"); library(expm);
P0 %*% (P %^% h)
```

─── MATLAB ───────

```
P   = [ .64   .32   .04
        .40   .50   .10
        .25   .50   .25 ];

P0  = [ 1/3  1/3  1/3 ];
h   = 60;
P0 * P^h
```

Recall that percentage signs "%" are used for matrix operations in R, as opposed to elementwise operations; also see Appendix Section A.5.

Taking a power of a large matrix is computationally expensive for large h. One would really like to take a limit of P_h as $h \to \infty$ instead of a tedious computation of $P_0 \cdot P^h$. We learn how to do it in the next section.

6.2.3   Steady-state distribution

This section discusses the distribution of states of a Markov chain after a large number of transitions.

DEFINITION 6.8 ─────

A collection of limiting probabilities

$$\pi_x = \lim_{h \to \infty} P_h(x)$$

is called a **steady-state distribution** of a Markov chain $X(t)$.

When this limit exists, it can be used as a forecast of the distribution of X after *many* transitions. A fast system (say, a central processing unit with a clock rate of several gigahertz), will go through a very large number of transitions very quickly. Then, its distribution of states at virtually any time is the steady-state distribution.

Computing the steady-state distribution

When a steady-state distribution π *exists*, it can be computed as follows. We notice that π is a limit of not only P_h but also P_{h+1}. The latter two are related by the formula

$$P_h P = P_0 P^h P = P_0 P^{h+1} = P_{h+1}.$$

Taking the limit of P_h and P_{h+1}, as $h \to \infty$, we obtain

$$\pi P = \pi. \tag{6.5}$$

Then, solving (6.5) for π, we get the steady-state distribution of a Markov chain with a transition probability matrix P.

The system of steady-state equations (6.5) consists of n equations with n unknowns which are probabilities of n states. However, this system is *singular*; it has infinitely many solutions.

That is because both sides of the system (6.5) can be multiplied by any constant C, and thus, any multiple of π is also a solution of (6.5).

There is yet one more condition that we have not used. Being a distribution, all the probabilities π_x must add to 1,

$$\pi_1 + \ldots + \pi_n = 1.$$

Only one solution of (6.5) can satisfy this condition, and this is the steady-state distribution.

Steady-state distribution

$$\pi = \lim_{h \to \infty} P_h$$

is computed as a solution of

$$\begin{cases} \pi P = \pi \\ \displaystyle\sum_x \pi_x = 1 \end{cases}$$

Example 6.11 (WEATHER, CONTINUED). In Example 6.7 on p. 140, the transition probability matrix of sunny and rainy days is

$$P = \begin{pmatrix} 0.7 & 0.3 \\ 0.4 & 0.6 \end{pmatrix}.$$

The steady-state equation for this Markov chain is

$$\pi P = \pi,$$

or

$$(\pi_1,\ \pi_2) = (\pi_1,\ \pi_2) \begin{pmatrix} 0.7 & 0.3 \\ 0.4 & 0.6 \end{pmatrix} = (0.7\pi_1 + 0.4\pi_2,\ 0.3\pi_1 + 0.6\pi_2).$$

We obtain a system of equations,

$$\begin{cases} 0.7\pi_1 + 0.4\pi_2 = \pi_1 \\ 0.3\pi_1 + 0.6\pi_2 = \pi_2 \end{cases} \Leftrightarrow \begin{cases} 0.4\pi_2 = 0.3\pi_1 \\ 0.3\pi_1 = 0.4\pi_2 \end{cases} \Leftrightarrow \pi_2 = \frac{3}{4}\pi_1.$$

We see that two equations in our system reduced to one. This will always happen: *one equation will follow from the others*, and this is because the system $\pi P = \pi$ is singular. It remains to use the *normalizing equation* $\sum \pi_x = 1$,

$$\pi_1 + \pi_2 = \pi_1 + \frac{3}{4}\pi_1 = \frac{7}{4}\pi_1 = 1,$$

from where

$$\underline{\pi_1 = 4/7} \quad \text{and} \quad \underline{\pi_2 = 3/7}.$$

In a long history of this city, $4/7 \approx 57\%$ of days are sunny, and $3/7 \approx 43\%$ are rainy. $\quad \diamond$

Example 6.12 (SHARED DEVICE, CONTINUED). In Example 6.9 on p. 143, the transition probability matrix for the number of concurrent users is

$$P = \begin{pmatrix} .64 & .32 & .04 \\ .40 & .50 & .10 \\ .25 & .50 & .25 \end{pmatrix}.$$

Let us find the steady-state distribution. It will automatically serve as our forecast for the number of users next day or next month. Thus, it will answer the question posed in the end of Section 6.2.2.

The steady-state equations are:

$$\begin{cases} .64\pi_0 + .40\pi_1 + .25\pi_2 &= \pi_0 \\ .32\pi_0 + .50\pi_1 + .50\pi_2 &= \pi_1 \\ .04\pi_0 + .10\pi_1 + .25\pi_2 &= \pi_2 \end{cases} ; \quad \begin{cases} -.36\pi_0 + .40\pi_1 + .25\pi_2 &= 0 \\ .32\pi_0 - .50\pi_1 + .50\pi_2 &= 0 \\ .04\pi_0 + .10\pi_1 - .75\pi_2 &= 0 \end{cases}$$

Solving this system by method of elimination, we express π_2 from the first equation and substitute into the other equations,

$$\begin{cases} \pi_2 = 1.44\pi_0 - 1.6\pi_1 \\ .32\pi_0 - .50\pi_1 + .50(1.44\pi_0 - 1.6\pi_1) &= 0 \\ .04\pi_0 + .10\pi_1 - .75(1.44\pi_0 - 1.6\pi_1) &= 0 \end{cases} ; \quad \begin{cases} \pi_2 = 1.44\pi_0 - 1.6\pi_1 \\ 1.04\pi_0 - 1.3\pi_1 &= 0 \\ -1.04\pi_0 + 1.3\pi_1 &= 0 \end{cases}$$

The last two equations are equivalent. This was anticipated; one equation should always follow from the others, so we are "probably" on a right track.

Express π_1 from the second equation and substitute into the first one,

$$\begin{cases} \pi_2 &= 1.44\pi_0 - 1.6(.8\pi_0) \\ \pi_1 &= .8\pi_0 \end{cases} ; \quad \begin{cases} \pi_2 &= .16\pi_0 \\ \pi_1 &= .8\pi_0 \end{cases}$$

Finally, use the normalizing equation,

$$\pi_0 + \pi_1 + \pi_2 = \pi_0 + .8\pi_0 + .16\pi_0 = 1.96\pi_0 = 1,$$

from where we compute the answer,

$$\begin{cases} \pi_0 &= 1/1.96 &= .5102 \\ \pi_1 &= .8(.5102) &= .4082 \\ \pi_2 &= .16(.5102) &= .0816 \end{cases}$$

This is a direct but a little longer way of finding the steady-state distribution π. For this particular problem, a shorter solution is offered in Exercise 6.25.

$\Diamond$

The limit of P^h

Then, what will the h-step transition probabilities be for large h? It turns out that matrix $P^{(h)}$ also has a limit, as $h \to \infty$, and the limiting matrix has the form

$$\Pi = \lim_{h \to \infty} P^{(h)} = \begin{pmatrix} \pi_1 & \pi_2 & \cdots & \pi_n \\ \pi_1 & \pi_2 & \cdots & \pi_n \\ \vdots & \vdots & \cdots & \vdots \\ \pi_1 & \pi_2 & \cdots & \pi_n \end{pmatrix}.$$

All the rows of the limiting matrix Π are equal, and they consist of the steady-state probabilities π_x!

How can this phenomenon be explained? First, the forecast for some very remote future should not depend on the current state $X(0)$. Say, our weather forecast for the next century should not depend on the weather we have today. Therefore, $p_{ik} = p_{jk}$ for all i, j, k, and this is why all rows of Π coincide.

Second, our forecast, independent of $X(0)$, will just be given in terms of long-term proportions, that is, π_x. Indeed, if your next-year vacation is in August, then the best weather forecast you can get will likely be given as the historical average for this time of the year.

Steady state

What is the steady state of a Markov chain? Suppose the system has reached its steady state, so that the current distribution of states is $P_t = \pi$. A system makes one more transition, and the distribution becomes $P_{t+1} = \pi P$. But $\pi P = \pi$, and thus, $P_t = P_{t+1}$. We see that *in a steady state, transitions do not affect the distribution*. A system may go from one state to another, but the distribution of states does not change. In this sense, it is *steady*.

Existence of a steady-state distribution. Regular Markov chains

As we know from Calculus, there are situations when a limit simply does not exist. Similarly, there are Markov chains with no steady-state distribution.

Example 6.13 (PERIODIC MARKOV CHAIN, NO STEADY-STATE DISTRIBUTION). In chess, a knight can only move to a field of different color, from white to black, and from black to white. Then, the transition probability matrix of the color of its field is

$$P = \begin{pmatrix} 0 & 1 \\ 1 & 0 \end{pmatrix}.$$

Computing P^2, P^3, etc., we find that

$$P^{(h)} = P^h = \begin{cases} \begin{pmatrix} 1 & 0 \\ 0 & 1 \end{pmatrix} & \text{for all odd } h \\[2ex] \begin{pmatrix} 0 & 1 \\ 1 & 0 \end{pmatrix} & \text{for all even } h \end{cases}$$

Indeed, after any odd number of moves, the knight's field will change its color, and after any even number of moves, it will return to the initial color. Thus, there is no limit of P^h and $P^{(h)}$. $\diamond$

The Markov chain in Example 6.13 is *periodic* with period 2 because $X(t) = X(t+2)$ for all t with probability 1. Periodic Markov chains cannot be regular; from any state, some h-step transitions are possible and some are not, depending on whether or not h is divisible by the period.

There are other situations when steady-state probabilities cannot be found. What we need is a criterion for the existence of a steady-state distribution.

DEFINITION 6.9 ─────

A Markov chain is **regular** if

$$p_{ij}^{(h)} > 0$$

for some h and all i, j. That is, for some h, matrix $P^{(h)}$ has only non-zero entries, and h-step transitions from any state to any state are possible.

Any regular Markov chain has a steady-state distribution.

Example 6.14. Markov chains in Examples 6.7 and 6.9 are *regular* because all transitions are possible for $h = 1$ already, and matrix P does not contain any zeros.   ◊

Example 6.15. A Markov chain with transition probability matrix

$$P = \begin{pmatrix} 0 & 1 & 0 & 0 \\ 0 & 0 & 1 & 0 \\ 0 & 0 & 0 & 1 \\ 0.9 & 0 & 0 & 0.1 \end{pmatrix}.$$

is also regular. Matrix P contains zeros, and so do P^2, P^3, P^4, and P^5. The 6-step transition probability matrix

$$P^{(6)} = \begin{pmatrix} .009 & .090 & .900 & .001 \\ .001 & .009 & .090 & .900 \\ .810 & .001 & .009 & .180 \\ .162 & .810 & .001 & .027 \end{pmatrix}$$

contains no zeros and proves regularity of this Markov chain.

In fact, computation of all P^h up to $h = 6$ is not required in this problem. Regularity can also be seen from the transition diagram in Figure 6.4. Based on this diagram, any state j can be reached in 6 steps from any state i. Indeed, moving counterclockwise through this Figure, we can reach state 4 from any state i in ≤ 3 steps. Then, we can reach state j from state 4 in ≤ 3 additional steps, for the total of ≤ 6 steps. If we can reach state i from state j in fewer than 6 steps, we just use the remaining steps circling around state 4. For example, state 2 is reached from state 1 in 6 steps as follows:

$$1 \rightarrow 2 \rightarrow 3 \rightarrow 4 \rightarrow 4 \rightarrow 1 \rightarrow 2.$$

Notice that we don't have to compute $p_{ij}^{(h)}$. We only need to verify that they are all positive for some h.   ◊

Example 6.16 (IRREGULAR MARKOV CHAIN, ABSORBING STATES). When there is a state i with $p_{ii} = 1$, a Markov chain cannot be regular. There is no exit from state i; therefore, $p_{ij}^{(h)} = 0$ for all h and all $j \neq i$. Such a state is called *absorbing*. For example, state 4 in Figure 6.5a is absorbing, therefore, the Markov chain is irregular.

There may be several absorbing states or an entire absorbing zone, from which the remaining states can never be reached. For example, states 3, 4, and 5 in Figure 6.5b form an *absorbing zone*, some kind of a Bermuda triangle. When this process finds itself in the set $\{3, 4, 5\}$, there is no route from there to the set $\{1, 2\}$. As a result, probabilities $p_{31}^{(h)}$, $p_{52}^{(h)}$ and some

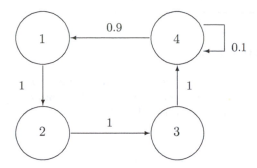

FIGURE 6.4: *Transition diagram for a regular Markov chain in Example 6.15.*

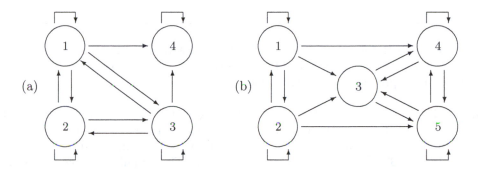

FIGURE 6.5: *Absorbing states and absorbing zones (Example 6.16).*

others equal 0 for all h. Although all p_{ij} may be less than 1, such Markov chains are still irregular.

Notice that both Markov chains do have steady-state distributions. The first process will eventually reach state 4 and will stay there for good. Therefore, the limiting distribution of $X(h)$ is $\pi = \lim P_h = (0, 0, 0, 1)$. The second Markov chain will eventually leave states 1 and 2 for good, thus its limiting (steady-state) distribution has the form $\pi = (0, 0, \pi_3, \pi_4, \pi_5)$. $\Diamond$

Conclusion

This section gives us an important method of analyzing rather complicated stochastic systems. Once the Markov property of a process is established, it only remains to find its one-step transition probabilities. Then, the steady-state distribution can be computed, and thus, we obtain the distribution of the process at any time, after a sufficient number of transitions.

This methodology will be our main working tool in Chapter 7, when we study queuing systems and evaluate their performance.

6.3   Counting processes

A large number of situations can be described by *counting processes*. As becomes clear from their name, they *count*. What they count differs greatly from one process to another. These may be counts of arrived jobs, completed tasks, transmitted messages, detected errors, scored goals, and so on.

DEFINITION 6.10 ——————————————————————————

A stochastic process X is **counting** if $X(t)$ is the number of items counted by the time t.

As time passes, one can count additional items; therefore, sample paths of a counting process are always *non-decreasing*. Also, counts are nonnegative integers, $X(t) \in \{0, 1, 2, 3, ...\}$. Hence, all counting processes are *discrete-state*.

Example 6.17 (E-MAILS AND ATTACHMENTS). Figure 6.6 shows sample paths of two counting process, $X(t)$ being the number of transmitted e-mails by the time t and $A(t)$ being the number of transmitted attachments. According to the graphs, e-mails were transmitted at $t = 8, 22, 30, 32, 35, 40, 41, 50, 52$, and 57 min. The e-mail counting process $X(t)$ increments by 1 at each of these times. Only 3 of these e-mails contained attachments. One attachment was sent at $t = 8$, five more at $t = 35$, making the total of $A(35) = 6$, and two more attachments at $t = 50$, making the total of $A(50) = 8$.   ◇

Two classes of counting processes will be discussed in detail, a discrete-time *Binomial process* and a continuous-time *Poisson process*.

6.3.1   Binomial process

We again consider a sequence of independent Bernoulli trials with probability of success p (Sections 3.4.1–3.4.4) and start counting "successes."

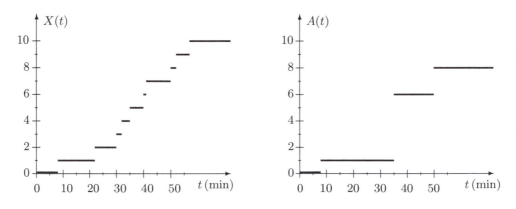

FIGURE 6.6: *Counting processes in Example 6.17.*

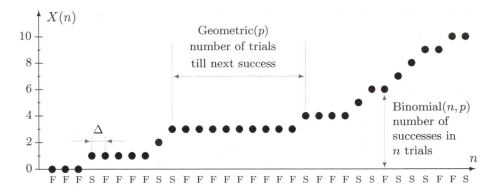

FIGURE 6.7: *Binomial process (sample path). Legend: S = success, F=failure.*

DEFINITION 6.11

> **Binomial process** $X(n)$ is the number of successes in the first n independent Bernoulli trials, where $n = 0, 1, 2, \ldots$.

It is a discrete-time discrete-space counting stochastic process. Moreover, it is *Markov*, and therefore, a Markov chain.

As we know from Sections 3.4.2–3.4.3, the distribution of $X(n)$ at any time n is Binomial(n, p), and the number of trials Y between two consecutive successes is Geometric(p) (Figure 6.7).

$$
\underline{\text{NOTATION}} \;\bigg\|\; \begin{array}{rcl} X(n) & = & \text{number of successes in } n \text{ trials} \\ Y & = & \text{number of trials between consecutive successes} \end{array} \;\bigg\|
$$

Relation to real time: frames

The "time" variable n actually measures the number of trials. It is not expressed in minutes or seconds. However, it can be related to real time.

Suppose that Bernoulli trials occur at equal time intervals, every Δ seconds. Then n trials occur during time $t = n\Delta$, and thus, the value of the process at time t has Binomial distribution with parameters $n = t/\Delta$ and p. The expected number of successes during t seconds is therefore

$$
\mathbf{E}\left\{ X\left(\frac{t}{\Delta}\right) \right\} = \frac{t}{\Delta}\, p,
$$

which amounts to

$$
\lambda = \frac{p}{\Delta}
$$

successes per second.

DEFINITION 6.12

> **Arrival rate** $\lambda = p/\Delta$ is the average number of successes per one unit of time. The time interval Δ of each Bernoulli trial is called a **frame**. The **interarrival time** is the time between successes.

These concepts, arrival rate and interarrival time, deal with modeling arrivals of jobs, messages, customers, and so on with a Binomial counting process, which is a common method in discrete-time queuing systems (Section 7.3). The key assumption in such models is that no more than 1 arrival is allowed during each Δ-second frame, so each frame is a Bernoulli trial. If this assumption appears unreasonable, and two or more arrivals can occur during the same frame, one has to model the process with a smaller Δ.

$$
\begin{array}{rcl}
\underline{\text{NOTATION}} \quad \lambda & = & \text{arrival rate} \\
\Delta & = & \text{frame size} \\
p & = & \text{probability of arrival (success)} \\
& & \text{during one frame (trial)} \\
X(t/\Delta) & = & \text{number of arrivals by the time } t \\
T & = & \text{interarrival time}
\end{array}
$$

The interarrival period consists of a Geometric number of frames Y, each frame taking Δ seconds. Hence, the interarrival time can be computed as

$$T = Y\Delta.$$

It is a *rescaled Geometric* random variable taking possible values Δ, 2Δ, 3Δ, etc. Its expectation and variance are

$$
\begin{aligned}
\mathbf{E}(T) &= \mathbf{E}(Y)\Delta = \frac{1}{p}\Delta = \frac{1}{\lambda}; \\
\mathrm{Var}(T) &= \mathrm{Var}(Y)\Delta^2 = (1-p)\left(\frac{\Delta}{p}\right)^2 \quad \text{or} \quad \frac{1-p}{\lambda^2}.
\end{aligned}
$$

$$
\begin{array}{rcl}
& \lambda & = & p/\Delta \\
& n & = & t/\Delta \\
\textbf{Binomial} & & & \\
\textbf{counting process} \quad & X(n) & = & \textit{Binomial}(n,p) \\
& Y & = & \textit{Geometric}(p) \\
& T & = & Y\Delta
\end{array}
$$

Example 6.18 (MAINFRAME COMPUTER). Jobs are sent to a mainframe computer at a rate of 2 jobs per minute. Arrivals are modeled by a Binomial counting process.

(a) Choose such a frame size that makes the probability of a new job during each frame equal 0.1.

(b) Using the chosen frames, compute the probability of more than 3 jobs received during one minute.

(c) Compute the probability of more than 30 jobs during 10 minutes.

(d) What is the average interarrival time, and what is the variance?

(e) Compute the probability that the next job does not arrive during the next 30 seconds.

Solution.

(a) We have $\lambda = 2 \text{ min}^{-1}$ and $p = 0.1$. Then

$$\Delta = \frac{p}{\lambda} = 0.05 \text{ min or 3 sec.}$$

(b) During $t = 1$ min, we have $n = t/\Delta = 20$ frames. The number of jobs during this time is Binomial$(n = 20, p = 0.1)$. From Table A2 (in the Appendix),

$$\boldsymbol{P}\{X(n) > 3\} = 1 - \boldsymbol{P}\{X(n) \leq 3\} = 1 - 0.8670 = 0.1330.$$

(c) Here $n = 10/0.05 = 200$ frames, and we use Normal approximation to Binomial(n, p) distribution (recall p. 94). With a proper continuity correction, and using Table A4,

$$
\begin{aligned}
\boldsymbol{P}\{X(n) > 30\} &= \boldsymbol{P}\{X(n) > 30.5\} \\
&= \boldsymbol{P}\left\{\frac{X(n) - np}{\sqrt{np(1-p)}} > \frac{30.5 - (200)(0.1)}{\sqrt{(200)(0.1)(1-0.1)}}\right\} \\
&= \boldsymbol{P}\{Z > 2.48\} = 1 - 0.9934 = 0.0066.
\end{aligned}
$$

Comparing questions (b) and (c), notice that 3 jobs during 1 minute is not the same as 30 jobs during 10 minutes!

(d) $\boldsymbol{E}(T) = 1/\lambda = 1/2$ min or 30 sec. Intuitively, this is rather clear because the jobs arrive at a rate of two per minute. The interarrival time has variance

$$\text{Var}(T) = \frac{1-p}{\lambda^2} = \frac{0.9}{2^2} = 0.225.$$

(e) For the interarrival time $T = Y\Delta = Y(0.05)$ and a Geometric variable Y,

$$
\begin{aligned}
\boldsymbol{P}\{T > 0.5 \text{ min}\} &= \boldsymbol{P}\{Y(0.05) > 0.5\} = \boldsymbol{P}\{Y > 10\} \\
&= \sum_{k=11}^{\infty} (1-p)^{k-1}p = (1-p)^{10} = (0.9)^{10} = 0.3138.
\end{aligned}
$$

Alternatively, this is also the probability of 0 arrivals during $n = t/\Delta = 0.5/0.05 = 10$ frames, which is also $(1-p)^{10}$. $\diamond$

Markov property

Binomial counting process is *Markov*, with transition probabilities

$$
p_{ij} = \begin{cases}
p & \text{if} \quad j = i+1 \\
1 - p & \text{if} \quad j = i \\
0 & \text{otherwise}
\end{cases}
$$

That is, during each frame, the number of successes X increments by 1 in case of a success or remains the same in case of a failure (see Figure 6.8). Transition probabilities are constant over time and independent of the past values of $X(n)$; therefore, it is a *stationary Markov chain*.

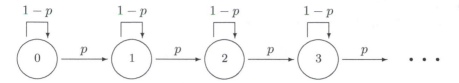

FIGURE 6.8: *Transition diagram for a Binomial counting process.*

This Markov chain is *irregular* because $X(n)$ is non-decreasing, thus $p_{10}^{(h)} = 0$ for all h. Once we see one success, the number of successes will never return to zero. Thus, this process has no steady-state distribution.

The h-step transition probabilities simply form a Binomial distribution. Indeed, $p_{ij}^{(h)}$ is the probability of going from i to j successes in h transitions, i.e.,

$$
\begin{aligned}
p_{ij}^{(h)} &= \boldsymbol{P}\left\{(j-i)\text{ successes in }h\text{ trials}\right\}\\[2mm]
&= \begin{cases} \dbinom{h}{j-i} p^{j-i}(1-p)^{h-j+i} & \text{if } \quad 0 \le j - i \le h \\[3mm] 0 & \text{otherwise} \end{cases}
\end{aligned}
$$

Notice that the transition probability matrix has ∞ rows and ∞ columns because $X(n)$ can reach any large value for sufficiently large n:

$$
P = \begin{pmatrix} 1-p & p & 0 & 0 & \cdots \\ 0 & 1-p & p & 0 & \cdots \\ 0 & 0 & 1-p & p & \cdots \\ 0 & 0 & 0 & 1-p & \cdots \\ \cdots & \cdots & \cdots & \cdots & \cdots \end{pmatrix}
$$

6.3.2 Poisson process

Continuous time

We now turn attention to *continuous-time* stochastic processes. The time variable t will run continuously through the whole interval, and thus, even during one minute there will be infinitely many moments when the process $X(t)$ may change. Then, how can one study such models?

Often a continuous-time process can be viewed as a limit of some discrete-time process whose frame size gradually decreases to zero, therefore allowing more frames during any fixed period of time,

$$\Delta \downarrow 0 \text{ and } n \uparrow \infty.$$

Example 6.19. For illustration, let us recall how movies are made. Although all motions on the screen seem continuous, we realize that an infinite amount of information could not be stored in a video cassette. Instead, we see a discrete sequence of exposures that run so fast that each motion seems continuous and smooth.

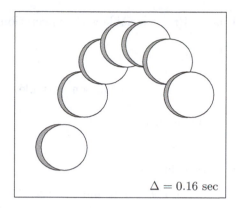 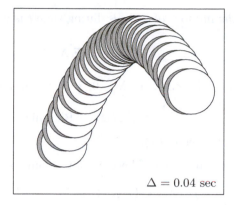

$\Delta = 0.16$ sec $\Delta = 0.04$ sec

FIGURE 6.9: *From discrete motion to continuous motion: reducing the frame size Δ.*

Early-age video cameras shot exposures rather slowly; the interval Δ between successive shots was pretty long (~ 0.2–0.5 sec). As a result, the quality of recorded video was rather low. Movies were "too discrete."

Modern camcorders can shoot up to 200 exposures per second attaining $\Delta = 0.005$. With such a small Δ, the resulting movie seems perfectly continuous. A shorter frame Δ results in a "more continuous" process (Figure 6.9). $\diamond$

Poisson process as the limiting case

Going from discrete time to continuous time, Poisson process is the limiting case of a Binomial counting process as $\Delta \downarrow 0$.

DEFINITION 6.13 ———————————————————

Poisson process is a continuous-time counting stochastic process obtained from a Binomial counting process when its frame size Δ decreases to 0 while the arrival rate λ remains constant.

We can now study a Poisson process by taking a Binomial counting process and letting its frame size decrease to zero.

Consider a Binomial counting process that counts arrivals or other events occurring at a rate λ. Let $X(t)$ denote the number of arrivals occurring until time t. What will happen with this process if we let its frame size Δ converge to 0?

The arrival rate λ remains constant. Arrivals occur at the same rate (say, messages arrive at a server) regardless of your choice of frame Δ.

The number of frames during time t increases to infinity,

$$n = \frac{t}{\Delta} \uparrow \infty \text{ as } \Delta \downarrow 0.$$

The probability of an arrival during each frame is proportional to Δ, so it also decreases to 0,

$$p = \lambda \Delta \downarrow 0 \text{ as } \Delta \downarrow 0.$$

Then, *the number of arrivals* during time t is a Binomial(n, p) variable with expectation

$$\mathbf{E}\, X(t) = np = \frac{tp}{\Delta} = \lambda t.$$

In the limiting case, as $\Delta \downarrow 0$, $n \uparrow \infty$, and $p \downarrow 0$, it becomes a Poisson variable with parameter $np = \lambda t$,

$$X(t) = \text{Binomial}(n, p) \to \text{Poisson}(\lambda)$$

(if in doubt, see p. 67).

The *interarrival time T* becomes a random variable with the c.d.f.

$$
\begin{aligned}
F_T(t) \quad &= \quad \boldsymbol{P}\{T \le t\} = \boldsymbol{P}\{Y \le n\} && \text{because } T = Y\Delta \text{ and } t = n\Delta \\
&= \quad 1 - (1 - p)^n && \text{Geometric distribution of } Y \\
&= \quad 1 - \left(1 - \frac{\lambda t}{n}\right)^n && \text{because } p = \lambda\Delta = \lambda t/n \\
&\to \quad 1 - e^{\lambda t}. && \text{This is the ``Euler limit'':} \\
& && (1 + x/n)^n \to e^x \text{ as } n \to \infty
\end{aligned}
$$

What we got is the c.d.f. of Exponential distribution! Hence, the interarrival time is Exponential with parameter λ.

Further, the time T_k of the k-th arrival is the sum of k Exponential interrarival times that has Gamma(k, λ) distribution. From this, we immediately obtain our familiar *Gamma-Poisson formula* (4.14),

$$\boldsymbol{P}\{T_k \le t\} = \boldsymbol{P}\{\,k\text{-th arrival before time } t\,\} = \boldsymbol{P}\{X(t) \ge k\}$$

where T_k is Gamma(k, λ) and $X(t)$ is Poisson(λt).

$$
\boxed{
\begin{array}{rcl}
X(t) &=& \text{Poisson}(\lambda t) \\
T &=& \text{Exponential}(\lambda) \\
T_k &=& \text{Gamma}(k, \lambda) \\
\boldsymbol{P}\{T_k \le t\} &=& \boldsymbol{P}\{X(t) \ge k\} \\
\boldsymbol{P}\{T_k > t\} &=& \boldsymbol{P}\{X(t) < k\}
\end{array}
}
$$

Poisson process

A sample path of some Poisson process is shown in Figure 6.10.

Example 6.20 (WEB SITE HITS). The number of hits to a certain web site follows a Poisson process with the intensity parameter $\lambda = 7$ hits per minute.

On the average, how much time is needed to get 10,000 hits? What is the probability that this will happen within 24 hours?

Solution. The time of the $10{,}000$-th hit T_k has Gamma distribution with parameters $k = 10{,}000$ and $\lambda = 7$ min^{-1}. Then, the expected time of the k-th hit is

$$\mu = \mathbf{E}(T_k) = \frac{k}{\lambda} = \underline{1{,}428.6 \text{ min or } 23.81 \text{ hrs.}}$$

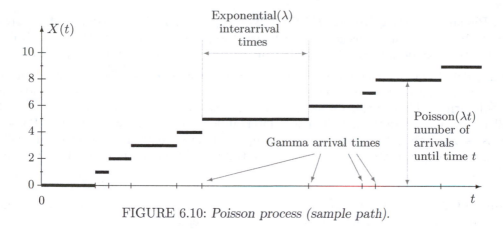

FIGURE 6.10: *Poisson process (sample path).*

Also,

$$\sigma = \text{Std}(T_k) = \frac{\sqrt{k}}{\lambda} = 14.3 \text{ min.}$$

By the Cental Limit Theorem of Section 4.3, we can use the Normal approximation to the Gamma distribution of T_k. The probability that the 10,000-th hit occurs within 24 hours (1440 min) is

$$\boldsymbol{P}\{T_k < 1440\} = \boldsymbol{P}\left\{\frac{T_k - \mu}{\sigma} < \frac{1440 - 1428.6}{14.3}\right\} = \boldsymbol{P}\{Z < 0.80\} = \underline{0.7881}.$$

$\Diamond$

Rare events and modeling

A more conventional definition of a Poisson process sounds like this.

DEFINITION 6.14 ————

Poisson process $X(t)$ is a continuous-time counting process with independent increments, such that
(a) $\boldsymbol{P}\{X(t+\Delta) - X(t) = 1\} = \lambda\Delta + o(\Delta)$ as $\Delta \to 0$;
(b) $\boldsymbol{P}\{X(t+\Delta) - X(t) > 1\} = o(\Delta)$ as $\Delta \to 0$.

In this definition, differences $X(t+\Delta) - X(t)$ are called *increments*. For a Poisson process, an increment is the number of arrivals during time interval $(t, t+\Delta]$.

Properties (a) and (b) imply the known fact that Binomial process probabilities differ from their limits (as $\Delta \to 0$) by a small quantity converging to 0 faster than Δ. This quantity is in general denoted by $o(\Delta)$;

$$\frac{o(\Delta)}{\Delta} \to 0 \text{ as } \Delta \to 0.$$

For a Binomial counting process, $P\{X(t + \Delta) - X(t) = 1\}$ is the probability of 1 arrival during 1 frame, and it equals $p = \lambda\Delta$, whereas the probability of more than 1 arrival is zero. For the Poisson process, these probabilities may be different, but only by a little.

The last definition describes formally what we called **rare events**. These events occur at random times; probability of a new event during a short interval of time is proportional to the length of this interval. Probability of more than 1 event during that time is much smaller comparing with the length of such interval. For such sequences of events a Poisson process is a suitable stochastic model. Examples of rare events include telephone calls, message arrivals, virus attacks, errors in codes, traffic accidents, natural disasters, network blackouts, and so on.

Example 6.21 (MAINFRAME COMPUTER, REVISITED). Let us now model the job arrivals in Example 6.18 on p. 154 with a *Poisson process*. Keeping the same arrival rate $\lambda = 2$ min^{-1}, this will remove a perhaps unnecessary assumption that no more than 1 job can arrive during any given frame.

Revisiting questions (b)–(e) in Example 6.18, we now obtain,

(b) The probability of more than 3 jobs during 1 minute is

$$P\{X(1) > 3\} = 1 - P\{X(1) \le 3\} = 1 - 0.8571 = 0.1429,$$

from Table A3 with parameter $\lambda t = (2)(1) = 2$.

(c) The probability of more than 30 jobs during 10 minutes is

$$P\{X(10) > 30\} = 1 - P\{X(10) \le 30\} = 1 - 0.9865 = 0.0135,$$

from Table A3 with parameter $\lambda t = (2)(10) = 20$.

(d) For the Exponential($\lambda = 2$) distribution of interarrival times, we get again that

$$\mathbf{E}(T) = \frac{1}{\lambda} = 0.5 \text{ min or } 30 \text{ sec}.$$

This is because the jobs arrive to the computer at the same rate regardless of whether we model their arrivals with a Binomial or Poisson process. Also,

$$\text{Var}(T) = \frac{1}{\lambda^2} = 0.25.$$

(e) The probability that the next job does not arrive during the next 30 seconds is

$$P\{T > 0.5 \text{ min}\} = e^{-\lambda(0.5)} = e^{-1} = 0.3679.$$

Alternatively, this is also the probability of 0 arrivals during 0.5 min, i.e.,

$$P\{X(0.5) = 0\} = 0.3679,$$

from Table A3 with parameter $\lambda t = (2)(0.5) = 1$.

We see that most of the results are somewhat different from Example 6.18, where the same events were modeled by a Binomial counting process. Noticeably, the variance of interarrival times increased. Binomial process introduces a restriction on the number of arrivals during each frame, therefore reducing variability. $\Diamond$

6.4 Simulation of stochastic processes

A number of important characteristics of stochastic processes require lengthy complex computations unless they are estimated by means of Monte Carlo methods. One may be interested to explore the time it takes a process to attain a certain level, the time the process spends above some level or above another process, the probability that one process reaches a certain value ahead of another process, etc. Also, it is often important to predict future behavior of a stochastic process.

Discrete-time processes

Sample paths of discrete-time stochastic processes are usually simulated *sequentially*. We start with $X(0)$, then simulate $X(1)$ from the conditional distribution of $X(1)$ given $X(0)$,

$$P_{X_1|X_0}(x_1|x_0) = \frac{P_{X_0,X_1}(x_0, x_1)}{P_{X_0}(x_0)}, \tag{6.6}$$

then $X(2)$ from the conditional distribution of $X(2)$ given $X(0)$ and X_1,

$$P_{X_2|X_0,X_1}(x_2|x_0, x_1) = \frac{P_{X_0,X_1,X_2}(x_0, x_1, x_2)}{P_{X_0,X_1}(x_0, x_1)}, \tag{6.7}$$

etc. In the case of *continuous-state* (but still, discrete-time) processes, the probability mass functions $P_{X_j}(x_j)$ in (6.6) and (6.7) will be replaced with densities $f_{X_j}(x_j)$.

Markov chains

For *Markov chains*, all conditional distributions in (6.6) and (6.7) have a simple form given by *one-step transition probabilities*. Then, the simulation algorithm reduces to the following.

Algorithm 6.1 *(Simulation of Markov chains)*

1. Initialize: generate $X(0)$ from the initial distribution $P_0(x)$.

2. Transition: having generated $X(t) = i$, generate $X(t+1)$ as a discrete random variables that takes value j with probability p_{ij}. See Algorithm 5.1 on p. 108.

3. Return to step 2 until a sufficiently long sample path is generated.

Example 6.22 (WEATHER FORECASTS AND WATER RESOURCES). In Example 6.8, the Markov chain of sunny and rainy days has initial distribution

$$\begin{cases} P_0(\text{sunny}) &= P_0(1) &= 0.2, \\ P_0(\text{rainy}) &= P_0(2) &= 0.8, \end{cases}$$

and transition probability matrix

$$P = \begin{pmatrix} 0.7 & 0.3 \\ 0.4 & 0.6 \end{pmatrix}.$$

For simplicity, sunny days are denoted by $X = 1$, and rainy days by $X = 2$. The following codes will generate the forecast for the next 100 days.

— R ——————— — MATLAB ———————

```
N <- 100                 N = 100;                  % length of sample path
X <- rep(0,N)            X = zeros(N,1);           % initialization
Q <- c(0.2,0.8)          Q = [0.2 0.8];            % initial distribution
prob <- c(.7,.4,.3,.6)                             % vector of probabilities
P <- matrix(prob,2,2)    P = [0.7 0.3; 0.4 0.6];   % transition prob. matrix
U <- runif(N)            U = rand(N,1);            % N Uniform variables
for (t in 1:N) {         for t=1:N;                % simulate X sequentially
  X[t] <- (U[t]<=Q[1])+    X(t) = (U(t)<=Q(1))+... % X(t)=1 with prob. Q(1)
      2*(U[t] > Q[1])          2*(U(t) > Q(1));    % X(t)=2 with prob. Q(2)
  Q <- P[ X[t], ]          Q = P(X(t),:);          % the X(t)-th row of P
                                                   %   is the pmf of X(t+1)
}; X                     end; X                    % resulting path of X
```

This program returns a sequence of states that looks like this:

2211122212222112111111112222222112121112222111111111112221211221...

Notice fairly long segments of sunny days ($X = 1$) and rainy days ($X = 2$), showing dependence among the generated variables. This is because a sunny day is more likely to be followed by another sunny day, and a rainy day is more likely to be followed by a rainy day.

Generating a large number of such sequences, we can estimate, say, the probability of 20 consecutive days without rain at least once during the next year, the average time between such droughts, the expected number of times this may happen during the same year, etc. As an immediate application, based on this, water resources can be allocated accordingly.

Binomial process

Simulation of a Binomial process is rather straightforward, as it is based on a sequence of independent Bernoulli trials. After simulating such a sequence, compute partial sums as follows.

— R ——————— — MATLAB ———————

```
N <- 100; p <- 0.4;     N = 100; p = 0.4;
X <- rep(0,N)           X = zeros(N,1);           % initialization
Y <- rbinom(N,1,p)      Y = binornd(1,p,N,1);     % sequence of Bernoulli(p)
X[1] <- Y[1]            X(1) = Y(1);              %    random variables
for (t in 2:N) {        for t = 2:N;              % X(t) is the number of
  X[t]<-X[t-1]+Y[t]       X(t)=X(t-1)+Y(t);       %    successes in the first
}; X                    end; X                    %    t Bernoulli trials
```

It is a good illustration to look at the simulated stochastic process in real time. Any stochastic process evolves in time, correct? Animation of the generated discrete-time Binomial process can be created as follows.

```
—— R ——————        —— MATLAB ——————
plot(1,X[1],             plot(1,X(1),'o');         % Start the plot and make
   xlim=c(0,N),          axis([0 N 0 max(X)]);     %    a box for all X(t)
   ylim=c(0,max(X)))     hold on;                  % Keep all the points
for (t in 2:N) {         for t=2:N;
   points(t,X[t])          plot(t,X(t),'o');        % Plot points with circles
   Sys.sleep(0.5)          pause(0.5);              % A half-second pause
}                        end; hold off             %    after each point
```

If you have access to R or MATLAB, try this code. You will see a real discrete-time process with half-second frames Δ.

Continuous-time processes

Simulation of continuous-time processes has a clear problem. The time t runs continuously through the time interval, taking infinitely many values in this range. However, we cannot store an infinite number of random variables in memory of our computer!

For most practical purposes, it suffices to generate a discrete-time process with a rather short frame Δ (discretization). For example, a two-dimensional *Brownian motion process*, a continuous-time process with continuous sample paths and independent bivariate Normal increments, can be simulated by the following codes.

```
—— R ——————————             —— MATLAB ——————
N<-10000                              N=10000;
X<-matrix(rep(0,N*2),N,2)             X=zeros(N,2);
Z<-matrix(rnorm(N*2),N,2)             Z=randn(N,2);
X[1,]<-Z[1,]                          X(1,:)=Z(1,:);
for (t in 2:N) { X[t,]<-X[t-1,]+Z[t,] }   for t=2:N;
Xlimits = c(min(X[,1]),max(X[,1]))      X(t,:)=X(t-1,:)+Z(t,:);
Ylimits = c(min(X[,2]),max(X[,2]))    end;
for (t in 2:N) { plot(X[1:t,],        comet(X(:,1),X(:,2));
  type='l', xlim=Xlimits, ylim=Ylimits) }
```

The last command, `comet`, creates a two-dimensional animation of the generated process. MATLAB users are invited to define a 3D Brownian motion process in a similar way and animate it with the `comet3` command.

Poisson process

Poisson processes can be generated without discretization. Indeed, although they are continuous-time, the value of $X(t)$ can change only a finite number of times during each interval. The process changes every time a new "rare event" or arrival occurs, which happens a Poisson(λt) number of times during an interval of length t.

Then, it suffices to generate these moments of arrival. As we know from Section 6.3.2, the first arrival time is Exponential(λ), and each interarrival time is Exponential(λ) too. Then, a segment of a Poisson process during a time interval $[0, M]$ can be generated and "animated" with the following codes.

—— R ——

```
M <- 1000; lambda <- 0.04;   # M is our time frame
S <- 0; T <- 0;              # T is a growing vector of arrival times
while (S <= M) {             # Loop ends when arrival time S exceeds M
  S <- S + rexp(1,lambda);   # Add Exponential interarrival time
  T <- c(T,S);      }        # Store the new arrival time in T
N <- length(T);             # Generated number of arrivals
X <- rep(1,M);              # Initialize the Poisson process
for (t in 1:M) {
  X[t] <- sum(T <= t)        # X[t] = number of arrivals by the time t
  plot(X[1:t],xlim=c(0,M),ylim=c(0,N))      # Path of X until time t
}
```

—— MATLAB ——

```
M=1000; lambda=0.04;         % M is our time frame
S=0; T=0;                    % T is a growing vector of arrival times
while S<=M;                  % Loop ends when arrival time S exceeds M
  S=S+exprnd(1/lambda);      % Add Exponential interarrival time
  T=[T S];                   % Vector of arrival times extends
end;                         % by one element
N=length(T);                 % Generated number of arrivals
X=zeros(M,1);                % Initialize the Poisson process
for t=1:M;
  X(t)=sum(T<=t);            % X(t) is the number of arrivals
end;                         % by the time t
comet(X);                    % Animation of the generated process!
```

Summary and conclusions

Stochastic processes are random variables that change, evolve, and develop in time. There are discrete- and continuous-time, discrete- and continuous-state processes, depending on their possible times and possible values.

Markov processes form an important class, where only the most recent value of the process is needed to predict its future probabilities. Then, a Markov chain is fully described by its initial distribution and transition probabilities. Its limiting behavior, after a large number of transitions, is determined by a steady-state distribution which we compute by solving a system of steady-state equations. Binomial and Poisson counting processes are both Markov, with discrete and continuous time, respectively.

Thus we developed an important tool for studying rather complex stochastic (involving uncertainty) systems. As long as the process is Markov, we compute its forecast, steady-state distribution, and other probabilities, expectations, and quantities of interest. Continuous-time processes can be viewed as limits of some discrete-time processes, when the frame size reduces to zero.

In the next chapter, we use this approach to evaluate performance of queuing systems.

Exercises

6.1. A small computer lab has 2 terminals. The number of students working in this lab is recorded at the end of every hour. A computer assistant notices the following pattern:

- If there are 0 or 1 students in a lab, then the number of students in 1 hour has a 50-50% chance to increase by 1 or remain unchanged.

- If there are 2 students in a lab, then the number of students in 1 hour has a 50-50% chance to decrease by 1 or remain unchanged.

(a) Write the transition probability matrix for this Markov chain.

(b) Is this a regular Markov chain? Justify your answer.

(c) Suppose there is nobody in the lab at 7 am. What is the probability of nobody working in the lab at 10 am?

6.2. A computer system can operate in two different modes. Every hour, it remains in the same mode or switches to a different mode according to the transition probability matrix

$$\begin{pmatrix} 0.4 & 0.6 \\ 0.6 & 0.4 \end{pmatrix}.$$

(a) Compute the 2-step transition probability matrix.

(b) If the system is in Mode I at 5:30 pm, what is the probability that it will be in Mode I at 8:30 pm on the same day?

6.3. Markov chains find direct applications in genetics. Here is an example.

An offspring of a black dog is black with probability 0.6 and brown with probability 0.4. An offspring of a brown dog is black with probability 0.2 and brown with probability 0.8.

(a) Write the transition probability matrix of this Markov chain.

(b) Rex is a brown dog. Compute the probability that his grandchild is black.

6.4. Every day, Eric takes the same street from his home to the university. There are 4 street lights along his way, and Eric has noticed the following Markov dependence. If he sees a green light at an intersection, then 60% of time the next light is also green, and 40% of time the next light is red. However, if he sees a red light, then 70% of time the next light is also red, and 30% of time the next light is green.

(a) Construct the transition probability matrix for the street lights.

(b) If the first light is green, what is the probability that the third light is red?

(c) Eric's classmate Jacob has *many* street lights between his home and the university. If the *first* street light is green, what is the probability that the *last* street light is red? (Use the steady-state distribution.)

6.5. The pattern of sunny and rainy days on planet Rainbow is a homogeneous Markov chain with two states. Every sunny day is followed by another sunny day with probability 0.8. Every rainy day is followed by another rainy day with probability 0.6. Compute the probability that April 1 next year is rainy on Rainbow.

6.6. A computer device can be either in a busy mode (state 1) processing a task, or in an idle mode (state 2), when there are no tasks to process. Being in a busy mode, it can finish a task and enter an idle mode any minute with the probability 0.2. Thus, with the probability 0.8 it stays another minute in a busy mode. Being in an idle mode, it receives a new task any minute with the probability 0.1 and enters a busy mode. Thus, it stays another minute in an idle mode with the probability 0.9. The initial state X_0 is idle. Let X_n be the state of the device after n minutes.

(a) Find the distribution of X_2.

(b) Find the steady-state distribution of X_n.

6.7. A Markov chain has the transition probability matrix

$$P = \begin{pmatrix} 0.3 & \dots & 0 \\ 0 & 0 & \dots \\ 1 & \dots & \dots \end{pmatrix}$$

(a) Fill in the blanks.

(b) Show that this is a regular Markov chain.

(c) Compute the steady-state probabilities.

6.8. A Markov chain has 3 possible states: A, B, and C. Every hour, it makes a transition to a *different* state. From state A, transitions to states B and C are equally likely. From state B, transitions to states A and C are equally likely. From state C, it always makes a transition to state A. Find the steady-state distribution of states.

6.9. Tasks are sent to a supercomputer at an average rate of 6 tasks per minute. Their arrivals are modeled by a Binomial counting process with 2-second frames.

(a) Compute the probability of more than 2 tasks sent during 10 seconds.

(b) Compute the probability of more than 20 tasks sent during 100 seconds. You may use a suitable approximation.

6.10. The number of electronic messages received by an interactive message center is modeled by Binomial counting process with 15-second frames. The average arrival rate is 1 message per minute. Compute the probability of receiving more than 3 messages during a 2-minute interval.

6.11. Jobs are sent to a printer at the average rate of 2 jobs per minute. Binomial counting process is used to model these jobs.

(a) What frame length Δ gives the probability 0.1 of an arrival during any given frame?

(b) With this value of Δ, compute the expectation and standard deviation for the number of jobs sent to the printer during a 1-hour period.

6.12. On the average, 2 airplanes per minute land at a certain international airport. We would like to model the number of landings by a Binomial counting process.

(a) What frame length should one use to guarantee that the probability of a landing during any frame does not exceed 0.1?

(b) Using the chosen frames, compute the probability of no landings during the next 5 minutes.

(c) Using the chosen frames, compute the probability of more than 100 landed airplanes during the next hour.

6.13. On the average, every 12 seconds a customer makes a call using a certain phone card. Calls are modeled by a Binomial counting process with 2-second frames. Find the mean and the variance for the time, in seconds, between two consecutive calls.

6.14. Customers of a certain internet service provider connect to the internet at the average rate of 3 customers per minute. Assuming Binomial counting process with 5-second frames, compute the probability of more than 10 new connections during the next 3 minutes. Compute the mean and the standard deviation of the number of seconds between connections.

6.15. Customers of an internet service provider connect to the internet at the average rate of 12 new connections per minute. Connections are modeled by a Binomial counting process.

(a) What frame length Δ gives the probability 0.15 of an arrival during any given frame?

(b) With this value of Δ, compute the expectation and standard deviation for the number of seconds between two consecutive connections.

6.16. Messages arrive at a transmission center according to a Binomial counting process with 30 frames per minute. The average arrival rate is 40 messages per hour. Compute the mean and standard deviation of the number of messages arrived between 10 am and 10:30 am.

6.17. Messages arrive at an electronic message center at random times, with an average of 9 messages per hour.

(a) What is the probability of receiving at least five messages during the next hour?

(b) What is the probability of receiving exactly five messages during the next hour?

6.18. Messages arrive at an interactive message center according to a Binomial counting process with the average interarrival time of 15 seconds. Choosing a frame size of 5 seconds, compute the probability that during 200 minutes of operation, no more than 750 messages arrive.

6.19. Messages arrive at an electronic mail server at the average rate of 4 messages every 5 minutes. Their number is modeled by a Binomial counting process.

(a) What frame length makes the probability of a new message arrival during a given frame equal 0.05?

(b) Suppose that 50 messages arrived during some 1-hour period. Does this indicate that the arrival rate is on the increase? Use frames computed in (a).

6.20. Power outages are unexpected rare events occurring according to a Poisson process with the average rate of 3 outages per month. Compute the probability of more than 5 power outages during three summer months.

6.21. Telephone calls to a customer service center occur according to a Poisson process with the rate of 1 call every 3 minutes. Compute the probability of receiving more than 5 calls during the next 12 minutes.

6.22. Network blackouts occur at an average rate of 5 blackouts per month.

(a) Compute the probability of more than 3 blackouts during a given month.

(b) Each blackout costs $1500 for computer assistance and repair. Find the expectation and standard deviation of the monthly total cost due to blackouts.

6.23. An internet service provider offers special discounts to every third connecting customer. Its customers connect to the internet according to a Poisson process with the rate of 5 customers per minute. Compute:

(a) the probability that no offer is made during the first 2 minutes

(b) expectation and variance of the time of the first offer

6.24. On the average, Mr. Z drinks and drives once in 4 years. He knows that

- Every time when he drinks and drives, he is caught by police.
- According to the laws of his state, the third time when he is caught drinking and driving results in the loss of his driver's license.
- Poisson process is the correct model for such "rare events" as drinking and driving.

What is the probability that Mr. Z will keep his driver's license for at least 10 years?

6.25. Refer to Example 6.9. Find the steady-state distribution for the number of users by writing $X(t)$ as a sum of two independent Markov chains,

$$X(t) = Y_1(t) + Y_2(t),$$

where $Y_i(t) = 1$ if user i is connected at time t and $Y_i(t) = 0$ if user i is not connected, for $i = 1, 2$. Find the steady-state distribution for each Y_i, then use it to find the steady-state distribution for X. Compare your result with Example 6.12.

6.26. (COMPUTER MINI-PROJECT) Generate 10,000 transitions of the Markov chain in Example 6.9 on p. 143. How many times do you find your generated Markov chain in each of its three states? Does it match the distribution found in Example 6.12?

6.27. (COMPUTER MINI-PROJECT) In this project, we explore the effect of Markov dependence.

Start with a sunny day and generate weather forecasts for the next 100 days according to the Markov chain in Example 6.7 on p. 140. Observe periods of sunny days and periods of rainy days.

Then consider another city where each day, sunny or rainy, is followed by a sunny day with probability $(4/7)$ and by a rainy day with probability $(3/7)$. As we know from Example 6.11 on p. 147, this is the steady-state distribution of sunny and rainy days, so the overall proportion of sunny days should be the same in both cities. Generate weather forecasts for 100 days and compare with the first city. What happened with periods of consecutive sunny and rainy days?

Each day in the first city, the weather depends on the previous day. In the second city, weather forecasts for different days are independent.

6.28. (COMPUTER MINI-PROJECT) Generate a 24-hour segment of a Poisson process of arrivals with the arrival rate $\lambda = 2$ hours^{-1}. Graph its trajectory. Do you observe several arrivals in a short period of time followed by relatively long periods with no arrivals at all? Why are the arrivals so unevenly distributed over the 24-hour period?

Chapter 7

Queuing Systems

We are now ready to analyze a broad range of *queuing systems* that play a crucial role in Computer Science and other fields.

DEFINITION 7.1

> A **queuing system** is a facility consisting of one or several servers designed to perform certain tasks or process certain jobs and a queue of jobs waiting to be processed.

Jobs arrive at the queuing system, wait for an available server, get processed by this server, and leave.

Examples of queuing systems are:

- a personal or shared computer executing tasks sent by its users;
- an internet service provider whose customers connect to the internet, browse, and disconnect;
- a printer processing jobs sent to it from different computers;
- a customer service with one or several representatives on duty answering calls from their customers;
- a TV channel viewed by many people at various times;
- a toll area on a highway, a fast food drive-through lane, or an automated teller machine (ATM) in a bank, where cars arrive, get the required service and depart;
- a medical office serving patients; and so on.

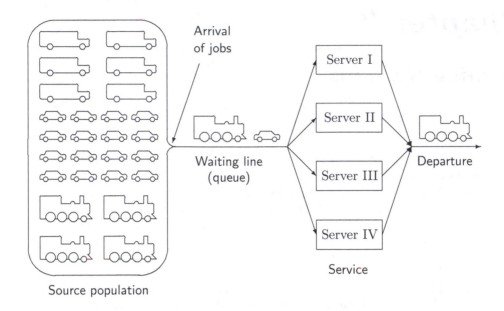

FIGURE 7.1: *Main components of a queuing system.*

7.1 Main components of a queuing system

How does a queuing system work? What happens with a job when it goes through a queuing system? The main stages are depicted in Figure 7.1.

Arrival

Typically, jobs arrive to a queuing system at random times. A *counting process* $A(t)$ tells the number of arrivals that occurred by the time t. In stationary queuing systems (whose distribution characteristics do not change over time), arrivals occur at *arrival rate*

$$\lambda_A = \frac{\mathbf{E}A(t)}{t}$$

for any $t > 0$, which is the expected number of arrivals per 1 unit of time. Then, the expected time between arrivals is

$$\mu_A = \frac{1}{\lambda_A}.$$

Queuing and routing to servers

Arrived jobs are typically processed according to the order of their arrivals, on a "first come–first serve" basis.

When a new job arrives, it may find the system in different states. If one server is available at that time, it will certainly take the new job. If several servers are available, the job may be randomized to one of them, or the server may be chosen according to some rules. For example, the fastest server or the least loaded server may be assigned to process the new job. Finally, if all servers are busy working on other jobs, the new job will join the queue, wait until all the previously arrived jobs are completed, and get routed to the next available server.

Various additional constraints may take place. For example, a queue may have a *buffer* that limits the number of waiting jobs. Such a queuing system will have *limited capacity*; the total number of jobs in it at any time is bounded by some constant C. If the capacity is full (for example, in a parking garage), a new job cannot enter the system until another job departs.

Also, jobs may leave the queue prematurely, say, after an excessively long waiting time. Servers may also open and close during the day as people need rest and servers need maintenance. Complex queuing systems with many extra conditions may be difficult to study analytically; however, we shall learn Monte Carlo methods for queuing systems in Section 7.6.

Service

Once a server becomes available, it immediately starts processing the next assigned job. In practice, service times are random because they depend on the amount of work required by each task. The average service time is μ_S. It may vary from one server to another as some computers or customer service representatives work faster than others. The *service rate* is defined as the average number of jobs processed by a continuously working server during one unit of time. It equals

$$\lambda_S = \frac{1}{\mu_S}.$$

Departure

When the service is completed, the job leaves the system.

The following parameters and random variables describe performance of a queuing system.

NOTATION	Parameters of a queuing system
λ_A =	arrival rate
λ_S =	service rate
μ_A =	$1/\lambda_A$ = mean interarrival time
μ_S =	$1/\lambda_S$ = mean service time
r =	$\lambda_A/\lambda_S = \mu_S/\mu_A$ = utilization, or arrival-to-service ratio

NOTATION	Random variables of a queuing system

$$
\begin{aligned}
X_s(t) &= \text{number of jobs receiving service at time } t \\
X_w(t) &= \text{number of jobs waiting in a queue at time } t \\
X(t) &= X_s(t) + X_w(t), \\
&\quad \text{the total number of jobs in the system at time } t \\[6pt]
S_k &= \text{service time of the } k\text{-th job} \\
W_k &= \text{waiting time of the } k\text{-th job} \\
R_k &= S_k + W_k, \text{ response time, the total time a job spends in the} \\
&\quad \text{system from its arrival until the departure}
\end{aligned}
$$

Utilization r is an important parameter. As we see in later sections, it shows whether or not a system can function under the current or even higher rate of arrivals, and how much the system is over- or underloaded.

A queuing system is *stationary* if the distributions of S_k, W_k, and R_k are independent of k. In this case, index k will often be omitted.

Most of the time, our goal will be finding the distribution of $X(t)$, the total number of jobs in the system. The other characteristics of a queuing system will be assessed from that. As a result, we shall obtain a comprehensive performance evaluation of a queuing system.

7.2 The Little's Law

The Little's Law gives a simple relationship between the expected number of jobs, the expected response time, and the arrival rate. It is valid for any stationary queuing system.

Little's Law

$$
\lambda_A \, \mathbf{E}(R) = \mathbf{E}(X)
$$

PROOF: For an elegant derivation of the Little's Law, calculate the shaded area in Figure 7.2. In this figure, rectangles represent the jobs, stretching between their arrival and departure times. Thus, the length of the k-th rectangle equals

$$
\text{Departure time} - \text{Arrival time} = R_k,
$$

and so does its area.

By the time T, there are $A(T)$ arrivals. Among them, $X(T)$ jobs remain in the system at time T. Only a portion of these jobs is completed by time T; the other portion (call it ε) will take place after time T. Then, the total shaded area equals

$$
\text{Shaded area} = \sum_{k=1}^{A(T)} R_k - \varepsilon. \tag{7.1}
$$

Alternatively, we can recall from Calculus that every area can be computed by integration. We let

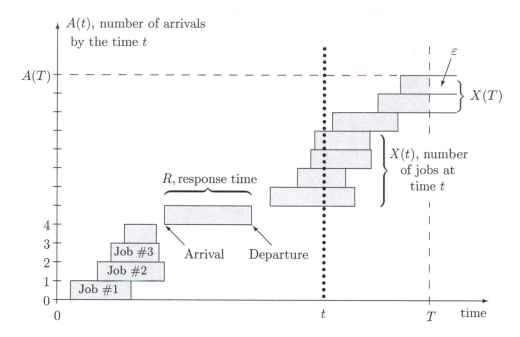

FIGURE 7.2: *Queuing system and an illustration to the Little's Law.*

t run from 0 to T and integrate the cross-section of the shaded region at t. As seen on the picture, the length of this cross-section is $X(t)$, the number of jobs in the system at time t. Hence,

$$\text{Shaded area} = \int_0^T X(t)dt. \tag{7.2}$$

Combining (7.1) and (7.2), we get

$$\sum_{k=1}^{A(T)} R_k - \varepsilon = \int_0^T X(t)\,dt. \tag{7.3}$$

It remains to take expectations, divide by T, and let $T \to \infty$. Then $\varepsilon/T \to 0$. In the left-hand side of (7.3), we get

$$\lim_{T \to \infty} \frac{1}{T} \mathbf{E}\left(\sum_{k=1}^{A(T)} R_k - \varepsilon \right) = \lim_{T \to \infty} \frac{\mathbf{E}(A(T))\,\mathbf{E}(R)}{T} - 0 = \lambda_A\,\mathbf{E}(R).$$

Recall that the arrival rate is $\lambda_A = \mathbf{E}(A(T))/T$.

In the right-hand side of (7.3), we get the average value of $X(t)$,

$$\lim_{T \to \infty} \frac{1}{T} \mathbf{E} \int_0^T X(t)\,dt = \mathbf{E}(X).$$

Therefore, $\lambda_A\,\mathbf{E}(R) = \mathbf{E}(X)$. $\qquad\qquad\qquad\qquad\qquad\qquad\qquad\qquad\qquad\qquad\qquad\square$

Example 7.1 (QUEUE IN A BANK). You walk into a bank at 10:00. Being there, you count a total of 10 customers and assume that this is the typical, average number. You also notice that on the average, customers walk in every 2 minutes. When should you expect to finish services and leave the bank?

<u>Solution.</u> We have $\mathbf{E}(X) = 10$ and $\mu_A = 2$ min. By the Little's Law,

$$\mathbf{E}(R) = \frac{\mathbf{E}(X)}{\lambda_A} = \mathbf{E}(X)\mu_A = (10)(2) = \underline{20 \text{ min}}.$$

That is, your expected response time is 20 minutes, and you should expect to leave at 10:20.

The Little's Law is universal, it applies to any stationary queuing system and even the system's components — the queue and the servers. Thus, we can immediately deduce the equations for the number of waiting jobs,

$$\mathbf{E}(X_w) = \lambda_A \, \mathbf{E}(W),$$

and for the number of jobs currently receiving service,

$$\mathbf{E}(X_s) = \lambda_A \, \mathbf{E}(S) = \lambda_A \mu_S = r.$$

We have obtained another important definition of *utilization*.

DEFINITION 7.2

> Utilization r is the expected number of jobs receiving service at any given time.

This Little's law is a fairly recent result. It was obtained by *John D. C. Little* who is currently an Institute Professor at Massachusetts Institute of Technology in the United States.

The Little's Law only relates *expectations* of the number of jobs and their response time. In the rest of this chapter, we evaluate the entire distribution of $X(t)$ that directs us to various probabilities and expectations of interest. These quantities will describe and predict performance of a queuing system.

DEFINITION 7.3

> The number of jobs in a queuing system, $X(t)$, is called a **queuing process**. In general, it is not a counting process because jobs arrive and depart, therefore, their number may increase and decrease whereas any counting process is non-decreasing.

FIGURE 7.3: *Transition diagram for a Bernoulli single-server queuing process.*

7.3 Bernoulli single-server queuing process

DEFINITION 7.4 —————————————————

Bernoulli single-server queuing process is a discrete-time queuing process with the following characteristics:

- one server

- unlimited capacity

- arrivals occur according to a Binomial process, and the probability of a new arrival during each frame is p_A

- the probability of a service completion (and a departure) during each frame is p_S provided that there is at least one job in the system at the beginning of the frame

- service times and interarrival times are independent

Everything learned in Section 6.3.1 about Binomial *counting* processes applies to arrivals of jobs. It also applies to service completions all the time when there is at least one job in the system. We can then deduce that

- there is a Geometric(p_A) number of frames between successive arrivals;
- each service takes a Geometric(p_S) number of frames;
- service of any job takes at least one frame;
- $p_A = \lambda_A \Delta$;
- $p_S = \lambda_S \Delta$.

Markov property

Moreover, Bernoulli single-server queuing process is a *homogeneous Markov chain* because probabilities p_A and p_S never change. The number of jobs in the system increments by 1 with each arrival and decrements by 1 with each departure (Figure 7.3). Conditions of a Binomial process guarantee that at most one arrival and at most one departure may occur during each frame. Then, we can compute all transition probabilities,

$$\begin{aligned} p_{00} &= \boldsymbol{P}\{\text{ no arrivals }\} &= 1 - p_A \\ p_{01} &= \boldsymbol{P}\{\text{ new arrival }\} &= p_A \end{aligned}$$

and for all $i \geq 1$,

$$
\begin{aligned}
p_{i,i-1} &= \boldsymbol{P}\{\text{ no arrivals } \cap \text{ one departure }\} &&= (1-p_A)p_S \\
p_{i,i} &= \boldsymbol{P}\{\text{ no arrivals } \cap \text{ no departures }\} \\
&\quad + \boldsymbol{P}\{\text{ one arrival } \cap \text{ one departure }\} &&= (1-p_A)(1-p_S)+p_Ap_S \\
p_{i,i+1} &= \boldsymbol{P}\{\text{ one arrival } \cap \text{ no departures }\} &&= p_A(1-p_S)
\end{aligned}
$$

The transition probability matrix (of an interesting size $\infty \times \infty$) is three-diagonal,

$$
P = \begin{pmatrix}
1-p_A & p_A & 0 & \cdots \\
(1-p_A)p_S & \begin{matrix}(1-p_A)(1-p_S)\\+p_Ap_S\end{matrix} & p_A(1-p_S) & \cdots \\
0 & (1-p_A)p_S & \begin{matrix}(1-p_A)(1-p_S)\\+p_Ap_S\end{matrix} & \cdots \\
0 & 0 & (1-p_A)p_S & \ddots \\
\vdots & \vdots & \ddots & \ddots
\end{pmatrix} \tag{7.4}
$$

All the other transition probabilities equal 0 because the number of jobs cannot change by more than one during any single frame.

This transition probability matrix may be used, for example, to simulate this queuing system and study its performance, as we did with general Markov chains in Section 6.4. One can also compute k-step transition probabilities and predict the load of a server or the length of a queue at any time in future.

Example 7.2 (PRINTER). Any printer represents a single-server queuing system because it can process only one job at a time while other jobs are stored in a queue. Suppose the jobs are sent to a printer at the rate of 20 per hour, and that it takes an average of 40 seconds to print each job. Currently a printer is printing a job, and there is another job stored in a queue. Assuming Bernoulli single-server queuing process with 20-second frames,

(a) what is the probability that the printer will be idle in 2 minutes?
(b) what is the expected length of a queue in 2 minutes? Find the expected number of waiting jobs and the expected total number of jobs in the system.

Solution. We are given:

$$
\begin{aligned}
\lambda_A &= 20 \text{ hrs}^{-1} = 1/3 \text{ min}^{-1}, \\
\lambda_S &= 1/\mu_S = 1/40 \text{ sec}^{-1} = 1.5 \text{ min}^{-1}, \\
\Delta &= 1/3 \text{ min}.
\end{aligned}
$$

Compute

$$
\begin{aligned}
p_A &= \lambda_A \Delta &&= 1/9 \\
p_S &= \lambda_S \Delta &&= 1/2,
\end{aligned}
$$

and all the transition probabilities for the number of jobs X,

$$
\begin{aligned}
p_{00} &= 1 - p_A = 8/9 = 0.889 \\
p_{01} &= p_A = 1/9 = 0.111
\end{aligned}
$$

and for $i \geq 1$,

$$
\begin{aligned}
p_{i,i-1} &= (1 - p_A)p_S = 4/9 = 0.444 \\
p_{i,i+1} &= p_A(1 - p_S) = 1/18 = 0.056 \\
p_{i,i} &= 1 - 0.444 - 0.056 = 0.5
\end{aligned}
\tag{7.5}
$$

Notice the "shortcut" used in computing $p_{i,i}$. Indeed, the row sum in a transition probability matrix is always 1; therefore, $p_{i,i} = 1 - p_{i,i-1} - p_{i,i+1}$.

We got the following transition probability matrix,

$$
P = \begin{pmatrix}
.889 & .111 & 0 & 0 & \cdots \\
.444 & .5 & .056 & 0 & \cdots \\
0 & .444 & .5 & .056 & \cdots \\
0 & 0 & .444 & .5 & \ddots \\
\vdots & \vdots & \vdots & \ddots & \ddots
\end{pmatrix}
$$

Since there are 2 min$/\Delta = 6$ frames in 2 minutes, we need a distribution of X after 6 frames, which is

$$
P_6 = P_0 P^6,
$$

as we know from Section 6.2, formula (6.4). Here there is an interesting problem. How do we deal with matrix P that has infinitely many rows and columns?

Fortunately, we only need a small portion of this matrix. There are 2 jobs currently in the system, so the initial distribution is

$$
P_0 = (0\ 0\ 1\ 0\ 0\ 0\ 0\ 0).
$$

In a course of 6 frames, their number can change by 6 at most (look at Figure 7.3), and thus, it is sufficient to consider the the first 9 rows and 9 columns of P only, corresponding to states 0, 1, ..., 8. Then, we compute the distribution after 6 frames,

$$
P_6 = P_0 P^6 = (0\ 0\ 1\ 0\ 0\ 0\ 0\ 0)
\begin{pmatrix}
.889 & .111 & 0 & \cdots & 0 & 0 \\
.444 & .5 & .056 & \cdots & 0 & 0 \\
0 & .444 & .5 & \cdots & 0 & 0 \\
\vdots & \vdots & \vdots & \vdots & \vdots & \vdots \\
0 & 0 & 0 & \cdots & .5 & .056 \\
0 & 0 & 0 & \cdots & .444 & .5
\end{pmatrix}^6
$$

$$
= (.644\ .250\ .080\ .022\ .004\ .000\ .000\ .000\ .000).
$$

Based on this distribution,

(a) The probability that the printer is idle after 2 minutes is the probability of no jobs in the system at that time,

$$
P_6(0) = \underline{0.644}.
$$

(b) The expected number of jobs in the system is

$$\mathbf{E}(X) = \sum_{x=0}^{8} x P_6(x) = \underline{0.494 \text{ jobs.}}$$

Out of these jobs, X_w are waiting in a queue and X_s are getting service. However, the server processes at most 1 job at a time; therefore, X_s equals 0 and 1. It has *Bernoulli* distribution, and its expectation is

$$\mathbf{E}(X_s) = \boldsymbol{P} \{ \text{ printer is busy } \} = 1 - \boldsymbol{P} \{ \text{ printer is idle } \} = 1 - 0.644 = 0.356.$$

Therefore, the expected length of a queue equals

$$\mathbf{E}(X_w) = \mathbf{E}(X) - \mathbf{E}(X_s) = 0.494 - 0.356 = \underline{0.138 \text{ waiting jobs.}}$$

$\Diamond$

Matrix computations in Example 7.2 are rather lengthy if you do them by hand; however, they present no problem for R, MATLAB, matrix calculators, or any other tool that can handle matrices. Here are R and MATLAB codes for computing $P_6 = P_0 P^6$ and $\mathbf{E}(X_6) = \sum x P_6(x)$. Notice that in R, a power of a matrix like P^6 can be computed in a package "expm".

— R —————

```
P0 <- rep(0,9)              % Initial distribution with P_0(2) = 1
P0[3] <- 1                  % P_0(2) is the 3rd element of vector P0
P <- matrix(0,9,9)          % Define a 9x9 transition probability matrix
P[1,1] <- 8/9; P[1,2] <- 1/9; P[9,8] <- 4/9; P[9,9] <- 0.5;
for (k in 2:8) { P[k,k-1]<-4/9; P[k,k]<-0.5; P[k,k+1]<-1/18; }
install.packages("expm")    % Use matrix operations from R package "expm"
library(expm)               %
P6 <- P0 %*% (P%^%6)        % Distribution after 6 frames
EX <- (0:8) %*% t(P6)       % Computes E(X_6) as a product of two vectors
```

— MATLAB ———

```
P0 = zeros(1,9);            % Initial distribution with P_0(2) = 1
P0(3) = 1;                  % P_0(2) is the 3rd element of vector P0
P = zeros(9);               % Define a 9x9 transition probability matrix
P(1,1) = 8/9; P(1,2) = 1/9; P(9,8) = 4/9; P(9,9) = .5;
for k=2:8; P(k,k-1)=4/9; P(k,k)=.5; P(k,k+1)=1/18; end;
P6 = P0 * P^6               % Distribution after 6 frames
EX = (0:8) * P6'            % Computes E(X_6) as a product of two vectors
```

Steady-state distribution

Bernoulli single-server queuing process is an *irregular Markov chain*. Indeed, any k-step transition probability matrix contains zeros because a k-step transition from 0 to $(k+1)$ is impossible as it requires at least $(k+1)$ arrivals, and this cannot happen by the conditions of the Binomial process of arrivals.

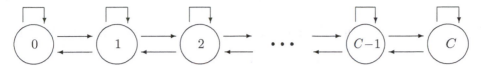

FIGURE 7.4: *Transition diagram for a Bernoulli single-server queuing process with limited capacity.*

Nevertheless, any system whose service rate exceeds the arrival rate (that is, jobs can be served faster than they arrive, so there is no overload),

$$\lambda_S > \lambda_A,$$

does have a steady-state distribution. Its computation is possible, despite the infinite dimension of P, but a little cumbersome. Instead, we shall compute the steady-state distribution for a continuous-time queuing process, taking, as usual, the limit of P as $\Delta \to 0$. Computations will simplify significantly.

7.3.1 Systems with limited capacity

As we see, the number of jobs in a Bernoulli single-server queuing system may potentially reach any number. However, many systems have limited resources for storing jobs. Then, there is a maximum number of jobs C that can possibly be in the system simultaneously. This number is called *capacity*.

How does limited capacity change the behavior of a queuing system? Until the capacity C is reached, the system operates without any limitation, as if $C = \infty$. All transition probabilities are the same as in (7.4).

The situation changes only when $X = C$. At this time, the system is full; it can accept new jobs into its queue only if some job departs. As before, the number of jobs decrements by 1 if there is a departure and no new arrival,

$$p_{C,C-1} = (1 - p_A)p_S.$$

In *all* other cases, the number of jobs remains at $X = C$. If there is no departure during some frame, and a new job arrives, this job cannot enter the system. Hence,

$$p_{C,C} = (1 - p_A)(1 - p_S) + p_A p_S + p_A(1 - p_S) = 1 - (1 - p_A)p_S.$$

This Markov chain has states 0, 1, ..., C (Figure 7.4), its transition probability matrix is finite, and any state can be reached in C steps; hence, the Markov chain is regular, and its steady-state distribution is readily available.

Example 7.3 (TELEPHONE WITH TWO LINES). Having a telephone with 2 lines, a customer service representative can talk to a customer and have another one "on hold." This is a system with limited capacity $C = 2$. When the capacity is reached and someone tries to call, (s)he will get a busy signal or voice mail.

Suppose the representative gets an average of 10 calls per hour, and the average phone conversation lasts 4 minutes. Modeling this by a Bernoulli single-server queuing process with limited capacity and 1-minute frames, compute the steady-state distribution and interpret it.

<u>Solution</u>. We have $\lambda_A = 10$ hrs$^{-1} = 1/6$ min^{-1}, $\lambda_S = 1/4$ min^{-1}, and $\Delta = 1$ min. Then

$$
\begin{aligned}
p_A &= \lambda_A \Delta &= 1/6, \\
p_S &= \lambda_S \Delta &= 1/4.
\end{aligned}
$$

The Markov chain $X(t)$ has 3 states, $X = 0$, $X = 1$, and $X = 2$. The transition probability matrix is

$$
P = \begin{pmatrix}
1-p_A & p_A & 0 \\
(1-p_A)p_S & (1-p_A)(1-p_S)+p_A p_S & p_A(1-p_S) \\
0 & (1-p_A)p_S & 1-(1-p_A)p_S
\end{pmatrix}
$$

$$
= \begin{pmatrix}
5/6 & 1/6 & 0 \\
5/24 & 2/3 & 1/8 \\
0 & 5/24 & 19/24
\end{pmatrix}. \tag{7.6}
$$

Next, we solve the steady-state equations

$$
\pi P = \pi \Rightarrow
\begin{cases}
\dfrac{5}{6}\pi_0 + \dfrac{5}{24}\pi_1 = \pi_0 \\[2mm]
\dfrac{1}{6}\pi_0 + \dfrac{2}{3}\pi_1 + \dfrac{5}{24}\pi_2 = \pi_1 \\[2mm]
\dfrac{1}{8}\pi_1 + \dfrac{19}{24}\pi_2 = \pi_2
\end{cases}
\Rightarrow
\begin{cases}
\dfrac{5}{24}\pi_1 = \dfrac{1}{6}\pi_0 \\[2mm]
\dfrac{1}{6}\pi_0 + \dfrac{2}{3}\pi_1 + \dfrac{5}{24}\pi_2 = \pi_1 \\[2mm]
\dfrac{1}{8}\pi_1 = \dfrac{5}{24}\pi_2
\end{cases}
$$

As we expect, the second equation follows from the others. After substitution, it becomes an identity

$$
\frac{5}{24}\pi_1 + \frac{2}{3}\pi_1 + \frac{1}{8}\pi_1 = \pi_1.
$$

We use this as a double-check and turn to the normalizing equation

$$
\pi_0 + \pi_1 + \pi_2 = \frac{5}{4}\pi_1 + \pi_1 + \frac{3}{5}\pi_1 = \frac{57}{20}\pi_1 = 1,
$$

from where

$$
\pi_0 = 25/57 = \underline{0.439}, \quad \pi_1 = 20/57 = \underline{0.351}, \quad \pi_2 = 12/57 = \underline{0.210}.
$$

Interpreting this result, 43.9% of time the representative is not talking on the phone, 35.1% of time (s)he talks but has the second line open, and 21.0% of time both lines are busy and new calls don't get through. ◊

7.4　M/M/1 system

We now turn our attention to *continuous-time* queuing processes. Our usual approach is to move from discrete time to continuous time gradually by reducing the frame size Δ to zero.

First, let us explain what the notation "M/M/1" actually means.

NOTATION	A queuing system can be denoted as A/S/n/C, where

A   denotes the distribution of interarrival times
S   denotes the distribution of service times
n   is the number of servers
C   is the capacity

Default capacity is $C = \infty$ (unlimited capacity)

Letter M denotes *Exponential distribution* because it is *memoryless*, and the resulting process is *Markov*.

DEFINITION 7.5

An **M/M/1 queuing process** is a continuous-time Markov queuing process with the following characteristics,

- one server;
- unlimited capacity;
- Exponential interarrival times with the arrival rate λ_A;
- Exponential service times with the service rate λ_S;
- service times and interarrival times are independent.

From Section 6.3.2, we know that Exponential interarrival times imply a *Poisson process* of arrivals with parameter λ_A. This is a very popular model for telephone calls and many other types of arriving jobs.

M/M/1 as a limiting case of a Bernoulli queuing process

We study M/M/1 systems by considering a Bernoulli single-server queuing process and letting its frame Δ go to zero. Our goal is to derive the steady-state distribution and other quantities of interest that measure the system's performance.

When the frame Δ gets small, its square Δ^2 becomes practically negligible, and transition probabilities for a Bernoulli single-server queuing process can be written as

$$
\begin{aligned}
p_{00} &= 1 - p_A &= 1 - \lambda_A \Delta \\
p_{10} &= p_A &= \lambda_A \Delta
\end{aligned}
$$

and for all $i \geq 1$,

$$
\begin{aligned}
p_{i,i-1} &= (1 - p_A)p_S &= (1 - \lambda_A \Delta)\lambda_S \Delta &\approx \lambda_S \Delta \\
p_{i,i+1} &= p_A(1 - p_S) &= \lambda_A \Delta(1 - \lambda_S \Delta) &\approx \lambda_A \Delta \\
p_{i,i} &= (1 - p_A)(1 - p_S) + p_A p_S &\approx 1 - \lambda_A \Delta - \lambda_S \Delta
\end{aligned}
$$

Remark: To be rigorous, these probabilities are written up to a small term of order $O(\Delta^2)$, as $\Delta \to 0$. These terms with Δ^2 will eventually cancel out in our derivation.

We have obtained the following transition probability matrix,

$$P \approx \begin{pmatrix} 1 - \lambda_A \Delta & \lambda_A \Delta & 0 & 0 & \cdots \\ \lambda_S \Delta & 1 - \lambda_A \Delta - \lambda_S \Delta & \lambda_A \Delta & 0 & \cdots \\ 0 & \lambda_S \Delta & 1 - \lambda_A \Delta - \lambda_S \Delta & \lambda_A \Delta & \cdots \\ 0 & 0 & \lambda_S \Delta & 1 - \lambda_A \Delta - \lambda_S \Delta & \ddots \\ \vdots & \vdots & \vdots & \ddots & \ddots \end{pmatrix} \tag{7.7}$$

Steady-state distribution for an M/M/1 system

As we solve the standard steady-state system of equations (infinitely many equations with infinitely many unknowns, actually),

$$\begin{cases} \pi P = \pi \\ \sum \pi_i = 1 \end{cases}$$

a rather simple, almost familiar final answer may come as a surprise!

The solution can be broken into little steps. Start by multiplying $\pi = (\pi_0, \pi_1, \pi_2, \ldots)$ by the first column of P and get

$$\pi_0(1 - \lambda_A \Delta) + \pi_1 \lambda_S \Delta = \pi_0 \quad \Rightarrow \quad \lambda_A \Delta \pi_0 = \lambda_S \Delta \pi_1 \quad \Rightarrow \quad \boxed{\lambda_A \pi_0 = \lambda_S \pi_1}.$$

This is *the first balance equation*.

Remark: We divided both sides of the equation by Δ. At this point, if we kept the Δ^2 terms, we would have gotten rid of them anyway by taking the limit as $\Delta \to 0$.

Next, multiplying π by the second column of P, we get

$$\pi_0 \lambda_A \Delta + \pi_1(1 - \lambda_A \Delta - \lambda_S \Delta) + \pi_2 \lambda_S \Delta = \pi_1 \quad \Rightarrow \quad (\lambda_A + \lambda_S)\pi_1 = \lambda_A \pi_0 + \lambda_S \pi_2.$$

Thanks to the first balance equation, $\lambda_A \pi_0$ and $\lambda_S \pi_1$ cancel each other, and we obtain the *second balance equation*,

$$\boxed{\lambda_A \pi_1 = \lambda_S \pi_2}.$$

This trend of balance equations will certainly continue because every next column of matrix P is just the same as the previous column, only shifted down by 1 position. Thus, the *general balance equation* looks like

$$\boxed{\lambda_A \pi_{i-1} = \lambda_S \pi_i} \quad \text{or} \quad \boxed{\pi_i = r \, \pi_{i-1}} \tag{7.8}$$

where $r = \lambda_A / \lambda_S$ is *utilization, or arrival-to-service ratio*.

Repeatedly applying (7.8) for i, $i - 1$, $i - 2$, etc., we express each π_i via π_0,

$$\pi_i = r \, \pi_{i-1} = r^2 \pi_{i-2} = r^3 \pi_{i-3} = \ldots = r^i \pi_0.$$

Finally, we apply the normalizing condition $\sum \pi_i = 1$, recognizing along the way that we deal with the geometric series,

$$\sum_{i=0}^{\infty} \pi_i = \sum_{i=0}^{\infty} r^i \pi_0 = \frac{\pi_0}{1 - r} = 1 \quad \Rightarrow \quad \begin{cases} \pi_0 &= 1 - r \\ \pi_1 &= r\pi_0 = r(1 - r) \\ \pi_2 &= r^2 \pi_0 = r^2(1 - r) \\ & \quad \text{etc.} \end{cases}$$

Does this distribution look familiar? Every probability equals the previous probability multiplied by the same constant r.

This distribution of $X(t)$ is *Shifted Geometric*, because $Y = X + 1$ has the standard Geometric distribution with parameter $p = 1 - r$,

$$\boldsymbol{P}\{Y = y\} = \boldsymbol{P}\{X = y - 1\} = \pi_{y-1} = r^{y-1}(1 - r) = (1 - p)^{y-1}p \ \text{ for } \ y \geq 1,$$

as in Section 3.4.3. This helps us compute the expected number of jobs in the system at any time and its variance,

$$\mathbf{E}(X) = \mathbf{E}(Y - 1) = \mathbf{E}(Y) - 1 = \frac{1}{1 - r} - 1 = \frac{r}{1 - r}$$

and

$$\text{Var}(X) = \text{Var}(Y - 1) = \text{Var}(Y) = \frac{r}{(1 - r)^2}$$

(for Geometric expectation and variance, see (3.10) on p. 62).

M/M/1 system:
steady-state distribution
of the number of jobs

$$\pi_x = \boldsymbol{P}\{X = x\} = r^x(1 - r)$$
$$\text{for } x = 0, 1, 2, \ldots$$

$$\mathbf{E}(X) = \frac{r}{1 - r}$$

$$\text{Var}(X) = \frac{r}{(1 - r)^2}$$

where $r = \lambda_A/\lambda_S = \mu_S/\mu_A$

(7.9)

7.4.1 Evaluating the system's performance

Many important system characteristics can be obtained directly from the distribution (7.9).

Utilization

We now see the actual meaning of the *arrival-to-service ratio*, or *utilization*. $r = \lambda_A/\lambda_S$. According to (7.9), it equals

$$r = 1 - \pi_0 = \boldsymbol{P}\{X > 0\},$$

which is the probability that there are jobs in the system, and therefore, the server is busy processing a job. Thus,

$$\boldsymbol{P}\{\text{ server is busy }\} = r$$
$$\boldsymbol{P}\{\text{ server is idle }\} = 1 - r$$

We can also say that r is the proportion of time when the server is put to work. In other words, r (utilization!) shows how much the server is utilized.

The system is functional if $r < 1$. In fact, our derivation of the distribution of X is only possible when $r < 1$, otherwise the geometric series used there diverges.

If $r \geq 1$, the system gets *overloaded*. Arrivals are too frequent comparing with the service rate, and the system cannot manage the incoming flow of jobs. The number of jobs in the system will accumulate in this case (unless, of course, it has a limited capacity).

Waiting time

When a new job arrives, it finds the system with X jobs in it. While these X jobs are being served, the new job awaits its turn in a queue. Thus, its waiting time consists of service times of X earlier jobs,

$$W = S_1 + S_2 + S_3 + \ldots + S_X.$$

Perhaps, the first job in this queue has already started its service. Magically, this possibility does not affect the distribution of W. Recall that service times in M/M/1 systems are *Exponential*, and this distribution has a *memoryless property*. At any moment, the remaining service time of the first job still has Exponential(λ_S) distribution regardless of how long it is already being served!

We can then conclude that the new job has *expected waiting time*

$$\mathbf{E}(W) = \mathbf{E}(S_1 + \ldots + S_X) = \mathbf{E}(S)\,\mathbf{E}(X) = \frac{\mu_S\, r}{1 - r} \quad \text{or} \quad \frac{r}{\lambda_S(1 - r)}.$$

(By writing the product of expectations, we actually used the fact that service times are independent of the number of jobs in the system at that time).

Remark: Notice that W is a rare example of a variable whose distribution is neither discrete nor continuous. On one hand, it has a probability mass function at 0 because $\mathbf{P}\{W = 0\} = 1 - r$ is the probability that the server is idle and available and there is no waiting time for a new job. On the other hand, W has a density $f(x)$ for all $x > 0$. Given any positive number of jobs $X = n$, the waiting time is the sum of n Exponential times which is Gamma(n, λ_S). Such a distribution of W is *mixed*.

Response time

Response time is the time a job spends in the system, from its arrival until its departure. It consists of waiting time (if any) and service time. The *expected response time* can then be computed as

$$\mathbf{E}(R) = \mathbf{E}(W) + \mathbf{E}(S) = \frac{\mu_S\, r}{1 - r} + \mu_S = \frac{\mu_S}{1 - r} \quad \text{or} \quad \frac{1}{\lambda_S(1 - r)}.$$

Queue

The length of a queue is the number of waiting jobs,

$$X_w = X - X_s.$$

The number of jobs X_s getting service at any time is 0 or 1. It is therefore a Bernoulli variable with parameter

$$P\{ \text{ server is busy } \} = r.$$

Hence, the *expected queue length* is

$$\mathbf{E}(X_w) = \mathbf{E}(X) - \mathbf{E}(X_s) = \frac{r}{1-r} - r = \frac{r^2}{1-r}.$$

Little's Law revisited

The Little's Law certainly applies to the M/M/1 queuing system and its components, the queue and the server. Assuming the system is functional ($r < 1$), all the jobs go through the entire system, and thus, each component is subject to the same arrival rate λ_A. The Little's Law then guarantees that

$$
\begin{aligned}
\lambda_A \, \mathbf{E}(R) &= \mathbf{E}(X), \\
\lambda_A \, \mathbf{E}(S) &= \mathbf{E}(X_s), \\
\lambda_A \, \mathbf{E}(W) &= \mathbf{E}(X_w).
\end{aligned}
$$

Using our results in (7.10), it wouldn't be a problem for you to verify all three equations, would it?

M/M/1:
main
performance
characteristics

$$
\begin{aligned}
\mathbf{E}(R) &= \frac{\mu_S}{1-r} = \frac{1}{\lambda_S(1-r)} \\[2mm]
\mathbf{E}(W) &= \frac{\mu_S\, r}{1-r} = \frac{r}{\lambda_S(1-r)} \\[2mm]
\mathbf{E}(X) &= \frac{r}{1-r} \\[2mm]
\mathbf{E}(X_w) &= \frac{r^2}{1-r} \\[2mm]
P\{\text{server is busy}\} &= r \\
P\{\text{server is idle}\} &= 1-r
\end{aligned}
$$

(7.10)

Example 7.4 (MESSAGE TRANSMISSION WITH A SINGLE CHANNEL). Messages arrive to a communication center at random times with an average of 5 messages per minute. They are transmitted through a single channel in the order they were received. On average, it takes 10 seconds to transmit a message. Conditions of an M/M/1 queue are satisfied. Compute the main performance characteristics for this center.

<u>Solution</u>. The arrival rate $\lambda_A = 5 \text{ min}^{-1}$ and the expected service time $\mu_S = 10$ sec or $(1/6)$ min are given. Then, the utilization is

$$r = \lambda_A/\lambda_S = \lambda_A \mu_S = \underline{5/6}.$$

This also represents the proportion of time when the channel is busy and the probability of a non-zero waiting time.

The average number of messages stored in the system at any time is

$$\mathbf{E}(X) = \frac{r}{1-r} = \underline{5}.$$

Out of these, an average of

$$\mathbf{E}(X_w) = \frac{r^2}{1-r} = \underline{4.17}$$

messages are waiting, and

$$\mathbf{E}(X_s) = r = \underline{0.83}$$

are being transmitted.

When a message arrives to the center, its waiting time until its transmission begins averages

$$\mathbf{E}(W) = \frac{\mu_S r}{1-r} = \underline{50 \text{ seconds}},$$

whereas the total amount of time since its arrival until the end of its transmission has an average of

$$\mathbf{E}(R) = \frac{\mu_S}{1-r} = \underline{1 \text{ minute}}.$$

$\diamond$

Example 7.5 (FORECAST). Let's continue Example 7.4. Suppose that next year the customer base of our transmission center is projected to increase by 10%, and thus, the intensity of incoming traffic λ_A increases by 10% too. How will this affect the center's performance?

Solution. Recompute the main performance characteristics under the new arrival rate

$$\lambda_A^{\text{NEW}} = (1.1)\lambda_A^{\text{OLD}} = 5.5 \text{ min}^{-1}.$$

Now the utilization equals $r = 11/12$, getting dangerously close to 1 where the system gets overloaded. For high values of r, various parameters of the system increase rapidly. A 10% increase in the arrival rate will result in rather significant changes in other variables. Using (7.10), we now get

$$\begin{array}{rcl}
\mathbf{E}(X) & = & 11 \text{ jobs}, \\
\mathbf{E}(X_w) & = & 10.08 \text{ jobs}, \\
\mathbf{E}(W) & = & 110 \text{ seconds, and} \\
\mathbf{E}(R) & = & 2 \text{ minutes}.
\end{array}$$

We see that the response time, the waiting time, the average number of stored messages, and therefore, the average required amount of memory more than doubled when the number of customers increased by a mere 10%. $\diamond$

When a system gets nearly overloaded

As we observed in Example 7.5, the system slowed down significantly as a result of a 10% increase in the intensity of incoming traffic, projected for the next year. One may try to forecast the two-year future of the system, assuming a 10% increase of a customer base each year. It will appear that during the second year the utilization will exceed 1, and the system will be unable to function.

What is a practical solution to this problem? Another channel or two may be added to the center to help the existing channel handle all the arriving messages! The new system will then have more than one channel-server, and it will be able to process more arriving jobs.

Such systems are analyzed in the next section.

7.5 Multiserver queuing systems

We now turn our attention to queuing systems with several servers. We assume that each server can perform the same range of services; however, in general, some servers may be faster than others. Thus, the service times for different servers may potentially have different distributions.

When a job arrives, it either finds all servers busy serving jobs, or it finds one or several available servers. In the first case, the job will wait in a queue for its turn whereas in the second case, it will be routed to one of the idle servers. A mechanism assigning jobs to available servers may be random, or it may be based on some rule. For example, some companies follow the equity principle and make sure that each call to their customer service is handled by the least loaded customer service representative.

The number of servers may be finite or infinite. A system with infinitely many servers can afford an unlimited number of concurrent users. For example, any number of people can watch a TV channel simultaneously. A job served by a system with $k = \infty$ servers will never wait; it will always find available idle servers.

As in previous sections, here is our plan for analyzing multiserver systems:

- first, verify if the number of jobs in the system at time t is a Markov process and write its transition probability matrix P. For continuous-time processes, we select a "very short" frame Δ and eventually let Δ converge to zero;

- next, compute the steady-state distribution π, and

- finally, use π to evaluate the system's long-term performance characteristics.

We treat a few common and analytically simple cases in detail. Advanced theory goes further; in this book we shall analyze more complex and non-Markov queuing systems by Monte Carlo methods in Section 7.6.

Notice that utilization r no longer has to be less than 1. A system with k servers can handle k times the traffic of a single-server system; therefore, it will function with any $r < k$.

As always, we start with a discrete-time process that can be described on a language of Bernoulli trials.

7.5.1 Bernoulli k-server queuing process

DEFINITION 7.6 ————————————————————————

> **Bernoulli k-server queuing process** is a discrete-time queuing process with the following characteristics:
>
> – k servers
>
> – unlimited capacity
>
> – arrivals occur according to a Binomial counting process; the probability of a new arrival during each frame is p_A
>
> – during each frame, each *busy* server completes its job with probability p_S independently of the other servers and independently of the process of arrivals

---

Markov property

As a result of this definition, all interarrival times and all service times are independent *Geometric* random variables (multiplied by the frame length Δ) with parameters p_A and p_S, respectively. This is similar to a single-server process in Section 7.3. Geometric variables have a *memoryless property*, they forget the past, and therefore, again our process is Markov.

The new feature is that now several jobs may finish during the same frame.

Suppose that $X_s = n$ jobs are currently getting service. During the next frame, each of them may finish and depart, independently of the other jobs. Then the number of departures is the number of successes in n independent Bernoulli trials, and therefore, it has *Binomial* distribution with parameters n and p_S.

This will help us compute the transition probability matrix.

Transition probabilities

Let's compute transition probabilities

$$p_{ij} = \boldsymbol{P}\left\{X(t + \Delta) = j \mid X(t) = i\right\}.$$

To do this, suppose there are i jobs in a k-server system. Then, the number n of busy servers is *the smaller* of the number of jobs i and the total number of servers k,

$$n = \min\left\{i, k\right\}.$$

- For $i \leq k$, the number of servers is sufficient for the current jobs, all jobs are getting service, and the number of departures X_d during the next frame is Binomial(i, p_S).

- For $i > k$, there are more jobs than servers. Then all k servers are busy, and the number of departures X_d during the next frame is Binomial(k, p_S).

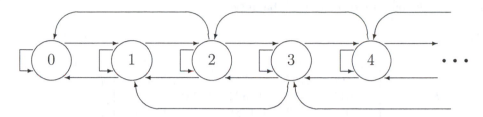

FIGURE 7.5: *Transition diagram for a Bernoulli queuing system with 2 servers.*

In addition, a new job arrives during the next frame with probability p_A.

Accounting for all the possibilities, we get:

$$
\begin{aligned}
p_{i,i+1} &= \boldsymbol{P}\{\,1\text{ arrival, }0\text{ departures}\,\} = p_A \cdot (1-p_S)^n; \\[2mm]
p_{i,i} &= \boldsymbol{P}\{\,1\text{ arrival, }1\text{ departure}\,\} + \boldsymbol{P}\{\,0\text{ arrivals, }0\text{ departures}\,\} \\[2mm]
&= p_A \cdot \binom{n}{1} p_S(1-p_S)^{n-1} + (1-p_A) \cdot (1-p_S)^n; \\[2mm]
p_{i,i-1} &= \boldsymbol{P}\{\,1\text{ arrival, }2\text{ departures}\,\} + \boldsymbol{P}\{\,0\text{ arrivals, }1\text{ departure}\,\} \\[2mm]
&= p_A \cdot \binom{n}{2} p_S^2(1-p_S)^{n-2} + (1-p_A) \cdot \binom{n}{1} p_S(1-p_S)^{n-1}; \\[2mm]
\cdots \quad &\cdots \quad \cdots \ \cdots \ \cdots \\[2mm]
p_{i,i-n} &= \boldsymbol{P}\{\,0\text{ arrivals, }n\text{ departures}\,\} = (1-p_A) \cdot p_S^n.
\end{aligned}
$$

(7.11)

Seemingly long, these formulas need nothing but the computation of Binomial probabilities for the number of departing jobs X_d.

A transition diagram for a 2-server system is shown in Figure 7.5. The number of concurrent jobs can make transitions from i to $i-2$, $i-1$, i, and $i+1$.

Example 7.6 (CUSTOMER SERVICE AS A TWO-SERVER SYSTEM WITH LIMITED CAPACITY). There are two customer service representatives on duty answering customers' calls. When both of them are busy, two more customers may be "on hold," but other callers will receive a "busy signal." Customers call at the rate of 1 call every 5 minutes, and the average service takes 8 minutes. Assuming a two-server Bernoulli queuing system with limited capacity and 1-minute frames, compute

(a) the steady-state distribution of the number of concurrent jobs;

(b) the proportion of callers who get a "busy signal";

(c) the percentage of time each representative is busy, if each of them takes 50% of all calls.

<u>Solution.</u> This system has $k = 2$ servers, capacity $C = 4$, $\lambda_A = 1/5$ min^{-1}, $\lambda_S = 1/8$ min^{-1}, and $\Delta = 1$ min. Then

$$p_A = \lambda_A \Delta = 0.2 \quad \text{and} \quad p_S = \lambda_S \Delta = 0.125.$$

Using (7.11), we compute transition probabilities:

$$P \;=\; \begin{pmatrix} 0.8000 & 0.2000 & 0 & 0 & 0 \\ 0.1000 & 0.7250 & 0.1750 & 0 & 0 \\ 0.0125 & 0.1781 & 0.6562 & 0.1531 & 0 \\ 0 & 0.0125 & 0.1781 & 0.6562 & 0.1531 \\ 0 & 0 & 0.0125 & 0.1781 & 0.8094 \end{pmatrix} \begin{array}{l} \text{From} \\ \text{state:} \\ 0 \\ 1 \\ 2 \\ 3 \\ 4 \end{array}$$

	To state:	0	1	2	3	4

Frankly, we used the following computer code to compute this matrix,

—— R ——————

```
k <- 2; C <- 4; pa <- 0.2; ps <- 0.125;
P <- matrix(0,C+1,C+1)          # (C+1) states from 0 to C
for (i in 0:C) { n <- min(i,k)  # n = number of busy servers
  P[i+1,i-n+1] <- (1-pa)*ps^n;  # transition when all n jobs depart
  for (j in 0:(n-1)) {
    P[i+1,i-j+1] <-                 # transition i -> i-j
      pa * choose(n,j+1)*(1-ps)^(n-j-1)*ps^(j+1) +
      (1-pa) * choose(n,j)*(1-ps)^(n-j)*ps^j;
  }
  if (i<C) {                    # capacity is not reached
    P[i+1,i+2] <- pa * (1-ps)^n;
    } else {                    # capacity is reached
    P[C+1,C+1] <- P[C+1,C+1] + pa * (1-ps)^n
  }
}; P
```

—— Matlab ————

```
k=2; C=4; pa=0.2; ps=0.125;
P = zeros(C+1,C+1);            % (C+1) states from 0 to C
  for i=0:C; n = min(i,k)      % number of busy servers
  P(i+1,i-n+1)=(1-pa)*ps^n;    % transition when all n jobs depart
  for j=0:n-1;
    P(i+1,i-j+1) = ...         % transition i -> i-j
      pa * nchoosek(n,j+1)*(1-ps)^(n-j-1)*ps^(j+1)...
      + (1-pa) * nchoosek(n,j)*(1-ps)^(n-j)*ps^j;
  end;
  if i<C;                      % capacity is not reached
    P(i+1,i+2) = pa * (1-ps)^n;
    else                       % capacity is reached
    P(C+1,C+1) = P(C+1,C+1) + pa * (1-ps)^n;
  end;
end; P
```

And now we can answers all the questions on page 191.

(a) The steady-state distribution for this system is

$$
\begin{cases}
\pi_0 & = & 0.1527 \\
\pi_1 & = & 0.2753 \\
\pi_2 & = & 0.2407 \\
\pi_3 & = & 0.1837 \\
\pi_4 & = & 0.1476
\end{cases}
$$

(b) The proportion of callers who hear a "busy signal" is the probability that the system is full when a job arrives, which is

$$
\boldsymbol{P}\{X = C\} = \pi_4 = \underline{0.1476}.
$$

(c) Each customer service rep is busy when there are two, three, or four jobs in the system, plus a half of the time when there is one job (because there is a 50% chance that the other rep handles this job). This totals

$$
\pi_2 + \pi_3 + \pi_4 + 0.5\pi_1 = \underline{0.7090} \text{ or } \underline{70.90\%}.
$$

$\Diamond$

7.5.2   M/M/k systems

An M/M/k system is a multiserver extention of M/M/1. According to the general "A/S/n/C" notation in Section 7.4, it means the following.

DEFINITION 7.7 ——————————————————————

An **M/M/k queuing process** is a continuous-time Markov queuing process with

- k servers
- unlimited capacity
- Exponential interarrival times with the arrival rate λ_A
- Exponential service time for each server with the service rate λ_S, independent of all the arrival times and the other servers

---

Once again, we move from the discrete-time Bernoulli multiserver process to the continuous-time M/M/k by letting the frame Δ go to 0. For very small Δ, transition probabilities (7.11) simplify,

$$
\begin{aligned}
p_{i,i+1} &= \lambda_A \Delta \cdot (1 - \lambda_S \Delta)^n \approx \lambda_A \Delta = p_A \\[2mm]
p_{i,i} &= \lambda_A \Delta \cdot n \lambda_S \Delta (1 - \lambda_S \Delta)^{n-1} + (1 - \lambda_A \Delta) \cdot (1 - \lambda_S \Delta)^n \\
&\approx 1 - \lambda_A \Delta - n \lambda_S \Delta = 1 - p_A - n p_S \\[2mm]
p_{i,i-1} &\approx n \lambda_S \Delta = n p_S \\[2mm]
p_{i,j} &= 0 \text{ for all other } j.
\end{aligned}
\tag{7.12}
$$

Again, $n = \min \{i, k\}$ is the number of jobs receiving service among the total of i jobs in the system.

For example, for $k = 3$ servers, the transition probability matrix is

$$P \approx \begin{pmatrix} 1 - p_A & p_A & 0 & 0 & 0 & \ddots \\ p_S & 1 - p_A - p_S & p_A & 0 & 0 & \ddots \\ 0 & 2p_S & 1 - p_A - 2p_S & p_A & 0 & \ddots \\ 0 & 0 & 3p_S & 1 - p_A - 3p_S & p_A & \ddots \\ 0 & 0 & 0 & 3p_S & 1 - p_A - 3p_S & \ddots \\ 0 & 0 & 0 & 0 & 3p_S & \ddots \\ \ddots & \ddots & \ddots & \ddots & \ddots & \ddots \end{pmatrix} \qquad (7.13)$$

Let's try to understand these probabilities intuitively. Recall that Δ is very small, so we ignored terms proportional to Δ^2, Δ^3, etc. Then, no more than one event, arrival or departure, may occur during each frame. Probability of more than one event is of the order $O(\Delta^2)$. Changing the number of jobs by 2 requires at least 2 events, and thus, such changes cannot occur during one frame.

At the same time, transition from i to $i - 1$ may be caused by a departure of any one of n currently served jobs. This is why we see the departure probability p_S multiplied by n.

Steady-state distribution

From matrix P, it would not be hard for us to find π, the steady-state distribution for the number of concurrent jobs X, and further performance characteristics. We already went through similar steps for a special case $k = 1$ in Section 7.4.

Again, we solve the system

$$\begin{cases} \pi P = \pi \\ \sum_i \pi_i = 1 \end{cases}$$

Multiplying π by the first column of P gives our familiar *balance equation*

$$\pi_0(1 - p_A) + \pi_1 p_S = \pi_0 \quad \Rightarrow \quad \pi_0 p_A = \pi_1 p_S \quad \Rightarrow \quad \pi_0 p_A = \pi_1 p_S \quad \Rightarrow \quad \boxed{\pi_1 = r\pi_0},$$

where

$$r = \frac{p_A}{p_S} = \frac{\lambda_A}{\lambda_S}$$

is *utilization*. Things become different from the second balance equation, affected by multiple servers,

$$\pi_0 p_A + \pi_1(1 - p_A - p_S) + 2\pi_2 p_S = \pi_1 \quad \Rightarrow \quad \pi_1 p_A = 2\pi_2 p_S \quad \Rightarrow \quad \boxed{\pi_2 = 2r\pi_1}.$$

Continuing, we get all the balance equations,

$$
\begin{cases}
\pi_1 &= r\pi_0 \\
\pi_2 &= r\pi_1/2 &= r^2\pi_0/2! \\
\pi_3 &= r\pi_2/3 &= r^3\pi_0/3! \\
&\cdots \quad \cdots \quad \cdots \quad \cdots \quad \cdots \quad \cdots \\
\pi_k &= r\pi_{k-1}/k &= r^k\pi_0/k!
\end{cases}
$$

until the number of jobs k occupies all the servers, after which

$$
\begin{cases}
\pi_{k+1} &= (r/k)\pi_k &= (r/k)\,r^k\pi_0/k! \\
\pi_{k+2} &= (r/k)\pi_{k+1} &= (r/k)^2\,r^k\pi_0/k! \\
& \text{etc.}
\end{cases}
$$

The normalizing equation $\sum \pi_i = 1$ returns

$$
\begin{aligned}
1 &= \pi_0 + \pi_1 + \dots \\
&= \pi_0\left(1 + r + \frac{r^2}{2!} + \frac{r^3}{3!} + \dots + \frac{r^k}{k!} + \frac{r^k}{k!}(r/k) + \frac{r^k}{k!}(r/k)^2 + \dots\right) \\
&= \pi_0\left(\sum_{i=0}^{k-1}\frac{r^i}{i!} + \frac{r^k}{(1-r/k)k!}\right),
\end{aligned}
$$

using in the last line the formula for a geometric series with the ratio (r/k).

We have obtained the following result.

M/M/k system:
steady-state
distribution
of the number
of jobs

$$
\pi_x = P\{X = x\} = \begin{cases}
\dfrac{r^x}{x!}\,\pi_0 & \text{for } x \le k \\[2ex]
\dfrac{r^k}{k!}\,\pi_0\left(\dfrac{r}{k}\right)^{x-k} & \text{for } x > k
\end{cases}
$$

where

$$
\pi_0 = P\{X = 0\} = \frac{1}{\displaystyle\sum_{i=0}^{k-1}\frac{r^i}{i!} + \frac{r^k}{(1-r/k)k!}}
$$

and $r = \lambda_A/\lambda_S$

(7.14)

Example 7.7 (A MULTICHANNEL MESSAGE TRANSMISSION CENTER). In Example 7.5 on p. 188, we worried about the system's incapability to handle the increasing stream of messages. That was a single-server queuing system.

Suppose now that in view of our forecast, 2 additional channels are built with the same parameters as the first channel. As a result, we have an M/M/3 system. Now it can easily handle even twice as much traffic!

Indeed, suppose now that the arrival rate λ_A has doubled since Example 7.4. Now it equals 10 min^{-1}. The service rate is still 6 min^{-1} for each server. The system's utilization is

$$r = \lambda_A/\lambda_S = 10/6 = 1.67 > 1,$$

which is fine. Having 3 servers, the system will function with any $r < 3$.

What percentage of messages will be sent immediately, with no waiting time?

Solution. A message does not wait at all times when there is an idle server (channel) to transmit it. The latter happens when the number of jobs in the system is less than the number of servers. Hence,

$$\boldsymbol{P}\{W = 0\} = \boldsymbol{P}\{X < 3\} = \pi_0 + \pi_1 + \pi_2 = \underline{0.70},$$

where the steady-state probabilities are computed by formula (7.14) as

$$\pi_0 = \cfrac{1}{1 + 1.67 + \cfrac{(1.67)^2}{2!} + \cfrac{(1.67)^3}{(1 - 1.67/3)3!}} = \frac{1}{5.79} = 0.17,$$

$$\pi_1 = (1.67)\pi_0 = 0.29, \quad \text{and} \quad \pi_2 = \frac{(1.67)^2}{2!}\pi_0 = 0.24.$$

$\Diamond$

7.5.3 Unlimited number of servers and M/M/∞

An unlimited number of servers completely eliminates the waiting time. Whenever a job arrives, there will always be servers available to handle it, and thus, the response time R consists of the service time only. In other words,

$$X = X_s, \quad R = S, \quad X_w = 0, \quad \text{and} \quad W = 0.$$

Have we seen such queuing systems?

Clearly, nobody can physically build an *infinite* number of devices. Having an unlimited number of servers simply means that any number of concurrent users can be served simultaneously. For example, most internet service providers and most long-distance telephone companies allow virtually any number of concurrent connections; an unlimited number of people can watch a TV channel, listen to a radio station, or take a sun bath.

Sometimes a model with infinitely many servers is a reasonable approximation for a system where jobs typically don't wait and get their service immediately. This may be appropriate for a computer server, a grocery store, or a street with practically no traffic jams.

M/M/∞ queueing system

The definition and all the theory of M/M/k systems applies to M/M/∞ when we substitute $k = \infty$ or take a limit as the number of servers k goes to infinity. The number of jobs will always be less than the number of servers ($i < k$), and therefore, we always have $n = i$. That is, with i jobs in the system, exactly i servers are busy.

The number of jobs X in an M/M/∞ system has a transition probability matrix

$$
P = \begin{pmatrix}
1 - p_A & p_A & 0 & 0 & \ddots \\
p_S & 1 - p_A - p_S & p_A & 0 & \ddots \\
0 & 2p_S & 1 - p_A - 2p_S & p_A & \ddots \\
0 & 0 & 3p_S & 1 - p_A - 3p_S & \ddots \\
0 & 0 & 0 & 4p_S & \ddots \\
\ddots & \ddots & \ddots & \ddots & \ddots
\end{pmatrix} \tag{7.15}
$$

and its steady-state distribution is ... *doesn't it look familiar?*

Taking a limit of (7.14) as $k \to \infty$, we get

$$
\pi_0 = \frac{1}{\displaystyle\sum_{i=0}^{\infty} \frac{r^i}{i!} + \lim_{k \to \infty} \frac{r^k}{(1 - r/k)k!}} = \frac{1}{e^r + 0} = e^{-r},
$$

and for all $x \geq 0$,

$$
\pi_x = \frac{r^x}{x!} \pi_0 = e^{-r} \frac{r^x}{x!}.
$$

We certainly recognize *Poisson distribution*! Yes, the number of concurrent jobs in an M/M/∞ system is Poisson with parameter $r = \lambda_A/\lambda_S$.

M/M/∞ system: steady-state distribution of the number of jobs	The number of jobs is **Poisson**(r) $\pi_x = P\{X = x\} = e^{-r} \dfrac{r^x}{x!}$ $\mathbf{E}(X) = \mathrm{Var}(X) = r$

$$\tag{7.16}$$

Example 7.8 (A POWERFUL SERVER). A certain powerful server can afford practically any number of concurrent users. The users connect to a server at random times, every 3 minutes, on the average, according to a Poisson counting process. Each user spends an Exponential amount of time on the server with an average of 1 hour and disconnects from it, independently of other users.

Such a description fits an M/M/∞ queuing system with

$$
r = \mu_S/\mu_A = 60 \, \text{min} \, / 3 \, \text{min} = 20.
$$

The number of concurrent users is Poisson(20). We can use Table A3 in the Appendix to find out that

$$
P\{X = 0\} = 0.0000,
$$

that is, the server practically never becomes idle.

Also, if an urgent message is sent to all the users, then $\mathbf{E}(X) = 20$ users, on the average, will see it immediately. Fifteen or more users will receive this message immediately with probability

$$P\{X \geq 15\} = 1 - F(14) = 1 - 0.1049 = \underline{0.8951}.$$

$\Diamond$

7.6 Simulation of queuing systems

Let us summarize what we have achieved in this chapter. We developed a theory and understood how to analyze and evaluate rather basic queuing systems: Bernoulli and M/M/k. We have covered the cases of one server and several servers, limited and unlimited capacity, but never considered such complications as customers' premature dropout, service interruption, non-Markov processes of arrivals and/or services, and distinction between different servers.

Most of our results were obtained from the Markov property of the considered queuing processes. For these systems, we derived a steady-state distribution of the number of concurrent jobs and computed the vital performance characteristics from it.

The only general result is the Little's Law of Section 7.2, which can be applied to any queuing system.

In practice, however, many queuing systems have a rather complex structure. Jobs may arrive according to a non-Poisson process; often the rate of arrivals changes during the day (there is a rush hour on highways and on the internet). Service times may have different distributions, and they are not always memoryless, thus the Markov property may not be satisfied. The number of servers may also change during the day (additional servers may turn on during rush hours). Some customers may get dissatisfied with a long waiting time and quit in the middle of their queue. And so on.

Queuing theory does not cover all the possible situations. On the other hand, we can simulate the behavior of almost any queuing system and study its properties by *Monte Carlo methods*.

Markov case

A queuing system is *Markov* only when its interarrival and service times are memoryless. Then the future can be predicted from the present without relying on the past (Section 6.2).

If this is the case, we compute the transition probability matrix as we did in (7.4), (7.6), (7.7), (7.11), (7.12), (7.13), or (7.15), and simulate the Markov chain according to Algorithm 6.1 on p. 161. To study long-term characteristics of a queuing system, the initial distribution of X_0 typically does not matter, so we may start this algorithm with 0 jobs in the system and immediately "switch on" the servers.

Even when the system is Markov, some interesting characteristics do not follow from its steady-state distribution directly. They can now be estimated from a Monte Carlo study by simulated long-run proportions and averages. For example, we may be interested in estimating the percentage of dissatisfied customers, the expected number of users during a period of time from 3 pm till 5 pm, etc. We may also compute various forecasts (Section 5.3.3).

General case

Monte Carlo methods of Chapter 5 let us simulate and evaluate rather complex queuing systems far beyond Bernoulli and M/M/k. As long as we know the distributions of inter-arrival and service times, we can generate the processes of arrivals and services. To assign jobs to servers, we keep track of servers that are available each time when a new job arrives. When all the servers are busy, the new job will enter a queue.

As we simulate the work of a queuing system, we keep records of events and variables that are of interest to us. After a large number of Monte Carlo runs, we average our records in order to estimate probabilities by long-run proportions and expected values by long-run averages.

Example: simulation of a multiserver queuing system

Let us simulate one day, from 8 am till 10 pm, of a queuing system that has

- four servers;
- Gamma distributed service times with parameters given in the table,

Server	α	λ
I	6	0.3 min^{-1}
II	10	0.2 min^{-1}
III	7	0.7 min^{-1}
IV	5	1.0 min^{-1}

- a Poisson process of arrivals with the rate of 1 arrival every 4 min, independent of service times;
- random assignment of servers, when more than 1 server is available.

In addition, suppose that after 15 minutes of waiting, jobs withdraw from a queue if their service has not started.

For an ordinary day of work of this system, we are interested to estimate the expected values of:

- the total time each server is busy with jobs;
- the total number of jobs served by each server;
- the average waiting time;
- the longest waiting time;
- the number of withdrawn jobs;
- the number of times a server was available immediately (this is also the number of jobs with no waiting time);
- the number of jobs remaining in the system at 10 pm.

We start by entering parameters of the system.

```
— R ———————          — MATLAB ————
k   <- 4                  k  = 4;                  % number of servers
mu  <- 4                  mu = 4;                  % mean interarrival time
alpha <- c(6,10,7,5)      alpha = [6 10 7 5];      % parameters of service
lambda <- c(.3,.2,.7,1.0) lambda = [.3 .2 .7 1.0]; % times
```

Then we initialize variables and start keeping track of arriving jobs. These are empty arrays so far. They get filled with each arriving job.

```
— R ———————          — MATLAB ————
arrival <- c()            arrival = [ ];           % arrival time
start   <- c()            start   = [ ];           % service starts
finish  <- c()            finish  = [ ];           % departure time
server  <- c()            server  = [ ];           % assigned server
j <- 0                    j = 0;                   % job number
T <- 0                    T = 0;                   % arrival of a new job
A <- rep(0,k)             A = zeros(1,k);          % A is an array of times
           % when each of the servers becomes available to take the next job
```

The queuing system is ready to work! We start a "while"-loop over the number of arriving jobs. It will run until the end of the day, when arrival time T reaches 14 hours, or 840 minutes. The length of this loop, the total number of arrived jobs, is random.

```
— R ———————          — MATLAB ————
while (T < 840) {         while T < 840;           % until end of the day
  j <- j+1                  j = j+1;               % next job
  T <- T+rexp(1,1/mu)       T = T-mu*log(rand);    % arrival time of job j
  arrival<-c(arrival,T)     arrival=[arrival T];
```

Generated in accordance with Example 5.10 on p. 111, the arrival time T is obtained by incrementing the previous arrival time by an Exponential interarrival time (R and MAT-LAB users can take advantage of built-in functions **rexp** and **exprnd**). Generated within this while-loop, the last job actually arrives after 10 pm. You may either accept it or delete it.

Next, we need to assign the new job j to a server, following the rule of random assignment. There are two cases here: either all servers are busy at the arrival time T, or some servers are available.

```
— R ———————
Nfree <- sum( A < T )               # number of free servers at time T
u <- 1                              # u = server that will take job j
if (Nfree == 0) {                   # If all servers are busy at time T ...
  for (v in 2:k) {
    if (A[v] < A[u]) {              # Find the server that gets available
      u <- v;                       # first and assign job j to it
    }
  }
}
```

```
  if (A[u]-T > 15) {          # If job j waits more than 15 min,
    start <- c(start, T+15)   # then it withdraws at time T+15.
    finish <- c(finish, T+15)
    u <- 0                    # No server is assigned.
  } else {                    # If job j doesn't withdraw ...
    start <- c(start, A[u])
  }
} else {                      # If there are servers available ...
  u <- ceiling(runif(1)*k)    # Generate a random server, from 1 to k
  while (A[u] > T) {          # Random search for an available server
    u <- ceiling(runif(1)*k)
  }
start <- c(start, T)          # Service starts immediately at T
}
server <- c(server, u)
```

— MATLAB ——————

```
Nfree = sum( A < T );         % number of free servers at time T
u = 1;                        % u = server that will take job j

if Nfree == 0;                % If all servers are busy at time T ...
  for v=2:k;
    if A(v)<A(u);             % Find the server that gets available
      u = v;                  % first and assign job j to it
    end;
  end;

  if A(u)-T > 15;             % If job j waits more than 15 min,
    start=[start T+15];       % then it withdraws at time T+15.
    finish=[finish T+15];
    u = 0;                    % No server is assigned.
  else                        % If job j doesn't withdraw ...
    start = [start A(u)];
  end;

else                          % If there are servers available ...
  u = ceil(rand*k);           % Generate a random server, from 1 to k
  while A(u) > T;             % Random search for an available server
    u = ceil(rand*k);
  end;
start = [start T];            % Service starts immediately at T
end;
server = [server u];
```

The server (u) for the new job (j) is determined. We can now generate its service time S from the suitable Gamma distribution and update our records. Again, we'll generate the Gamma service times from a Uniform random number generator although nothing prevents R and MATLAB users from making use of built-in functions `rgamma` and `gammarnd`. Our program continues...

— R ——————

```
if (u > 0) {                        # if job j doesn't withdraw ...
  S <- sum( -1/lambda[u] * log(runif(alpha[u])) )
  finish <- c(finish, start[j]+S)
  A[u] <- start[j] + S;             # This is the time when server u
}                                   # will now become available
}                                   # End of the while-loop
job = 1:j; print( data.frame(job, arrival, start, finish, server) )
```

— MATLAB ——————

```
if u > 0;                           % if job j doesn't withdraw ...
  S = sum( -1/lambda(u) * log(rand(alpha(u),1) ) );
  finish = [finish start(j)+S];
  A(u)   = start(j) + S;            % This is the time when server u
  end;                              % will now become available
end;                                % End of the while-loop
disp([(1:j)' arrival' start' finish' server'])
```

The day (our while-loop) has ended. The last command displays the following records,

Job	Arrival time	Service starts	Service ends	Server
1	.....	.....	.....	.....
2	.....	.....	.....	.....
.....	.....	.....	.....	.....
.....	.....	.....	.....	.....

From this master table, all variables of our interest can be computed.

— R ——————

```
Twork = rep(NA,k); Njobs = rep(NA,k);
for (u in 1:k) {                        # Twork(u) is the total working
  Twork[u] <- sum((server == u)*(finish-start))    # time for server u
  Njobs(u) <- sum(server == u)          # number of jobs served by u
}
Wmean <- mean(start-arrival)            # average waiting time
Wmax <- max(start-arrival)              # the longest waiting time
Nwithdr <- sum(server == 0)             # number of withdrawn jobs
Nav <- sum(start-arrival < 0.00001)     # number of jobs that did not wait
Nat10 <- sum(finish > 840)              # number of jobs at 10 pm
```

— MATLAB ——————

```
for u=1:k;                              % Twork(u) is the total working
  Twork(u) = sum((server==u).*(finish-start));   % time for server u
  Njobs(u) = sum(server==u);            % number of jobs served by u
```

```
end;
Wmean = mean(start-arrival);      % average waiting time
Wmax = max(start-arrival);        % the longest waiting time
Nwithdr = sum(server==0);         % number of withdrawn jobs
Nav = sum(start-arrival < 0.00001);  % number of jobs that did not wait
Nat10 = sum(finish > 840);        % number of jobs at 10 pm
```

We have computed all the required quantities for one day. Put this code into a do-loop and simulate a large number of days. The required expected values are then estimated by averaging variables `Wmean`, `Nwithdr`, etc., over all days.

Summary and conclusions

Queuing systems are service facilities designed to serve randomly arriving jobs.

Several classes of queuing systems were studied in this chapter. We discussed discrete-time and continuous-time, Markov and non-Markov, one-server and multiserver queuing processes with limited and unlimited capacity.

Detailed theory was developed for Markov systems, where we derived the distribution of the number of concurrent jobs and obtained analytic expressions for a number of vital characteristics. These results are used to evaluate performance of a queuing system, forecast its future efficiency when parameters change, and see if it is still functional under the new conditions.

Performance of more complicated and advanced queuing systems can be evaluated by Monte Carlo methods. One needs to simulate arrivals of jobs, assignment of servers, and service times and to keep track of all variables of interest. A sample computer code for such a simulation is given.

Exercises

7.1. Customers arrive at the ATM at the rate of 10 customers per hour and spend 2 minutes, on average, on all the transactions. This system is modeled by the Bernoulli single-server queuing process with 10-second frames. Write the transition probability matrix for the number of customers at the ATM at the end of each frame.

7.2. Consider Bernoulli single-server queuing process with an arrival rate of 2 jobs per minute, a service rate of 4 jobs per minute, frames of 0.2 minutes, and a capacity limited by 2 jobs. Compute the steady-state distribution of the number of jobs in the system.

7.3. Performance of a car wash center is modeled by the Bernoulli single-server queuing process with 2-minute frames. The cars arrive every 10 minutes, on the average. The average service time is 6 minutes. Capacity is unlimited. If there are no cars at the center at 10 am, compute the probability that one car will be washed and another car will be waiting at 10:04 am.

7.4. Masha has a telephone with two lines that allows her to talk with a friend and have at most one other friend on hold. On the average, she gets 10 calls every hour, and an average conversation takes 2 minutes. Assuming a single-server limited-capacity Bernoulli queuing process with 1-minute frames, compute the fraction of time Masha spends using her telephone.

7.5. A customer service representative can work with one person at a time and have at most one other customer waiting. Compute the steady-state distribution of the number of customers in this queuing system at any time, assuming that customers arrive according to a Binomial counting process with 3-minute frames and the average interarrival time of 10 minutes, and the average service takes 15 minutes.

7.6. Performance of a telephone with 2 lines is modeled by the Bernoulli single-server queuing process with limited capacity $(C = 2)$. If both lines of a telephone are busy, the new callers receive a busy signal and cannot enter the queue. On the average, there are 5 calls per hour, and the average call takes 20 minutes. Compute the steady-state probabilities using four-minute frames.

7.7. Jobs arrive at the server at the rate of 8 jobs per hour. The service takes 3 minutes, on the average. This system is modeled by the Bernoulli single-server queuing process with 5-second frames and capacity limited by 3 jobs. Write the transition probability matrix for the number of jobs in the system at the end of each frame.

7.8. For an M/M/1 queuing process with the arrival rate of 5 min^{-1} and the average service time of 4 seconds, compute

(a) the proportion of time when there are exactly 2 customers in the system;

(b) the expected response time (the expected time from the arrival till the departure).

7.9. Jobs are sent to a printer at random times, according to a Poisson process of arrivals, with a rate of 12 jobs per hour. The time it takes to print a job is an Exponential random variable, independent of the arrival time, with the average of 2 minutes per job.

(a) A job is sent to a printer at noon. When is it expected to be printed?

(b) How often does the total number of jobs in a queue and currently being printed exceed 2?

7.10. A vending machine is modeled as an M/M/1 queue with the arrival rate of 20 customers per hour and the average service time of 2 minutes.

(a) A customer arrives at 8 pm. What is the expected waiting time?

(b) What is the probability that nobody is using the vending machine at 3 pm?

7.11. Customers come to a teller's window according to a Poisson process with a rate of 10 customers every 25 minutes. Service times are Exponential. The average service takes 2 minutes. Compute

  (a) the average number of customers in the system and the average number of customers waiting in a line;

  (b) the fraction of time when the teller is busy with a customer;

  (c) the fraction of time when the teller is busy and at least five other customers are waiting in a line.

7.12. For an M/M/1 queuing system with the average interarrival time of 5 minutes and the average service time of 3 minutes, compute

  (a) the expected response time;

  (b) the fraction of time when there are fewer than 2 jobs in the system;

  (c) the fraction of customers who have to wait before their service starts.

7.13. Jobs arrive at the service facility according to a Poisson process with the average interarrival time of 4.5 minutes. A typical job spends a Gamma distributed time with parameters $\alpha = 12$, $\lambda = 5$ min^{-1} in the system and leaves.

  (a) Compute the average number of jobs in the system at any time.

  (b) Suppose that only 20 jobs arrived during the last three hours. Is this an evidence that the expected interarrival time has increased?

7.14. Trucks arrive at a weigh station according to a Poisson process with the average rate of 1 truck every 10 minutes. Inspection time is Exponential with the average of 3 minutes. When a truck is on scale, the other arrived trucks stay in a line waiting for their turn. Compute

  (a) the expected number of trucks in the line at any time;

  (b) the proportion of time when the weigh station is empty;

  (c) the expected time each truck spends at the station, from arrival till departure.

7.15. Consider a hot-line telephone that has no second line. When the telephone is busy, the new callers get a busy signal. People call at the average rate of 2 calls per minute. The average duration of a telephone conversation is 1 minute. The system behaves like a Bernoulli single-server queuing process with a frame size of 1 second.

  (a) Compute the steady-state distribution for the number of concurrent jobs.

  (b) What is the probability that more than 150 people attempted to call this number between 2 pm and 3 pm?

7.16. On the average, every 6 minutes a customer arrives at an M/M/k queuing system, spends an average of 20 minutes there, and departs. What is the mean number of customers in the system at any given time?

7.17. Verify the Little's Law for the M/M/1 queuing system and for its components – waiting and service.

7.18. A metered parking lot with two parking spaces represents a Bernoulli two-server queuing system with capacity limited by two cars and 30-second frames. Cars arrive at the rate of one car every 4 minutes. Each car is parked for 5 minutes, on the average.

(a) Compute the transition probability matrix for the number of parked cars.

(b) Find the steady-state distribution for the number of parked cars.

(c) What fraction of the time are both parking spaces vacant?

(d) What fraction of arriving cars will not be able to park?

(e) Every 2 minutes of parking costs 25 cents. Assuming that the drivers use all the parking time they pay for, what revenue is the owner of this parking lot expected to get every 24 hours?

7.19. (*This exercise may require a computer or at least a calculator.*) A walk-in hairdressing saloon has two hairdressers and two extra chairs for waiting customers. We model this system as a Bernoulli queuing process with two servers, 1-minute frames, capacity limited by 4 customers, arrival rate of 3 customers per hour, and the average service time of 45 minutes, not including the waiting time.

(a) Compute the transition probability matrix for the number of customers in the saloon at any time and find the steady-state distribution.

(b) Use this steady-state distribution to compute the expected number of customers in the saloon at any time, the expected number of customers waiting for a hairdresser, and the fraction of customers who found all the available seats already occupied.

(c) Each hairdresser comes for an eight-hour working day. How does it split between their working time and their resting time?

(d) How will the performance characteristics in (b,c) change if they simply put two more chairs into the waiting area?

7.20. Two tellers are now on duty in a bank from Exercise 7.11, and they work as an M/M/2 queuing system with the arrival rate of 0.4 min^{-1} and the mean service time of 2 min.

(a) Compute the steady state probabilities $\pi_0, ..., \pi_{10}$ and use them to approximate the expected number of customers in the bank as

$$\mathbf{E}(X) \approx \sum_{x=0}^{10} x\pi_x.$$

(b) Use the Little's Law to find the expected response time.

(c) From this, derive the expected waiting time and the expected number of waiting customers.

(d*) Derive the exact expected number of customers $\mathbf{E}(X)$ without the approximation in (a) and use it to recalculate the answers in (b) and (c).

Hint: $\mathbf{E}(X)$ is computed as a series; relate it to the expected number of jobs in the M/M/1 system.

7.21. A toll area on a highway has three toll booths and works as an M/M/3 queuing system. On the average, cars arrive at the rate of one car every 5 seconds, and it takes 12 seconds to pay the toll, not including the waiting time. Compute the fraction of time when there are ten or more cars waiting in the line.

7.22. Sports fans tune to a local sports talk radio station according to a Poisson process with the rate of three fans every two minutes and listen it for an Exponential amount of time with the average of 20 minutes.

(a) What queuing system is the most appropriate for this situation?

(b) Compute the expected number of concurrent listeners at any time.

(b) Find the fraction of time when 40 or more fans are tuned to this station.

7.23. Internet users visit a certain web site according to an M/M/∞ queuing system with the arrival rate of 2 min^{-1} and the expected time of 5 minutes spent at the site. Compute

(a) the expected number of visitors of the web site at any time;

(b) the fraction of time when nobody is browsing this web site.

7.24. (MINI-PROJECT) Messages arrive at an electronic mail server according to a Poisson process with the average frequency of 5 messages per minute. The server can process only one message at a time, and messages are processed on a "first come – first serve" basis. It takes an Exponential amount of time U to process any text message, *plus* an Exponential amount of time V, independent of U, to process attachments (if there are any), with $\mathbf{E}(U) = 2$ seconds and $\mathbf{E}(V) = 7$ seconds. Forty percent of messages contain attachments. Estimate the expected response time of this server by the Monte Carlo method.

(Notice that because of the attachments, the overall service times are not Exponential, so the system is not M/M/1.).

7.25. (MINI-PROJECT) A doctor scheduled her patients to arrive at equal 15-minute intervals. Patients are then served in the order of their arrivals, and each of them needs a Gamma time with the doctor that has parameters $\alpha = 4$ and $\lambda = 0.3$ min^{-1}. Use Monte Carlo simulations to estimate

(a) the expected response time;

(b) the expected waiting time;

(c) the probability that a patient has to wait before seeing the doctor.

(*This system is not M/M/1 because the interarrival times are not random.*).

7.26. (COMPUTER PROJECT) A multiserver system (computer lab, customer service, telephone company) consists of $n = 4$ servers (computers, customer service representatives, telephone cables). Every server is able to process any job, but some of them work faster than the others. The service times are distributed according to the table.

Server	Distribution	Parameters
I	Gamma	$\alpha = 7, \lambda = 3 \text{ min}^{-1}$
II	Gamma	$\alpha = 5, \lambda = 2 \text{ min}^{-1}$
III	Exponential	$\lambda = 0.3 \text{ min}^{-1}$
IV	Uniform	$a = 4 \text{ min}, b = 9 \text{ min}$

The jobs (customers, telephone calls) arrive to the system at random times, independently of each other, according to a Poisson process. The average interarrival time is 2 minutes. If a job arrives, and there are free servers available, then the job is *equally likely* to be processed by any of the available servers. If no servers are available at the time of arrival, the job enters a queue. After waiting for 6 minutes, if the service has not started, the job leaves the system. The system works 10 hours a day, from 8 am till 6 pm.

Run at least 1000 Monte Carlo simulations and estimate the following quantities:

(a) the expected waiting time for a randomly selected job;

(b) the expected response time;

(c) the expected length of a queue (excluding the jobs receiving service), when a new job arrives;

(d) the expected *maximum* waiting time during a 10-hour day;

(e) the expected *maximum* length of a queue during a 10-hour day;

(f) the probability that at least one server is available, when a job arrives;

(g) the probability that at least two servers are available, when a job arrives;

(h) the expected number of jobs processed by each server;

(i) the expected time each server is idle during the day;

(j) the expected number of jobs still remaining in the system at 6:03 pm;

(k) the expected percentage of jobs that left the queue prematurely.

7.27. (COMPUTER PROJECT) An IT support help desk represents a queuing system with five assistants taking calls from customers. The calls occur according to a Poisson process with the average rate of one call every 45 seconds. The service times for the 1st, 2nd, 3rd, 4th, and 5th assistants are all Exponential random variables with parameters $\lambda_1 = 0.1$, $\lambda_2 = 0.2$, $\lambda_3 = 0.3$, $\lambda_4 = 0.4$, and $\lambda_5 = 0.5 \text{ min}^{-1}$, respectively (the jth help desk assistant has $\lambda_k = k/10$ min^{-1}). Besides the customers who are being assisted, up to ten other customers can be placed on hold. At times when this capacity is reached, the new callers receive a busy signal.

Use the Monte Carlo methods to estimate the following performance characteristics,

(a) the fraction of customers who receive a busy signal;

(b) the expected response time;

(c) the average waiting time;

(d) the portion of customers served by each help desk assistant;

(e) all of the above if the 6th help desk assistant has been hired to help the first five, and $\lambda_6 = 0.6$. How has this assistant, the fastest on the team, improved performance of the queuing system?

Questions (b-d) refer to those customers who got into the system and did not receive a busy signal.

Part III

Statistics

Chapter 8

Introduction to Statistics

The first seven chapters of this book taught us to analyze problems and systems involving uncertainty, to find probabilities, expectations, and other characteristics for a variety of situations, and to produce forecasts that may lead to important decisions.

What was given to us in all these problems? Ultimately, *we needed to know the distribution and its parameters*, in order to compute probabilities or at least to estimate them by means of Monte Carlo. Often the distribution may not be given, and we learned how to fit the suitable model, say, Binomial, Exponential, or Poisson, given the type of variables we deal with. In any case, parameters of the fitted distribution had to be reported to us explicitly, or they had to follow directly from the problem.

This, however, is rarely the case in practice. Only sometimes the situation may be under our control, where, for example, produced items have predetermined specifications, and therefore, one knows parameters of their distribution.

Much more often *parameters are not known*. Then, how can one apply the knowledge of Chapters 1–7 and compute probabilities? The answer is simple: *we need to collect data*. A properly collected sample of data can provide rather sufficient information about parameters of the observed system. In the next sections and chapters, we learn how to use this sample

- to visualize data, understand the patterns, and make quick statements about the system's behavior;

- to characterize this behavior in simple terms and quantities;

- to estimate the distribution parameters;

- to assess reliability of our estimates;

– to test statements about parameters and the entire system;

– to understand relations among variables;

– to fit suitable models and use them to make forecasts.

8.1   Population and sample, parameters and statistics

Data collection is a crucially important step in Statistics. We use the collected and observed *sample* to make statements about a much larger set — the *population*.

DEFINITION 8.1

> A **population** consists of all units of interest. Any numerical characteristic of a population is a **parameter**. A **sample** consists of observed units collected from the population. It is used to make statements about the population. Any function of a sample is called **statistic**.

In real problems, we would like to make statements about the population. To compute probabilities, expectations, and make optimal decisions under uncertainty, we need to know the population *parameters*. However, the only way to know these parameters is to measure the entire population, i.e., to conduct a *census*.

Instead of a census, we may collect data in a form of a random sample from a population (Figure 8.1). This is our data. We can measure them, perform calculations, and *estimate* the unknown parameters of the population up to a certain *measurable* degree of accuracy.

$$
\text{NOTATION} \quad \left\| \begin{array}{rcl} \theta &=& \text{population parameter} \\ \widehat{\theta} &=& \text{its estimator, obtained from a sample} \end{array} \right\|
$$

Example 8.1 (CUSTOMER SATISFACTION). For example, even if 80% of all users are satisfied with their internet connection, it does not mean that exactly 8 out of 10 customers in your observed sample are satisfied. As we can see from Table A2 in the Appendix, with probability 0.0328, only five out of ten sampled customers are satisfied. In other words, there is a 3% chance for a random sample to suggest that contrary to the claimed population parameter, no more than 50% of users are satisfied.   ◊

This example shows that a sample may sometimes give a rather misleading information about the population although this happens with a low probability. *Sampling errors cannot be excluded.*

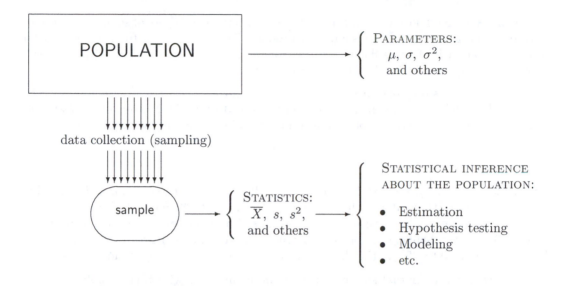

FIGURE 8.1: *Population parameters and sample statistics.*

Sampling and non-sampling errors

Sampling and non-sampling errors refer to any discrepancy between a collected sample and a whole population.

Sampling errors are caused by the mere fact that only a sample, a portion of a population, is observed. For most of reasonable statistical procedures, sampling errors decrease (and converge to zero) as the sample size increases.

Non-sampling errors are caused by inappropriate sampling schemes or wrong statistical techniques. Often no wise statistical techniques can rescue a poorly collected sample of data.

Look at some examples of wrong sampling practices.

Example 8.2 (SAMPLING FROM A WRONG POPULATION). To evaluate the work of a Windows help desk, a survey of *social science students of some university* is conducted. This sample poorly represents the whole population of *all Windows users*. For example, computer science students and especially computer professionals may have a totally different opinion about the Windows help desk. $\Diamond$

Example 8.3 (DEPENDENT OBSERVATIONS). Comparing two brands of notebooks, a senior manager asks all employees of her group to state which notebook they like and generalizes the obtained responses to conclude which notebook is better. Again, these employees are not randomly selected from the population of all users of these notebooks. Also, their opinions are likely to be *dependent*. Working together, these people often communicate, and their points of view affect each other. Dependent observations do not necessarily cause non-sampling errors, if they are handled properly. The fact is, in such cases, we cannot assume independence. $\Diamond$

Example 8.4 (NOT EQUALLY LIKELY). A survey among passengers of some airline is conducted in the following way. A sample of random flights is selected from a list, and ten passengers on each of these flights are also randomly chosen. Each sampled passenger is asked to fill a questionnaire. Is this a representative sample?

Suppose Mr. X flies only once a year whereas Ms. Y has business trips twice a month. Obviously, Ms. Y has a much higher chance to be sampled than Mr. X. Unequal probabilities have to be taken into account, otherwise a non-sampling error will inevitably occur.   ◊

Example 8.5 (PRESIDENTIAL ELECTION OF 1936). A popular weekly magazine The Literary Digest correctly predicted the winners of 1920, 1924, 1928, and 1932 U.S. Presidential Elections. However, it failed to do so in 1936! Based on a survey of ten million people, it predicted an overwhelming victory of Governor Alfred Landon. Instead, Franklin Delano Roosevelt received 98.49% of the electoral vote, won 46 out of 48 states, and was re-elected.

So, what went wrong in that survey? At least two main issues with their sampling practice caused this prediction error. First, the sample was based on the population of subscribers of The Literary Digest that was dominated by Republicans. Second, responses were voluntary, and 77% of mailed questionnaires were not returned, introducing further bias. These are classical examples of non-sampling errors.   ◊

In this book, we focus on *simple random sampling*, which is one way to avoid non-sampling errors.

DEFINITION 8.2 ————————————————————————————

> **Simple random sampling** is a sampling design where units are collected from the entire population independently of each other, all being equally likely to be sampled.

Observations collected by means of a simple random sampling design are **iid** (independent, identically distributed) random variables.

Example 8.6. To evaluate its customers' satisfaction, a bank makes a list of all the accounts. A Monte Carlo method is used to choose a random number between 1 and N, where N is the total number of bank accounts. Say, we generate a Uniform$(0,N)$ variable X and sample an account number $\lceil X \rceil$ from the list. Similarly, we choose the second account, uniformly distributed among the remaining $N - 1$ accounts, etc., until we get a sample of the desired size n. This is a simple random sample.   ◊

Obtaining a good, representative random sample is rather important in Statistics. Although we have only a portion of the population in our hands, a rigorous sampling design followed by a suitable statistical inference allows to estimate parameters and make statements with a certain measurable degree of confidence.

8.2 Descriptive statistics

Suppose a good random sample

$$\mathcal{S} = (X_1, X_2, \ldots, X_n)$$

has been collected. For example, to evaluate effectiveness of a processor for a certain type of tasks, we recorded the CPU time in seconds for $n = 30$ randomly chosen jobs (data set CPU),

$$
\begin{array}{cccccccccc}
70 & 36 & 43 & 69 & 82 & 48 & 34 & 62 & 35 & 15 \\
59 & 139 & 46 & 37 & 42 & 30 & 55 & 56 & 36 & 82 \\
38 & 89 & 54 & 25 & 35 & 24 & 22 & 9 & 56 & 19
\end{array}
\tag{8.1}
$$

What information do we get from this collection of numbers?

We know that X, the CPU time of a random job, is a random variable, and its value does not have to be among the observed thirty. We'll use the collected data to describe the distribution of X.

Simple **descriptive statistics** measuring the location, spread, variability, and other characteristics can be computed immediately. In this section, we discuss the following statistics,

- **mean**, measuring the average value of a sample;

- **median**, measuring the central value;

- **quantiles** and **quartiles**, showing where certain portions of a sample are located;

- **variance**, **standard deviation**, and **interquartile range**, measuring variability and spread of data.

Each statistic is a random variable because it is computed from random data. It has a so-called *sampling distribution*.

Each statistic estimates the corresponding population parameter and adds certain information about the distribution of X, the variable of interest.

We used similar methods in Section 5.3.2, where we estimated parameters from Monte Carlo samples obtained by computer simulations. Here we estimate parameters and make conclusions based on real, not simulated, data.

8.2.1 Mean

Sample mean $\overline{X}$ estimates the population mean $\mu = \mathbf{E}(X)$.

DEFINITION 8.3

Sample mean $\overline{X}$ is the arithmetic average,

$$\overline{X} = \frac{X_1 + \ldots + X_n}{n}$$

Naturally, being the average of sampled observations, $\overline{X}$ estimates the average value of the whole distribution of X. Computed from random data, $\overline{X}$ does not necessarily equal μ; however, we would expect it to converge to μ when a large sample is collected.

Sample means possess a number of good properties. They are *unbiased, consistent,* and *asymptotically Normal.*

Remark: This is true if the population has finite mean and variance, which is the case for almost all the distributions in this book (see, however, Example 3.20 on p. 62).

Unbiasedness

DEFINITION 8.4 —————

> An estimator $\widehat{\theta}$ is **unbiased** for a parameter θ if its expectation equals the parameter,
> $$\mathbf{E}(\widehat{\theta}) = \theta$$
> for all possible values of θ.
>
> **Bias** of $\widehat{\theta}$ is defined as $\mathrm{Bias}(\widehat{\theta}) = \mathbf{E}(\widehat{\theta} - \theta)$.

Unbiasedness means that in a long run, collecting a large number of samples and computing $\widehat{\theta}$ from each of them, on the average we hit the unknown parameter θ exactly. In other words, in a long run, unbiased estimators neither underestimate nor overestimate the parameter.

Sample mean estimates μ unbiasedly because its expectation is

$$\mathbf{E}(\overline{X}) = \mathbf{E}\left(\frac{X_1 + \ldots + X_n}{n}\right) = \frac{\mathbf{E}X_1 + \ldots + \mathbf{E}X_n}{n} = \frac{n\mu}{n} = \mu.$$

Consistency

DEFINITION 8.5 —————

> An estimator $\widehat{\theta}$ is **consistent** for a parameter θ if the probability of its sampling error of any magnitude converges to 0 as the sample size increases to infinity. Stating it rigorously,
>
> $$\boldsymbol{P}\left\{|\widehat{\theta} - \theta| > \varepsilon\right\} \to 0 \text{ as } n \to \infty$$
>
> for any $\varepsilon > 0$. That is, when we estimate θ from a large sample, the estimation error $|\widehat{\theta} - \theta|$ is unlikely to exceed ε, and it does it with smaller and smaller probabilities as we increase the sample size.

Consistency of $\overline{X}$ follows directly from Chebyshev's inequality on p. 54.

To use this inequality, we find the variance of $\overline{X}$,

$$\text{Var}(\overline{X}) = \text{Var}\left(\frac{X_1 + \ldots + X_n}{n}\right) = \frac{\text{Var}X_1 + \ldots + \text{Var}X_n}{n^2} = \frac{n\sigma^2}{n^2} = \frac{\sigma^2}{n}. \tag{8.2}$$

Then, using Chebyshev's inequality for the random variable $\overline{X}$, we get

$$\boldsymbol{P}\left\{|\overline{X} - \mu| > \varepsilon\right\} \leq \frac{\text{Var}(\overline{X})}{\varepsilon^2} = \frac{\sigma^2/n}{\varepsilon^2} \to 0,$$

as $n \to \infty$.

Thus, a sample mean is *consistent*. Its sampling error will be small with a higher and higher probability, as we collect larger and larger samples.

Asymptotic Normality

By the Central Limit Theorem, the sum of observations, and therefore, the sample mean have approximately Normal distribution if they are computed from a large sample. That is, the distribution of

$$Z = \frac{\overline{X} - \mathbf{E}\overline{X}}{\text{Std}\overline{X}} = \frac{\overline{X} - \mu}{\sigma\sqrt{n}}$$

converges to Standard Normal as $n \to \infty$. This property is called **Asymptotic Normality**.

Example 8.7 (CPU TIMES). Looking at the CPU data on p. 217 (data set CPU), we estimate the average (expected) CPU time μ by

$$\overline{X} = \frac{70 + 36 + \ldots + 56 + 19}{30} = \frac{1447}{30} = 48.2333.$$

We may conclude that the mean CPU time of *all* the jobs is "near" 48.2333 seconds.

$\diamond$

NOTATION			
μ	$=$	population mean	
$\overline{X}$	$=$	sample mean, estimator of μ	
σ	$=$	population standard deviation	
s	$=$	sample standard deviation, estimator of σ	
σ^2	$=$	population variance	
s^2	$=$	sample variance, estimator of σ	

8.2.2 Median

One disadvantage of a sample mean is its *sensitivity to extreme observations*. For example, if the first job in our sample is unusually heavy, and it takes 30 minutes to get processed instead of 70 seconds, this one extremely large observation shifts the sample mean from 48.2333 sec to 105.9 sec. Can we call such an estimator "reliable"?

Another simple measure of location is a *sample median*, which estimates the *population median*. It is much less sensitive than the sample mean.

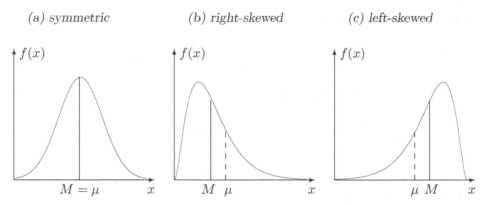

FIGURE 8.2: *A mean μ and a median M for distributions of different shapes.*

DEFINITION 8.6 ────────

Median means a "central" value.

Sample median $\widehat{M}$ is a number that is exceeded by at most a half of observations and is preceded by at most a half of observations.

Population median M is a number that is exceeded with probability no greater than 0.5 and is preceded with probability no greater than 0.5. That is, M is such that

$$\begin{cases} \boldsymbol{P}\{X > M\} & \leq & 0.5 \\ \boldsymbol{P}\{X < M\} & \leq & 0.5 \end{cases}$$

Understanding the shape of a distribution

Comparing the mean μ and the median M, one can tell whether the distribution of X is right-skewed, left-skewed, or symmetric (Figure 8.2):

$$\begin{array}{lcl} \text{Symmetric distribution} & \Rightarrow & M = \mu \\ \text{Right-skewed distribution} & \Rightarrow & M < \mu \\ \text{Left-skewed distribution} & \Rightarrow & M > \mu \end{array}$$

Computation of a population median

For continuous distributions, computing a population median reduces to solving one equation:

$$\begin{cases} \boldsymbol{P}\{X > M\} & = & 1 - F(M) & \leq & 0.5 \\ \boldsymbol{P}\{X < M\} & = & F(M) & \leq & 0.5 \end{cases} \Rightarrow \quad F(M) = 0.5.$$

Example 8.8 (UNIFORM, FIGURE 8.3A). Uniform(a, b) distribution has a cdf

$$F(x) = \frac{x - a}{b - a} \quad \text{for } a < x < b.$$

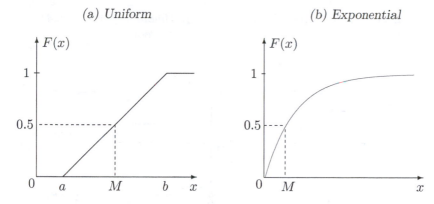

FIGURE 8.3: *Computing medians of continuous distributions.*

Solving the equation $F(M) = (M - a)/(b - a) = 0.5$, we get

$$M = \frac{a+b}{2}.$$

It coincides with the mean because the Uniform distribution is symmetric. ◊

Example 8.9 (EXPONENTIAL, FIGURE 8.3B). Exponential(λ) distribution has a cdf

$$F(x) = 1 - e^{-\lambda x} \text{ for } x > 0.$$

Solving the equation $F(M) = 1 - e^{-\lambda x} = 0.5$, we get

$$M = \frac{\ln 2}{\lambda} = \frac{0.6931}{\lambda}.$$

We know that $\mu = 1/\lambda$ for Exponential distribution. Here the median is smaller than the mean because Exponential distribution is right-skewed. ◊

For discrete distributions, equation $F(x) = 0.5$ has either a whole interval of roots or no roots at all (see Figure 8.4).

In the first case, any number in this interval, excluding the ends, is a median. Notice that the median in this case is not unique (Figure 8.4a). Often the middle of this interval is reported as the median.

In the second case, the smallest x with $F(x) \geq 0.5$ is the median. It is the value of x where the cdf jumps over 0.5 (Figure 8.4b).

Example 8.10 (SYMMETRIC BINOMIAL, FIGURE 8.4A). Consider Binomial distribution with $n = 5$ and $p = 0.5$. From Table A2, we see that for all $2 < x < 3$,

$$\begin{cases} P\{X < x\} &= F(2) &= 0.5 \\ P\{X > x\} &= 1 - F(2) &= 0.5 \end{cases}$$

By Definition 8.6, any number of the interval (2,3) is a median.

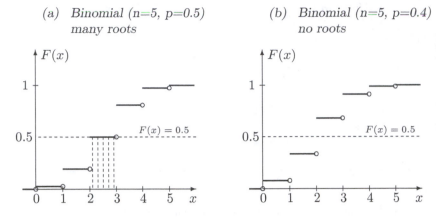

FIGURE 8.4: *Computing medians of discrete distributions.*

This result agrees with our intuition. With $p = 0.5$, successes and failures are equally likely. Pick, for example, $x = 2.4$ in the interval $(2,3)$. Having fewer than 2.4 successes (i.e., at most two) has the same chance as having fewer than 2.4 failures (i.e., at least 3 successes). Therefore, $X < 2.4$ with the same probability as $X > 2.4$, which makes $x = 2.4$ a central value, a median. We can say that $x = 2.4$ (and any other x between 2 and 3) splits the distribution into two equal parts. Then, it is a median. $\Diamond$

Example 8.11 (ASYMMETRIC BINOMIAL, FIGURE 8.4B). For the Binomial distribution with $n = 5$ and $p = 0.4$,

$$F(x) < 0.5 \quad \text{for} \quad x < 2$$
$$F(x) > 0.5 \quad \text{for} \quad x \geq 2$$

but there is no value of x where $F(x) = 0.5$. Then, $M = 2$ is the median.

Seeing a value on either side of $x = 2$ has probability less than 0.5, which makes $x = 2$ a center value. $\Diamond$

Computing sample medians

A sample is always discrete, it consists of a finite number of observations. Then, computing a sample median is similar to the case of discrete distributions.

In simple random sampling, all observations are equally likely, and thus, equal probabilities on each side of a median translate into an equal number of observations.

Again, there are two cases, depending on the sample size n.

Sample median	If n is odd, the $\left(\dfrac{n+1}{2}\right)$-th smallest observation is a median. If n is even, any number between the $\left(\dfrac{n}{2}\right)$-th smallest and the $\left(\dfrac{n+2}{2}\right)$-th smallest observations is a median.

Example 8.12 (MEDIAN CPU TIME). Let's compute the median of $n = 30$ CPU times from the data on p. 217 (data set CPU).

First, order the data,

$$
\begin{array}{cccccccccc}
9 & 15 & 19 & 22 & 24 & 25 & 30 & 34 & 35 & 35 \\
36 & 36 & 37 & 38 & \mathbf{42} & \mathbf{43} & 46 & 48 & 54 & 55 \\
56 & 56 & 59 & 62 & 69 & 70 & 82 & 82 & 89 & 139
\end{array}
\tag{8.3}
$$

Next, since $n = 30$ is even, find $n/2 = 15$-th smallest and $(n + 2)/2 = 16$-th smallest observations. These are 42 and 43. Any number between them is a sample median (typically reported as 42.5). ◊

We see why medians are not sensitive to extreme observations. If in the previous example, the first CPU time happens to be 30 minutes instead of 70 seconds, it does not affect the sample median at all!

Sample medians are easy to compute. In fact, no computations are needed, only the ordering. If you are driving (and only if you find it safe!), here is a simple experiment that you can conduct yourself.

Example 8.13 (MEDIAN SPEED ON A HIGHWAY). How can you measure the median speed of cars on a multilane road without a radar? It's very simple. Adjust your speed so that a half of cars overtake you, and you overtake the other half. Then you are driving with the median speed! ◊

8.2.3 Quantiles, percentiles, and quartiles

Generalizing the notion of a median, we replace 0.5 in Definition 8.6 by some $0 < p < 1$.

DEFINITION 8.7 —————————————————

A p-**quantile** of a population is such a number x that solves equations

$$
\begin{cases}
\boldsymbol{P}\{X < x\} & \leq & p \\
\boldsymbol{P}\{X > x\} & \leq & 1 - p
\end{cases}
$$

A **sample p-quantile** is any number that exceeds at most $100p\%$ of the sample, and is exceeded by at most $100(1 - p)\%$ of the sample.

A γ-**percentile** is (0.01γ)-quantile.

First, second, and third **quartiles** are the 25th, 50th, and 75th percentiles. They split a population or a sample into four equal parts.

A **median** is at the same time a 0.5-quantile, 50th percentile, and 2nd quartile.

NOTATION

$$
\begin{aligned}
q_p &= \text{population } p\text{-quantile} \\
\widehat{q}_p &= \text{sample } p\text{-quantile, estimator of } q_p \\[4pt]
\pi_\gamma &= \text{population } \gamma\text{-percentile} \\
\widehat{\pi}_\gamma &= \text{sample } \gamma\text{-percentile, estimator of } \pi_\gamma \\[4pt]
Q_1, Q_2, Q_3 &= \text{population quartiles} \\
\widehat{Q}_1, \widehat{Q}_2, \widehat{Q}_3 &= \text{sample quartiles, estimators of } Q_1, Q_2, \text{ and } Q_3 \\[4pt]
M &= \text{population median} \\
\widehat{M} &= \text{sample median, estimator of } M
\end{aligned}
$$

Quantiles, quartiles, and percentiles are related as follows.

Quantiles, quartiles, percentiles

$$
q_p = \pi_{100p}
$$
$$
Q_1 = \pi_{25} = q_{1/4} \qquad Q_3 = \pi_{75} = q_{3/4}
$$
$$
M = Q_2 = \pi_{50} = q_{1/2}
$$

Sample statistics are of course in a similar relation.

Computing quantiles is very similar to computing medians.

Example 8.14 (SAMPLE QUARTILES). Let us compute the 1st and 3rd quartiles of CPU times. Again, we look at the ordered sample (data set CPU)

$$
\begin{array}{cccccccccc}
9 & 15 & 19 & 22 & 24 & 25 & 30 & \mathbf{34} & 35 & 35 \\
36 & 36 & 37 & 38 & 42 & 43 & 46 & 48 & 54 & 55 \\
56 & 56 & \mathbf{59} & 62 & 69 & 70 & 82 & 82 & 89 & 139
\end{array}
$$

First quartile $\widehat{Q}_1$. For $p = 0.25$, we find that 25% of the sample equals $np = 7.5$, and 75% of the sample is $n(1-p) = 22.5$ observations. From the ordered sample, we see that only the 8th element, 34, has no more than 7.5 observations to the left and no more than 22.5 observations to the right of it. Hence, $\widehat{Q}_1 = 34$.

Third quartile $\widehat{Q}_3$. Similarly, the third sample quartile is the 23rd smallest element, $\widehat{Q}_3 = 59$.

◊

Example 8.15 (CALCULATING FACTORY WARRANTIES FROM POPULATION PERCENTILES). A computer maker sells extended warranty on the produced computers. It agrees to issue a warranty for x years if it knows that only 10% of computers will fail before the warranty expires. It is known from past experience that lifetimes of these computers have Gamma distribution with $\alpha = 60$ and $\lambda = 5$ years^{-1}. Compute x and advise the company on the important decision under uncertainty about possible warranties.

Solution. We just need to find the tenth percentile of the specified Gamma distribution and let

$$
x = \pi_{10}.
$$

As we know from Section 4.3, being a sum of Exponential variables, a Gamma variable is approximately Normal for large $\alpha = 60$. Using (4.12), compute

$$
\begin{aligned}
\mu &= \alpha/\lambda = 12, \\
\sigma &= \sqrt{\alpha/\lambda^2} = 1.55.
\end{aligned}
$$

From Table A4, the 10th percentile of a standardized variable

$$
Z = \frac{X - \mu}{\sigma}
$$

equals (-1.28) (find the probability closest to 0.10 in the table and read the corresponding value of z). Unstandardizing it, we get

$$
x = \mu + (-1.28)\sigma = 12 - (1.28)(1.55) = \underline{10.02}.
$$

Thus, the company can issue a 10-year warranty rather safely.

Remark: Of course, one does not have to use Normal approximation in the last example. A number of computer packages have built-in commands for the exact evaluation of probabilities and quantiles. For example, the 10th percentile of Gamma($\alpha = 60, \lambda = 5$) distribution can be obtained by the command `qgamma(0.10,60,5)` in R and by `gaminv(0.10, 60, 1/5)` in MATLAB. ◇

8.2.4 Variance and standard deviation

Statistics introduced in the previous sections showed where the average value and certain percentages of a population are located. Now we are going to measure *variability* of our variable, how unstable the variable can be, and how much the actual value can differ from its expectation. As a result, we'll be able to assess reliability of our estimates and accuracy of our forecasts.

DEFINITION 8.8 ————

For a sample $(X_1, X_2, \ldots, X_n)$, a **sample variance** is defined as

$$
s^2 = \frac{1}{n-1} \sum_{i=1}^{n} \left(X_i - \overline{X} \right)^2. \tag{8.4}
$$

It measures variability among observations and estimates the population variance $\sigma^2 = \text{Var}(X)$.

Sample standard deviation is a square root of a sample variance,

$$
s = \sqrt{s^2}.
$$

It measures variability in the same units as X and estimates the population standard deviation $\sigma = \text{Std}(X)$.

Both population and sample variances are measured in squared units (in^2, sec^2, \$^{2}, etc.). Therefore, it is convenient to have standard deviations that are comparable with our variable of interest, X.

The formula for s^2 follows the same idea as that for σ^2. It is also the average squared deviation from the mean, this time computed for a sample. Like σ^2, sample variance measures how far the actual values of X are from their average.

Computation

Often it is easier to compute the sample variance using another formula,

$$\textbf{Sample variance} \qquad \boxed{\; s^2 = \dfrac{\displaystyle\sum_{i=1}^{n} X_i^2 - n\overline{X}^2}{n-1}. \;} \qquad (8.5)$$

Remark: Expressions (8.4) and (8.5) are equivalent because

$$\sum \left(X_i - \overline{X}\right)^2 = \sum X_i^2 - 2\overline{X}\sum X_i + \sum \overline{X}^2 = \sum X_i^2 - 2\overline{X}\left(n\overline{X}\right) + n\overline{X}^2 = \sum X_i^2 - n\overline{X}^2.$$

When $X_1, \ldots, X_n$ are integers, but $(X_1 - \overline{X}), \ldots, (X_n - \overline{X})$ are fractions, it may be easier to use (8.5). However, $(X_n - \overline{X})$ are generally smaller in magnitude, and thus, we'd rather use (8.4) if $X_1, \ldots, X_n$ are rather large numbers.

Example 8.16 (CPU TIME, CONTINUED). For the data in (8.1) on p. 217 (data set CPU), we have computed $\overline{X} = 48.2333$. Following Definition 8.8, we can compute the sample variance as

$$s^2 = \frac{(70 - 48.2333)^2 + \ldots + (19 - 48.2333)^2}{30 - 1} = \frac{20,391}{29} = 703.1506 \ (\text{sec}^2).$$

Alternatively, using (8.5),

$$s^2 = \frac{70^2 + \ldots + 19^2 - (30)(48.2333)^2}{30 - 1} = \frac{90,185 - 69,794}{29} = 703.1506 \ (\text{sec}^2).$$

The sample standard deviation is

$$s = \sqrt{703.1506} = 26.1506 \ (\text{sec}^2).$$

We can use these results, for example, as follows. Since $\overline{X}$ and s estimate the population mean and standard deviation, we can make a claim that at least 8/9 of all tasks require less than

$$\overline{X} + 3s = 127.78 \text{ seconds} \qquad (8.6)$$

of CPU time. We used Chebyshev's inequality (3.8) to derive this (also see Exercise 8.3). $\lozenge$

A seemingly strange coefficient $\left(\frac{1}{n-1}\right)$ ensures that s^2 is an **unbiased** estimator of σ^2.

PROOF: Let us prove the unbiasedness of s^2.

<u>Case 1</u>. Suppose for a moment that the population mean $\mu = \mathbf{E}(X) = 0$. Then

$$\mathbf{E}X_i^2 = \operatorname{Var}X_i = \sigma^2,$$

and by (8.2),

$$\mathbf{E}\overline{X}^2 = \mathrm{Var}\overline{X} = \sigma^2/n.$$

Then,

$$\mathbf{E}s^2 = \frac{\mathbf{E}\sum X_i^2 - n\,\mathbf{E}\overline{X}^2}{n-1} = \frac{n\sigma^2 - \sigma^2}{n-1} = \sigma^2.$$

<u>Case 2</u>. If $\mu \neq 0$, consider auxiliary variables $Y_i = X_i - \mu$. Variances don't depend on constant shifts (see (3.7), p. 53), therefore, Y_i have the same variance as X_i. Their sample variances are equal too,

$$s_Y^2 = \frac{\sum (Y_i - \overline{Y})^2}{n-1} = \frac{\sum (X_i + \mu - (\overline{X} - \mu))^2}{n-1} = \frac{\sum (X_i - \overline{X})^2}{n-1} = s_X^2.$$

Since $\mathbf{E}(Y_i) = 0$, Case 1 applies to these variables. Thus,

$$\mathbf{E}(s_X^2) = \mathbf{E}(s_Y^2) = \sigma_Y^2 = \sigma_X^2.$$

$\square$

Similarly to $\overline{X}$, it can be shown that under rather mild assumptions, sample variance and sample standard deviation are **consistent** and **asymptotically Normal**.

8.2.5 Standard errors of estimates

Besides the population variances and standard deviations, it is helpful to evaluate variability of computed statistics and especially parameter estimators.

DEFINITION 8.9 ———

Standard error of an estimator $\widehat{\theta}$ is its standard deviation, $\sigma(\widehat{\theta}) = \mathrm{Std}(\widehat{\theta})$.

$$\underline{\text{NOTATION}} \left\| \begin{array}{rcl} \sigma(\widehat{\theta}) & = & \text{standard error of estimator } \widehat{\theta} \text{ of parameter } \theta \\ s(\widehat{\theta}) & = & \text{estimated standard error } = \widehat{\sigma}(\widehat{\theta}) \end{array} \right\|$$

As a measure of variability, standard errors show precision and reliability of estimators. They show how much estimators of the same parameter θ can vary if they are computed from different samples. Ideally, we would like to deal with unbiased or nearly unbiased estimators that have low standard error (Figure 8.5).

Example 8.17 (STANDARD ERROR OF A SAMPLE MEAN). Parameter $\theta = \mu$, the population mean, is estimated from the sample of size n by the sample mean $\widehat{\theta} = \overline{X}$. We already know that the standard error of this estimator is $\sigma(\overline{X}) = \sigma/\sqrt{n}$, and it can be estimated by $s(\overline{X}) = s/\sqrt{n}$. $\diamondsuit$

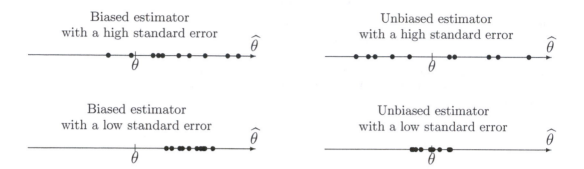

FIGURE 8.5: *Bias and standard error of an estimator. In each case, the dots represent parameter estimators $\widehat{\theta}$ obtained from 10 different random samples.*

8.2.6 Interquartile range

Sample mean, variance, and standard deviation are *sensitive to outliers*. If an extreme observation (an **outlier**) erroneously appears in our data set, it can rather significantly affect the values of $\overline{X}$ and s^2.

In practice, outliers may be a real problem that is hard to avoid. To detect and identify outliers, we need measures of variability that are not very sensitive to them.

One such measure is an interquartile range.

DEFINITION 8.10 ————————————————

> An **interquartile range** is defined as the difference between the first and the third quartiles,
> $$IQR = Q_3 - Q_1.$$
>
> It measures variability of data. Not much affected by outliers, it is often used to detect them. IQR is estimated by the *sample interquartile range*
>
> $$\widehat{IQR} = \widehat{Q}_3 - \widehat{Q}_1.$$

Detection of outliers

A "rule of thumb" for identifying outliers is the rule of **1.5(IQR)**. Measure $1.5(\widehat{Q}_3 - \widehat{Q}_1)$ down from the first quartile and up from the third quartile. All the data points observed outside of this interval are assumed suspiciously far. They are the first candidates to be handled as outliers.

Remark: The rule of $1.5(IQR)$ originally comes from the assumption that the data are nearly normally distributed. If this is a valid assumption, then 99.3% of the population should appear within 1.5 interquartile ranges from quartiles (Exercise 8.4). It is so unlikely to see a value of X outside of this range that such an observation may be treated as an outlier.

Example 8.18 (ANY OUTLYING CPU TIMES?). Can we suspect that sample (8.1) (data set CPU) has outliers? Compute

$$\widehat{IQR} = \widehat{Q}_3 - \widehat{Q}_1 = 59 - 34 = 25$$

and measure 1.5 interquartile ranges from each quartile:

$$
\begin{aligned}
\widehat{Q}_1 - 1.5(\widehat{IQR}) &= 34 - 37.5 &= -3.5; \\
\widehat{Q}_3 + 1.5(\widehat{IQR}) &= 59 + 37.5 &= 96.5.
\end{aligned}
$$

In our data, one task took 139 seconds, which is well outside of the interval $[-3.5, 96.5]$. This may be an outlier. $\Diamond$

Handling of outliers

What should we do if the 1.5(IQR) rule suggests possible outliers in the sample?

Many people simply delete suspicious observations, keeping in mind that one outlier can significantly affect sample mean and standard deviation and therefore spoil our statistical analysis. However, deleting them immediately may not be the best idea.

It is rather important to track the history of outliers and understand the reason they appeared in the data set. There may be a pattern that a practitioner would want to be aware of. It may be a new trend that was not known before. Or, it may be an observation from a very special part of the population. Sometimes important phenomena are discovered by looking at outliers.

If it is confirmed that a suspected observation entered the data set by a mere mistake, it can be deleted.

8.3 Graphical statistics

Despite highly developed theory and methodology of Statistics, when it comes to analysis of real data, experienced statisticians will often follow a very simple advice:

> **Before you do anything with a data set,**
> **look at it!**

A quick look at a sample may clearly suggest

- a probability model, i.e., a family of distributions to be used;
- statistical methods suitable for the given data;

– presence or absence of outliers;

– presence or absence of heterogeneity;

– existence of time trends and other patterns;

– relation between two or several variables.

There is a number of simple and advanced ways to *visualize* data. This section introduces

- histograms,

- stem-and-leaf plots,

- boxplots,

- time plots, and

- scatter plots.

Each graphical method serves a certain purpose and discovers certain information about data.

8.3.1   Histogram

A **histogram** shows the shape of a pmf or a pdf of data, checks for homogeneity, and suggests possible outliers. To construct a histogram, we split the range of data into equal intervals, "bins," and count how many observations fall into each bin.

A **frequency histogram** consists of columns, one for each bin, whose height is determined by the *number* of observations in the bin.

A **relative frequency histogram** has the same shape but a different vertical scale. Its column heights represent the *proportion* of all data that appeared in each bin.

The sample of CPU times on p. 217 (data set CPU) stretches from 9 to 139 seconds. Choosing intervals [0,14], [14,28], [28,42], . . . as bins, we count

$$
\begin{array}{llllll}
1 & \text{observation} & \text{between} & 0 & \text{and} & 14 \\
5 & \text{observations} & " & 14 & " & 28 \\
9 & " & " & 28 & " & 42 \\
7 & " & " & 42 & " & 56 \\
4 & " & " & 56 & " & 70 \\
\end{array}
\qquad (8.7)
$$

....

Using this for column heights, a (frequency) histogram of CPU times is then constructed (Figure 8.6a). A relative frequency histogram (Figure 8.6b) is only different in the vertical scale. Each count is now divided by the sample size $n = 30$.

What information can we draw from these histograms?

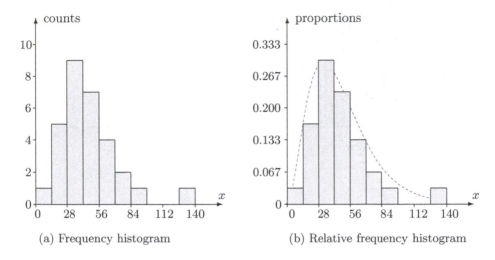

(a) Frequency histogram (b) Relative frequency histogram

FIGURE 8.6: *Histograms of CPU data.*

> *Histograms have a shape similar to the pmf or pdf of data,*
> *especially in large samples.*

Remark: To understand the last statement, let's imagine for a moment that the data are integers and all columns in a relative frequency histogram have a unit width. Then the height of a column above a number x equals the proportion of x's in a sample, and in large samples it approximates the probability $P(x)$, the pmf (probability is a long-run proportion, Chapter 2).

For continuous distributions, the height of a unit-width column equals its area. Probabilities are areas under the density curve (Chapter 4). Thus, we get approximately the same result either computing sample proportions or integrating the curve that connects the column tops on a relative frequency histogram.

Now, if columns have a non-unit (but still, equal) width, it will only change the horizontal scale but will not alter the shape of a histogram. In each case, this shape will be similar to the graph of the population pmf or pdf.

The following information can be drawn from the histograms shown in Figure 8.6:

- Continuous distribution of CPU times is not symmetric; it is right-skewed as we see 5 columns to the right of the highest column and only 2 columns to the left.

- Among continuous distributions in Chapter 4, only Gamma distribution has a similar shape; a Gamma family seems appropriate for CPU times. We sketched a suitable Gamma pdf with a dashed curve in Figure 8.6b. It is rescaled because our columns don't have a unit width.

- The time of 139 seconds stands alone suggesting that it is in fact an outlier.

- There is no indication of heterogeneity; all data points except $x = 139$ form a rather homogeneous group that fits the sketched Gamma curve.

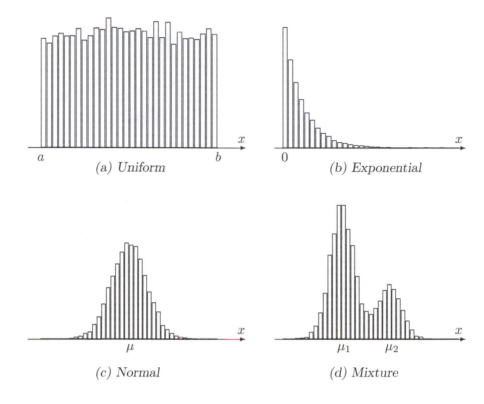

FIGURE 8.7: *Histograms of various samples.*

How else may histograms look like?

We saw a rather nice fit of a Gamma distribution in Figure 8.6b, except for one outlier. What other shapes of histograms can we see, and what other conclusions about the population can we make?

Certainly, histograms come in all shapes and sizes. Four examples are shown in Figure 8.7.

In Figure 8.7a, the distribution is almost symmetric, and columns have almost the same height. Slight differences can be attributed to the randomness of our sample, i.e., the *sampling error*. The histogram suggest a Uniform or Discrete Uniform distribution between a and b.

In Figure 8.7b, the distribution is heavily right-skewed, column heights decrease exponentially fast. This sample should come from an Exponential distribution, if variables are continuous, or from Geometric, if they are discrete.

In Figure 8.7c, the distribution is symmetric, with very quickly vanishing "tails." Its bell shape reminds a Normal density that, as we know from Section 4.2.4, decays at a rate of $\sim e^{-cx^2}$. We can locate the center μ of a histogram and conclude that this sample is likely to come from a Normal distribution with a mean close to μ.

Figure 8.7d presents a rather interesting case that deserves special attention.

Mixtures

Let us look at Figure 8.7d . We have not seen a distribution with two "humps" in the previous chapters. Most likely, here we deal with a **mixture of distributions**. Each observation comes from distribution F_1 with some probability p_1 and comes from distribution F_2 with probability $p_2 = 1 - p_1$.

Mixtures typically appear in heterogeneous populations that consist of several groups: females and males, graduate and undergraduate students, daytime and nighttime internet traffic, Windows, Unix, or Macintosh users, etc. In such cases, we can either study each group separately, or use the Law of Total Probability on p. 31, write the (unconditional) cdf as

$$F(x) = p_1 F_1(x) + p_2 F_2(x) + \ldots,$$

and study the whole population at once.

Bell shapes of both humps in Figure 8.7d suggest that the sample came from a mixture of two Normal distributions (with means around μ_1 and μ_2), with a higher probability of having mean μ_1, since the left hump is bigger.

The choice of bins

Experimenting with histograms, you can notice that their shape may depend on the choice of bins. One can hear various rules of thumb about a good choice of bins, but in general,

- there should not be too few or too many bins;

- their number may increase with a sample size;

- they should be chosen to make the histogram informative, so that we can see shapes, outliers, etc.

In Figure 8.6, we simply divided the range of CPU data into 10 equal intervals, 14 sec each, and apparently this was sufficient for drawing important conclusions.

As two extremes, consider histograms in Figure 8.8 constructed from the same CPU data.

The first histogram has too many columns; therefore, each column is short. Most bins have only 1 observation. This tells little about the actual shape of the distribution; however, we can still notice an outlier, $X = 139$.

The second histogram has only 3 columns. It is hard to guess the family of distributions here, although a flat Uniform distribution is already ruled out. The outlier is not seen; it merged with the rightmost bin.

Both histograms in Figure 8.8 can be made more informative by a better choice of bins.

8.3.2   Stem-and-leaf plot

Stem-and-leaf plots are similar to histograms although they carry more information. Namely, they also show how the data are distributed *within* columns.

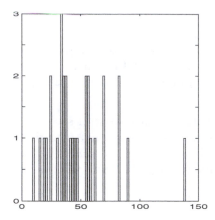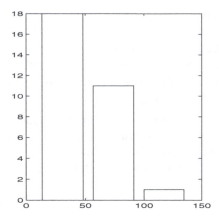

FIGURE 8.8: *Wrong choice of bins for CPU data: too many bins, too few bins.*

To construct a stem-and-leaf plot, we need to draw a stem and a leaf. The first one or several digits form a stem, and the next digit forms a leaf. Other digits are dropped; in other words, the numbers get rounded. For example, a number 239 can be written as

$$23 \mid 9$$

with 23 going to the stem and 9 to the leaf, or as

$$2 \mid 3$$

with 2 joining the stem, 3 joining the leaf, and digit 9 being dropped. In the first case, the *leaf unit* equals 1 while in the second case, the leaf unit is 10, showing that the (rounded) number is not 23 but 230.

For the CPU data on p. 217 (data set CPU), let the last digits form a leaf. The remaining digits go to the stem. Each CPU time is then written as

$$10 \text{ "stem"} + \text{"leaf"},$$

making the following stem-and-leaf plot,

```
                              0 | 9
       LEAF UNIT = 1          1 | 5  9
                              2 | 2  4  5
                              3 | 0  4  5  5  6  6  7  8
                              4 | 2  3  6  8
                              5 | 4  5  6  6  9
                              6 | 2  9
                              7 | 0
                              8 | 2  2  9
                              9 |
                             10 |
                             11 |
                             12 |
                             13 | 9
```

Turning this plot by 90 degrees counterclockwise, we get a *histogram* with 10-unit bins (because each stem unit equals 10). Thus, all the information seen on a histogram can be obtained here too. In addition, now we can see individual values within each column. We have the entire sample sorted and written in the form of a stem-and-leaf plot. If needed, we can even compute sample mean, median, quartiles, and other statistics from it.

Example 8.19 (COMPARISON). Sometimes stem-and-leaf plots are used to compare two samples. For this purpose, one can put two leaves on the same stem. Consider, for example, samples of round-trip transit times (known as pings) received from two locations (data set `Pings`).

Location I: 0.0156, 0.0396, 0.0355, 0.0480, 0.0419, 0.0335, 0.0543, 0.0350, 0.0280, 0.0210, 0.0308, 0.0327, 0.0215, 0.0437, 0.0483 seconds

Location II: 0.0298, 0.0674, 0.0387, 0.0787, 0.0467, 0.0712, 0.0045, 0.0167, 0.0661, 0.0109, 0.0198, 0.0039 seconds

Choosing a leaf unit of 0.001, a stem unit of 0.01, and dropping the last digit, we construct the following two stem-and-leaf plots, one to the left and one to the right of the stem.

```
LEAF UNIT = 0.001                          0 | 3  4
                                    5     1 | 0  6  9
                        1  1  8     2 |
              0  2  3  5  5  9     3 | 8
                    1  3  8  8     4 | 6
                             4     5 |
                                    6 | 1  6  7
                                    7 | 8
```

Looking at these two plots, one will see about the same average ping from the two locations. One will also realize that the first location has a more stable connection because its pings have lower variability and lower variance. For the second location, the fastest ping will be understood as

$$\{10(\text{leaf } 0) + \text{stem } 3\} \, (\text{leaf unit } 0.001) = 0.003,$$

and the slowest ping as

$$\{10(\text{leaf } 7) + \text{stem } 8\} \, (\text{leaf unit } 0.001) = 0.078.$$

8.3.3 Boxplot

The main descriptive statistics of a sample can be represented graphically by a **boxplot**. To construct a boxplot, we draw a box between the first and the third quartiles, a line inside a box for a median, and extend whiskers to the smallest and the largest observations, thus representing a so-called *five-point summary*:

$$\text{five-point summary} = \left(\min X_i, \ \widehat{Q}_1, \ \widehat{M}, \ \widehat{Q}_3, \ \max X_i \right).$$

Often a sample mean $\overline{X}$ is also depicted with a dot or a cross. Observations further than 1.5 interquartile ranges are usually drawn separately from whiskers, indicating the possibility of outliers. This is in accordance with the 1.5(IQR) rule (see Section 8.2.6).

The mean and five-point summary of CPU times were found in Examples 8.7, 8.12, and 8.14,

$$\overline{X} = 48.2333; \ \min X_i = 9, \ \widehat{Q}_1 = 34, \ \widehat{M} = 42.5, \ \widehat{Q}_3 = 59, \ \max X_i = 139.$$

We also know that $X = 139$ is more than $1.5(\widehat{IQR})$ away from the third quartile, and we suspect that it may be an outlier.

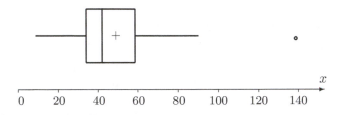

FIGURE 8.9: *Boxplot of CPU time data.*

A boxplot is drawn in Figure 8.9. The mean is depicted with a "+," and the right whisker extends till the second largest observation $X = 89$ because $X = 139$ is a suspected outlier (depicted with a little circle).

From this boxplot, one can conclude:

- The distribution of CPU times is right skewed because (1) the mean exceeds the median, and (2) the right half of the box is larger than the left half.

- Each half of a box and each whisker represents approximately 25% of the population. For example, we expect about 25% of all CPU times to fall between 42.5 and 59 seconds.

Parallel boxplots

Boxplots are often used to compare different populations or parts of the same population. For such a comparison, samples of data are collected from each part, and their boxplots are drawn on the same scale next to each other.

For example, seven parallel boxplots in Figure 8.10 represent the amount of internet traffic handled by a certain center during a week. We can see the following general patterns:

- The heaviest internet traffic occurs on Fridays.

- Fridays also have the highest variability.

- The lightest traffic is seen on weekends, with an increasing trend from Saturday to Monday.

- Each day, the distribution is right-skewed, with a few outliers on each day except Saturday. Outliers indicate occurrences of unusually heavy internet traffic.

Trends can also be seen on scatter plots and time plots.

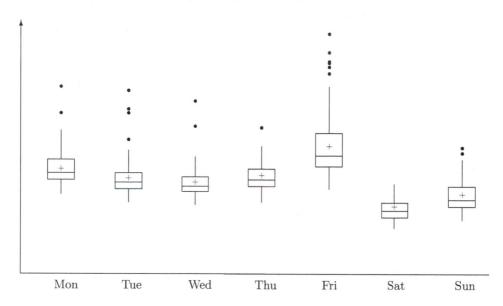

FIGURE 8.10: *Parallel boxplots of internet traffic.*

8.3.4   Scatter plots and time plots

Scatter plots are used to see and understand a relationship between two variables. These can be temperature and humidity, experience and salary, age of a network and its speed, number of servers and the expected response time, and so on.

To study the relationship, both variables are measured on each sampled item. For example, temperature and humidity during each of n days, age and speed of n networks, or experience and salary of n randomly chosen computer scientists are recorded. Then, a **scatter plot** consists of n points on an (x, y)-plane, with x- and y-coordinates representing the two recorded variables.

Example 8.20 (ANTIVIRUS MAINTENANCE).   Protection of a personal computer largely depends on the frequency of running antivirus software on it. One can set to run it every day, once a week, once a month, etc.

During a scheduled maintenance of computer facilities, a computer manager records the number of times the antivirus software was launched on each computer during 1 month (variable X) and the number of detected worms (variable Y). The data for 30 computers are in the table (data set `Antivirus`).

X	30	30	30	30	30	30	30	30	30	30	30	15	15	15	10
Y	0	0	1	0	0	0	1	1	0	0	0	0	1	1	0

X	10	10	6	6	5	5	5	4	4	4	4	4	1	1	1
Y	0	2	0	4	1	2	0	2	1	0	1	0	6	3	1

Is there a connection between the frequency of running antivirus software and the number of worms in the system? A scatter plot of these data is given in Figure 8.11a. It clearly shows that the number of worms reduces, in general, when the antivirus is employed more frequently. This relationship, however, is not certain because no worm was detected on some "lucky" computers although the antivirus software was launched only once a week on them.

$\Diamond$

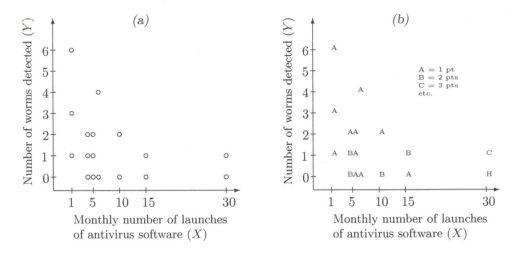

FIGURE 8.11: *Scatter plots for Examples 8.20 and 8.21.*

Example 8.21 (PLOTTING IDENTICAL POINTS). Looking at the scatter plot in Figure 8.11a, the manager in Example 8.20 realized that a portion of data is hidden there because there are identical observations. For example, no worms were detected on 8 computers where the antivirus software is used daily (30 times a month). Then, Figure 8.11a may be misleading.

When the data contain identical pairs of observations, the points on a scatter plot are often depicted with either numbers or letters ("A" for 1 point, "B" for two identical points, "C" for three, etc.). You can see the result in Figure 8.11b. ◇

When we study time trends and development of variables over time, we use **time plots**. These are scatter plots with x-variable representing time.

Example 8.22 (WORLD POPULATION). For example, here is how the world population increased between 1950 and 2019 (Figure 8.12). We can clearly see that the population increases at an almost steady rate. The actual data are given in Table 11.1 on p. 376 and in data set `PopulationWorld`. Later, in Chapter 11, we'll learn how to estimate trends seen on time plots and scatter plots and even make forecasts for the future. ◇

R notes

In R, `mean`, `median`, `var`, `sd`, `min`, `max`, `range`, `quantile`, and `length` are used to determine the sample mean, median, variance, standard deviation, minimum, maximum, range, quantiles, and the sample size of an observed variable. A `summary` command calculates the whole five-point summary for a boxplot – minimum, maximum, three quartiles, and additionally, the mean. The boxplot itself can be drawn simply by typing `boxplot(X)` for an observed variable X. Try this for the whole data set, `boxplot(Dataset)`, and you will obtain beautiful parallel boxplots (but notice that they will appear all on the same scale).

Similarly, `summary` can also be applied to the entire data set, and this is often used to gain quick information, as the first acquaintance with data.

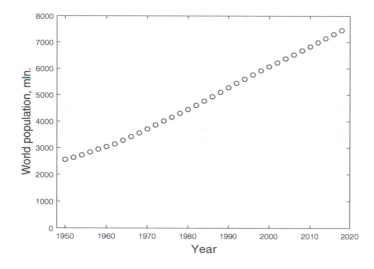

FIGURE 8.12: *Time plot of the world population in 1950–2019.*

For graphical statistics, `hist` is used for histograms, `stem` for stem-and-leaf plots, and `plot` for scatter plots. In these plots, you can optionally set a color and thickness. For example, `plot(X,Y,col="blue",lwd=3)` will produce a plot of Y vs X in blue points that are three times thicker (and therefore, will look brighter) than those without an `lwd` option. To plot several variables on the same scatter plot, you can add plots to an existing one by a command `points`. For example, `points(X,Z,col="green")` will superimpose a green scatter plot of Z vs X on an earlier blue scatter plot of Y vs X. Applying this command to the whole data set, `plot(Dataset)`, will produce a *scatter plot matrix*. That is, R will draw a square matrix of scatter plots of all the variables against each other variable.

MATLAB notes

For the standard descriptive statistics, MATLAB has ready-to-go commands `mean`, `median`, `std`, `var`, `min`, and `max`. Also, `quantile(X,p)` and `prctile(X,100*p)` give sample p-quantile $\hat{q}_p$ which is also the $100p$-percentile $\hat{\pi}_{100p}$, and `iqr(X)` returns the interquartile range.

Graphing data sets is also possible in one command. To see a histogram, write `hist(x)` or `hist(x,n)` where x is the data, and n is the desired number of bins. For a boxplot, enter `boxplot(X)`. By the way, if X is a matrix, then you will obtain parallel boxplots. To draw a scatter plot or a time plot, write `scatter(X,Y)` or just `plot(X,Y)`. You can nicely choose the color and symbol in this command. For example, `plot(X,Y,'r*',X,Z,'bd')` will plot variables Y and Z against variable X, with Y-points marked by red stars and Z-points marked by blue diamonds.

Summary and conclusions

Methods discussed in this chapter provide a quick description of a data set, general conclusions about the population, and help to formulate conjectures about its features.

Sample mean, median, variance, moments, and quantiles estimate and give us an idea about the corresponding population parameters. The interquartile range is used to detect outliers.

Before we start working with a data set, we look at it! There is a number of methods to display data: histograms, stem-and-leaf plots, boxplots, scatter plots, time plots. Good data displays show the shape of the distribution, the type of skewness or symmetry, interrelations, general trends, and possible outliers.

More advanced and accurate statistical methods are introduced in the next chapter.

Exercises

8.1. The numbers of blocked intrusion attempts on each day during the first two weeks of the month were

$$56, 47, 49, 37, 38, 60, 50, 43, 43, 59, 50, 56, 54, 58$$

After the change of firewall settings, the numbers of blocked intrusions during the next 20 days were

$$53, 21, 32, 49, 45, 38, 44, 33, 32, 43, 53, 46, 36, 48, 39, 35, 37, 36, 39, 45.$$

Comparing the number of blocked intrusions before and after the change,

(a) construct side-by-side stem-and-leaf plots;

(b) compute the five-point summaries and construct parallel boxplots;

(c) comment on your findings.

These data are available in data set `Intrusions`.

8.2. A network provider investigates the load of its network. The number of concurrent users is recorded at fifty locations (thousands of people),

17.2	22.1	18.5	17.2	18.6	14.8	21.7	15.8	16.3	22.8
24.1	13.3	16.2	17.5	19.0	23.9	14.8	22.2	21.7	20.7
13.5	15.8	13.1	16.1	21.9	23.9	19.3	12.0	19.9	19.4
15.4	16.7	19.5	16.2	16.9	17.1	20.2	13.4	19.8	17.7
19.7	18.7	17.6	15.9	15.2	17.1	15.0	18.8	21.6	11.9

These data are available in data set `ConcurrentUsers`.

(a) Compute the sample mean, variance, and standard deviation of the number of concurrent users.

(b) Estimate the standard error of the sample mean.

(c) Compute the five-point summary and construct a boxplot.

(d) Compute the interquartile range. Are there any outliers?

(e) It is reported that the number of concurrent users follows approximately Normal distribution. Does the histogram support this claim?

8.3. Verify the use of Chebyshev's inequality in (8.6) of Example 8.16. Show that if the population mean is indeed 48.2333 and the population standard deviation is indeed 26.5170, then at least 8/9 of all tasks require less than 127.78 seconds of CPU time.

8.4. Use Table A4 to compute the probability for *any* Normal random variable to take a value within 1.5 interquartile ranges from population quartiles.

8.5. The following data set shows population of the United States (in million) since 1790,

Year	1790	1800	1810	1820	1830	1840	1850	1860	1870	1880	1890	1900
Population	3.9	5.3	7.2	9.6	12.9	17.1	23.2	31.4	38.6	50.2	63.0	76.2

Year	1910	1920	1930	1940	1950	1960	1970	1980	1990	2000	2010
Population	92.2	106.0	123.2	132.2	151.3	179.3	203.3	226.5	248.7	281.4	308.7

Construct a time plot for the U.S. population. What kind of trend do you see? What information can be extracted from this plot?

These data are available in data set `PopulationUSA`.

8.6. Refer to Exercise 8.5 (data set `PopulationUSA`). Compute 10-year *increments* of the population growth $x_1 = 5.3 - 3.9$, $x_2 = 7.2 - 5.3$, etc.

(a) Compute sample mean, median, and variance of 10-year increments. Discuss how the U.S. population changes during a decade.

(b) Construct a time plot of 10-year increments and discuss the observed pattern.

8.7. Refer to Exercise 8.5 (data set `PopulationUSA`). Compute 10-year *relative population change* $y_1 = (5.3 - 3.9)/3.9$, $y_2 = (7.2 - 5.3)/5.3$, etc.

(a) Compute sample mean, median, and variance of the relative population change.

(b) Construct a time plot of the relative population change. What trend do you see now?

(c) Comparing the time plots in Exercises 8.6 and 8.7, what kind of correlation between x_i and y_i would you expect? Verify by computing the sample correlation coefficient

$$r = \frac{\sum (x_i - \overline{x})(y_i - \overline{y})/(n-1)}{s_x s_y}.$$

What can you conclude? How would you explain this phenomenon?

8.8. Consider three data sets (also, in data set `Symmetry`).

(1) 19, 24, 12, 19, 18, 24, 8, 5, 9, 20, 13, 11, 1, 12, 11, 10, 22, 21, 7, 16, 15, 15, 26, 16, 1, 13, 21, 21, 20, 19

(2) 17, 24, 21, 22, 26, 22, 19, 21, 23, 11, 19, 14, 23, 25, 26, 15, 17, 26, 21, 18, 19, 21, 24, 18, 16, 20, 21, 20, 23, 33

(3) 56, 52, 13, 34, 33, 18, 44, 41, 48, 75, 24, 19, 35, 27, 46, 62, 71, 24, 66, 94, 40, 18, 15, 39, 53, 23, 41, 78, 15, 35

(a) For each data set, draw a histogram and determine whether the distribution is right-skewed, left-skewed, or symmetric.

(b) Compute sample means and sample medians. Do they support your findings about skewness and symmetry? How?

8.9. The following data set represents the number of new computer accounts registered during ten consecutive days.

$$43, 37, 50, 51, 58, 105, 52, 45, 45, 10.$$

(a) Compute the mean, median, quartiles, and standard deviation.

(b) Check for outliers using the 1.5(IQR) rule.

(c) Delete the detected outliers and compute the mean, median, quartiles, and standard deviation again.

(d) Make a conclusion about the effect of outliers on basic descriptive statistics.

These data are available in data set Accounts.

Chapter 9

Statistical Inference I

After taking a general look at the data, we are ready for more advanced and more informative statistical analysis.

In this chapter, we learn how

- *to estimate parameters* of the distribution. Methods of Chapter 8 mostly concern measure of location (mean, median, quantiles) and variability (variance, standard deviation, interquartile range). As we know, this does not cover all possible parameters, and thus, we still lack a general methodology of estimation.

- *to construct confidence intervals.* Any estimator, computed from a collected random sample instead of the whole population, is understood as only an approximation of the corresponding parameter. Instead of one estimator that is subject to a *sampling error*, it is often more reasonable to produce an interval that will contain the true population parameter with a certain known high probability.

- *to test hypotheses.* That is, we shall use the collected sample to verify statements and claims about the population. As a result of each test, a statement is either rejected on basis of the observed data or accepted (not rejected). Sampling error in this analysis results in a possibility of wrongfully accepting or rejecting the hypothesis; however, we can design tests to control the probability of such errors.

Results of such statistical analysis are used for making decisions under uncertainty, developing optimal strategies, forecasting, evaluating and controlling performance, and so on.

9.1   Parameter estimation

By now, we have learned a few elementary ways to determine the *family of distributions*. We take into account the nature of our data, basic description, and range; propose a suitable family of distributions; and support our conjecture by looking at a histogram.

In this section, we learn how to estimate parameters of distributions. As a result, a large family will be reduced to just one distribution that we can use for performance evaluation, forecasting, etc.

Example 9.1 (POISSON). For example, consider a sample of computer chips with a certain type of rare defects. The number of defects on each chip is recorded. This is the number of rare events, and thus, it should follow a Poisson distribution with *some* parameter λ.

We know that $\lambda = \mathbf{E}(X)$ is the expectation of a Poisson variable (Section 3.4.5). Then, should we estimate it with a sample mean $\overline{X}$? Or, should we use a sample variance s^2 because λ also equals $\mathrm{Var}(X)$? $\diamond$

Example 9.2 (GAMMA). Suppose now that we deal with a Gamma(α, λ) family of distributions. Its parameters α and λ do not represent the mean, variance, standard deviation, or any other measures discussed in Chapter 8. What would the estimation algorithm be this time? $\diamond$

Questions raised in these examples do not have unique answers. Statisticians developed a number of estimation techniques, each having certain optimal properties.

Two rather popular methods are discussed in this section:

– method of moments, and

– method of maximum likelihood.

Several other methods are introduced later: bootstrap in Section 10.3, Bayesian parameter estimation in Section 10.4, and least squares estimation in Chapter 11.

9.1.1   Method of moments

Moments

First, let us define the moments.

DEFINITION 9.1 ——————————

The k-th **population moment** is defined as

$$\mu_k = \mathbf{E}(X^k).$$

The k-th **sample moment**

$$m_k = \frac{1}{n} \sum_{i=1}^{n} X_i^k$$

estimates μ_k from a sample $(X_1, \ldots, X_n)$.

The first sample moment is the sample mean $\overline{X}$.

Central moments are computed similarly, after centralizing the data, that is, subtracting the mean.

DEFINITION 9.2 ——————————

For $k \geq 2$, the k-th **population central moment** is defined as

$$\mu_k' = \mathbf{E}(X - \mu_1)^k.$$

The k-th **sample central moment**

$$m_k' = \frac{1}{n} \sum_{i=1}^{n} \left(X_i - \overline{X}\right)^k$$

estimates μ_k from a sample $(X_1, \ldots, X_n)$.

Remark: The second population central moment is variance $\mathrm{Var}(X)$. The second sample central moment is sample variance, although $(n-1)$ in its denominator is now replaced by n. We mentioned that estimation methods are not unique. For unbiased estimation of $\sigma^2 = \mathrm{Var}(X)$, we use

$$s^2 = \frac{1}{n-1}\sum_{i=1}^{n}(X_i - \overline{X})^2;$$

however, method of moments and method of maximum likelihood produce a different version,

$$S^2 = m_2' = \frac{1}{n}\sum_{i=1}^{n}(X_i - \overline{X})^2.$$

And this is not all! We'll see other estimates of σ^2 as well.

Estimation

Method of moments is based on a simple idea. Since our sample comes from a family of distributions $\{F(\theta)\}$, we choose such a member of this family whose properties are close to properties of our data. Namely, we shall match the *moments*.

To estimate k parameters, equate the first k population and sample moments,

$$\begin{cases} \mu_1 &= & m_1 \\ \cdots & \cdots & \cdots \\ \mu_k &= & m_k \end{cases}$$

The left-hand sides of these equations depend on the distribution parameters. The right-hand sides can be computed from data. The **method of moments estimator** is the solution of this system of equations.

Example 9.3 (POISSON). To estimate parameter λ of Poisson(λ) distribution, we recall that

$$\mu_1 = \mathbf{E}(X) = \lambda.$$

There is only one unknown parameter, hence we write one equation,

$$\mu_1 = \lambda = m_1 = \overline{X}.$$

"Solving" it for λ, we obtain

$$\widehat{\lambda} = \overline{X},$$

the method of moments estimator of λ. $\diamondsuit$

This does not look difficult, does it? Simplicity is the main attractive feature of the method of moments.

If it is easier, one may opt to equate *central moments*.

Example 9.4 (GAMMA DISTRIBUTION OF CPU TIMES). The histogram in Figure 8.6 suggested that CPU times have Gamma distribution with some parameters α and λ. To estimate them, we need two equations. From data on p. 217, we compute

$$m_1 = \overline{X} = 48.2333 \quad \text{and} \quad m_2' = S^2 = 679.7122.$$

and write two equations,

$$\begin{cases} \mu_1 = \mathbf{E}(X) = \alpha/\lambda = m_1 \\ \mu_2' = \mathrm{Var}(X) = \alpha/\lambda^2 = m_2'. \end{cases}$$

It is convenient to use the second *central* moment here because we already know the expression for the variance $m_2' = \mathrm{Var}(X)$ of a Gamma variable.

Solving this system in terms of α and λ, we get the method of moment estimates

$$\begin{cases} \widehat{\alpha} = m_1^2/m_2' = 3.4227 \\ \widehat{\lambda} = m_1/m_2' = 0.0710. \end{cases}$$

$\Diamond$

Of course, we solved these two examples so quickly because we already knew the moments of Poisson and Gamma distributions from Sections 3.4.5 and 4.2.3. When we see a new distribution for us, we'll have to compute its moments.

Consider, for example, *Pareto distribution* that plays an increasingly vital role in modern internet modeling due to very heavy internet traffic nowadays.

Example 9.5 (PARETO). A two-parameter *Pareto distribution* has a cdf

$$F(x) = 1 - \left(\frac{x}{\sigma}\right)^{-\theta} \quad \text{for } x > \sigma.$$

How should we compute method of moments estimators of σ and θ?

We have not seen Pareto distribution in this book so far, so we'll have to compute its first two moments.

We start with the density

$$f(x) = F'(x) = \frac{\theta}{\sigma}\left(\frac{x}{\sigma}\right)^{-\theta-1} = \theta\sigma^\theta x^{-\theta-1}$$

and use it to find the expectation

$$\mu_1 = \mathbf{E}(X) = \int_\sigma^\infty x\, f(x)\, dx = \theta\sigma^\theta \int_\sigma^\infty x^{-\theta} dx$$

$$= \theta\sigma^\theta \left.\frac{x^{-\theta+1}}{-\theta+1}\right|_{x=\sigma}^{x=\infty} = \frac{\theta\sigma}{\theta-1}, \quad \text{for } \theta > 1,$$

and the second moment

$$\mu_2 = \mathbf{E}(X^2) = \int_\sigma^\infty x^2 f(x)\, dx = \theta\sigma^\theta \int_\sigma^\infty x^{-\theta+1} dx = \frac{\theta\sigma^2}{\theta-2}, \quad \text{for } \theta > 2.$$

For $\theta \le 1$, a Pareto variable has an infinite expectation, and for $\theta \le 2$, it has an infinite second moment.

Then we solve the method of moments equations

$$\begin{cases} \mu_1 = \dfrac{\theta\sigma}{\theta-1} = m_1 \\ \mu_2 = \dfrac{\theta\sigma^2}{\theta-2} = m_2 \end{cases}$$

and find that

$$\widehat{\theta} = \sqrt{\frac{m_2}{m_2 - m_1^2} + 1} \quad \text{and} \quad \widehat{\sigma} = \frac{m_1(\widehat{\theta} - 1)}{\widehat{\theta}}. \tag{9.1}$$

When we collect a sample from Pareto distribution, we can compute sample moments m_1 and m_2 and estimate parameters by (9.1). ◇

On rare occasions, when k equations are not enough to estimate k parameters, we'll consider higher moments.

Example 9.6 (NORMAL). Suppose we already know the mean μ of a Normal distribution and would like to estimate the variance σ^2. Only one parameter σ^2 is unknown; however, the first method of moments equation

$$\mu_1 = m_1$$

does not contain σ^2 and therefore does not produce its estimate. We then consider the second equation, say,

$$\mu_2' = \sigma^2 = m_2' = S^2,$$

which gives us the method of moments estimate immediately, $\widehat{\sigma}^2 = S^2$. ◇

Method of moments estimates are typically easy to compute. They can serve as a quick tool for estimating parameters of interest.

9.1.2   Method of maximum likelihood

Another interesting idea is behind the method of *maximum likelihood estimation*.

Since the sample $\boldsymbol{X} = (X_1, \ldots, X_n)$ has already been observed, we find such parameters that maximize the probability (likelihood) for this to happen. In other words, we make the event that has already happened to be as likely as possible. This is yet another way to make the chosen distribution consistent with the observed data.

DEFINITION 9.3

> **Maximum likelihood estimator** is the parameter value that maximizes the likelihood of the observed sample. For a discrete distribution, we maximize the joint pmf of data $P(X_1, \ldots, X_n)$. For a continuous distribution, we maximize the joint density $f(X_1, \ldots, X_n)$.

Both cases, discrete and continuous, are explained below.

Discrete case

For a discrete distribution, the probability of a given sample is the joint pmf of data,

$$P\{\boldsymbol{X} = (X_1, \ldots, X_n)\} = P(\boldsymbol{X}) = P(X_1, \ldots, X_n) = \prod_{i=1}^{n} P(X_i),$$

because in a simple random sample, all observed X_i are independent.

To maximize this likelihood, we consider the critical points by taking derivatives with respect to all unknown parameters and equating them to 0. The maximum can only be attained at such parameter values θ where the derivative $\frac{\partial}{\partial \theta} P(\boldsymbol{X})$ equals 0, where it does not exist, or at the boundary of the set of possible values of θ (to review this, see Section A.4.4).

A nice computational shortcut is to take logarithms first. Differentiating the sum

$$\ln \prod_{i=1}^{n} P(X_i) = \sum_{i=1}^{n} \ln P(X_i)$$

is easier than differentiating the product $\prod P(X_i)$. Besides, logarithm is an increasing function, so the likelihood $P(\boldsymbol{X})$ and the log-likelihood $\ln P(\boldsymbol{X})$ are maximized by exactly the same parameters.

Example 9.7 (POISSON). The pmf of Poisson distribution is

$$P(x) = e^{-\lambda} \frac{\lambda^x}{x!},$$

and its logarithm is

$$\ln P(x) = -\lambda + x \ln \lambda - \ln(x!).$$

Thus, we need to maximize

$$\ln P(\boldsymbol{X}) = \sum_{i=1}^{n} (-\lambda + X_i \ln \lambda) + C = -n\lambda + \ln \lambda \sum_{i=1}^{n} X_i + C,$$

where $C = -\sum \ln(x!)$ is a constant that does not contain the unknown parameter λ.

Find the critical point(s) of this log-likelihood. Differentiating it and equating its derivative to 0, we get

$$\frac{\partial}{\partial \lambda} \ln P(\boldsymbol{X}) = -n + \frac{1}{\lambda} \sum_{i=1}^{n} X_i = 0.$$

This equation has only one solution

$$\widehat{\lambda} = \frac{1}{n} \sum_{i=1}^{n} X_i = \overline{X}.$$

Since this is the only critical point, and since the likelihood vanishes (converges to 0) as $\lambda \downarrow 0$ or $\lambda \uparrow \infty$, we conclude that $\widehat{\lambda}$ is the maximizer. Therefore, it is the maximum likelihood estimator of λ.

For the Poisson distribution, the method of moments and the method of maximum likelihood returned the same estimator, $\widehat{\lambda} = \overline{X}$. $\Diamond$

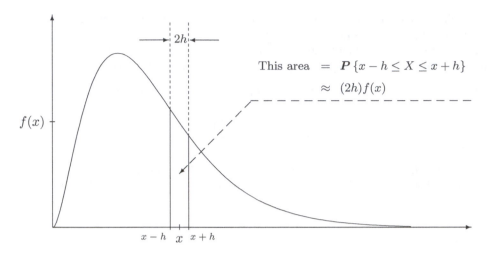

FIGURE 9.1: *Probability of observing "almost" $X = x$.*

Continuous case

In the continuous case, the probability to observe exactly the given number $X = x$ is 0, as we know from Chapter 4. Instead, the method of maximum likelihood will maximize the probability of observing "almost" the same number.

For a very small h,

$$\boldsymbol{P}\{x - h < X < x + h\} = \int_{x-h}^{x+h} f(y)dy \approx (2h)f(x).$$

That is, the probability of observing a value close to x is proportional to the density $f(x)$ (see Figure 9.1). Then, for a sample $\boldsymbol{X} = (X_1, \ldots, X_n)$, the maximum likelihood method will maximize the joint density $f(X_1, \ldots, X_n)$.

Example 9.8 (EXPONENTIAL). The Exponential density is

$$f(x) = \lambda e^{-\lambda x},$$

so the log-likelihood of a sample can be written as

$$\ln f(\boldsymbol{X}) = \sum_{i=1}^{n} \ln\left(\lambda e^{-\lambda X_i}\right) = \sum_{i=1}^{n} (\ln \lambda - \lambda X_i) = n \ln \lambda - \lambda \sum_{i=1}^{n} X_i.$$

Taking its derivative with respect to the unknown parameter λ, equating it to 0, and solving for λ, we get

$$\frac{\partial}{\partial \lambda} \ln f(\boldsymbol{X}) = \frac{n}{\lambda} - \sum_{i=1}^{n} X_i = 0,$$

resulting in

$$\widehat{\lambda} = \frac{n}{\sum X_i} = \frac{1}{\overline{X}}.$$

Again, this is the only critical point, and the likelihood $f(\boldsymbol{X})$ vanishes as $\lambda \downarrow 0$ or $\lambda \uparrow \infty$. Thus, $\widehat{\lambda} = \overline{X}$ is the maximum likelihood estimator of λ. This time, it also coincides with the method of moments estimator (Exercise 9.3b).   ◊

Sometimes the likelihood has no critical points inside its domain, then it is maximized at the boundary.

Example 9.9 (UNIFORM). Based on a sample from Uniform$(0, b)$ distribution, how can we estimate the parameter b?

The Uniform$(0, b)$ density is

$$f(x) = \frac{1}{b} \quad \text{for } 0 \le x \le b.$$

It is decreasing in b, and therefore, it is maximized at the the smallest possible value of b, which is x.

For a sample $(X_1, \ldots, X_n)$, the joint density

$$f(X_1, \ldots, X_n) = \left(\frac{1}{b}\right)^n \quad \text{for } 0 \le X_1, \ldots, X_n \le b$$

also attains its maximum at the smallest possible value of b which is now the largest observation. Indeed, $b \ge X_i$ for all i only if $b \ge \max(X_i)$. If $b < \max(X_i)$, then $f(\boldsymbol{X}) = 0$, and this cannot be the maximum value.

Therefore, the maximum likelihood estimator is $\widehat{b} = \max(X_i)$. ◇

When we estimate more than 1 parameter, all the partial derivatives should be equal 0 at the critical point. If no critical points exist, the likelihood is again maximized on the boundary.

Example 9.10 (PARETO). For the Pareto distribution in Example 9.5, the log-likelihood is

$$\ln f(\boldsymbol{X}) = \sum_{i=1}^{n} \ln \left(\theta \sigma^\theta X_i^{-\theta-1}\right) = n \ln \theta + n\theta \ln \sigma - (\theta + 1) \sum_{i=1}^{n} \ln X_i$$

for $X_1, \ldots, X_n \ge \sigma$. Maximizing this function over both σ and θ, we notice that it always increases in σ. Thus, we estimate σ by its largest possible value, which is the smallest observation,

$$\widehat{\sigma} = \min(X_i).$$

We can substitute this value of σ into the log-likelihood and maximize with respect to θ,

$$\frac{\partial}{\partial \theta} \ln f(\boldsymbol{X}) = \frac{n}{\theta} + n \ln \widehat{\sigma} - \sum_{i=1}^{n} \ln X_i = 0;$$

$$\widehat{\theta} = \frac{n}{\sum \ln X_i - n \ln \widehat{\sigma}} = \frac{n}{\sum \ln (X_i/\widehat{\sigma})}.$$

The maximum likelihood estimates of σ and θ are

$$\widehat{\sigma} = \min(X_i) \quad \text{and} \quad \widehat{\theta} = \frac{n}{\sum \ln (X_i/\widehat{\sigma})}.$$

◇

Maximum likelihood estimators are rather popular because of their nice properties. Under mild conditions, these estimators are consistent, and for large samples, they have an approximately Normal distribution. Often in complicated problems, finding a good estimation scheme may be challenging whereas the maximum likelihood method always gives a reasonable solution.

9.1.3 Estimation of standard errors

How good are the estimators that we learned in Sections 9.1.1 and 9.1.2? Standard errors can serve as measures of their accuracy. To estimate them, we derive an expression for the standard error and estimate all the unknown parameters in it.

Example 9.11 (ESTIMATION OF THE POISSON PARAMETER). In Examples 9.3 and 9.7, we found the method of moments and maximum likelihood estimators of the Poisson parameter λ. Both estimators appear to be equal the sample mean $\widehat{\lambda} = \overline{X}$. Let us now estimate the *standard error* of $\widehat{\lambda}$.

<u>Solution</u>. There are at least two ways to do it.

On one hand, $\sigma = \sqrt{\lambda}$ for the Poisson(λ) distribution, so $\sigma(\widehat{\lambda}) = \sigma(\overline{X}) = \sigma/\sqrt{n} = \sqrt{\lambda/n}$, as we know from (8.2) on p. 219. Estimating λ by $\overline{X}$, we obtain

$$ s_1(\widehat{\lambda}) = \sqrt{\frac{\overline{X}}{n}} = \frac{\sqrt{\sum X_i}}{n}. $$

On the other hand, we can use the sample standard deviation and estimate the standard error of the sample mean as in Example 8.17,

$$ s_2(\widehat{\lambda}) = \frac{s}{\sqrt{n}} = \sqrt{\frac{\sum (X_i - \overline{X})^2}{n(n-1)}}. $$

Apparently, we can estimate the standard error of $\widehat{\lambda}$ by two good estimators, s_1 and s_2. ◊

Example 9.12 (ESTIMATION OF THE EXPONENTIAL PARAMETER). Derive the standard error of the maximum likelihood estimator in Example 9.8 and estimate it, assuming a sample size $n \geq 3$.

<u>Solution</u>. This requires some integration work. Fortunately, we can take a shortcut because we know that the integral of any Gamma density is one, i.e.,

$$ \int_0^\infty \frac{\lambda^\alpha}{\Gamma(\alpha)} x^{\alpha-1} e^{-\lambda x} dx = 1 \quad \text{for any } \alpha > 0, \ \lambda > 0. $$

Now, notice that $\widehat{\lambda} = 1/\overline{X} = n/\sum X_i$, where $\sum X_i$ has Gamma (n, λ) distribution because each X_i is Exponential(λ).

Therefore, the k-th moment of $\widehat{\lambda}$ equals

$$
\begin{aligned}
\mathbf{E}(\widehat{\lambda}^k) &= \mathbf{E}\left(\frac{n}{\sum X_i}\right)^k = \int_0^\infty \left(\frac{n}{x}\right)^k \frac{\lambda^n}{\Gamma(n)} x^{n-1} e^{-\lambda x} dx = \frac{n^k \lambda^n}{\Gamma(n)} \int_0^\infty x^{n-k-1} e^{-\lambda x} dx \\
&= \frac{n^k \lambda^n}{\Gamma(n)} \frac{\Gamma(n-k)}{\lambda^{n-k}} \int_0^\infty \frac{\lambda^{n-k}}{\Gamma(n-k)} x^{n-k-1} e^{-\lambda x} dx \\
&= \frac{n^k \lambda^n}{\Gamma(n)} \frac{\Gamma(n-k)}{\lambda^{n-k}} \cdot 1 = \frac{n^k \lambda^k (n-k-1)!}{(n-1)!}.
\end{aligned}
$$

Substituting $k = 1$, we get the first moment,

$$\mathbf{E}(\widehat{\lambda}) = \frac{n\lambda}{n-1}.$$

Substituting $k = 2$, we get the second moment,

$$\mathbf{E}(\widehat{\lambda}^2) = \frac{n^2\lambda^2}{(n-1)(n-2)}.$$

Then, the standard error of $\widehat{\lambda}$ is

$$\sigma(\widehat{\lambda}) = \sqrt{\operatorname{Var}(\widehat{\lambda})} = \sqrt{\mathbf{E}(\widehat{\lambda}^2) - \mathbf{E}^2(\widehat{\lambda})} = \sqrt{\frac{n^2\lambda^2}{(n-1)(n-2)} - \frac{n^2\lambda^2}{(n-1)^2}} = \frac{n\lambda}{(n-1)\sqrt{n-2}}.$$

We have just estimated λ by $\widehat{\lambda} = 1/\overline{X}$; therefore, we can estimate the standard error $\sigma(\widehat{\lambda})$ by

$$s(\widehat{\lambda}) = \frac{n}{\overline{X}(n-1)\sqrt{n-2}} \quad \text{or} \quad \frac{n^2}{\sum X_i(n-1)\sqrt{n-2}}.$$

$\Diamond$

Although this was not too long, estimation of standard errors can become much harder for just slightly more complex estimators. In some cases, a nice analytic formula for $\sigma(\widehat{\theta})$ may not exist. Then, a modern method of *bootstrap* will be used, and we discuss it in Section 10.3.

Computer notes

R package "MASS" (Modern Applied Statistics with S) includes maximum likelihood estimation for almost all the distributions discussed in this book. You just specify the observed variable and the family of distributions in R command `fitdistr`. For example, `fitdistr(X,'normal')`, `fitdistr(X,'Poisson')`, or `fitdistr(X,'geometric')`. Here is an R solution to Example 9.4.

— R ——————

```
install.packages("MASS")              # Invoke package "MASS"
library(MASS)
x<-c(70,36,43,69,82,48,34,            # Enter the data from Example 9.4
62,35,15,59,139,46,37,42,30,55,56,36,82,38,89,54,25,35,24,22,9,56,19);
fitdistr(x,'gamma')
```

As a result, R returns parameter estimates $\widehat{\alpha}$ (shape) and $\widehat{\lambda}$ (frequency or rate) along with their estimated standard errors in parentheses.

shape	rate
3.63007913	0.07526080
(0.89719441)	(0.01994748)

MATLAB has a similar tool...

— MATLAB ——————

```
x=[70,36,43,69,82,48,34,62,35,15,59,139,46,37,42,...
30,55,56,36,82,38,89,54,25,35,24,22,9,56,19);
fitdist(x,'gamma')
```

... with a slightly different output.

```
Gamma distribution
    a = 3.63007   [2.23591, 5.89356]
    b = 13.2871   [7.90162, 22.3433]
```

MATLAB understands the second parameter of Gamma distribution as a *scale parameter* β instead of a *frequency parameter* λ. They are directly related, $\beta = 1/\lambda$. Also, instead of standard errors, it attaches *confidence intervals* of both parameters. Confidence intervals? We learn them in the next section.

9.2 Confidence intervals

When we report an estimator $\widehat{\theta}$ of a population parameter θ, we know that most likely

$$\widehat{\theta} \neq \theta$$

due to a sampling error. We realize that we have estimated θ *up to some error*. Likewise, nobody understands the internet connection of 11 megabytes per second as exactly 11 megabytes going through the network every second, and nobody takes a meteorological forecast as the promise of exactly the predicted temperature.

Then how much can we trust the reported estimator? How far can it be from the actual parameter of interest? What is the probability that it will be reasonably close? And if we observed an estimator $\widehat{\theta}$, then what can the actual parameter θ be?

To answer these questions, statisticians use *confidence intervals*, which contain parameter values that deserve some confidence, given the observed data.

DEFINITION 9.4 ——————

> An interval $[a, b]$ is a $(1 - \alpha)100\%$ **confidence interval** for the parameter θ if it contains the parameter with probability $(1 - \alpha)$,
>
> $$P\{a \leq \theta \leq b\} = 1 - \alpha.$$
>
> The **coverage probability** $(1 - \alpha)$ is also called a **confidence level**.

Let us take a moment to think about this definition. The probability of a random event $\{a \leq \theta \leq b\}$ has to be $(1 - \alpha)$. What randomness is involved in this event?

The population parameter θ is not random. It is a population feature, independent of any random sampling procedure, and therefore, it remains constant. On the other hand,

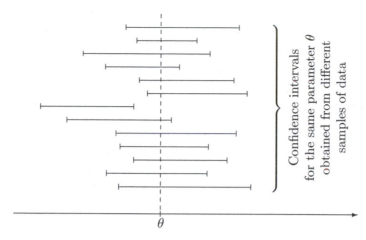

Confidence intervals for the same parameter θ obtained from different samples of data

θ

FIGURE 9.2: *Confidence intervals and coverage of parameter θ.*

the interval is computed from random data, and therefore, it is random. The *coverage probability* refers to the chance that our interval covers a constant parameter θ.

This is illustrated in Figure 9.2. Suppose that we collect many random samples and produce a confidence interval from each of them. If these are $(1 - \alpha)100\%$ confidence intervals, then we expect $(1 - \alpha)100\%$ of them to cover θ and $100\alpha\%$ of them to miss it. In Figure 9.2, we see one interval that does not cover θ. No mistake was made in data collection and construction of this interval. It missed the parameter only due to a *sampling error*.

It is therefore *wrong* to say, *"I computed a 90% confidence interval, it is [3, 6]. Parameter belongs to this interval with probability 90%."* The parameter is constant; it either belongs to the interval [3, 6] (with probability 1) or does not. In this case, 90% refers to the proportion of confidence intervals that contain the unknown parameter in a long run.

9.2.1 Construction of confidence intervals: a general method

Given a sample of data and a desired confidence level $(1 - \alpha)$, how can we construct a confidence interval $[a, b]$ that will satisfy the coverage condition

$$\boldsymbol{P} \{a \leq \theta \leq b\} = 1 - \alpha$$

in Definition 9.4?

We start by estimating parameter θ. *Assume there is an unbiased estimator $\widehat{\theta}$ that has a Normal distribution.* When we standardize it, we get a Standard Normal variable

$$Z = \frac{\widehat{\theta} - \mathbf{E}(\widehat{\theta})}{\sigma(\widehat{\theta})} = \frac{\widehat{\theta} - \theta}{\sigma(\widehat{\theta})}, \tag{9.2}$$

where $\mathbf{E}(\widehat{\theta}) = \theta$ because $\widehat{\theta}$ is unbiased, and $\sigma(\widehat{\theta}) = \sigma(\widehat{\theta})$ is its standard error.

This variable falls between the Standard Normal quantiles $q_{\alpha/2}$ and $q_{1-\alpha/2}$, denoted by

$$
\begin{aligned}
-z_{\alpha/2} &= q_{\alpha/2} \\
z_{\alpha/2} &= q_{1-\alpha/2}
\end{aligned}
$$

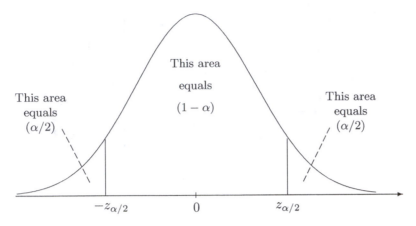

FIGURE 9.3: *Standard Normal quantiles* $\pm z_{\alpha/2}$ *and partition of the area under the density curve.*

with probability $(1 - \alpha)$, as you can see in Figure 9.3.

Then,

$$P\left\{ -z_{\alpha/2} \leq \frac{\widehat{\theta} - \theta}{\sigma(\widehat{\theta})} \leq z_{\alpha/2} \right\} = 1 - \alpha.$$

Solving the inequality inside $\{\ldots\}$ for θ, we get

$$P\left\{ \widehat{\theta} - z_{\alpha/2} \cdot \sigma(\widehat{\theta}) \leq \theta \leq \widehat{\theta} - z_{\alpha/2} \cdot \sigma(\widehat{\theta}) \right\} = 1 - \alpha.$$

The problem is solved! We have obtained two numbers

$$
\begin{aligned}
a &= \widehat{\theta} - z_{\alpha/2} \cdot \sigma(\widehat{\theta}) \\
b &= \widehat{\theta} + z_{\alpha/2} \cdot \sigma(\widehat{\theta})
\end{aligned}
$$

such that

$$P\left\{ a \leq \theta \leq b \right\} = 1 - \alpha.$$

Confidence interval, Normal distribution	If parameter θ has an unbiased, Normally distributed estimator $\widehat{\theta}$, then $$\widehat{\theta} \pm z_{\alpha/2} \cdot \sigma(\widehat{\theta}) = \left[\widehat{\theta} - z_{\alpha/2} \cdot \sigma(\widehat{\theta}),\ \widehat{\theta} + z_{\alpha/2} \cdot \sigma(\widehat{\theta}) \right]$$ is a $(1 - \alpha)100\%$ confidence interval for θ. If the distribution of $\widehat{\theta}$ is *approximately* Normal, we get an *approximately* $(1 - \alpha)100\%$ confidence interval.

(9.3)

In this formula, $\widehat{\theta}$ is the **center of the interval**, and $z_{\alpha/2} \cdot \sigma(\widehat{\theta})$ is the **margin**. The margin of error is often reported along with poll and survey results. In newspapers and press releases, it is usually computed for a 95% confidence interval.

We have seen quantiles $\pm z_{\alpha/2}$ in inverse problems (Example 4.12 on p. 91). Now, in confidence estimation, and also, in the next section on hypothesis testing, they will play a crucial role as we'll need to attain the desired confidence level α. The most commonly used values are

$$z_{0.10} = 1.282, \qquad z_{0.05} = 1.645, \qquad z_{0.025} = 1.960,$$
$$z_{0.01} = 2.326, \qquad z_{0.005} = 2.576.$$

(9.4)

NOTATION $\begin{Vmatrix} z_\alpha = q_{1-\alpha} = \Phi^{-1}(1-\alpha) \text{ is the value of a Standard} \\ \text{Normal variable } Z \text{ that is exceeded with probability } \alpha \end{Vmatrix}$

Several important applications of this general method are discussed below. In each problem, we

(a) find an unbiased estimator of θ,

(b) check if it has a Normal distribution,

(c) find its standard error $\sigma(\widehat{\theta}) = \text{Std}(\widehat{\theta})$,

(d) obtain quantiles $\pm z_{\alpha/2}$ from the table of Normal distribution (Table A4 in the Appendix), and finally,

(e) apply the rule (9.3).

9.2.2 Confidence interval for the population mean

Let us construct a confidence interval for the population mean

$$\theta = \mu = \mathbf{E}(X).$$

Start with an estimator,

$$\widehat{\theta} = \overline{X} = \frac{1}{n} \sum_{i=1}^{n} X_i.$$

The rule (9.3) is applicable in two cases.

1. If a sample $\boldsymbol{X} = (X_1, \ldots, X_n)$ comes from Normal distribution, then $\overline{X}$ is also Normal, and rule (9.3) can be applied.

2. If a sample comes from *any* distribution, but the sample size n is large, then $\overline{X}$ has an approximately Normal distribution according to the Central Limit Theorem on p. 93. Then rule (9.3) gives an approximately $(1-\alpha)100\%$ confidence interval.

In Section 8.2.1, we derived

$$\mathbf{E}(\overline{X}) \;=\; \mu \qquad \text{(thus, it is an unbiased estimator);}$$
$$\sigma(\overline{X}) \;=\; \sigma/\sqrt{n}.$$

Then, (9.3) reduces to the following $(1-\alpha)100\%$ confidence interval for μ.

$$\boxed{\;\overline{X} \pm z_{\alpha/2}\dfrac{\sigma}{\sqrt{n}}\;}$$

Confidence interval for the mean; σ is known (9.5)

Example 9.13. Construct a 95% confidence interval for the population mean based on a sample of measurements

$$2.5,\; 7.4,\; 8.0,\; 4.5,\; 7.4,\; 9.2$$

if measurement errors have Normal distribution, and the measurement device guarantees a standard deviation of $\sigma = 2.2$.

Solution. This sample has size $n = 6$ and sample mean $\overline{X} = 6.50$. To attain a confidence level of

$$1 - \alpha = 0.95,$$

we need $\alpha = 0.05$ and $\alpha/2 = 0.025$. Hence, we are looking for quantiles

$$q_{0.025} = -z_{0.025} \qquad \text{and} \qquad q_{0.975} = z_{0.025}.$$

From (9.4) or Table A4, we find that $q_{0.975} = 1.960$. Substituting these values into (9.5), we obtain a 95% confidence interval for μ,

$$\overline{X} \pm z_{\alpha/2}\frac{\sigma}{\sqrt{n}} = 6.50 \pm (1.960)\frac{2.2}{\sqrt{6}} = 6.50 \pm 1.76 \text{ or } [4.74,\; 8.26].$$

$\Diamond$

The only situation when method (9.3) cannot be applied is when the sample size is small and the distribution of data is not Normal. Special methods for the given distribution of X are required in this case.

9.2.3   Confidence interval for the difference between two means

Under the same conditions as in the previous section,

– Normal distribution of data or
– sufficiently large sample size,

we can construct a confidence interval for the *difference* between two means.

This problem arises when we compare two populations. It may be a comparison of two materials, two suppliers, two service providers, two communication channels, two labs, etc. From each population, a sample is collected (Figure 9.4),

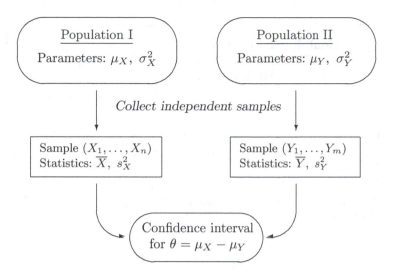

FIGURE 9.4: *Comparison of two populations.*

$$\boldsymbol{X} = (X_1, \ldots, X_n) \quad \text{from one population,}$$
$$\boldsymbol{Y} = (Y_1, \ldots, Y_m) \quad \text{from the other population.}$$

Suppose that the two samples are collected **independently** of each other.

To construct a confidence interval for the difference between population means

$$\theta = \mu_X - \mu_Y,$$

we complete the usual steps (a)–(e) below.

(a) Propose an estimator of θ,
$$\widehat{\theta} = \overline{X} - \overline{Y}.$$

It is natural to come up with this estimator because $\overline{X}$ estimates μ_X and $\overline{Y}$ estimates μ_Y.

(b) Check that $\widehat{\theta}$ is unbiased. Indeed,

$$\mathbf{E}(\widehat{\theta}) = \mathbf{E}\left(\overline{X} - \overline{Y}\right) = \mathbf{E}\left(\overline{X}\right) - \mathbf{E}\left(\overline{Y}\right) = \mu_X - \mu_Y = \theta.$$

(c) Check that $\widehat{\theta}$ has a Normal or approximately Normal distribution. This is true if the observations are Normal or *both* sample sizes m and n are large.

(d) Find the standard error of $\widehat{\theta}$ (using independence of $\boldsymbol{X}$ and $\boldsymbol{Y}$),

$$\sigma(\widehat{\theta}) = \sqrt{\text{Var}\left(\overline{X} - \overline{Y}\right)} = \sqrt{\text{Var}\left(\overline{X}\right) + \text{Var}\left(\overline{Y}\right)} = \sqrt{\frac{\sigma_X^2}{n} + \frac{\sigma_Y^2}{m}}.$$

(e) Find quantiles $\pm z_{\alpha/2}$ and compute the confidence interval according to (9.3). This results in the following formula.

Confidence interval for the difference of means; known standard deviations	$\overline{X} - \overline{Y} \pm z_{\alpha/2} \sqrt{\dfrac{\sigma_X^2}{n} + \dfrac{\sigma_Y^2}{m}}$	(9.6)

Example 9.14 (EFFECT OF AN UPGRADE). A manager evaluates effectiveness of a major hardware upgrade by running a certain process 50 times before the upgrade and 50 times after it. Based on these data, the average running time is 8.5 minutes before the upgrade, 7.2 minutes after it. Historically, the standard deviation has been 1.8 minutes, and presumably it has not changed. Construct a 90% confidence interval showing how much the mean running time reduced due to the hardware upgrade.

Solution. We have $n = m = 50$, $\sigma_X = \sigma_Y = 1.8$, $\overline{X} = 8.5$, and $\overline{Y} = 7.2$. Also, the confidence level $(1 - \alpha)$ equals 0.9, hence $\alpha/2 = 0.05$, and $z_{\alpha/2} = 1.645$.

The distribution of times may not be Normal; however, due to large sample sizes, the estimator

$$\widehat{\theta} = \overline{X} - \overline{Y}$$

is approximately Normal by the Central Limit Theorem. Thus, formula (9.6) is applicable, and a 90% confidence interval for the difference of means $(\mu_X - \mu_Y)$ is

$$8.5 - 7.2 \pm (1.645)\sqrt{1.8^2 \left(\frac{1}{50} + \frac{1}{50}\right)} = \underline{1.3 \pm 0.6} \text{ or } \underline{[0.7, 1.9]}.$$

We can say that the hardware upgrade resulted in a 1.3-minute reduction of the mean running time, with a 90% confidence margin of 0.6 minutes. $\diamond$

9.2.4   Selection of a sample size

Formula (9.3) describes a confidence interval as

$$\text{center} \pm \text{margin}$$

where

$$\begin{aligned} \text{center} &= \widehat{\theta}, \\ \text{margin} &= z_{\alpha/2} \cdot \sigma(\widehat{\theta}). \end{aligned}$$

We can revert the problem and ask a very practical question: *How large a sample should be collected to provide a certain desired precision of our estimator?*

In other words, what sample size n guarantees that the margin of a $(1 - \alpha)100\%$ confidence interval does not exceed a specified limit Δ?

To answer this question, we only need to solve the inequality

$$\text{margin} \leq \Delta \qquad (9.7)$$

in terms of n. Typically, parameters are estimated more accurately based on larger samples, so that the standard error $\sigma(\widehat{\theta})$ and the margin are decreasing functions of sample size n. Then, (9.7) must be satisfied for sufficiently large n.

9.2.5 Estimating means with a given precision

When we estimate a population mean, the margin of error is

$$\text{margin} = z_{\alpha/2} \cdot \sigma/\sqrt{n}.$$

Solving inequality (9.7) for n results in the following rule.

Sample size for a given precision

> In order to attain a margin of error Δ for estimating a population mean with a confidence level $(1 - \alpha)$,
>
> a sample of size $\quad n \geq \left(\dfrac{z_{\alpha/2} \cdot \sigma}{\Delta} \right)^2 \quad$ is required.

(9.8)

When we compute the expression in (9.8), it will most likely be a fraction. Notice that we can only *round it up* to the nearest integer sample size. If we round it down, our margin will exceed Δ.

Looking at (9.8), we see that a large sample will be necessary

- to attain a narrow margin (small Δ);

- to attain a high confidence level (small α); and

- to control the margin under high variability of data (large σ).

In particular, we need to quadruple the sample size in order to half the margin of the interval.

Example 9.15. In Example 9.13, we constructed a 95% confidence with the center 6.50 and margin 1.76 based on a sample of size 6. Now, that was too wide, right? How large a sample do we need to estimate the population mean with a margin of at most 0.4 units with 95% confidence?

Solution. We have $\Delta = 0.4$, $\alpha = 0.05$, and from Example 9.13, $\sigma = 2.2$. By (9.8), we need a sample of

$$n \geq \left(\frac{z_{0.05/2} \cdot \sigma}{\Delta} \right)^2 = \left(\frac{(1.960)(2.2)}{0.4} \right)^2 = 116.2.$$

Keeping in mind that this is the minimum sample size that satisfies Δ, and we are only allowed to round it up, we need a sample of at least 117 observations.

$\Diamond$

9.3 Unknown standard deviation

A rather heavy condition was assumed when we constructed all the confidence intervals. We assumed a *known standard deviation* σ and used it in all the derived formulas.

Sometimes this assumption is perfectly valid. We may know the variance from a large archive of historical data, or it may be given as precision of a measuring device.

Much more often, however, the population variance is unknown. We'll then estimate it from data and see if we can still apply methods of the previous section.

Two broad situations will be considered:

– large samples from any distribution,

– samples of any size from a Normal distribution.

In the only remaining case, a small non-Normal sample, a confidence interval will be constructed by special methods. A popular modern approach called *bootstrap* is discussed in Section 10.3.3.

9.3.1 Large samples

A large sample should produce a rather accurate estimator of a variance. We can then replace the true standard error $\sigma(\widehat{\theta})$ in (9.3) by its estimator $s(\widehat{\theta})$, and obtain an approximate confidence interval

$$\widehat{\theta} \pm z_{\alpha/2} \cdot s(\widehat{\theta}).$$

Example 9.16 (DELAYS AT NODES). Internet connections are often slowed by delays at nodes. Let us determine if the delay time increases during heavy-volume times.

Five hundred packets are sent through the same network between 5 pm and 6 pm (sample $\boldsymbol{X}$), and three hundred packets are sent between 10 pm and 11 pm (sample $\boldsymbol{Y}$). The early sample has a mean delay time of 0.8 sec with a standard deviation of 0.1 sec whereas the second sample has a mean delay time of 0.5 sec with a standard deviation of 0.08 sec. Construct a 99.5% confidence interval for the difference between the mean delay times.

Solution. We have $n = 500, \overline{X} = 0.8, s_X = 0.1; m = 300, \overline{Y} = 0.5, s_Y = 0.08$. Large sample sizes allow us to replace unknown population standard deviations by their estimates and use an approximately Normal distribution of sample means.

For a confidence level of $1 - \alpha = 0.995$, we need

$$z_{\alpha/2} = z_{0.0025} = q_{0.9975}.$$

Look for the *probability* 0.9975 in the body of Table A4 and find the corresponding value of z,

$$z_{0.0025} = 2.81.$$

Then, a 99.5% confidence interval for the difference of mean execution times is

$$\overline{X} - \overline{Y} \;\pm\; z_{0.0025}\sqrt{\frac{s_X^2}{n} + \frac{s_Y^2}{m}} = (0.8 - 0.5) \pm (2.81)\sqrt{\frac{(0.1)^2}{500} + \frac{(0.08)^2}{300}}$$
$$= \;\; 0.3 \pm 0.018 \;\; \text{or} \;\; [0.282, \, 0.318].$$

$\Diamond$

9.3.2 Confidence intervals for proportions

In particular, we surely don't know the variance when we estimate a population proportion.

DEFINITION 9.5 ———————

We assume a subpopulation A of items that have a certain *attribute*. By the **population proportion** we mean the probability

$$p = \boldsymbol{P}\{i \in A\}$$

for a randomly selected item i to have this attribute.

A **sample proportion**

$$\widehat{p} = \frac{\text{number of sampled items from } A}{n}$$

is used to estimate p.

Let us use the *indicator* variables

$$X_i = \begin{cases} 1 & \text{if} \quad i \in A \\ 0 & \text{if} \quad i \notin A \end{cases}$$

Each X_i has Bernoulli distribution with parameter p. In particular,

$$\boldsymbol{E}(X_i) = p \quad \text{and} \quad \text{Var}(X_i) = p(1-p).$$

Also,

$$\widehat{p} = \frac{1}{n}\sum_{i=1}^{n} X_i$$

is nothing but a sample mean of X_i.

Therefore,

$$\boldsymbol{E}(\widehat{p}) = p \quad \text{and} \quad \text{Var}(\widehat{p}) = \frac{p(1-p)}{n},$$

as we know from properties of sample means on p. 219.

We conclude that

1. a sample proportion $\widehat{p}$ is unbiased for the population proportion p;

2. it has approximately Normal distribution for large samples, because it has a form of a sample mean;

3. when we construct a confidence interval for p, we do not know the standard deviation $\text{Std}(\widehat{p})$.

Indeed, knowing the standard deviation is equivalent to knowing p, and if we know p, why would we need a confidence interval for it?

Thus, we estimate the unknown standard error

$$\sigma(\widehat{p}) = \sqrt{\frac{p(1-p)}{n}}$$

by

$$s(\widehat{p}) = \sqrt{\frac{\widehat{p}(1-\widehat{p})}{n}}$$

and use it in the general formula

$$\widehat{p} \pm z_{\alpha/2} \cdot s(\widehat{p})$$

to construct an approximate $(1-\alpha)100\%$ confidence interval.

Confidence interval for a population proportion	$\widehat{p} \pm z_{\alpha/2}\sqrt{\dfrac{\widehat{p}(1-\widehat{p})}{n}}$

Similarly, we can construct a confidence interval for the *difference between two proportions*. In two populations, we have proportions p_1 and p_2 of items with an attribute. Independent samples of size n_1 and n_2 are collected, and both parameters are estimated by sample proportions $\widehat{p}_1$ and $\widehat{p}_2$.

Summarizing, we have

$$\text{Parameter of interest:} \quad \theta = p_1 - p_2$$

$$\text{Estimated by:} \quad \widehat{\theta} = \widehat{p}_1 - \widehat{p}_2$$

$$\text{Its standard error:} \quad \sigma(\widehat{\theta}) = \sqrt{\frac{p_1(1-p_1)}{n_1} + \frac{p_2(1-p_2)}{n_2}}$$

$$\text{Estimated by:} \quad s(\widehat{\theta}) = \sqrt{\frac{\widehat{p}_1(1-\widehat{p}_1)}{n_1} + \frac{\widehat{p}_2(1-\widehat{p}_2)}{n_2}}$$

Confidence interval for the difference of proportions	$\widehat{p}_1 - \widehat{p}_2 \pm z_{\alpha/2}\sqrt{\dfrac{\widehat{p}_1(1-\widehat{p}_1)}{n_1} + \dfrac{\widehat{p}_2(1-\widehat{p}_2)}{n_2}}$

Example 9.17 (PRE-ELECTION POLL). A candidate prepares for the local elections. During his campaign, 42 out of 70 randomly selected people in town A and 59 out of 100 randomly selected people in town B showed they would vote for this candidate. Estimate the difference in support that this candidate is getting in towns A and B with 95% confidence. Can we state affirmatively that the candidate gets a stronger support in town A?

<u>Solution</u>. We have $n_1 = 70$, $n_2 = 100$, $\widehat{p}_1 = 42/70 = 0.6$, and $\widehat{p}_2 = 59/100 = 0.59$. For the confidence interval, we have

$$\text{center} = \widehat{p}_1 - \widehat{p}_2 = 0.01,$$

and

$$\begin{aligned} \text{margin} &= z_{0.05/2}\sqrt{\frac{\widehat{p}_1(1-\widehat{p}_1)}{n_1} + \frac{\widehat{p}_2(1-\widehat{p}_2)}{n_2}} \\ &= (1.960)\sqrt{\frac{(0.6)(0.4)}{70} + \frac{(0.59)(0.41)}{100}} = 0.15. \end{aligned}$$

Then

$$0.01 \pm 0.15 = [\text{-0.14, 0.16}]$$

is a 95% confidence interval for the difference in support $(p_1 - p_2)$ in the two towns.

So, is the support stronger in town A? On one hand, the estimator $\widehat{p}_1 - \widehat{p}_2 = 0.01$ suggests that the support is 1% higher in town A than in town B. On the other hand, the difference could appear positive just because of a sampling error. As we see, the 95% confidence interval includes a large range of negative values too. Therefore, the obtained data does *not* indicate affirmatively that the support in town A is stronger.

In fact, we will test in Example 9.33 if there is any difference between the two towns and will conclude that there is no evidence for it or against it. A formal procedure for testing statements like that will be introduced in Section 9.4. ◇

9.3.3 Estimating proportions with a given precision

Our confidence interval for a population proportion has a margin

$$\text{margin} = z_{\alpha/2}\sqrt{\frac{\widehat{p}(1-\widehat{p})}{n}}.$$

A standard way of finding the sample size that provides the desired margin Δ is to solve the inequality

$$\text{margin} \le \Delta \quad \text{or} \quad n \ge \widehat{p}(1-\widehat{p})\left(\frac{z_{\alpha/2}}{\Delta}\right)^2.$$

However, this inequality includes $\widehat{p}$. To know $\widehat{p}$, we first need to collect a sample, but to know the sample size, we first need to know $\widehat{p}$!

A way out of this circle is shown in Figure 9.5. As we see, the function $\widehat{p}(1-\widehat{p})$ never exceeds 0.25. Therefore, we can replace the unknown value of $\widehat{p}(1-\widehat{p})$ by 0.25 and find a sample size n, perhaps larger than we actually need, that will ensure that we estimate $\widehat{p}$ with a margin not exceeding Δ. That is, choose a sample size

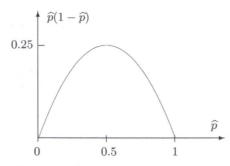

FIGURE 9.5: *Function $\widehat{p}(1 - \widehat{p})$ attains its maximum at $\widehat{p} = 0.5$.*

$$n \geq 0.25 \left(\frac{z_{\alpha/2}}{\Delta} \right)^2.$$

It will automatically be at least as large as the required $\widehat{p}(1 - \widehat{p})(z_{\alpha/2}/\Delta)^2$, regardless of the unknown value of $\widehat{p}$.

Example 9.18. A sample of size

$$n \geq 0.25 \left(\frac{1.960}{0.1} \right)^2 = 96.04$$

(that is, at least 97 observations) always guarantees that a population proportion is estimated with an error of at most 0.1 with a 95% confidence. $\Diamond$

9.3.4   Small samples: Student's t distribution

Having a small sample, we can no longer pretend that a sample standard deviation s is an accurate estimator of the population standard deviation σ. Then, how should we adjust the confidence interval when we replace σ by s, or more generally, when we replace the standard error $\sigma(\widehat{\theta})$ by its estimator $s(\widehat{\theta})$?

A famous solution was proposed by *William Gosset* (1876–1937), known by his pseudonym *Student*. Working for the Irish brewery Guinness, he derived the T-distribution for the quality control problems in brewing.

Student followed the steps similar to our derivation of a confidence interval on p. 255. Then he replaced the true but unknown standard error of $\widehat{\theta}$ by its estimator $s(\widehat{\theta})$ and concluded that the **T-ratio**

$$t = \frac{\widehat{\theta} - \theta}{s(\widehat{\theta})},$$

the *ratio* of two random variables, no longer has a Normal distribution!

Student figured the distribution of a T-ratio. For the problem of estimating the mean based on n Normal observations $X_1, \ldots, X_n$, this was **T-distribution** with $(n - 1)$ *degrees of* *freedom*. Table A5 gives critical values t_α of the T-distribution that we'll use for confidence intervals.

So, using T-distribution instead of Standard Normal and estimated standard error instead of the unknown true one, we obtain the confidence interval for the population mean.

<table>
<tr>
<td>Confidence
interval
for the mean;
σ is unknown</td>
<td>$$\overline{X} \pm t_{\alpha/2}\frac{s}{\sqrt{n}}$$

where $t_{\alpha/2}$ is a critical value from T-distribution
with $n-1$ degrees of freedom</td>
<td>(9.9)</td>
</tr>
</table>

Example 9.19 (UNAUTHORIZED USE OF A COMPUTER ACCOUNT). If an unauthorized person accesses a computer account with the correct username and password (stolen or cracked), can this intrusion be detected? Recently, a number of methods have been proposed to detect such unauthorized use. The time between keystrokes, the time a key is depressed, the frequency of various keywords are measured and compared with those of the account owner. If there are significant differences, an intruder is detected.

The following times between keystrokes were recorded when a user typed the username and password (data set `Keystrokes`):

.24, .22, .26, .34, .35, .32, .33, .29, .19, .36, .30, .15, .17, .28, .38, .40, .37, .27 seconds

As the first step in detecting an intrusion, let's construct a 99% confidence interval for the mean time between keystrokes assuming Normal distribution of these times.

Solution. The sample size is $n = 18$, the sample mean time is $\overline{X} = 0.29$ sec, and the sample standard deviation is $s = 0.074$. The critical value of t distribution with $n - 1 = 17$ degrees of freedom is $t_{\alpha/2} = t_{0.005} = 2.898$. Then, the 99% confidence interval for the mean time is

$$0.29 \pm (2.898)\frac{0.074}{\sqrt{18}} = 0.29 \pm 0.05 = [0.24; 0.34]$$

Example 9.28 on p. 282 will show whether this result signals an intrusion. ◊

The density of Student's T-distribution is a bell-shaped symmetric curve that can be easily confused with Normal. Comparing with the Normal density, its peak is lower and its tails are thicker. Therefore, a larger number t_α is generally needed to cut area α from the right tail. That is,

$$t_\alpha > z_\alpha$$

for small α. As a consequence, the confidence interval (9.9) is wider than the interval (9.5) for the case of known σ. This wider margin is the price paid for not knowing the standard deviation σ. When we lack a certain piece of information, we cannot get a more accurate estimator.

However, we see in Table A5 that

$$t_\alpha \to z_\alpha,$$

as the number of degrees of freedom ν tends to infinity. Indeed, having a large sample (hence, large $\nu = n - 1$), we can count on a very accurate estimator of σ, and thus, the confidence interval is almost as narrow as if we knew σ in this case.

Degrees of freedom ν is the parameter of T-distribution controlling the shape of the T-density curve. Its meaning is the *dimension* of a vector used to estimate the variance. Here we estimate σ^2 by a sample variance

$$s^2 = \frac{1}{n-1} \sum_{i=1}^{n} (X_i - \overline{X})^2,$$

and thus, we use a vector

$$\boldsymbol{X}' = \left(X_1 - \overline{X}, \ldots, X_n - \overline{X} \right).$$

The initial vector $\boldsymbol{X} = (X_1, \ldots, X_n)$ has dimension n; therefore, it has n degrees of freedom. However, when the sample mean $\overline{X}$ is subtracted from each observation, there appears a linear relation among the elements,

$$\sum_{i=1}^{n} (X_i - \overline{X}) = 0.$$

We lose 1 degree of freedom due to this constraint; the vector $\boldsymbol{X}'$ belongs to an $(n-1)$-dimensional hyperplane, and this is why we have only $\nu = n - 1$ degrees of freedom.

In many similar problems, degrees of freedom can be computed as

$$\begin{array}{c} \text{number of} \\ \text{degrees of freedom} \end{array} = \text{sample size} - \begin{array}{c} \text{number of estimated} \\ \text{location parameters} \end{array} \qquad (9.10)$$

9.3.5 Comparison of two populations with unknown variances

We now construct a confidence interval for the difference of two means $\mu_X - \mu_Y$, comparing the population of X's with the population of Y's.

Again, independent random samples are collected,

$$\boldsymbol{X} = (X_1, \ldots, X_n) \quad \text{and} \quad \boldsymbol{Y} = (Y_1, \ldots, Y_m),$$

one from each population, as in Figure 9.4 on p. 259. This time, however, population variances σ_X^2 and σ_Y^2 are unknown to us, and we use their estimates.

Two important cases need to be considered here. In one case, there exists an exact and simple solution based on T-distribution. The other case suddenly appears to be a famous *Behrens–Fisher problem*, where no exact solution exists, and only approximations are available.

Case 1. Equal variances

Suppose there are reasons to assume that the two populations have equal variances,

$$\sigma_X^2 = \sigma_Y^2 = \sigma^2.$$

For example, two sets of data are collected with the same measurement device, thus, measurements have different means but the same precision.

In this case, there is only one variance σ^2 to estimate instead of two. We should use both samples $\mathbf{X}$ and $\mathbf{Y}$ to estimate their common variance. This estimator of σ^2 is called a **pooled sample variance**, and it is computed as

$$s_p^2 = \frac{\sum_{i=1}^{n}(X_i - \overline{X})^2 + \sum_{i=1}^{m}(Y_i - \overline{Y})^2}{n+m-2} = \frac{(n-1)s_X^2 + (m-1)s_Y^2}{n+m-2}. \tag{9.11}$$

Substituting this variance estimator in (9.6) for σ_X^2 and σ_Y^2, we get the following confidence interval.

Confidence interval for the difference of means; equal, unknown standard deviations	$\overline{X} - \overline{Y} \pm t_{\alpha/2}\, s_p \sqrt{\dfrac{1}{n} + \dfrac{1}{m}}$ where s_p is the *pooled standard deviation*, a root of the pooled variance in (9.11) and $t_{\alpha/2}$ is a critical value from T-distribution with $(n+m-2)$ degrees of freedom

Example 9.20 (CD WRITER AND BATTERY LIFE). CD writing is energy consuming; therefore, it affects the battery lifetime on laptops. To estimate the effect of CD writing, 30 users are asked to work on their laptops until the "low battery" sign comes on.

Eighteen users without a CD writer worked an average of 5.3 hours with a standard deviation of 1.4 hours. The other twelve, who used their CD writer, worked an average of 4.8 hours with a standard deviation of 1.6 hours. Assuming Normal distributions with equal population variances ($\sigma_X^2 = \sigma_Y^2$), construct a 95% confidence interval for the battery life reduction caused by CD writing.

Solution. Effect of the CD writer is measured by the reduction of the mean battery life. We have $n = 12$, $\overline{X} = 4.8$, $s_X = 1.6$ for users with a CD writer and $m = 18$, $\overline{Y} = 5.3$, $s_Y = 1.4$ for users without it. The pooled standard deviation is

$$s_p = \sqrt{\frac{(n-1)s_X^2 + (m-1)s_Y^2}{n+m-2}} = \sqrt{\frac{(11)(1.6)^2 + (17)(1.4)^2}{28}} = 1.4818$$

(check: it has to be between s_X and s_Y). The critical value is $t_{0.025} = 2.048$ (use 28 d.f.). The 95% confidence interval for the difference between the mean battery lives is

$$(4.8 - 5.3) \pm (2.048)(1.4818)\sqrt{\frac{1}{18} + \frac{1}{12}} = -0.5 \pm 1.13 = [-1.63;\ 0.63]. \qquad \Diamond$$

Remark: Let's discuss formula (9.11). First, notice that different sample means $\overline{X}$ and $\overline{Y}$ are used for X-terms and Y-terms. Indeed, our two populations may have different means. As we know, variance of any variable measures its deviation from its mean. Thus, from each observation we subtract its own mean-estimate.

Second, we lose 2 degrees of freedom due to the estimation of two means. Two constraints,

$$\sum_{i=1}^{n}(X_i - \overline{X}) = 0 \quad \text{and} \quad \sum_{i=1}^{m}(Y_i - \overline{Y}) = 0,$$

show that the number of **degrees of freedom** is only $(n + m - 2)$ instead of $(n + m)$. We see this coefficient in the denominator, and it makes s_p^2 an unbiased estimator of σ^2 (see Exercise 9.19).

Case 2. Unequal variances

The most difficult case is when both variances are unknown and unequal. Confidence estimation of $\mu_X - \mu_Y$ in this case is known as the *Behrens–Fisher problem*. Certainly, we can replace unknown variances σ_X^2, σ_Y^2 by their estimates s_X^2, s_Y^2 and form a T-ratio

$$t = \frac{(\overline{X} - \overline{Y}) - (\mu_X - \mu_Y)}{\sqrt{\dfrac{s_X^2}{n} + \dfrac{s_Y^2}{m}}}.$$

However, it won't have a T-distribution.

An approximate solution was proposed in the 1940s by *Franklin E. Satterthwaite*, who worked for General Electric Company at that time. Satterthwaite used the method of moments to estimate degrees of freedom ν of a T-distribution that is "closest" to this T-ratio. This number depends on unknown variances. Estimating them by sample variances, he obtained the formula that is now known as *Satterthwaite approximation*,

$$\nu = \frac{\left(\dfrac{s_X^2}{n} + \dfrac{s_Y^2}{m}\right)^2}{\dfrac{s_X^4}{n^2(n - 1)} + \dfrac{s_Y^4}{m^2(m - 1)}}. \tag{9.12}$$

This number of degrees of freedom often appears non-integer. There are T-distributions with non-integer ν, see Section A.2.1. To use Table A5, just take the closest ν that is given in that table.

Formula (9.12) is widely used for t-intervals and t-tests.

Confidence interval for the difference of means; unequal, unknown standard deviations	$\overline{X} - \overline{Y} \pm t_{\alpha/2} \sqrt{\dfrac{s_X^2}{n} + \dfrac{s_Y^2}{m}}$ where $t_{\alpha/2}$ is a critical value from T-distribution with ν degrees of freedom given by formula (9.12)

Example 9.21 (COMPARISON OF TWO SERVERS). An account on server A is more expensive than an account on server B. However, server A is faster. To see if it's optimal to go with the faster but more expensive server, a manager needs to know how much faster it is. A certain computer algorithm is executed 30 times on server A and 20 times on server B with the following results,

	Server A	Server B
Sample mean	6.7 min	7.5 min
Sample standard deviation	0.6 min	1.2 min

Construct a 95% confidence interval for the difference $\mu_1 - \mu_2$ between the mean execution times on server A and server B, assuming that the observed times are approximately Normal.

Solution. We have $n = 30$, $m = 20$, $\overline{X} = 6.7$, $\overline{Y} = 7.5$, $s_X = 0.6$, and $s_Y = 1.2$. The second standard deviation is twice larger than the first one; therefore, equality of population variances can hardly be assumed. We use the method for unknown, unequal variances.

Using Satterthwaite approximation (9.12), we find degrees of freedom:

$$\nu = \frac{\left(\dfrac{(0.6)^2}{30} + \dfrac{(1.2)^2}{20} \right)^2}{\dfrac{(0.6)^4}{30^2(29)} + \dfrac{(1.2)^4}{20^2(19)}} = 25.4.$$

To use Table A5, we round this ν to 25 and find $t_{0.025} = 2.060$. Then, the confidence interval is

$$\overline{X} - \overline{Y} \pm t_{\alpha/2} \sqrt{\frac{s_X^2}{n} + \frac{s_Y^2}{m}} = 6.7 - 7.5 \pm (2.060) \sqrt{\frac{(0.6)^2}{30} + \frac{(1.2)^2}{20}}$$

$$= -0.8 \pm 0.6 \quad \text{or} \quad [-1.4, -0.2].$$

$\Diamond$

9.4 Hypothesis testing

A vital role of Statistics is in verifying statements, claims, conjectures, and in general - *testing hypotheses*. Based on a random sample, we can use Statistics to verify whether

- a system has not been infected,

- a hardware upgrade was efficient,

- the average number of concurrent users increased by 2000 this year,

- the average connection speed is 54 Mbps, as claimed by the internet service provider,

- the proportion of defective products is at most 3%, as promised by the manufacturer,

- service times have Gamma distribution,

- the number of errors in software is independent of the manager's experience,

- etc.

Testing statistical hypotheses has wide applications far beyond Computer Science. These methods are used to prove efficiency of a new medical treatment, safety of a new automobile brand, innocence of a defendant, and authorship of a document; to establish cause-and-effect relationships; to identify factors that can significantly improve the response; to fit stochastic models; to detect information leaks; and so forth.

9.4.1 Hypothesis and alternative

To begin, we need to state exactly what we are testing. These are *hypothesis* and *alternative*.

$$\text{\underline{Notation}} \quad \left\| \begin{array}{lcl} H_0 & = & \text{hypothesis (the null hypothesis)} \\ H_A & = & \text{alternative (the alternative hypothesis)} \end{array} \right\|$$

H_0 and H_A are simply two mutually exclusive statements. Each test results either in acceptance of H_0 or its rejection in favor of H_A.

A null hypothesis is always an equality, absence of an effect or relation, some "normal," usual statement that people have believed in for years. In order to overturn the common belief and to reject the hypothesis, we need *significant evidence*. Such evidence can only be provided by data. Only when such evidence is found, and when it strongly supports the alternative H_A, can the hypothesis H_0 be rejected in favor of H_A.

Based on a random sample, a statistician cannot tell whether the hypothesis is true or the alternative. We need to see the entire population to tell that. The purpose of each test is to determine whether the data provides sufficient evidence against H_0 in favor of H_A.

This is similar to a criminal trial. Typically, the jury cannot tell whether the defendant committed a crime or not. It is not their task. They are only required to determine if the presented evidence against the defendant is sufficient and convincing. By default, called *presumption of innocence*, insufficient evidence leads to acquittal.

Example 9.22. To verify that the the average connection speed is 54 Mbps, we test the hypothesis $H_0 : \mu = 54$ against the *two-sided alternative* $H_A : \mu \neq 54$, where μ is the average speed of all connections.

However, if we worry about a *low* connection speed only, we can conduct a one-sided test of

$$H_0 : \mu = 54 \quad \text{vs} \quad H_A : \mu < 54.$$

In this case, we only measure the amount of evidence supporting the *one-sided alternative* $H_A : \mu < 54$. In the absence of such evidence, we gladly accept the null hypothesis. $\Diamond$

DEFINITION 9.6 ─────────────

Alternative of the type $H_A : \mu \neq \mu_0$ covering regions on both sides of the hypothesis ($H_0 : \mu = \mu_0$) is a **two-sided alternative**.

Alternative $H_A : \mu < \mu_0$ covering the region to the left of H_0 is **one-sided, left-tail**.

Alternative $H_A : \mu > \mu_0$ covering the region to the right of H_0 is **one-sided, right-tail**.

Example 9.23. To verify whether the average number of concurrent users increased by 2000, we test

$$H_0 : \mu_2 - \mu_1 = 2000 \quad \text{vs} \quad H_A : \mu_2 - \mu_1 \neq 2000,$$

where μ_1 is the average number of concurrent users last year, and μ_2 is the average number of concurrent users this year. Depending on the situation, we may replace the *two-sided* alternative $H_A : \mu_2 - \mu_1 \neq 2000$ with a one-sided alternative $H_A^{(1)} : \mu_2 - \mu_1 < 2000$ or $H_A^{(2)} : \mu_2 - \mu_1 > 2000$. The test of H_0 against $H_A^{(1)}$ evaluates the amount of evidence that the mean number of concurrent users changed by fewer than 2000. Testing against $H_A^{(2)}$, we see if there is sufficient evidence to claim that this number increased by more than 2000. $\Diamond$

Example 9.24. To verify if the proportion of defective products is at most 3%, we test

$$H_0 : p = 0.03 \quad \text{vs} \quad H_A : p > 0.03,$$

where p is the proportion of defects in the whole shipment.

Why do we choose the *right-tail alternative* $H_A : p > 0.03$? That is because we reject the shipment only if significant evidence supporting this alternative is collected. If the data suggest that $p < 0.03$, the shipment will still be accepted. $\Diamond$

9.4.2 Type I and Type II errors: level of significance

When testing hypotheses, we realize that all we see is a random sample. Therefore, with all the best statistics skills, our decision to accept or to reject H_0 may still be wrong. That would be a *sampling error* (Section 8.1).

Four situations are possible,

	Result of the test	
	Reject H_0	**Accept** H_0
H_0 **is true**	*Type I error*	correct
H_0 **is false**	correct	*Type II error*

In two of the four cases, the test results in a *correct decision*. Either we accepted a true hypothesis, or we rejected a false hypothesis. The other two situations are sampling errors.

DEFINITION 9.7 ———————————————————————————

A **type I error** occurs when we reject the true null hypothesis.

A **type II error** occurs when we accept the false null hypothesis.

Each error occurs with a certain probability that we hope to keep small. A good test results in an erroneous decision only if the observed data are somewhat extreme.

A type I error is often considered more dangerous and undesired than a type II error. Making a type I error can be compared with convicting an innocent defendant or sending a patient to a surgery when (s)he does not need one.

For this reason, we shall design tests that bound the probability of type I error by a pre-assigned small number α. Under this condition, we may want to minimize the probability of type II error.

DEFINITION 9.8 ———————————————————————————

Probability of a type I error is the **significance level** of a test,

$$\alpha = P \left\{ \text{reject } H_0 \mid H_0 \text{ is true} \right\}.$$

Probability of rejecting a false hypothesis is the **power of the test**,

$$p(\theta) = P \left\{ \text{reject } H_0 \mid \theta; \, H_A \text{ is true} \right\}.$$

It is usually a function of the parameter θ because the alternative hypothesis includes a set of parameter values. Also, the power is the probability to avoid a Type II error.

Typically, hypotheses are tested at significance levels as small as 0.01, 0.05, or 0.10, although there are exceptions. Testing at a low level of significance means that only a large amount of evidence can force rejection of H_0. Rejecting a hypothesis at a very low level of significance is done with a lot of confidence that this decision is right.

9.4.3   Level α tests: general approach

A standard algorithm for a level α test of a hypothesis H_0 against an alternative H_A consists of 3 steps.

Step 1. Test statistic

Testing hypothesis is based on a **test statistic** T, a quantity computed from the data that has some known, tabulated distribution F_0 if the hypothesis H_0 is true.

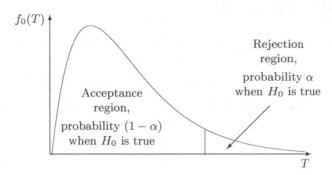

FIGURE 9.6: *Acceptance and rejection regions.*

Test statistics are used to discriminate between the hypothesis and the alternative. When we verify a hypothesis about some parameter θ, the test statistic is usually obtained by a suitable transformation of its estimator $\widehat{\theta}$.

Step 2. Acceptance region and rejection region

Next, we consider the **null distribution** F_0. This is the distribution of test statistic T when the hypothesis H_0 is true. If it has a density f_0, then the whole area under the density curve is 1, and we can always find a portion of it whose area is α, as shown in Figure 9.6. It is called **rejection region** ($\mathfrak{R}$).

The remaining part, the complement of the rejection region, is called **acceptance region** ($\mathfrak{A} = \overline{\mathfrak{R}}$). By the complement rule, its area is $(1 - \alpha)$.

These regions are selected in such a way that the values of test statistic T in the rejection region provide a stronger support of H_A than the values $T \in \mathfrak{A}$. For example, suppose that T is expected to be large if H_A is true. Then the rejection region corresponds to the right tail of the null distribution F_0 (Figure 9.6).

As another example, look at Figure 9.3 on p. 256. If the null distribution of T is *Standard Normal*, then the area between $(-z_{\alpha/2})$ and $z_{\alpha/2}$ equals exactly $(1 - \alpha)$. The interval

$$\mathfrak{A} = (-z_{\alpha/2}, z_{\alpha/2})$$

can serve as a level α acceptance region for a two-sided test of $H_0 : \theta = \theta_0$ vs $H_A : \theta \neq \theta_0$. The remaining part consists of two symmetric tails,

$$\mathfrak{R} = \overline{\mathfrak{A}} = (-\infty, -z_{\alpha/2}] \cup [z_{\alpha/2}, +\infty);$$

this is the rejection region.

Areas under the density curve are probabilities, and we conclude that

$$\boldsymbol{P}\{T \in \text{ acceptance region } \mid H_0\} = 1 - \alpha$$

and

$$\boldsymbol{P}\{T \in \text{ rejection region } \mid H_0\} = \alpha.$$

Step 3: Result and its interpretation

Accept the hypothesis H_0 if the test statistic T belongs to the acceptance region. Reject H_0 in favor of the alternative H_A if T belongs to the rejection region.

Our acceptance and rejection regions guarantee that the significance level of our test is

$$\begin{aligned} \text{Significance level} &= \boldsymbol{P}\{\text{ Type I error }\} \\ &= \boldsymbol{P}\{\text{ Reject } \mid H_0\} \\ &= \boldsymbol{P}\{T \in \mathfrak{R} \mid H_0\} \\ &= \alpha. \end{aligned} \qquad (9.13)$$

Therefore, indeed, we have a level α test!

The interesting part is to interpret our result correctly. Notice that conclusions like *"My level α test accepted the hypothesis. Therefore, the hypothesis is true with probability $(1 - \alpha)$"* are *wrong*! Statements H_0 and H_A are about a non-random population, and thus, the hypothesis can either be true with probability 1 or false with probability 1.

If the test rejects the hypothesis, all we can state is that the data provides sufficient evidence against H_0 and in favor of H_A. It may either happen because H_0 is not true, or because our sample is too extreme. The latter, however, can only happen with probability α.

If the test accepts the hypothesis, it only means that the evidence obtained from the data is not sufficient to reject it. In the absence of sufficient evidence, by default, we accept the null hypothesis.

Notation			
α	=	level of significance, probability of type I error
$p(\theta)$	=	power
T	=	test statistic
F_0, f_0	=	null distribution of T and its density
$\mathfrak{A}$	=	acceptance region
$\mathfrak{R}$	=	rejection region

9.4.4 Rejection regions and power

Our construction of the rejection region guaranteed the desired significance level α, as we proved in (9.13). However, one can choose many regions that will also have probability α (see Figure 9.7). Among them, which one is the best choice?

To avoid *type II errors*, we choose such a rejection region that will likely cover the test statistic T in case if the *alternative H_A* is true. This maximizes the *power* of our test because we'll rarely accept H_0 in this case.

Then, we look at our test statistic T under the alternative. Often

  (a) a *right-tail alternative* forces T to be large,

  (b) a *left-tail alternative* forces T to be small,

  (c) a *two-sided alternative* forces T to be either large or small

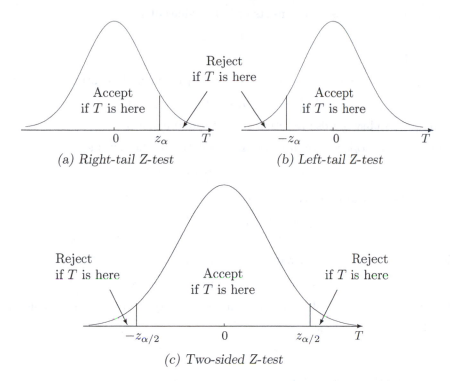

(a) *Right-tail Z-test*

(b) *Left-tail Z-test*

(c) *Two-sided Z-test*

FIGURE 9.7: *Acceptance and rejection regions for a Z-test with (a) a one-sided right-tail alternative; (b) a one-sided left-tail alternative; (c) a two-sided alternative.*

(although it certainly depends on how we choose T). If this is the case, it tells us exactly when we should reject the null hypothesis:

(a) For a **right-tail alternative**, the rejection region $\mathfrak{R}$ should consist of large values of T. Choose $\mathfrak{R}$ on the right, $\mathfrak{A}$ on the left (Figure 9.7a).

(b) For a **left-tail alternative**, the rejection region $\mathfrak{R}$ should consist of small values of T. Choose $\mathfrak{R}$ on the left, $\mathfrak{A}$ on the right (Figure 9.7b).

(c) For a **two-sided alternative**, the rejection region $\mathfrak{R}$ should consist of very small and very large values of T. Let $\mathfrak{R}$ consist of two extreme regions, while $\mathfrak{A}$ covers the middle (Figure 9.7c).

9.4.5 Standard Normal null distribution (Z-test)

An important case, in terms of a large number of applications, is when the null distribution of the test statistic is *Standard Normal.*

The test in this case is called a **Z-test**, and the test statistic is usually denoted by Z.

(a) A level α test with a **right-tail alternative** should

$$\begin{cases} \text{reject } H_0 & \text{if} \quad Z \geq z_\alpha \\ \text{accept } H_0 & \text{if} \quad Z < z_\alpha \end{cases} \qquad (9.14)$$

The rejection region in this case consists of large values of Z only,

$$\mathfrak{R} = [z_\alpha, +\infty), \qquad \mathfrak{A} = (-\infty, z_\alpha)$$

(see Figure 9.7a).

Under the null hypothesis, Z belongs to $\mathfrak{A}$ and we reject the null hypothesis with probability

$$\boldsymbol{P}\{T \geq z_\alpha \mid H_0\} = 1 - \Phi(z_\alpha) = \alpha,$$

making the probability of false rejection (type I error) equal α .

For example, we use this acceptance region to test the population mean,

$$H_0 : \ \mu = \mu_0 \ \text{ vs } \ H_A : \ \mu > \mu_0.$$

(b) With a **left-tail alternative**, we should

$$\begin{cases} \text{reject } H_0 & \text{if} \quad Z \leq -z_\alpha \\ \text{accept } H_0 & \text{if} \quad Z > -z_\alpha \end{cases} \tag{9.15}$$

The rejection region consists of small values of Z only,

$$\mathfrak{R} = (-\infty, -z_\alpha], \qquad \mathfrak{A} = (-z_\alpha, +\infty).$$

Similarly, $\boldsymbol{P}\{Z \in \mathfrak{R}\} = \alpha$ under H_0; thus, the probability of type I error equals α.

For example, this is how we should test

$$H_0 : \ \mu = \mu_0 \ \text{ vs } \ H_A : \ \mu < \mu_0.$$

(c) With a **two-sided alternative**, we

$$\begin{cases} \text{reject } H_0 & \text{if} \quad |Z| \geq z_{\alpha/2} \\ \text{accept } H_0 & \text{if} \quad |Z| < z_{\alpha/2} \end{cases} \tag{9.16}$$

The rejection region consists of very small and very large values of Z,

$$\mathfrak{R} = (-\infty, z_{\alpha/2}] \cup [z_{\alpha/2}, +\infty), \qquad A = (-z_{\alpha/2}, z_{\alpha/2}).$$

Again, the probability of type I error equals α in this case.

For example, we use this test for

$$H_0 : \ \mu = \mu_0 \ \text{ vs } \ H_A : \ \mu \neq \mu_0.$$

This is easy to remember:

- for a two-sided test, divide α by two and use $z_{\alpha/2}$;

- for a one-sided test, use z_α keeping in mind that the rejection region consists of just one piece.

Now consider testing a hypothesis about a population parameter θ. Suppose that its estimator $\widehat{\theta}$ has Normal distribution, at least approximately, and we know $\boldsymbol{E}(\widehat{\theta})$ and $\text{Var}(\widehat{\theta})$ if the hypothesis is true.

Then the test statistic

$$Z = \frac{\widehat{\theta} - \boldsymbol{E}(\widehat{\theta})}{\sqrt{\text{Var}(\widehat{\theta})}} \tag{9.17}$$

has Standard Normal distribution, and we can use (9.14), (9.15), and (9.16) to construct acceptance and rejection regions for a level α test. We call Z a **Z-statistic**.

Examples of Z-tests are in the next section.

9.4.6 Z-tests for means and proportions

As we already know,

- sample means have Normal distribution when the distribution of data is Normal;

- sample means have approximately Normal distribution when they are computed from large samples (the distribution of data can be arbitrary);

- sample proportions have approximately Normal distribution when they are computed from large samples;

- this extends to differences between means and between proportions

(see Sections 8.2.1 and 9.2.2–9.3.2).

For all these cases, we can use a Z-statistic (9.17) and rejection regions (9.14)–(9.16) to design powerful level α tests.

Z-tests are summarized in Table 9.1. You can certainly derive the test statistics without our help; see Exercise 9.6. The last row of Table 9.1 is explained in detail in Section 9.4.7.

Example 9.25 (Z-TEST ABOUT A POPULATION MEAN). The number of concurrent users for some internet service provider has always averaged 5000 with a standard deviation of 800. After an equipment upgrade, the average number of users at 100 randomly selected moments of time is 5200. Does it indicate, at a 5% level of significance, that the mean number of concurrent users has increased? Assume that the standard deviation of the number of concurrent users has not changed.

<u>Solution.</u> We test the null hypothesis $H_0 : \mu = 5000$ against a *one-sided right-tail alternative* $H_A : \mu > 5000$, because we are only interested to know if the mean number of users μ has increased.

Step 1: Test statistic. We are given: $\sigma = 800$, $n = 100$, $\alpha = 0.05$, $\mu_0 = 5000$, and from the sample, $\overline{X} = 5200$. The test statistic is

$$ Z = \frac{\overline{X} - \mu_0}{\sigma/\sqrt{n}} = \frac{5200 - 5000}{800/\sqrt{100}} = 2.5. $$

Step 2: Acceptance and rejection regions. The critical value is

$$ z_\alpha = z_{0.05} = 1.645 $$

(don't divide α by 2 because it is a one-sided test). With the right-tail alternative, we

$$ \begin{cases} \text{reject } H_0 & \text{if} \quad Z \geq 1.645 \\ \text{accept } H_0 & \text{if} \quad Z < 1.645 \end{cases} $$

Step 3: Result. Our test statistic $Z = 2.5$ belongs to the *rejection region*; therefore, we *reject the null hypothesis*. The data (5200 users, on the average, at 100 times) provided sufficient evidence in favor of the alternative hypothesis that the mean number of users has increased.

$\Diamond$

Null hypothesis	Parameter, estimator	If H_0 is true:		Test statistic
H_0	$\theta, \widehat{\theta}$	$\mathbf{E}(\widehat{\theta})$	$\mathrm{Var}(\widehat{\theta})$	$Z = \dfrac{\widehat{\theta} - \theta_0}{\sqrt{\mathrm{Var}(\widehat{\theta})}}$
One-sample Z-tests for means and proportions, based on a sample of size n				
$\mu = \mu_0$	$\mu, \overline{X}$	μ_0	$\dfrac{\sigma^2}{n}$	$\dfrac{\overline{X} - \mu_0}{\sigma/\sqrt{n}}$
$p = p_0$	$p, \widehat{p}$	p_0	$\dfrac{p_0(1-p_0)}{n}$	$\dfrac{\widehat{p} - p_0}{\sqrt{\frac{p_0(1-p_0)}{n}}}$
Two-sample Z-tests comparing means and proportions of two populations, based on independent samples of size n and m				
$\mu_X - \mu_Y = D$	$\mu_X - \mu_Y,$ $\overline{X} - \overline{Y}$	D	$\dfrac{\sigma_X^2}{n} + \dfrac{\sigma_Y^2}{m}$	$\dfrac{\overline{X} - \overline{Y} - D}{\sqrt{\frac{\sigma_X^2}{n} + \frac{\sigma_Y^2}{m}}}$
$p_1 - p_2 = D$	$p_1 - p_2,$ $\widehat{p}_1 - \widehat{p}_2$	D	$\dfrac{p_1(1-p_1)}{n} + \dfrac{p_2(1-p_2)}{m}$	$\dfrac{\widehat{p}_1 - \widehat{p}_2 - D}{\sqrt{\frac{\widehat{p}_1(1-\widehat{p}_1)}{n} + \frac{\widehat{p}_2(1-\widehat{p}_2)}{m}}}$
$p_1 = p_2$	$p_1 - p_2,$ $\widehat{p}_1 - \widehat{p}_2$	0	$p(1-p)\left(\dfrac{1}{n} + \dfrac{1}{m}\right),$ where $p = p_1 = p_2$	$\dfrac{\widehat{p}_1 - \widehat{p}_2}{\sqrt{\widehat{p}(1-\widehat{p})\left(\frac{1}{n} + \frac{1}{m}\right)}}$ where $\widehat{p} = \dfrac{n\widehat{p}_1 + m\widehat{p}_2}{n + m}$

TABLE 9.1: *Summary of Z-tests.*

Example 9.26 (TWO-SAMPLE Z-TEST OF PROPORTIONS). A quality inspector finds 10 defective parts in a sample of 500 parts received from manufacturer A. Out of 400 parts from manufacturer B, she finds 12 defective ones. A computer-making company uses these parts in their computers and claims that the quality of parts produced by A and B is the same. At the 5% level of significance, do we have enough evidence to disprove this claim?

Solution. We test $H_0 : p_A = p_B$, or $H_0 : p_A - p_B = 0$, against $H_A : p_A \neq p_B$. This is a two-sided test because no direction of the alternative has been indicated. We only need to verify whether or not the proportions of defective parts are equal for manufacturers A and B.

Step 1: Test statistic. We are given: $\widehat{p}_A = 10/500 = 0.02$ from a sample of size $n = 500$; $\widehat{p}_B = 12/400 = 0.03$ from a sample of size $m = 400$. The tested value is $D = 0$.

As we know, for these Bernoulli data, the variance depends on the unknown parameters p_A and p_B which are estimated by the sample proportions $\widehat{p}_A$ and $\widehat{p}_B$.

The test statistic then equals

$$Z = \frac{\widehat{p}_A - \widehat{p}_B - D}{\sqrt{\dfrac{\widehat{p}_A(1 - \widehat{p}_A)}{n} + \dfrac{\widehat{p}_B(1 - \widehat{p}_B)}{m}}} = \frac{0.02 - 0.03}{\sqrt{\dfrac{(0.02)(0.98)}{500} + \dfrac{(0.03)(0.97)}{400}}} = -0.945.$$

Step 2: Acceptance and rejection regions. This is a two-sided test; thus we divide α by 2, find $z_{0.05/2} = z_{0.025} = 1.96$, and

$$\begin{cases} \text{reject } H_0 & \text{if} \quad |Z| \geq 1.96; \\ \text{accept } H_0 & \text{if} \quad |Z| < 1.96. \end{cases}$$

Step 3: Result. The evidence against H_0 is insufficient because $|Z| < 1.96$. Although *sample proportions* of defective parts are unequal, the difference between them appears too small to claim that *population proportions* are different. $\diamond$

9.4.7 Pooled sample proportion

The test in Example 9.26 can be conducted differently and perhaps, more efficiently.

Indeed, we standardize the estimator $\widehat{\theta} = \widehat{p}_A - \widehat{p}_B$ using its expectation $\mathbf{E}(\widehat{\theta})$ and variance $\text{Var}(\widehat{\theta})$ under the null distribution, i.e., when H_0 is true. However, under the null hypothesis $p_A = p_B$. Then, when we standardize $(\widehat{p}_A - \widehat{p}_B)$, instead of estimating two proportions in the denominator, we only need to estimate one.

First, we estimate the common population proportion by the overall proportion of defective parts,

$$\widehat{p}(\text{pooled}) = \frac{\text{number of defective parts}}{\text{total number of parts}} = \frac{n\widehat{p}_A + m\widehat{p}_B}{n + m}.$$

Then we estimate the common variance as

$$\widehat{\text{Var}}(\widehat{p}_A - \widehat{p}_B) = \frac{\widehat{p}(1 - \widehat{p})}{n} + \frac{\widehat{p}(1 - \widehat{p})}{m} = \widehat{p}(1 - \widehat{p})\left(\frac{1}{n} + \frac{1}{m}\right)$$

and use it for the Z-statistic,

$$Z = \frac{\widehat{p}_A - \widehat{p}_B}{\sqrt{\widehat{p}(1 - \widehat{p})\left(\frac{1}{n} + \frac{1}{m}\right)}}.$$

Example 9.27 (EXAMPLE 9.26, CONTINUED). Here the pooled proportion equals

$$\widehat{p} = \frac{10 + 12}{500 + 400} = 0.0244,$$

so that

$$Z = \frac{0.02 - 0.03}{\sqrt{(0.0244)(0.9756)\left(\frac{1}{500} + \frac{1}{400}\right)}} = -0.966.$$

This does not affect our result. We obtained a different value of Z-statistic, but it also belongs to the acceptance region. We still don't have a significant evidence against the equality of two population proportions. $\diamond$

9.4.8 Unknown σ: T-tests

As we decided in Section 9.3, when we don't know the population standard deviation, we estimate it. The resulting *T-statistic* has the form

$$t = \frac{\widehat{\theta} - \mathbf{E}(\widehat{\theta})}{s(\widehat{\theta})} = \frac{\widehat{\theta} - \mathbf{E}(\widehat{\theta})}{\sqrt{\widehat{\mathrm{Var}(\theta)}}}.$$

In the case *when the distribution of* $\widehat{\theta}$ *is Normal*, the test is based on *Student's T-distribution* with acceptance and rejection regions according to the direction of H_A:

(a) For a **right-tail alternative**,

$$\begin{cases} \text{reject } H_0 & \text{if } \quad t \geq t_\alpha \\ \text{accept } H_0 & \text{if } \quad t < t_\alpha \end{cases} \tag{9.18}$$

(b) For a **left-tail alternative**,

$$\begin{cases} \text{reject } H_0 & \text{if } \quad t \leq -t_\alpha \\ \text{accept } H_0 & \text{if } \quad t > -t_\alpha \end{cases} \tag{9.19}$$

(c) For a **two-sided alternative**,

$$\begin{cases} \text{reject } H_0 & \text{if } \quad |t| \geq t_{\alpha/2} \\ \text{accept } H_0 & \text{if } \quad |t| < t_{\alpha/2} \end{cases} \tag{9.20}$$

Quantiles t_α and $t_{\alpha/2}$ are given in Table A5. As in Section 9.3.4, the number of degrees of freedom depends on the problem and the sample size, see Table 9.2 and formula (9.10).

As in Section 9.3.4, the **pooled sample variance**

$$s_p^2 = \frac{\displaystyle\sum_{i=1}^{n}(X_i - \overline{X})^2 + \sum_{i=1}^{m}(Y_i - \overline{Y})^2}{n + m - 2} = \frac{(n-1)s_X^2 + (m-1)s_Y^2}{n + m - 2}$$

is computed for the case of equal unknown variances. When variances are not equal, degrees of freedom are computed by Satterthwaite approximation (9.12).

Example 9.28 (UNAUTHORIZED USE OF A COMPUTER ACCOUNT, CONTINUED). A longtime authorized user of the account makes 0.2 seconds between keystrokes. One day, the data in Example 9.19 on p. 267 (data set `Keystrokes`) are recorded as someone typed the correct username and password. At a 5% level of significance, is this an evidence of an unauthorized attempt?

Hypothesis H_0	Conditions	Test statistic t	Degrees of freedom
$\mu = \mu_0$	Sample size n; unknown σ	$t = \dfrac{\overline{X} - \mu_0}{s/\sqrt{n}}$	$n - 1$
$\mu_X - \mu_Y = D$	Sample sizes n, m; unknown but equal standard deviations, $\sigma_X = \sigma_Y$	$t = \dfrac{\overline{X} - \overline{Y} - D}{s_p\sqrt{\frac{1}{n} + \frac{1}{m}}}$	$n + m - 2$
$\mu_X - \mu_Y = D$	Sample sizes n, m; unknown, unequal standard deviations, $\sigma_X \neq \sigma_Y$	$t = \dfrac{\overline{X} - \overline{Y} - D}{\sqrt{\frac{s_X^2}{n} + \frac{s_Y^2}{m}}}$	Satterthwaite approximation, formula (9.12)

TABLE 9.2: *Summary of T-tests.*

Let us test

$$H_0 : \mu = 0.2 \quad \text{vs} \quad H_A : \mu \neq 0.2$$

at a significance level $\alpha = 0.01$. From Example 9.19, we have sample statistics $n = 18$, $\overline{X} = 0.29$ and $s = 0.074$. Compute the T-statistic,

$$t = \frac{\overline{X} - 0.2}{s/\sqrt{n}} = \frac{0.29 - 0.2}{0.074/\sqrt{18}} = 5.16.$$

The rejection region is $\Re = (-\infty, -2.11] \cup [2.11, \infty)$, where we used T-distribution with $18 - 1 = 17$ degrees of freedom and $\alpha/2 = 0.025$ because of the two-sided alternative.

Since $t \in \Re$, we reject the null hypothesis and conclude that *there is a significant evidence of an unauthorized use of that account.* $\diamond$

Example 9.29 (CD WRITER AND BATTERY LIFE). Does a CD writer consume extra energy, and therefore, does it reduce the battery life on a laptop?

Example 9.20 on p. 269 provides data on battery lives for laptops with a CD writer (sample X) and without a CD writer (sample Y):

$$n = 12, \ \overline{X} = 4.8, \ s_X = 1.6; \ m = 18, \ \overline{Y} = 5.3, \ s_Y = 1.4; \ s_p = 1.4818.$$

Testing

$$H_0 : \mu_X = \mu_Y \quad \text{vs} \quad H_A : \mu_X < \mu_Y$$

at $\alpha = 0.05$, we obtain

$$t = \frac{\overline{X} - \overline{Y}}{s_p\sqrt{\frac{1}{n} + \frac{1}{m}}} = \frac{4.8 - 5.3}{(1.4818)\sqrt{\frac{1}{18} + \frac{1}{12}}} = -0.9054.$$

The rejection region for this left-tail test is $(-\infty, -z_\alpha] = (-\infty, -1.645]$. Since $t \notin \mathfrak{R}$, we accept H_0 concluding that there is *no evidence that laptops with a CD writer have a shorter battery life.* $\Diamond$

Example 9.30 (COMPARISON OF TWO SERVERS, CONTINUED). Is server A faster in Example 9.21 on p. 271? Formulate and test the hypothesis at a level $\alpha = 0.05$.

Solution. To see if server A is faster, we need to test

$$H_0 : \mu_X = \mu_Y \text{ vs } H_A : \mu_X < \mu_Y.$$

This is the case of unknown, unequal standard deviations. In Example 9.21, we used Satterthwaite approximation for the number of degrees of freedom and obtained $\nu = 25.4$. We should reject the null hypothesis if $t \leq -1.708$. Since

$$t = \frac{6.7 - 7.5}{\sqrt{\frac{(0.6)^2}{30} + \frac{(1.2)^2}{20}}} = -2.7603 \in \mathfrak{R},$$

we reject H_0 and conclude that there is evidence that *server A is faster.* $\Diamond$

When the distribution of $\hat{\theta}$ is not Normal, the Student's *T-distribution* cannot be used. The distribution of a T-statistic and all its probabilities will be different from Student's T, and as a result, our test may not have the desired significance level.

9.4.9 Duality: two-sided tests and two-sided confidence intervals

An interesting fact can be discovered if we look into our derivation of tests and confidence intervals. It turns out that we can conduct two-sided tests using nothing but the confidence intervals!

A level α Z-test of $H_0 : \theta = \theta_0$ vs $H_A : \theta \neq \theta_0$
accepts the null hypothesis

if and only if   (9.21)

a symmetric $(1 - \alpha)100\%$ confidence Z-interval for θ contains θ_0.

PROOF: The null hypothesis H_0 is accepted if and only if the Z-statistic belongs to the acceptance region, i.e.,

$$\left| \frac{\hat{\theta} - \theta_0}{\sigma(\hat{\theta})} \right| \leq z_{\alpha/2}.$$

This is equivalent to

$$\left| \hat{\theta} - \theta_0 \right| \leq z_{\alpha/2} \sigma(\hat{\theta}).$$

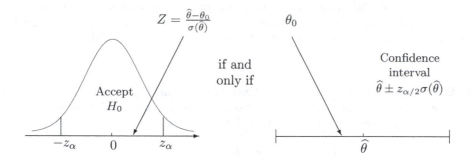

FIGURE 9.8: *Duality of tests and confidence intervals.*

We see that the distance from θ_0 to the center of Z-interval $\widehat{\theta}$ does not exceed its margin, $z_{\alpha/2}\sigma(\widehat{\theta})$ (see (9.3) and Figure 9.8). In other words, θ_0 belongs to the Z-interval. □

In fact, any two-sided test can be conducted this way. Accept $H_0 : \theta = \theta_0$ whenever a $(1-\alpha)100\%$ confidence interval for θ covers θ_0. Under θ_0, this test will accept the null hypothesis as often as the interval will cover θ_0, i.e., with probability $(1-\alpha)$. Thus, we have a level α test.

Rule (9.21) applies *only* when

- we are testing against a two-sided alternative (notice that our confidence intervals are two-sided too);

- significance level α of the test matches confidence level $(1-\alpha)$ of the confidence interval. For example, a two-sided 3% level test can be conducted using a 97% confidence interval.

Example 9.31. A sample of 6 measurements

$$2.5, 7.4, 8.0, 4.5, 7.4, 9.2$$

is collected from a Normal distribution with mean μ and standard deviation $\sigma = 2.2$. Test whether $\mu = 6$ against a two-sided alternative $H_A : \mu \neq 6$ at the 5% level of significance.

Solution. Solving Example 9.13 on p. 258, we have already constructed a 95% confidence interval for μ,

$$[4.74, \ 8.26].$$

The value of $\mu_0 = 6$ belongs to it; therefore, at the 5% level, the null hypothesis is accepted. ◇

Example 9.32. Use data in Example 9.31 to test whether $\mu = 7$.

Solution. The interval $[4.74, \ 8.26]$ contains $\mu_0 = 7$ too; therefore, the hypothesis $H_0 : \mu = 7$ is accepted as well. ◇

In the last two examples, how could we possibly accept both hypotheses, $\mu = 6$ and $\mu = 7$? Obviously, μ cannot be equal 6 and 7 at the same time! This is true. By accepting both null hypotheses, we only acknowledge that sufficient evidence against either of them is not found in the given data.

Example 9.33 (PRE-ELECTION POLL). In Example 9.17 on p. 265, we computed a 95% confidence interval for the difference of proportions supporting a candidate in towns A and B: $[-0.14, 0.16]$. This interval contains 0, therefore, the test of

$$H_0 : p_1 = p_2 \quad \text{vs} \quad H_A : p_1 \neq p_2$$

accepts the null hypothesis at the 5% level. Apparently, there is no evidence of unequal support of this candidate in the two towns. $\Diamond$

Example 9.34 (HARDWARE UPGRADE). In Example 9.14, we studied effectiveness of the hardware upgrade. We constructed a 90% confidence interval for the difference $(\mu_X - \mu_Y)$ in mean running times of a certain process: $[0.7, 1.9]$.

So, can we conclude that the upgrade was successful? Ineffective upgrade corresponds to a null hypothesis $H_0 : \mu_X = \mu_Y$, or $\mu_X - \mu_Y = 0$. Since the interval $[0.7, 1.9]$ does not contain 0, the no-effect hypothesis should be rejected at the 10% level of significance. $\Diamond$

Example 9.35 (WAS THE UPGRADE SUCCESSFUL? ONE-SIDED TEST). Let's look at Example 9.34 again. On second thought, we can only use Rule (9.21) to test the **two-sided alternative** $H_A : \mu_X \neq \mu_Y$, right? At the same time, the hardware upgrade is successful only when the running time reduces, i.e., $\mu_X > \mu_Y$. Then, we should judge effectiveness of the upgrade by a **one-sided, right-tail test** of

$$H_0 : \mu_X = \mu_Y \quad \text{vs} \quad H_A : \mu_X > \mu_Y. \tag{9.22}$$

Let us try to use the interval $[0.7, 1.9]$ for this test too. The null hypothesis in Example 9.34 is rejected at the 10% level in favor of a two-sided alternative, thus

$$|Z| > z_{\alpha/2} = z_{0.05}.$$

Then, either $Z < -z_{0.05}$ or $Z > z_{0.05}$. The first case is ruled out because the interval $[0.7, 1.9]$ consists of positive numbers, hence it cannot possibly support a left-tail alternative.

We conclude that $Z > z_{0.05}$, hence the test (9.22) results in rejection of H_0 at the 5% level of significance.

<u>Conclusion</u>. Our 90% confidence interval for $(\mu_X - \mu_Y)$ shows significant evidence, at the 5% level of significance, that the hardware upgrade was successful. $\Diamond$

Similarly, for the case of unknown variance(s).

> A level α T-test of $H_0 : \theta = \theta_0$ vs $H_A : \theta \neq \theta_0$
> accepts the null hypothesis
>
> if and only if
>
> a symmetric $(1 - \alpha)100\%$ confidence T-interval for θ contains θ_0.

Example 9.36 (UNAUTHORIZED USE OF A COMPUTER ACCOUNT, CONTINUED). A 99% confidence interval for the mean time between keystrokes is

$$[0.24; \ 0.34]$$

(Example 9.19 on p. 267 and data set `Keystrokes`). Example 9.28 on p. 282 tests whether the mean time is 0.2 seconds, which would be consistent with the speed of the account owner. The interval does not contain 0.2. Therefore, at a 1% level of significance, we have significant evidence that *the account was used by a different person.* $\diamond$

9.4.10 P-value

How do we choose α?

So far, we were testing hypotheses by means of acceptance and rejection regions. In the last section, we learned how to use confidence intervals for two-sided tests. Either way, we need to know the *significance level* α in order to conduct a test. Results of our test depend on it.

How do we choose α, the probability of making type I sampling error, rejecting the true hypothesis? Of course, when it seems too dangerous to reject true H_0, we choose a low significance level. How low? Should we choose $\alpha = 0.01$? Perhaps, 0.001? Or even 0.0001?

Also, if our *observed* test statistic $Z = Z_{\text{obs}}$ belongs to a rejection region but it is "too close to call" (see, for example, Figure 9.9), then how do we report the result? Formally, we should reject the null hypothesis, but practically, we realize that a slightly different significance level α could have expanded the acceptance region just enough to cover Z_{obs} and force us to accept H_0.

Suppose that the result of our test is crucially important. For example, the choice of a business strategy for the next ten years depends on it. In this case, can we rely so heavily on the choice of α? And if we rejected the true hypothesis just because we chose $\alpha = 0.05$ instead of $\alpha = 0.04$, then how do we explain to the chief executive officer that the situation was marginal? What is the statistical term for "too close to call"?

P-value

Using a P-value approach, we try not to rely on the level of significance. In fact, let us try to test a hypothesis using *all levels of significance!*

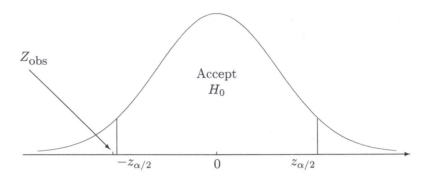

FIGURE 9.9: *This test is "too close to call": formally we reject the null hypothesis although the Z-statistic is almost at the boundary.*

Considering all levels of significance (between 0 and 1 because α is a probability of Type I error), we notice:

Case 1. If a level of significance is *very low*, we *accept* the null hypothesis (see Figure 9.10a). A low value of

$$\alpha = \boldsymbol{P}\left\{\text{ reject the null hypothesis when it is true }\right\}$$

makes it very unlikely to reject the hypothesis because it yields a very small rejection region. The right-tail area above the rejection region equals α.

Case 2. On the other extreme end, a *high significance level* α makes it likely to reject the null hypothesis and corresponds to a large rejection region. A sufficiently large α will produce such a large rejection region that will cover our test statistic, forcing us to *reject* H_0 (see Figure 9.10b).

Conclusion: there exists a boundary value between α-to-accept (case 1) and α-to-reject (case 2). This number is a *P-value* (Figure 9.11).

DEFINITION 9.9 ————————————————————————

> **P-value** is the lowest significance level α that forces rejection of the null hypothesis.
>
> **P-value** is also the highest significance level α that forces acceptance of the null hypothesis.

Testing hypotheses with a P-value

Once we know a P-value, we can indeed test hypotheses at *all* significance levels. Figure 9.11 clearly shows that for all $\alpha < P$ we accept the null hypothesis, and for all $\alpha > P$, we reject it.

Usual significance levels α lie in the interval $[0.01, 0.1]$ (although there are exceptions). Then, a P-value greater than 0.1 exceeds all natural significance levels, and the null hypothesis should be accepted. Conversely, if a P-value is less than 0.01, then it is smaller than all natural significance levels, and the null hypothesis should be rejected. Notice that we did not even have to specify the level α for these tests!

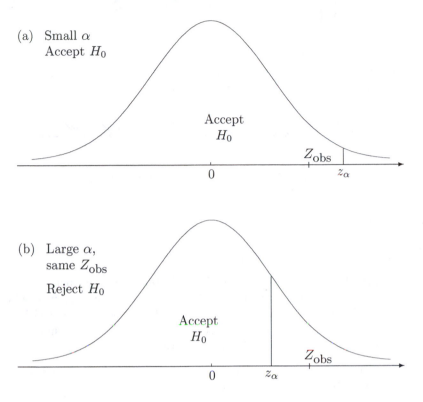

FIGURE 9.10: *(a) Under a low level of significance α, we accept the null hypothesis. (b) Under a high level of significance, we reject it.*

Only if the P-value happens to fall between 0.01 and 0.1, we really have to think about the level of significance. This is the "marginal case," "too close to call." When we report the conclusion, accepting or rejecting the hypothesis, we should always remember that with a slightly different α, the decision could have been reverted. When the matter is crucially important, a good decision is to collect more data until a more definitive answer can be obtained.

Testing H_0 with a P-value	For $\alpha < P$,	accept H_0
	For $\alpha > P$,	reject H_0
	Practically,	
	If $P < 0.01$,	reject H_0
	If $P > 0.1$,	accept H_0

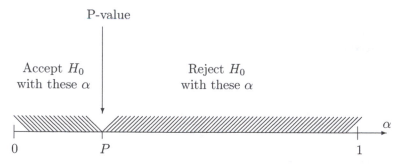

FIGURE 9.11: *P-value separates α-to-accept and α-to-reject.*

Computing P-values

Here is how a P-value can be computed from data.

Let us look at Figure 9.10 again. Start from Figure 9.10a, gradually increase α, and keep your eye at the vertical bar separating the acceptance and rejection region. It will move to the left until it hits the observed test statistic Z_{obs}. At this point, our decision changes, and we switch from case 1 (Figure 9.10a) to case 2 (Figure 9.10b). Increasing α further, we pass the Z-statistic and start accepting the null hypothesis.

What happens at the border of α-to-accept and α-to-reject? Definition 9.9 says that this borderline α is the **P-value**,

$$P = \alpha.$$

Also, at this border our observed Z-statistic coincides with the critical value z_α,

$$Z_{\text{obs}} = z_\alpha,$$

and thus,

$$P = \alpha = \boldsymbol{P}\{Z \geq z_\alpha\} = \boldsymbol{P}\{Z \geq Z_{\text{obs}}\}.$$

In this formula, Z is any Standard Normal random variable, and Z_{obs} is our observed test statistic, which is a concrete number, computed from data. First, we compute Z_{obs}, then use Table A4 to calculate

$$\boldsymbol{P}\{Z \geq Z_{\text{obs}}\} = 1 - \Phi(Z_{\text{obs}}).$$

Hypothesis H_0	Alternative H_A	P-value	Computation
$\theta = \theta_0$	right-tail $\theta > \theta_0$	$\boldsymbol{P}\{Z \geq Z_{\text{obs}}\}$	$1 - \Phi(Z_{\text{obs}})$
	left-tail $\theta < \theta_0$	$\boldsymbol{P}\{Z \leq Z_{\text{obs}}\}$	$\Phi(Z_{\text{obs}})$
	two-sided $\theta \neq \theta_0$	$\boldsymbol{P}\{|Z| \geq |Z_{\text{obs}}|\}$	$2(1 - \Phi(|Z_{\text{obs}}|))$

TABLE 9.3: *P-values for Z-tests.*

Hypothesis H_0	Alternative H_A	P-value	Computation
$\theta = \theta_0$	right-tail $\theta > \theta_0$	$P\{t \geq t_{\text{obs}}\}$	$1 - F_\nu(t_{\text{obs}})$
	left-tail $\theta < \theta_0$	$P\{t \leq t_{\text{obs}}\}$	$F_\nu(t_{\text{obs}})$
	two-sided $\theta \neq \theta_0$	$P\{|t| \geq |t_{\text{obs}}|\}$	$2(1 - F_\nu(|t_{\text{obs}}|))$

TABLE 9.4: *P-values for T-tests (F_ν is the cdf of T-distribution with the suitable number ν of degrees of freedom).*

P-values for the left-tail and for the two-sided alternatives are computed similarly, as given in Table 9.3.

This table applies to all the Z-tests in this chapter. It can be directly extended to the case of unknown standard deviations and T-tests (Table 9.4).

Understanding P-values

Looking at Tables 9.3 and 9.4, we see that *P-value* is the probability of observing a test statistic *at least as extreme as* Z_{obs} or t_{obs}. Being "extreme" is determined by the alternative. For a right-tail alternative, large numbers are extreme; for a left-tail alternative, small numbers are extreme; and for a two-sided alternative, both large and small numbers are extreme. In general, the more extreme test statistic we observe, the stronger support of the alternative it provides.

This creates another interesting definition of a P-value.

DEFINITION 9.10 ——————————————————————————

P-value is the probability of observing a test statistic that is as extreme as or more extreme than the test statistic computed from a given sample.

The following philosophy can be used when we test hypotheses by means of a P-value.

We are deciding between the null hypothesis H_0 and the alternative H_A. Observed is a test statistic Z_{obs}. If H_0 were true, how likely would it be to observe such a statistic? In other words, are the observed data consistent with H_0?

A high P-value tells that this or even more extreme value of Z_{obs} is quite possible under H_0, and therefore, we see no contradiction with H_0. The null hypothesis is not rejected.

Conversely, a low P-value signals that such an extreme test statistic is unlikely if H_0 is true. However, we really observed it. Then, our data are not consistent with the hypothesis, and we should reject H_0.

For example, if $P = 0.0001$, there is only 1 chance in 10,000 to observe what we really observed. The evidence supporting the alternative is highly significant in this case.

Example 9.37 (HOW SIGNIFICANT WAS THE UPGRADE?). Refer to Examples 9.14 and 9.34. At the 5% level of significance, we know that the hardware upgrade was successful. Was it marginally successful or very highly successful? Let us compute the P-value.

Start with computing a Z-statistic,

$$Z = \frac{\overline{X} - \overline{Y}}{\sqrt{\frac{\sigma_X^2}{n} + \frac{\sigma_Y^2}{m}}} = \frac{8.5 - 7.2}{\sqrt{\frac{1.8^2}{50} + \frac{1.8^2}{50}}} = 3.61.$$

From Table A4, we find that the P-value for the right-tail alternative is

$$P = \boldsymbol{P}\{Z \geq Z_{\text{obs}}\} = \boldsymbol{P}\{Z \geq 3.61\} = 1 - \Phi(3.61) = 0.0002.$$

The P-value is very low; therefore, we can reject the null hypothesis not only at the 5%, but also at the 1% and even 0.05% level of significance! We see now that the hardware upgrade was extremely successful. ◇

Example 9.38 (QUALITY INSPECTION). In Example 9.26, we compared the quality of parts produced by two manufacturers by a two-sided test. We obtained a test statistic

$$Z_{\text{obs}} = -0.94.$$

The P-value for this test equals

$$P = \boldsymbol{P}\{|Z| \geq |-0.94|\} = 2(1 - \Phi(0.94)) = 2(1 - 0.8264) = 0.3472.$$

This is a rather high P-value (greater than 0.1), and the null hypothesis is not rejected. Given H_0, there is a 34% chance of observing what we really observed. No contradiction with H_0, and therefore, no evidence that the quality of parts is not the same. ◇

Table A5 is not as detailed as Table A4. Often we can only use it to bound the P-value from below and from above. Typically, it suffices for hypothesis testing.

Example 9.39 (UNAUTHORIZED USE OF A COMPUTER ACCOUNT, CONTINUED). How significant is the evidence in Examples 9.28 and 9.36 on pp. 282, 287 that the account was used by an unauthorized person?

Under the null hypothesis, our T-statistic has T-distribution with 17 degrees of freedom. In the previous examples, we rejected H_0 first at the 5% level, then at the 1% level. Now, comparing $t = 5.16$ from Example 9.28 with the entire row 17 of Table A5, we find that it exceeds all the critical values given in the table until $t_{0.0001}$. Therefore, a two-sided test rejects the null hypothesis at a very low level $\alpha = 0.0002$, and the P-value is $P < 0.0002$. *The evidence of an unauthorized use is very strong!*

◇

9.5 Inference about variances

In this section, we'll derive confidence intervals and tests for the population variance $\sigma^2 = \text{Var}(X)$ and for the comparison of two variances $\sigma_X^2 = \text{Var}(X)$ and $\sigma_Y^2 = \text{Var}(Y)$. This will be a *new type of inference* for us because
(a) variance is a scale and not a location parameter,
(b) the distribution of its estimator, the sample variance, is not symmetric.

Variance often needs to be estimated or tested for quality control, in order to assess stability and accuracy, evaluate various risks, and also, for tests and confidence intervals for the population means when variance is unknown.

Recall that comparing two means in Section 9.3.5, we had to distinguish between the cases of equal and unequal variances. We no longer have to guess! In this section, we'll see how to test the null hypothesis $H_0 : \sigma_X^2 = \sigma_Y^2$ against the alternative $H_A : \sigma_X^2 \neq \sigma_Y^2$ and decide whether we should use the pooled variance (9.11) or the Satterthwaite approximation (9.12).

9.5.1 Variance estimator and Chi-square distribution

We start by estimating the population variance $\sigma^2 = \text{Var}(X)$ from an observed sample $\boldsymbol{X} = (X_1, \ldots, X_n)$. Recall from Section 8.2.4 that σ^2 is estimated *unbiasedly* and *consistently* by the sample variance

$$s^2 = \frac{1}{n-1} \sum_{i=1}^{n} \left(X_i - \overline{X}\right)^2 .$$

The summands $\left(X_i - \overline{X}\right)^2$ are not quite independent, as the Central Limit Theorem on p. 93 requires, because they all depend on $\overline{X}$. Nevertheless, the distribution of s^2 is approximately Normal, under mild conditions, when the sample is large.

For small to moderate samples, the distribution of s^2 is not Normal at all. It is not even symmetric. Indeed, why should it be symmetric if s^2 is always non-negative!

Distribution of the sample variance	When observations $X_1, \ldots, X_n$ are independent and Normal with $\text{Var}(X_i) = \sigma^2$, the distribution of $$\frac{(n-1)s^2}{\sigma^2} = \sum_{i=1}^{n} \left(\frac{X_i - \overline{X}}{\sigma}\right)^2$$ is *Chi-square with* $(n-1)$ *degrees of freedom*

Chi-square distribution, or χ^2, is a continuous distribution with density

$$f(x) = \frac{1}{2^{\nu/2}\Gamma(\nu/2)} \, x^{\nu/2-1} e^{-x/2}, \qquad x > 0,$$

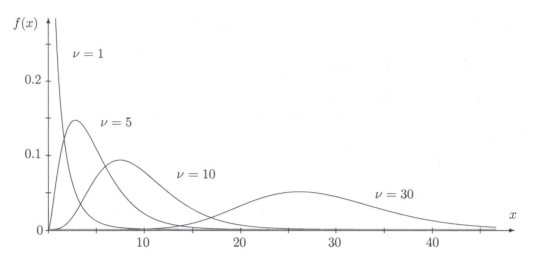

FIGURE 9.12: *Chi-square densities with $\nu = 1$, 5, 10, and 30 degrees of freedom. Each distribution is right-skewed. For large ν, it is approximately Normal.*

where $\nu > 0$ is a parameter that is called *degrees of freedom* and has the same meaning as for the Student's T-distribution (Figure 9.12).

Comparing this density with (4.7) on p. 85, we see that Chi-square distribution is a special case of Gamma,

$$\text{Chi-square}(\nu) = \text{Gamma}(\nu/2, 1/2),$$

and in particular, the Chi-square distribution with $\nu = 2$ degrees of freedom is Exponential(1/2).

We already know that Gamma(α, λ) distribution has expectation $\mathbf{E}(X) = \alpha/\lambda$ and Var$(X) = \alpha/\lambda^2$. Substituting $\alpha = \nu/2$ and $\lambda = 1/2$, we get the Chi-square moments,

$$\mathbf{E}(X) = \nu \quad \text{and} \quad \text{Var}(X) = 2\nu.$$

Chi-square distribution (χ^2)

ν	$=$	degrees of freedom
$f(x)$	$=$	$\dfrac{1}{2^{\nu/2}\Gamma(\nu/2)}\, x^{\nu/2-1} e^{-x/2}, \quad x > 0$
$\mathbf{E}(X)$	$=$	ν
Var(X)	$=$	2ν

(9.23)

Chi-square distribution was introduced around 1900 by a famous English mathematician *Karl Pearson* (1857-1936) who is regarded as a founder of the entire field of *Mathematical Statistics*. By the way, Pearson was a teacher and collaborator of William Gosset, which is why Student was Gosset's pseudonym.

Table A6 in the Appendix contains critical values of the Chi-square distribution.

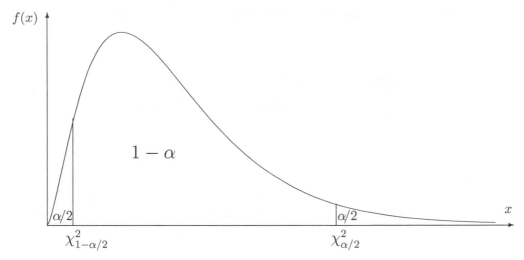

FIGURE 9.13: *Critical values of the Chi-square distribution.*

9.5.2 Confidence interval for the population variance

Let us construct a $(1 - \alpha)100\%$ confidence interval for the population variance σ^2, based on a sample of size n.

As always, we start with the estimator, the sample variance s^2. However, since the distribution of s^2 is not symmetric, our confidence interval won't have the form "estimator $\pm$ margin" as before.

Instead, we use Table A6 to find *the critical values* $\chi^2_{1-\alpha/2}$ and $\chi^2_{\alpha/2}$ of the Chi-square distribution with $\nu = n-1$ degrees of freedom. These critical values chop the areas of $(\alpha/2)$ on the right and on the left sides of the region under the Chi-square density curve, as on Figure 9.13. This is similar to $\pm z_{\alpha/2}$ and $\pm t_{\alpha/2}$ in the previous sections, although these Chi-square quantiles are no longer symmetric. Recall that $\chi^2_{\alpha/2}$ denotes the $(1 - \alpha/2)$-quantile, $q_{1-\alpha/2}$.

Then, the area between these two values is $(1 - \alpha)$.

A rescaled sample variance $(n - 1)s^2/\sigma^2$ has χ^2 density like the one on Figure 9.13, so

$$P\left\{\chi^2_{1-\alpha/2} \leq \frac{(n-1)s^2}{\sigma^2} \leq \chi^2_{\alpha/2}\right\} = 1 - \alpha.$$

Solving the inequality for the unknown parameter σ^2, we get

$$P\left\{\frac{(n-1)s^2}{\chi^2_{\alpha/2}} \leq \sigma^2 \leq \frac{(n-1)s^2}{\chi^2_{1-\alpha/2}}\right\} = 1 - \alpha.$$

A $(1 - \alpha)100\%$ confidence interval for the population variance is obtained!

$$\boxed{\textbf{Confidence interval} \atop \textbf{for the variance} \quad \left[\frac{(n-1)s^2}{\chi^2_{\alpha/2}}, \ \frac{(n-1)s^2}{\chi^2_{1-\alpha/2}}\right]} \qquad (9.24)$$

A confidence interval for the population standard deviation $\sigma = \sqrt{\sigma^2}$ is just one step away (Exercise 9.21).

$$
\textbf{Confidence interval for the standard deviation} \qquad \left[\sqrt{\frac{(n-1)s^2}{\chi^2_{\alpha/2}}}, \ \sqrt{\frac{(n-1)s^2}{\chi^2_{1-\alpha/2}}} \right] \qquad (9.25)
$$

Example 9.40. In Example 9.31 on p. 285, we relied on the reported parameters of the measurement device and assumed the known standard deviation $\sigma = 2.2$. Let us now rely on the data only and construct a 90% confidence interval for the standard deviation. The sample contained $n = 6$ measurements, 2.5, 7.4, 8.0, 4.5, 7.4, and 9.2.

<u>Solution.</u> Compute the sample mean and then the sample variance,

$$
\overline{X} = \frac{1}{6}(2.5 + \ldots + 9.2) = 6.5;
$$

$$
s^2 = \frac{1}{6-1}\left\{(2.5 - 6.5)^2 + \ldots + (9.2 - 6.5)^2\right\} = \frac{31.16}{5} = 6.232.
$$

(actually, we only need $(n-1)s^2 = 31.16$).

From Table A6 of Chi-square distribution with $\nu = n - 1 = 5$ degrees of freedom, we find the critical values $\chi^2_{1-\alpha/2} = \chi^2_{0.95} = 1.15$ and $\chi^2_{\alpha/2} = \chi^2_{0.05} = 11.1$. Then,

$$
\left[\sqrt{\frac{(n-1)s^2}{\chi^2_{\alpha/2}}}, \ \sqrt{\frac{(n-1)s^2}{\chi^2_{1-\alpha/2}}} \right] = \left[\sqrt{\frac{31.16}{11.1}}, \ \sqrt{\frac{31.16}{1.15}} \right] = \underline{[1.68, \ 5.21]}.
$$

is a 90% confidence interval for the population standard deviation (and by the way, $[1.68^2, 5.21^2] = [2.82, 27.14]$ is a 90% confidence interval for the variance). $\Diamond$

9.5.3 Testing variance

Suppose now that we need *to test* the population variance, for example, to make sure that the actual variability, uncertainty, volatility, or risk does not exceed the promised value. We'll derive a level α test based on the Chi-square distribution of the rescaled sample variance.

Level α test

Let $X_1, \ldots, X_n$ be a sample from the Normal distribution with the unknown population variance σ^2. For testing the null hypothesis

$$
H_0 : \sigma^2 = \sigma_0^2,
$$

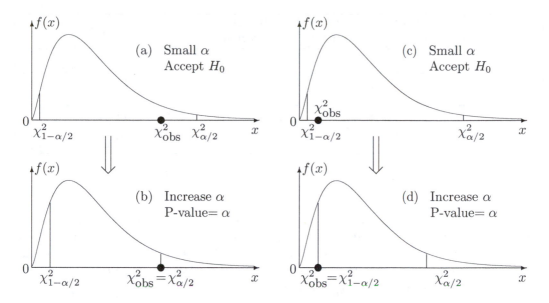

FIGURE 9.14: *P-value for a Chi-square test against a two-sided alternative.*

compute the χ^2-statistic

$$\chi^2_{\text{obs}} = \frac{(n-1)s^2}{\sigma_0^2}.$$

As we know, it follows the χ^2 distribution with $(n-1)$ degrees of freedom if H_0 is true and σ_0^2 is indeed the correct population variance. Thus, it only remains to compare χ^2_{obs} with the critical values from Table A6 of χ^2 distribution, using $\nu = n - 1$.

Testing against the *right-tail* alternative $H_A : \sigma^2 > \sigma_0^2$, reject H_0 if $\chi^2_{\text{obs}} \geq \chi^2_{\alpha}$.

Testing against the *left-tail* alternative $H_A : \sigma^2 < \sigma_0^2$, reject H_0 if $\chi^2_{\text{obs}} \leq \chi^2_{1-\alpha}$.

Testing against the *two-sided* alternative $H_A : \sigma^2 \neq \sigma_0^2$, reject H_0 if either $\chi^2_{\text{obs}} \geq \chi^2_{\alpha/2}$ or $\chi^2_{\text{obs}} \leq \chi^2_{1-\alpha/2}$.

As an exercise, please verify that in each case, the probability of type I error is exactly α.

P-value

For one-sided χ^2-tests, the P-value is computed the same way as in Z-tests and T-tests. It is always *the probability of the same or more extreme value of the test statistic than the one that was actually observed.* That is,

$$
\begin{aligned}
\text{P-value} &= \boldsymbol{P}\left\{\chi^2 \geq \chi^2_{\text{obs}}\right\} = 1 - F(\chi^2_{\text{obs}}) & \text{for a right-tail test,} \\
\text{P-value} &= \boldsymbol{P}\left\{\chi^2 \leq \chi^2_{\text{obs}}\right\} = F(\chi^2_{\text{obs}}) & \text{for a left-tail test,}
\end{aligned}
$$

where F is a cdf of χ^2 distribution with $\nu = n - 1$ degrees of freedom.

But how to compute the P-value for the *two-sided* alternative? Which values of χ^2 are considered "more extreme"? For example, do you think $\chi^2 = 3$ is more extreme than $\chi^2 = 1/3$?

Null Hypothesis	Alternative Hypothesis	Test statistic	Rejection region	P-value
$\sigma^2 = \sigma_0^2$	$\sigma^2 > \sigma_0^2$	$\dfrac{(n-1)s^2}{\sigma_0^2}$	$\chi^2_{\text{obs}} > \chi^2_{\alpha}$	$\boldsymbol{P}\left\{\chi^2 \geq \chi^2_{\text{obs}}\right\}$
	$\sigma^2 < \sigma_0^2$		$\chi^2_{\text{obs}} < \chi^2_{\alpha}$	$\boldsymbol{P}\left\{\chi^2 \leq \chi^2_{\text{obs}}\right\}$
	$\sigma^2 \neq \sigma_0^2$		$\chi^2_{\text{obs}} \geq \chi^2_{\alpha/2}$ or $\chi^2_{\text{obs}} \leq \chi^2_{1-\alpha/2}$	$2\min\left(\boldsymbol{P}\left\{\chi^2 \geq \chi^2_{\text{obs}}\right\},\ \boldsymbol{P}\left\{\chi^2 \leq \chi^2_{\text{obs}}\right\}\right)$

TABLE 9.5: χ^2-tests for the population variance.

We can no longer claim that the value further away from 0 is more extreme, as we did earlier for Z- and T-tests! Indeed, the χ^2-statistic is always positive, and in two-sided tests, its very small or very large values should be considered extreme. It should be fair to say that smaller values of χ^2 are more extreme than the observed one if χ^2_{obs} itself is small, and larger values are more extreme if χ^2_{obs} is large.

To make this idea rigorous, let us recall (from Section 9.4.10) that P-value equals the highest significance level α that yields acceptance of H_0. Start with a very small α as on Figure 9.14a. The null hypothesis H_0 is still accepted because $\chi^2_{\text{obs}} \in [\chi^2_{1-\alpha/2}, \chi^2_{\alpha/2}]$. Slowly increase α until the boundary between acceptance and rejection regions hits the observed test statistic χ^2_{obs}. At this moment, α equals to the P-value (Figure 9.14b), hence

$$P = 2\left(\frac{\alpha}{2}\right) = 2\boldsymbol{P}\left\{\chi^2 \geq \chi^2_{\text{obs}}\right\} = 2\left\{1 - F(\chi^2_{\text{obs}})\right\}. \tag{9.26}$$

It can also happen that the lower rejection boundary hits χ^2_{obs} first, as on Figure 9.14cd. In this case,

$$P = 2\left(\frac{\alpha}{2}\right) = 2\boldsymbol{P}\left\{\chi^2 \leq \chi^2_{\text{obs}}\right\} = 2F(\chi^2_{\text{obs}}). \tag{9.27}$$

So, the P-value is either given by (9.26) or by (9.27), depending on which one of them is smaller and which boundary hits χ^2_{obs} first. We can write this in one equation as

$$P = 2\min\left(\boldsymbol{P}\left\{\chi^2 \geq \chi^2_{\text{obs}}\right\}, \boldsymbol{P}\left\{\chi^2 \leq \chi^2_{\text{obs}}\right\}\right) = 2\min\left\{F(\chi^2_{\text{obs}}),\ 1 - F(\chi^2_{\text{obs}})\right\},$$

where F is the cdf of χ^2 distribution with $\nu = n - 1$ degrees of freedom.

Testing procedures that we have just derived are summarized in Table 9.5. The same tests can also be used for the *standard deviation* because testing $\sigma^2 = \sigma_0^2$ is equivalent to testing $\sigma = \sigma_0$.

Example 9.41. Refer to Example 9.40 on p. 296. The 90% confidence interval constructed there contains the suggested value of $\sigma = 2.2$. Then, by *duality* between confidence intervals and tests, there should be no evidence against this value of σ. Measure the amount of evidence against it by computing the suitable P-value.

<u>Solution.</u> The null hypothesis $H_0 : \sigma = 2.2$ is tested against $H_A : \sigma^2 \neq 2.2$. This is a two-sided test because we only need to know whether the standard deviation equals $\sigma_0 = 2.2$ or not.

Compute the test statistic from the data in Example 9.40,

$$\chi^2_{\text{obs}} = \frac{(n-1)s^2}{\sigma_0^2} = \frac{(5)(6.232)}{2.2^2} = 6.438.$$

Using Table A6 with $\nu = n - 1 = 5$ degrees of freedom, we see that $\chi^2_{0.80} < \chi^2_{\text{obs}} < \chi^2_{0.20}$. Therefore,

$$\boldsymbol{P}\left\{\chi^2 \geq \chi^2_{\text{obs}}\right\} > 0.2 \text{ and } \boldsymbol{P}\left\{\chi^2 \leq \chi^2_{\text{obs}}\right\} > 0.2,$$

hence,

$$P = 2 \min\left(\boldsymbol{P}\left\{\chi^2 \geq \chi^2_{\text{obs}}\right\}, \boldsymbol{P}\left\{\chi^2 \leq \chi^2_{\text{obs}}\right\}\right) \geq 0.4.$$

The evidence against $\sigma = 2.2$ is very weak; at all typical significance levels, $H_0 : \sigma = 2.2$ should be accepted. $\diamond$

Example 9.42 (REMAINING BATTERY LIFE). Creators of a new software claim that it measures the remaining notebook battery life with a standard deviation as low as 5 minutes. To test this claim, a fully charged notebook was disconnected from the power supply and continued on its battery. The experiment was repeated 50 times, and every time the predicted battery life was recorded. The sample standard deviation of these 50 normally distributed measurements was equal 5.6 minutes. At the 1% level of significance, do these data provide evidence that the actual standard deviation is greater than 5 min?

Solution. Test $H_0 : \sigma = 5$ against a right-tail $H_A : \sigma > 5$. From Table A6 with $\nu = n-1 = 49$ degrees of freedom, $\chi^2_{0.01} = 74.9$. The null hypothesis will be rejected if $\chi^2_{\text{obs}} > 74.9$.

Compute the test statistic,

$$\chi^2_{\text{obs}} = \frac{(n-1)s^2}{\sigma_0^2} = \frac{(49)(5.6)^2}{5^2} = 61.5.$$

Accept $H_0 : \sigma = 5$. The evidence against it is not significant at the 1% level of significance.

9.5.4 Comparison of two variances. F-distribution

In this section, we deal with two populations whose variances need to be compared. Such inference is used for the comparison of accuracy, stability, uncertainty, or risks arising in two populations.

Example 9.43 (EFFICIENT UPGRADE). A data channel has the average speed of 180 Megabytes per second. A hardware upgrade is supposed to improve *stability* of the data transfer while maintaining the same average speed. Stable data transfer rate implies low standard deviation. How can we estimate the relative change in the standard deviation of the transfer rate with 90% confidence? $\diamond$

Example 9.44 (CONSERVATIVE INVESTMENT). Two mutual funds promise the same expected return; however, one of them recorded a 10% higher *volatility* over the last 15 days. Is this a significant evidence for a conservative investor to prefer the other mutual fund? (Volatility is essentially the standard deviation of returns.)   ◊

Example 9.45 (WHICH METHOD TO USE?). For marketing purposes, a survey of users of two operating systems is conducted. Twenty users of operating system ABC record the average level of satisfaction of 77 on a 100-point scale, with a sample variance of 220. Thirty users of operating system DEF have the average satisfaction level 70 with a sample variance of 155. We already know from Section 9.4.8 how to compare the mean satisfaction levels. But what method should we choose? Should we assume equality of population variances, $\sigma_X^2 = \sigma_Y^2$ and use the pooled variance? Or we should allow for $\sigma_X^2 \neq \sigma_Y^2$ and use Satterthwaite approximation?   ◊

To compare variances or standard deviations, two independent samples $\boldsymbol{X} = (X_1, \ldots, X_n)$ and $\boldsymbol{Y} = (Y_1, \ldots, Y_m)$ are collected, one from each population, as on Figure 9.4 on p. 259. Unlike population means or proportions, variances are scale factors, and they are compared through their *ratio*

$$\theta = \frac{\sigma_X^2}{\sigma_Y^2}.$$

A natural estimator for the ratio of *population* variances $\theta = \sigma_X^2/\sigma_Y^2$ is the ratio of *sample* variances

$$\widehat{\theta} = \frac{s_X^2}{s_Y^2} = \frac{\sum(X_i - \overline{X})/(n-1)}{\sum(Y_i - \overline{Y})/(m-1)}. \tag{9.28}$$

The distribution of this statistic was obtained in 1918 by a famous English statistician and biologist *Sir Ronald Fisher* (1890-1962) and developed and formalized in 1934 by an American mathematician *George Snedecor* (1881-1974). Its standard form, after we divide each sample variance in formula (9.28) by the corresponding population variance, is therefore called the *Fisher–Snedecor distribution* or simply *F-distribution* with $(n-1)$ and $(m-1)$ degrees of freedom.

Distribution of the ratio of sample variances	For independent samples $X_1, \ldots, X_n$ from Normal (μ_X, σ_X) and $Y_1, \ldots, Y_m$ from Normal (μ_Y, σ_Y), the standardized ratio of variances $$F = \frac{s_X^2/\sigma_X^2}{s_Y^2/\sigma_Y^2} = \frac{\sum(X_i - \overline{X})^2/\sigma_X^2/(n-1)}{\sum(Y_i - \overline{Y})^2/\sigma_Y^2/(m-1)}$$ has *F-distribution with* $(n-1)$ and $(m-1)$ *degrees of freedom.*

$$\tag{9.29}$$

We know from Section 9.5.1 that for the Normal data, both s_X^2/σ_X^2 and s_Y^2/σ_Y^2 follow χ^2-distributions. We can now conclude that *the ratio of two independent* χ^2 *variables, each divided by its degrees of freedom, has F-distribution.* A ratio of two non-negative continuous random variables, any F-distributed variable is also non-negative and continuous.

F-distribution has two parameters, the *numerator degrees of freedom* and the *denominator degrees of freedom*. These are degrees of freedom of the sample variances in the numerator and denominator of the F-ratio (9.29).

Critical values of F-distribution are in Table A7, and we'll use them to construct confidence intervals and test hypotheses comparing two variances.

One question though... Comparing two variances, σ_X^2 and σ_Y^2, should we divide s_X^2 by s_Y^2 or s_Y^2 by s_X^2? Of course, both ratios are ok to use, but we have to keep in mind that in the first case we deal with $F(n-1, m-1)$ distribution, and in the second case with $F(m-1, n-1)$. This leads us to an important general conclusion –

$$\text{If } F \text{ has } F(\nu_1, \nu_2) \text{ distribution, then the distribution of } \frac{1}{F} \text{ is } F(\nu_2, \nu_1). \tag{9.30}$$

9.5.5 Confidence interval for the ratio of population variances

Here we construct a $(1-\alpha)100\%$ confidence interval for the parameter $\theta = \sigma_X^2/\sigma_Y^2$. This is about the sixth time we derive a formula for a confidence interval, so we are well familiar with the method, aren't we?

Start with the estimator, $\widehat{\theta} = s_X^2/s_Y^2$. Standardizing it to

$$F = \frac{s_X^2/\sigma_X^2}{s_Y^2/\sigma_Y^2} = \frac{s_X^2/s_Y^2}{\sigma_X^2/\sigma_Y^2} = \frac{\widehat{\theta}}{\theta},$$

we get an F-variable with $(n-1)$ and $(m-1)$ degrees of freedom. Therefore,

$$P\left\{ F_{1-\alpha/2}(n-1, m-1) \leq \frac{\widehat{\theta}}{\theta} \leq F_{\alpha/2}(n-1, m-1) \right\} = 1 - \alpha,$$

as on Figure 9.15. Solving the double inequality for the unknown parameter θ, we get

$$P\left\{ \frac{\widehat{\theta}}{F_{\alpha/2}(n-1, m-1)} \leq \theta \leq \frac{\widehat{\theta}}{F_{1-\alpha/2}(n-1, m-1)} \right\} = 1 - \alpha.$$

Therefore,

$$
\left[\frac{\widehat{\theta}}{F_{\alpha/2}(n-1, m-1)}, \; \frac{\widehat{\theta}}{F_{1-\alpha/2}(n-1, m-1)} \right]
$$
$$
= \left[\frac{s_X^2/s_Y^2}{F_{\alpha/2}(n-1, m-1)}, \; \frac{s_X^2/s_Y^2}{F_{1-\alpha/2}(n-1, m-1)} \right] \tag{9.31}
$$

is a $(1-\alpha)100\%$ confidence interval for $\theta = \sigma_X^2/\sigma_Y^2$.

The critical values $F_{1-\alpha/2}(n-1, m-1)$ and $F_{\alpha/2}(n-1, m-1)$ come from F-distribution with $(n-1)$ and $(m-1)$ degrees of freedom. However, our Table A7 has only small values of α. What can we do about $F_{1-\alpha/2}(n-1, m-1)$, a critical value with a large area on the right?

We can easily compute $F_{1-\alpha/2}(n-1, m-1)$ by making use of statement (9.30).

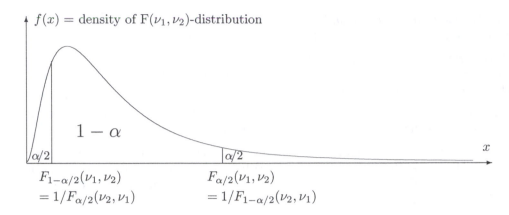

FIGURE 9.15: *Critical values of the F-distribution and their reciprocal property.*

Let $F(\nu_1, \nu_2)$ have F-distribution with ν_1 and ν_2 degrees of freedom, then its reciprocal $F(\nu_2, \nu_1) = 1/F(\nu_1, \nu_2)$ has ν_1 and ν_2 degrees of freedom. According to (9.30),

$$\alpha = P\left\{F(\nu_1, \nu_2) \le F_{1-\alpha}(\nu_1, \nu_2)\right\} = P\left\{F(\nu_2, \nu_1) \ge \frac{1}{F_{1-\alpha}(\nu_1, \nu_2)}\right\}$$

We see from here that $1/F_{1-\alpha}(\nu_1, \nu_2)$ is actually the α-critical value from $F(\nu_2, \nu_1)$ distribution because it cuts area α on the right; see Figure 9.15. We conclude that

**Reciprocal property
of F-distribution**

> The critical values of $F(\nu_1, \nu_2)$ and $F(\nu_2, \nu_1)$ distributions are related as follows,
>
> $$F_{1-\alpha}(\nu_1, \nu_2) = \frac{1}{F_\alpha(\nu_2, \nu_1)}$$

(9.32)

We can now obtain the critical values from Table A7 and formula (9.32), plug them into (9.31), and the confidence interval is ready.

**Confidence interval
for the ratio
of variances**

> $$\left[\frac{s_X^2}{s_Y^2 F_{\alpha/2}(n-1, m-1)}, \frac{s_X^2 F_{\alpha/2}(m-1, n-1)}{s_Y^2}\right]$$

(9.33)

Example 9.46 (EFFICIENT UPGRADE, CONTINUED). Refer to Example 9.43. After the upgrade, the instantaneous speed of data transfer, measured at 16 random instants, yields a standard deviation of 14 Mbps. Records show that the standard deviation was 22 Mbps before the upgrade, based on 27 measurements at random times. We are asked to construct a 90% confidence interval for the relative change in the standard deviation (assume Normal distribution of the speed).

Solution. From the data, $s_X = 14$, $s_Y = 22$, $n = 16$, and $m = 27$. For a 90% confidence interval, use $\alpha = 0.10$, $\alpha/2 = 0.05$. Find $F_{0.05}(15, 26) \approx 2.07$ and $F_{0.05}(26, 15) \approx 2.27$ from Table A7. Or, alternatively, use functions `qf(0.95,15,26)`, `qf(0.95,26,15)` in R or `finv(0.95,15,26)`, `finv(0.95,26,15)` in MATLAB to get the exact values, 2.0716 and 2.2722. Then, the 90% confidence interval for the *ratio of variances* $\theta = \sigma_X^2/\sigma_Y^2$ is

$$\left[\frac{14^2}{22^2 \cdot 2.07}, \frac{14^2 \cdot 2.27}{22^2}\right] = [0.20, 0.92].$$

For the *ratio of standard deviations* $\sigma_X/\sigma_Y = \sqrt{\theta}$, a 90% confidence interval is obtained by simply taking square roots,

$$\left[\sqrt{0.20}, \sqrt{0.92}\right] = [0.44, 0.96].$$

Thus, we can assert with a 90% confidence that the new standard deviation is between 44% and 96% of the old standard deviation. With this confidence level, *the relative reduction in the standard deviation of the data transfer rate (and therefore, the relative increase of stability) is between 4% and 56%* because this relative reduction is $(\sigma_Y - \sigma_X)/\sigma_Y = 1 - \sqrt{\theta}$.

$\Diamond$

Example 9.47 (EFFICIENT UPGRADE, CONTINUED AGAIN). Refer again to Examples 9.43 and 9.46. Can we infer that the channel became twice as stable as it was, if increase of stability is measured by the proportional reduction of standard deviation?

Solution. The 90% confidence interval obtained in Example 9.46 contains 0.5. Therefore, at the 10% level of significance, there is no evidence against $H_0 : \sigma_X/\sigma_Y = 0.5$, which is a two-fold reduction of standard deviation (recall Section 9.4.9 about the duality between confidence intervals and tests). This is all we can state - there is no evidence against the claim of a two-fold increase of stability. There is no "proof" that it actually happened. $\Diamond$

Testing hypotheses about the ratio of variances or standard deviations is in the next section.

9.5.6 F-tests comparing two variances

In this section, we test the null hypothesis about a *ratio of variances*

$$H_0 : \frac{\sigma_X^2}{\sigma_Y^2} = \theta_0 \tag{9.34}$$

against a one-sided or a two-sided alternative. Often we only need to know if two variances are equal, then we choose $\theta_0 = 1$. F-distribution is used to compare variances, so this test is called the **F-test**.

The test statistic for (9.34) is

$$F = \frac{s_X^2}{s_Y^2}/\theta_0,$$

Null Hypothesis $H_0 : \dfrac{\sigma_X^2}{\sigma_Y^2} = \theta_0$		Test statistic $F_{\text{obs}} = \dfrac{s_X^2}{s_Y^2}/\theta_0$
Alternative Hypothesis	Rejection region	P-value Use F$(n-1, m-1)$ distribution
$\dfrac{\sigma_X^2}{\sigma_Y^2} > \theta_0$	$F_{\text{obs}} \geq F_\alpha(n-1, m-1)$	$\boldsymbol{P}\{F \geq F_{\text{obs}}\}$
$\dfrac{\sigma_X^2}{\sigma_Y^2} < \theta_0$	$F_{\text{obs}} \leq F_\alpha(n-1, m-1)$	$\boldsymbol{P}\{F \leq F_{\text{obs}}\}$
$\dfrac{\sigma_X^2}{\sigma_Y^2} \neq \theta_0$	$F_{\text{obs}} \geq F_{\alpha/2}(n-1, m-1)$ or $F_{\text{obs}} < 1/F_{\alpha/2}(m-1, n-1)$	$2\min\left(\boldsymbol{P}\{F \geq F_{\text{obs}}\}, \boldsymbol{P}\{F \leq F_{\text{obs}}\}\right)$

TABLE 9.6: *Summary of F-tests for the ratio of population variances.*

which under the null hypothesis equals

$$F = \frac{s_X^2/\sigma_X^2}{s_Y^2/\sigma_Y^2}.$$

If $\boldsymbol{X}$ and $\boldsymbol{Y}$ are samples from Normal distributions, this *F-statistic* has F-distribution with $(n-1)$ and $(m-1)$ degrees of freedom.

Just like χ^2, F-statistic is also non-negative, with a non-symmetric right-skewed distribution. Level α tests and P-values are then developed similarly to χ^2, see Table 9.6.

Example 9.48 (WHICH METHOD TO USE? CONTINUED). In Example 9.45 on p. 300, $n = 20$, $\overline{X} = 77$, $s_X^2 = 220$; $m = 30$, $\overline{Y} = 70$, and $s_Y^2 = 155$. To compare the population means by a suitable method, we have to test whether the two population variances are equal or not.

<u>Solution.</u> Test $H_0 : \sigma_X^2 = \sigma_Y^2$ vs $H_A : \sigma_X^2 \neq \sigma_Y^2$ with the test statistic

$$F_{\text{obs}} = \frac{s_X^2}{s_Y^2} = 1.42.$$

For testing equality of variances, we let the tested ratio $\theta_0 = 1$. This is a two-sided test, so the P-value is

$$P = 2\min\left(\boldsymbol{P}\{F \geq 1.42\}, \boldsymbol{P}\{F \leq 1.42\}\right) = \ldots?$$

How to compute these probabilities for the F-distribution with $n - 1 = 19$ and $m - 1 = 29$ degrees of freedom? R and MATLAB, as always, can give us the exact answer. Typing `1-pf(1.42,19,29)` in R or `1-fcdf(1.42,19,29)` in MATLAB, we obtain $\boldsymbol{P}\{F \geq 1.42\} = 0.1926$. Then,

$$P = 2\min(0.1926, 1 - 0.1926) = \underline{0.3852}.$$

Table A7 can also be used, for an approximate but a completely satisfactory solution. This table does not have exactly 19 and 29 degrees of freedom and does not have a value $F_\alpha = 1.42$. However, looking at 15 and 20 d.f. for the numerator and 25 and 30 d.f. for the denominator, we see that 1.42 is always between $F_{0.25}$ and $F_{0.1}$. This will do it for us.

It implies that $P\{F \geq .42\} \in (0.1, 0.25)$, $P\{F \leq 1.42\} \in (0.75, 0.9)$, and therefore, the P-value is

$$P = 2P\{F \geq 1.42\} \in (0.2,\ 0.5)\ .$$

This is a high P-value showing no evidence of different variances. It should be ok to use the exact two-sample T-test with a pooled variance (according to which there is a mild evidence at a 4% level that the first operating system is better, $t = 1.80$, $P = 0.0388$). $\Diamond$

Example 9.49 (ARE ALL THE CONDITIONS MET?). In Example 9.44 on p. 300, we are asked to compare volatilities of two mutual funds and decide if one of them is more risky than the other. So, this is a one-sided test of

$$H_0 : \sigma_X = \sigma_Y \quad \text{vs} \quad H_A : \sigma_X > \sigma_Y.$$

The data collected over the period of 30 days show a 10% higher volatility of the first mutual fund, i.e., $s_X/s_Y = 1.1$. So, this is a standard F-test, right? A careless statistician would immediately proceed to the test statistic $F_{\text{obs}} = s_X^2/s_Y^2 = 1.21$ and the P-value $P = P\{F \geq F_{\text{obs}}\} > 0.25$ from Table A7 with $n - 1 = 29$ and $m - 1 = 29$ d.f., and jump to a conclusion that there is *no evidence that the first mutual fund carries a higher risk*.

Indeed, why not? Well, every statistical procedure has its assumptions, conditions under which our conclusions are valid. A careful statistician always checks the assumptions before reporting any results.

If we conduct an F-test and refer to the F-distribution, what conditions are required? We find the answer in (9.29) on p. 300. Apparently, for the F-statistic to have F-distribution under H_0, each of our two samples has to consist of independent and identically distributed Normal random variables, and the two samples have to be independent of each other.

Are these assumptions plausible, at the very least?

1. Normal distribution - may be. Returns on investments are typically not Normal but log-returns are.

2. Independent and identically distributed data within each sample - unlikely. Typically, there are economic trends, ups and downs, and returns on two days in a row should be dependent.

3. Independence between the two samples - it depends. If our mutual funds contain stocks from the same industry, their returns are surely dependent.

Actually, conditions 1-3 can be tested statistically, and for this we need to have the entire samples of data instead of the summary statistics.

The F-test is quite *robust*. It means that a mild departure from the assumptions 1-3 will not affect our conclusions severely, and we can treat our result as *approximate*. However, if the assumptions are not met even approximately, for example, the distribution of our data is asymmetric and far from Normal, then the P-value computed above is simply wrong. $\Diamond$

Discussion in Example 9.49 leads us to a very important practical conclusion.

> *Every statistical procedure is valid under certain assumptions.*
> *When they are not satisfied, the obtained results may be wrong and misleading.*
> *Therefore, unless there are reasons to believe that all the conditions are met,*
> *they have to be tested statistically.*

R notes

R has tools for tests and confidence intervals discussed in this chapter.

Testing means. To do one-sample T-test of $H_0 : \mu = \mu_0$ vs $H_A : \mu \neq \mu_0$ or to obtain a confidence interval for μ, type `t.test(X,mu=mu0)`. The output will include the sample mean, the test statistic, degrees of freedom, the p-value of this two-sided test, as well as the 95% confidence interval.

For a one-sided test, add an `alternative` option with values `"less"` or `"greater"`. However, notice that the confidence interval will also be one-sided in this case. The default confidence level is 95%. For other levels, add an option called `conf.level` option. For example, to get a 99% confidence interval for μ, type `t.test(X, conf.level=0.99)`, and for a one-sided, right-tail test of $H_0 : \mu = 10$ vs $H_A : \mu > 10$, use `t.test(X, mu=10, alternative="greater")`.

Two-sample tests and confidence intervals. The same command `t.test` can be employed for two-sample tests and confidence intervals. You only need to enter two variables instead of one. So, to test $H_0 : \mu_X - \mu_Y = 3$ vs $H_A : \mu_X - \mu_Y \neq 3$ and to construct a 90% confidence interval for the difference of means $(\mu_X - \mu_Y)$, enter `t.test(X, Y, mu=3, conf.level=0.90)`. Satterthwaite approximation will be applied by default. To conduct the equal-variances test with a pooled variance, give the option `var.equal = TRUE`. For a paired T-test, use `paired = TRUE`.

Testing variances. A two-sample F-test for variances requires a command `var.test`, whose syntax is rather similar to `t.test`. Say, to test $H_0 : \sigma_X^2 = \sigma_Y^2$ vs $H_A : \sigma_X^2 \neq \sigma_Y^2$, type `var.test(X, Y, ratio=1)`. This tool also has options `alternative` and `conf.level`, to let you choose a one-sided alternative and the desired confidence level.

Testing proportions. A special command `prop.test` is for the inference about population proportions. Variable X in this case is the number of successes, and it has to be followed by the number of trials n and the null proportion p_0 that we are testing. For example, `prop.test(60,100,0.5)` is for testing $H_0 : p = 0.5$ vs $H_A : p \neq 0.5$ based on a sample of $X = 60$ successes in $n = 100$ Bernoulli trials. This command also allows options `alternative` and `conf.level`.

For a two-sample test comparing two population proportions, a pair of counts and a pair of sample sizes should be written in place of X and n. For example, suppose that we need to conduct a two-sample test of proportions $H_0 : p_1 = p_2$ vs $H_A : p_1 < p_2$ based on samples of sizes $n_1 = 100$ and $n_2 = 120$. Thirty-three successes in the first sample and forty-eight in the second sample are observed. Command `prop.test( c(33,48), c(100,120), alternative="less" )` returns a p-value of 0.176, so the difference between proportions is not found statistically significant.

By the way, R applies a χ^2-test (Chi-square test) here although we know that it can also be done with a Z-test. The method used by R is more general. Doing it with a χ^2-test, R can handle more than two proportions with this command... as we'll see in the next chapter!

MATLAB notes

Some hypothesis tests are built in MATLAB. For a Z-test of $H_0 : \mu = \mu_0$ vs $H_A : \mu \neq \mu_0$, write `ztest(X,mu0,sigma,alpha)`, with the given tested value of μ_0, known standard deviation σ, and significance level α. For a one-sided test, add `'left'` or `'right'` after α. The test will return 0 if the null hypothesis is not rejected and 1 if it has to be rejected at level α. To see a P-value for this test, write `[h,p] = ztest(X,mu0,sigma)`.

T-tests can be done similarly, with a command `[h,p] = ttest(X,mu0)`. Of course, the standard deviation is not entered here; it is unknown.

Similar procedures are available for two-sample T-tests, as well as testing variances and ratios of variances. The corresponding commands are `ttest2(X,Y,alpha)`, `vartest(X,sigma0)`, and `vartest2(X,Y,alpha)`.

Confidence intervals for the mean, the difference of means, the variance, and the ratio of variances can be obtained simply by adding CI in `[h,p,CI]`. For example, `[h,p,CI] = vartest2(X,Y,0.01)` returns a 99% confidence interval for the ratio of variances σ_X^2/σ_Y^2, just as we derived it in Section 9.5.5.

Summary and conclusions

After taking a general look at the data by methods of Chapter 8, we proceeded with the more advanced and informative statistical inference described in this chapter.

There is a number of methods for estimating the unknown population parameters. Each method provides estimates with certain good and desired properties. We learned two general methods of **parameter estimation**.

Method of moments is based on matching the population and sample moments. It is relatively simple to implement, and it makes the estimated distribution "close" to the actual distribution of data in terms of their moments.

Maximum likelihood estimators maximize the probability of observing the sample that was really observed, thus making the actually occurring situation as likely as possible under the estimated distribution.

In addition to parameter estimates, **confidence intervals** show the margin of error and attach a confidence level to the result. A $(1 - \alpha)100\%$ confidence interval contains the unknown parameter with probability $(1 - \alpha)$. It means that $(1 - \alpha)100\%$ of all confidence intervals constructed from a large number of samples should contain the parameter, and only $100\alpha\%$ may miss it.

We can use data to verify statements and **test hypotheses**. Essentially, we measure the evidence provided by the data against the *null hypothesis* H_0. Then we decide whether it is sufficient for rejecting H_0.

Given *significance level* α, we can construct acceptance and rejection regions, compute a suitable *test statistic*, and make a decision depending on which region it belongs to. Alternatively, we may compute a P-value of the test. It shows how significant the evidence against H_0 is. Low P-values suggest rejection of the null hypothesis. P-value is the boundary between levels α-to reject and α-to-accept. It also represents the probability of observing the same or more extreme sample than the one that was actually observed.

Depending on what we are testing against, the *alternative hypothesis* may be two-sided or one-sided. We account for the direction of the alternative when we construct acceptance and rejection regions and when we compute a P-value.

We developed confidence intervals and tests of hypotheses for a number of parameters – population means, proportions, and variances, as well as the difference of means, the difference of proportions, and the ratio of variances. This covers the majority of practical problems.

Each developed statistical procedure is valid under certain assumptions. Verifying them is an important step in statistical inference.

Exercises

9.1. Estimate the unknown parameter θ from a sample

$$3, 3, 3, 3, 3, 7, 7, 7$$

drawn from a discrete distribution with the probability mass function

$$\begin{cases} P(3) & = & \theta \\ P(7) & = & 1 - \theta \end{cases}.$$

Compute two estimators of θ:

(a) the method of moments estimator;

(b) the maximum likelihood estimator.

Also,

(c) Estimate the standard error of each estimator of θ.

9.2. The number of times a computer code is executed until it runs without errors has a Geometric distribution with unknown parameter p. For 5 independent computer projects, a student records the following numbers of runs:

$$3 \quad 7 \quad 5 \quad 3 \quad 2$$

Estimate p

(a) by the method of moments;

(b) by the method of maximum likelihood.

9.3. Use method of moments and method of maximum likelihood to estimate

(a) parameters a and b if a sample from Uniform(a, b) distribution is observed;

(b) parameter λ if a sample from Exponential(λ) distribution is observed;

(c) parameter μ if a sample from Normal(μ, σ) distribution is observed, and we already know σ;

(d) parameter σ if a sample from Normal(μ, σ) distribution is observed, and we already know μ;

(e) parameters μ and σ if a sample from Normal(μ, σ) distribution is observed, and both μ and σ are unknown.

9.4. A sample of 3 observations $(X_1 = 0.4, \ X_2 = 0.7, \ X_3 = 0.9)$ is collected from a continuous distribution with density

$$f(x) = \begin{cases} \theta x^{\theta-1} & \text{if } 0 < x < 1 \\ 0 & \text{otherwise} \end{cases}$$

Estimate θ by *your favorite method.*

9.5. A sample $(X_1, ..., X_{10})$ is drawn from a distribution with a probability density function

$$\frac{1}{2}\left(\frac{1}{\theta}e^{-x/\theta} + \frac{1}{10}e^{-x/10}\right), \quad 0 < x < \infty$$

The sum of all 10 observations equals 150.

(a) Estimate θ by the method of moments.

(b) Estimate the standard error of your estimator in (a).

9.6. Verify columns 3-5 in Table 9.1 on p. 280. Section 9.4.7 will help you with the last row of the table.

9.7. In order to ensure efficient usage of a server, it is necessary to estimate the mean number of concurrent users. According to records, the average number of concurrent users at 100 randomly selected times is 37.7, with a standard deviation $\sigma = 9.2$.

(a) Construct a 90% confidence interval for the expectation of the number of concurrent users.

(b) At the 1% significance level, do these data provide significant evidence that the mean number of concurrent users is greater than 35?

9.8. Installation of a certain hardware takes random time with a standard deviation of 5 minutes.

(a) A computer technician installs this hardware on 64 different computers, with the average installation time of 42 minutes. Compute a 95% confidence interval for the population mean installation time.

(b) Suppose that the population mean installation time is 40 minutes. A technician installs the hardware on your PC. What is the probability that the installation time will be within the interval computed in (a)?

9.9. Salaries of entry-level computer engineers have Normal distribution with unknown mean and variance. Three randomly selected computer engineers have salaries (in $1000s):

$$30, 50, 70$$

(a) Construct a 90% confidence interval for the average salary of an entry-level computer engineer.

(b) Does this sample provide a significant evidence, at a 10% level of significance, that the average salary of all entry-level computer engineers is different from $80,000? Explain.

(c) Looking at this sample, one may think that the starting salaries have a great deal of variability. Construct a 90% confidence interval for the standard deviation of entry-level salaries.

9.10. We have to accept or reject a large shipment of items. For quality control purposes, we collect a sample of 200 items and find 24 defective items in it.

(a) Construct a 96% confidence interval for the proportion of defective items in the whole shipment.

(b) The manufacturer claims that at most one in 10 items in the shipment is defective. At the 4% level of significance, do we have sufficient evidence to disprove this claim? Do we have it at the 15% level?

9.11. Refer to Exercise 9.10. Having looked at the collected sample, we consider an alternative supplier. A sample of 150 items produced by the new supplier contains 13 defective items.

Is there significant evidence that the quality of items produced by the new supplier is higher than the quality of items in Exercise 9.10? What is the P-value?

9.12. An electronic parts factory produces resistors. Statistical analysis of the output suggests that resistances follow an approximately Normal distribution with a standard deviation of 0.2 ohms. A sample of 52 resistors has the average resistance of 0.62 ohms.

(a) Based on these data, construct a 95% confidence interval for the population mean resistance.

(b) If the actual population mean resistance is exactly 0.6 ohms, what is the probability that an average of 52 resistances is 0.62 ohms or higher?

9.13. Compute a P-value for the right-tail test in Example 9.25 on p. 279 and state your conclusion about a significant increase in the number of concurrent users.

9.14. Is there significant difference in speed between the two servers in Example 9.21 on p. 271?

(a) Use the confidence interval in Example 9.21 to conduct a two-sided test at the 5% level of significance.

(b) Compute a P-value of the two-sided test in (a).

(c) Is server A really faster? How strong is the evidence? Formulate the suitable hypothesis and alternative and compute the corresponding P-value.

State your conclusions in (a), (b), and (c).

9.15. According to Example 9.17 on p. 265, there is no significant difference, at the 5% level, between towns A and B in their support for the candidate. However, the level $\alpha = 0.05$ was chosen rather arbitrarily, and the candidate still does not know if he can trust the results when planning his campaign. Can we compare the two towns at *all* reasonable levels of significance? Compute the P-value of this test and state conclusions.

9.16. A sample of 250 items from lot A contains 10 defective items, and a sample of 300 items from lot B is found to contain 18 defective items.

(a) Construct a 98% confidence interval for the difference of proportions of defective items.

(b) At a significance level $\alpha = 0.02$, is there a significant difference between the quality of the two lots?

9.17. A news agency publishes results of a recent poll. It reports that a certain candidate has a 10-point stronger support in town A than in town B because 45% of the poll participants in town A and 35% of the poll participants in town B supported the candidate. What margin of error should the news agency report for each of the listed estimates, 45%, 35%, and 10%? Notice that 900 randomly selected registered voters participated in the poll in each town, and the reported margins of error correspond to 95% confidence intervals.

9.18. Consider the data about the number of blocked intrusions in Exercise 8.1, p. 240.

(a) Construct a 95% confidence interval for the difference between the average number of intrusion attempts per day before and after the change of firewall settings (assume equal variances).

(b) Can we claim a significant reduction in the rate of intrusion attempts? The number of intrusion attempts each day has approximately Normal distribution. Compute P-values and state your conclusions under the assumption of equal variances and without it. Does this assumption make a difference?

9.19. Consider two populations (X's and Y's) with different means but the same variance. Two independent samples, sizes n and m, are collected. Show that the pooled variance estimator

$$s_p^2 = \frac{\sum_{i=1}^{n}(X_i - \overline{X})^2 + \sum_{i=1}^{m}(Y_i - \overline{Y})^2}{n + m - 2}$$

estimates their common variance unbiasedly.

9.20. A manager questions the assumptions of Exercise 9.8. Her pilot sample of 40 installation times has a sample standard deviation of $s = 6.2$ min, and she says that it is significantly different from the assumed value of $\sigma = 5$ min. Do you agree with the manager? Conduct the suitable test of a standard deviation.

9.21. If $[a, b]$ is a $(1 - \alpha)100\%$ confidence interval for the population variance (with $a \geq 0$), prove that $[\sqrt{a}, \sqrt{b}]$ is a $(1 - \alpha)100\%$ confidence interval for the population standard deviation.

9.22. Recall Example 9.21 on p. 271, in which samples of 30 and 20 observations produced standard deviations of 0.6 min and 1.2 min, respectively. In this Example, we assumed unequal variances and used the suitable method only because the reported sample standard deviations *seemed* too different.

(a) Argue for or against the chosen method by testing equality of the population variances.

(b) Also, construct a 95% confidence interval for the ratio of the two population variances.

9.23. Anthony says to Eric that he is a stronger student because his average grade for the first six quizzes is higher. However, Eric replies that he is more stable because the variance of his grades is lower. The actual scores of the two friends (presumably, independent and normally distributed) are in the table.

	Quiz 1	Quiz 2	Quiz 3	Quiz 4	Quiz 5	Quiz 6
Anthony	85	92	97	65	75	96
Eric	81	79	76	84	83	77

(a) Is there significant evidence to support Anthony's claim? State H_0 and H_A. Test equality of variances and choose a suitable two-sample t-test. Then conduct the test and state conclusions.

(b) Is there significant evidence to support Eric's claim? State H_0 and H_A and conduct the test.

For each test, use the 5% level of significance.

9.24. Recall Exercise 9.23. Results essentially show that a sample of six quizzes was too small for Anthony to claim that he is a stronger student. We realize that each student has his own population of grades, with his own mean μ_i and variance σ_i^2. The observed quiz grades are sampled from this population, and they are different due to all the uncertainty and random factors involved when taking a quiz. Let us estimate the population parameters with some confidence.

(a) Construct a 90% confidence interval for the population mean score for each student.

(b) Construct a 90% confidence interval for the difference of population means. If you have not completed Exercise 9.23(a), start by testing equality of variances and choosing the appropriate method.

(c) Construct a 90% confidence interval for the population variance of scores for each student.

(d) Construct a 90% confidence interval for the ratio of population variances.

Chapter 10

Statistical Inference II

Statistical Inference journey continues. Methods covered in this chapter allow us to conduct new tests for independence and for the goodness of fit (sec. 10.1), test hypotheses without relying on a particular family of distributions (sec. 10.2), make full use of Monte Carlo methods for estimation and testing (sec. 10.3), and account for all the sources of information in addition to the real data (sec. 10.4).

10.1 Chi-square tests

Several important tests of statistical hypotheses are based on the *Chi-square distribution*. We have already used this distribution in Section 9.5 to study the population variance. This time, we will develop several tests based on the *counts* of our sampling units that fall in various categories. The general principle developed by Karl Pearson near year 1900 is to compare the *observed counts* against the *expected counts* via the *Chi-square statistic*

$$\text{Chi-square statistic} \quad \boxed{\chi^2 = \sum_{k=1}^{N} \frac{\{Obs(k) - Exp(k)\}^2}{Exp(k)}.} \quad (10.1)$$

Here the sum goes over N categories or groups of data defined depending on our testing problem; $Obs(k)$ is the actually observed number of sampling units in category k, and $Exp(k) = \mathbf{E}\{Obs(k) \mid H_0\}$ is the expected number of sampling units in category k if the null hypothesis H_0 is true.

This is always a *one-sided, right-tail* test. That is because only the low values of χ^2 show that the observed counts are close to what we expect them to be under the null hypotheses, and therefore, the data support H_0. On the contrary, large χ^2 occurs when Obs are far from Exp, which shows inconsistency of the data and the null hypothesis and does not support H_0.

Therefore, a level α rejection region for this Chi-square test is

$$R = [\chi^2_\alpha, +\infty),$$

and the P-value is always calculated as

$$P = \mathbf{P}\{\chi^2 \ge \chi^2_{\text{obs}}\}.$$

Pearson showed that the null distribution of χ^2 converges to the Chi-square distribution with $(N-1)$ degrees of freedom, as the sample size increases to infinity. This follows from a suitable version of the Central Limit Theorem. To apply it, we need to make sure the sample size is large enough. The rule of thumb requires an *expected count of at least 5 in each category*,

$$Exp(k) \ge 5 \quad \text{for all } k = 1, \dots, N.$$

If that is the case, then we can use the Chi-square distribution to construct rejection regions and compute P-values. If a count in some category is less than 5, then we should *merge* this category with another one, recalculate the χ^2 statistic, and then use the Chi-square distribution.

Here are several main applications of the Chi-square test.

10.1.1 Testing a distribution

The first type of applications focuses on testing whether the data belong to a particular distribution. For example, we may want to test whether a sample comes from the Normal distribution, whether interarrival times are Exponential and counts are Poisson, whether a random number generator returns high quality Standard Uniform values, or whether a die is unbiased.

In general, we observe a sample $(X_1, \dots, X_n)$ of size n from distribution F and test

$$H_0 : F = F_0 \quad \text{vs} \quad H_A : F \ne F_0 \quad (10.2)$$

for some given distribution F_0.

To conduct the test, we take all possible values of X under F_0, the *support* of F_0, and split them into N bins $B_1, \ldots, B_N$. A rule of thumb requires anywhere from 5 to 8 bins, which is quite enough to identify the distribution F_0 and at the same time have sufficiently high expected count in each bin, as it is required by the Chi-square test ($Exp \geq 5$).

The observed count for the k-th bin is the number of X_i that fall into B_k,

$$Obs(k) = \#\{i = 1, \ldots, n : X_i \in B_k\}.$$

If H_0 is true and all X_i have the distribution F_0, then $Obs(k)$, the number of "successes" in n trials, has Binomial distribution with parameters n and $p_k = F_0(B_k) = P\{X_i \in B_k \mid H_0\}$. Then, the corresponding expected count is the expected value of this Binomial distribution,

$$Exp(k) = np_k = nF_0(B_k).$$

After checking that all $Exp(k) \geq 5$, we compute the χ^2 statistic (10.1) and conduct the test.

Example 10.1 (IS THE DIE UNBIASED?). Suppose that after losing a large amount of money, an unlucky gambler questions whether the game was fair and the die was really unbiased. The last 90 tosses of this die gave the following results,

Number of dots on the die	1	2	3	4	5	6
Number of times it occurred	20	15	12	17	9	17

Let us test $H_0 : F = F_0$ vs $H_A : F \neq F_0$, where F is the distribution of the number of dots on the die, and F_0 is the *Discrete Uniform distribution*, under which

$$P(X = x) = \frac{1}{6} \quad \text{for} \quad x = 1, 2, 3, 4, 5, 6.$$

Observed counts are $Obs = 20, 15, 12, 17, 9$, and 17. The corresponding expected counts are

$$Exp(k) = np_k = (90)(1/6) = 15 \quad \text{(all more than 5)}.$$

Compute the Chi-square statistic

$$\chi^2_{obs} = \sum_{k=1}^{N} \frac{\{Obs(k) - Exp(k)\}^2}{Exp(k)}$$

$$= \frac{(20-15)^2}{15} + \frac{(15-15)^2}{15} + \frac{(12-15)^2}{15} + \frac{(17-15)^2}{15} + \frac{(9-15)^2}{15} + \frac{(17-15)^2}{15} = 5.2.$$

From Table A6 with $N - 1 = 5$ d.f., the P-value is

$$P = P\{\chi^2 \geq 5.2\} = \text{between 0.2 and 0.8}.$$

It means that there is no significant evidence to reject H_0, and therefore, no evidence that the die was biased. $\Diamond$

10.1.2   Testing a family of distributions

One more step. The Chi-square test can also be used to test the entire *model*. We are used to model the number of traffic accidents with Poisson, errors with Normal, and interarrival times with Exponential distribution. These are the models that are believed to fit the data well. Instead of just relying on this assumption, we can now test it and see if it is supported by the data.

We again suppose that the sample $(X_1, \ldots, X_n)$ is observed from a distribution F. It is desired to test whether F belongs to some *family of distributions* $\mathfrak{F}$,

$$H_0 : F \in \mathfrak{F} \qquad \text{vs} \qquad H_A : F \notin \mathfrak{F}. \tag{10.3}$$

Unlike test (10.2), the parameter θ of the tested family $\mathfrak{F}$ is not given; it is unknown. So, we have to estimate it by a consistent estimator $\widehat{\theta}$, to ensure $\widehat{\theta} \to \theta$ and to preserve the Chi-square distribution when $n \to \infty$. One can use the maximum likelihood estimator of θ.

Degrees of freedom of this Chi-square distribution will be reduced by the number of estimated parameters. Indeed, if θ is d-dimensional, then its estimation involves a system of d equations. These are d constraints which reduce the number of degrees of freedom by d.

It is often called a *goodness of fit test* because it measures how well the chosen model fits the data. Summarizing its steps,

- we find the maximum likelihood estimator $\widehat{\theta}$ and consider the distribution $F(\widehat{\theta}) \in \mathfrak{F}$;

- partition the support of $F(\widehat{\theta})$ into N bins $B_1, \ldots, B_N$, preferably with $N \in [5, 8]$;

- compute probabilities $p_k = \boldsymbol{P}\{X \in B_k\}$ for $k = 1, \ldots, N$ using $\widehat{\theta}$ as the parameter value;

- compute $Obs(k)$ from the data, $Exp(k) = np_k$, and the Chi-square statistic (10.1); if $np_k < 5$ for some k then merge B_k with another region;

- Compute the P-value or construct the rejection region using Chi-square distribution with $(N - d - 1)$ degrees of freedom, where d is the dimension of θ or the number of estimated parameters. State conclusions.

Example 10.2 (TRANSMISSION ERRORS). The number of transmission errors in communication channels is typically modeled by a Poisson distribution. Let us test this assumption. Among 170 randomly selected channels, 44 channels recorded no transmission error during a 3-hour period, 52 channels recorded one error, 36 recorded two errors, 20 recorded three errors, 12 recorded four errors, 5 recorded five errors, and one channel had seven errors.

Solution.   We are testing whether the unknown distribution F of the number of errors belongs to the Poisson family or not. That is,

$$H_0 : F = \text{Poisson}(\lambda) \text{ for some } \lambda \qquad \text{vs} \qquad H_A : F \neq \text{Poisson}(\lambda) \text{ for any } \lambda.$$

First, estimate parameter λ. Its maximum likelihood estimator equals

$$\widehat{\lambda} = \overline{X} = \frac{(44)(0) + (52)(1) + (36)(2) + (20)(3) + (12)(4) + (5)(5) + (1)(7)}{170} = 1.55$$

(see Example 9.7 on p. 249).

Next, the support of Poisson distribution is the set of all non-negative integers. Partition it into bins, let's say,

$$B_0 = \{0\}, \ B_1 = \{1\}, \ B_2 = \{2\}, \ B_3 = \{3\}, \ B_4 = \{4\}, \ B_5 = [5, \infty).$$

The observed counts for these bins are given: 44, 52, 36, 20, 12, and $5+1 = 6$. The expected counts are calculated from the Poisson pmf as

$$Exp(k) = np_k = ne^{-\widehat{\lambda}} \frac{\widehat{\lambda}^k}{k!} \qquad \text{for } k = 0, \ldots, 4, \ \ n = 170, \quad \text{and} \quad \widehat{\lambda} = 1.55.$$

The last expected count $Exp(5) = 170 - Exp(0) - \ldots - Exp(4)$ because $\sum p_k = 1$, and therefore, $\sum np_k = n$. Thus, we have

k	0	1	2	3	4	5
p_k	0.21	0.33	0.26	0.13	0.05	0.02
np_k	36.0	55.9	43.4	22.5	8.7	3.6

We notice that the last group has a count below five. So, let's combine it with the previous group, $B_4 = [4, \infty)$. For the new groups, we have

k	0	1	2	3	4
$Exp(k)$	36.0	55.9	43.4	22.5	12.3
$Obs(k)$	44.0	52.0	36.0	20.0	18.0

Then compute the Chi-square statistic

$$\chi^2_{\text{obs}} = \sum_k \frac{\{Obs(k) - Exp(k)\}^2}{Exp(k)} = 6.2$$

and compare it against the Chi-square distribution with $5 - 1 - 1 = 3$ d.f. in Table A6. With a P-value

$$P = \boldsymbol{P}\left\{\chi^2 \geq 6.2\right\} = \text{ between } 0.1 \text{ and } 0.2,$$

we conclude that there is *no evidence against a Poisson distribution* of the number of transmission errors. ◇

Testing families of continuous distributions is rather similar.

Example 10.3 (NETWORK LOAD). The data in Exercise 8.2 on p. 240 shows the number of concurrent users of a network in $n = 50$ locations. For further modeling and hypothesis testing, can we assume an approximately Normal distribution for the number of concurrent users?

Solution. For the distribution F of the number of concurrent users, let us test

$$H_0 : F = \text{Normal}(\mu, \sigma) \text{ for some } \mu \text{ and } \sigma \qquad \text{vs} \qquad H_A : F \neq \text{Normal}(\mu, \sigma) \text{ for any } \mu \text{ and } \sigma.$$

Maximum likelihood estimates of μ and σ are (see Exercise 9.3 on p. 309)

$$\widehat{\mu} = \overline{X} = 17.95 \quad \text{and} \quad \widehat{\sigma} = \sqrt{\frac{(n-1)s^2}{n}} = \sqrt{\frac{1}{n}\sum_{i=1}^{n}(X_i - \overline{X})^2} = 3.13.$$

Split the support $(-\infty, +\infty)$ (actually, only $[0, \infty)$ for the number of concurrent users) into bins, for example,

$$B_1 = (-\infty, 14), \; B_2 = [14, 16), \; B_3 = [16, 18), \; B_4 = [18, 20), \; B_5 = [20, 22), \; B_6 = [22, \infty)$$

(in thousands of users). While selecting these bins, we made sure that the expected counts in each of them will not be too low. Use Table A4 of Normal distribution to find the probabilities p_k,

$$p_1 = P(X \in B_1) = P(X \le 14) = P\left(Z \le \frac{14 - 17.95}{3.13}\right) = \Phi(-1.26) = 0.1038,$$

and similarly for $p_1, \ldots, p_6$. Then calculate $Exp(k) = np_k$ and count $Obs(k)$ (check!),

k	1	2	3	4	5	6
B_k	$(-\infty, 14)$	$[14, 16)$	$[16, 18)$	$[18, 20)$	$[20, 22)$	$[22, \infty)$
p_k	0.10	0.16	0.24	0.24	0.16	0.10
$Exp(k)$	5	8	12	12	8	5
$Obs(k)$	6	8	13	11	6	6

From this table, the test statistic is

$$\chi^2_{obs} = \sum_{k=1}^{N} \frac{\{Obs(k) - Exp(k)\}^2}{Exp(k)} = 1.07,$$

which has a high P-value

$$P = \boldsymbol{P}\{\chi^2 \ge 1.07\} > 0.2.$$

Chi-square distribution was used with $6 - 1 - 2 = 3$ d.f., where we lost 2 d.f. due to 2 estimated parameters. Hence, there is *no evidence against a Normal distribution* of the number of concurrent users.   ◊

R and MATLAB notes

For Chi-square tests, R has a command `chisq.test`. Here is how it solves Example 10.2.

— R ——————————

```
counts = c(44,52,36,20,12,6)               # Observed counts
prob = c( dpois(0:4,1.55), 1-ppois(4,1.55) )  # Poisson probabilities
                # of our chosen bins. These are P(0),...,P(4) and 1-F(4)
chisq.test(counts,p=prob)                   # Test for goodness of fit
```

As you notice, we test the observed counts against the Poisson distribution with parameter 1.55, the estimate derived in Example 10.2 by the method of maximum likelihood. It can also be computed in R using `fitdistr(x,'Poisson')`, the tool discussed in the end of Section 9.1.

A similar MATLAB command is `chi2gof`, a fancy abbreviation of a Chi-square test for the goodness of fit. Let's apply it to Example 10.3.

—— MATLAB ————

```
X = [17.2 22.1 18.5 17.2 18.6 14.8 21.7 15.8 16.3 22.8 24.1 13.3...
16.2 17.5 19.0 23.9 14.8 22.2 21.7 20.7 13.5 15.8 13.1 16.1 21.9...
23.9 19.3 12.0 19.9 19.4 15.4 16.7 19.5 16.2 16.9 17.1 20.2 13.4...
19.8 17.7 19.7 18.7 17.6 15.9 15.2 17.1 15.0 18.8 21.6 11.9];
[ decision, pvalue, stat ] = chi2gof(X, 'CDF', fitdist(X,'Normal'))
```

This gives us three pieces of information. New variables `pvalue` and `decision` are a P-value of the test and a decision based on it (1 if H_0 is rejected in favor of H_A meaning an evidence against Normal distribution, 0 otherwise). Also, the new object `stat` contains all the steps of this test - edges of chosen bins, observed and expected counts, and finally, the χ^2 statistic.

10.1.3   Testing independence

Many practical applications require testing independence of two factors. If there is a significant association between two features, it helps to understand the cause-and-effect relationships. For example, is it true that smoking causes lung cancer? Do the data confirm that drinking and driving increases the chance of a traffic accident? Does customer satisfaction with their PC depend on the operating system? And does the graphical user interface (GUI) affect popularity of a software?

Apparently, Chi-square statistics can help us test

H_0 : Factors A and B are independent   vs H_A : Factors A and B are dependent.

It is understood that each factor partitions the whole population $\mathcal{P}$ into two or more categories, $A_1, \ldots, A_k$ and $B_1, \ldots, B_m$, where $A_i \cap A_j = \varnothing$, $B_i \cap B_j = \varnothing$, for any $i \neq j$, and $\cup A_i = \cup B_i = \mathcal{P}$.

Independence of factors is understood just like independence of random variables in Section 3.2.2. Factors A and B are independent if any randomly selected unit x of the population belongs to categories A_i and B_j independently of each other. In other words, we are testing

$$H_0 : \boldsymbol{P}\{x \in A_i \cap B_j\} = \boldsymbol{P}\{x \in A_i\}\boldsymbol{P}\{x \in B_j\} \text{ for all } i,j$$
$$\text{vs} \tag{10.4}$$
$$H_A : \boldsymbol{P}\{x \in A_i \cap B_j\} \neq \boldsymbol{P}\{x \in A_i\}\boldsymbol{P}\{x \in B_j\} \text{ for some } i,j.$$

To test these hypotheses, we collect a sample of size n and count n_{ij} units that landed in the intersection of categories A_i and B_j. These are the *observed counts*, which can be nicely arranged in a *contingency table*,

	B_1	B_2	$\cdots$	B_m	row total
A_1	n_{11}	n_{12}	$\cdots$	n_{1m}	$n_{1\cdot}$
A_2	n_{21}	n_{22}	$\cdots$	n_{2m}	$n_{2\cdot}$
$\cdots$	$\cdots$	$\cdots$	$\cdots$	$\cdots$	$\cdots$
A_k	n_{k1}	n_{k2}	$\cdots$	n_{km}	$n_{k\cdot}$
column total	$n_{\cdot 1}$	$n_{\cdot 2}$	$\cdots$	$n_{\cdot m}$	$n_{\cdot\cdot} = n$

Notation $n_{i\cdot} = \sum_i n_{ij}$ and $n_{\cdot j} = \sum_j n_{ij}$ is quite common for the row totals and column totals.

Then we estimate all the probabilities in (10.4),

$$\widehat{P}\left\{x \in A_i \cap B_j\right\} = \frac{n_{ij}}{n}, \quad \widehat{P}\left\{x \in A_i\right\} = \sum_{j=1}^{m} \frac{n_{ij}}{n} = \frac{n_{i\cdot}}{n}, \quad \widehat{P}\left\{x \in B_j\right\} = \sum_{i=1}^{k} \frac{n_{ij}}{n} = \frac{n_{\cdot j}}{n}.$$

If H_0 is true, then we can also estimate the probabilities of intersection as

$$\widetilde{P}\left\{x \in A_i \cap B_j\right\} = \left(\frac{n_{i\cdot}}{n}\right)\left(\frac{n_{\cdot j}}{n}\right)$$

and estimate the *expected counts* as

$$\widehat{\mathrm{Exp}}(i,j) = n\left(\frac{n_{i\cdot}}{n}\right)\left(\frac{n_{\cdot j}}{n}\right) = \frac{(n_{i\cdot})(n_{\cdot j})}{n}.$$

This is the case when expected counts $\mathrm{Exp}(i,j) = \mathbf{E}\left\{Obs(i,j) \mid H_0\right\}$ are estimated under H_0. There is not enough information in the null hypothesis H_0 to compute them exactly.

After this preparation, we construct the usual *Chi-square statistic* comparing the observed and the estimated expected counts over the entire contingency table,

$$\chi_{\mathrm{obs}}^2 = \sum_{i=1}^{k} \sum_{j=1}^{m} \frac{\left\{Obs(i,j) - \widehat{\mathrm{Exp}}(i,j)\right\}^2}{\widehat{\mathrm{Exp}}(i,j)}. \tag{10.5}$$

This χ_{obs}^2 should now be compared against the Chi-square table. How many *degrees of freedom* does it have here? Well, since the table has k rows and m columns, wouldn't the number of degrees of freedom equal $(k \cdot m)$?

It turns out that the differences

$$d_{ij} = Obs(i,j) - \widehat{\mathrm{Exp}}(i,j) = n_{ij} - \frac{(n_{i\cdot})(n_{\cdot j})}{n}$$

in (10.5) have many constraints. For any $i = 1, \ldots, k$, the sum $d_{i\cdot} = \sum_j d_{ij} = n_{i\cdot} - \frac{(n_{i\cdot})(n_{\cdot\cdot})}{n} = 0$, and similarly, for any $j = 1, \ldots, m$, we have $d_{\cdot j} = \sum_i d_{ij} = n_{\cdot j} - \frac{(n_{\cdot\cdot})(n_{\cdot j})}{n} = 0$.

So, do we lose $(k+m)$ degrees of freedom due to these $(k+m)$ constraints? Yes, but there is also one constraint among these constraints! Whatever d_{ij} are, the equality $\sum_i d_{i\cdot} = \sum_j d_{\cdot j}$ always holds. So, if all the $d_{i\cdot}$ and $d_{\cdot j}$ equal zero except the last one, $d_{\cdot m}$, then $d_{\cdot m} = 0$ *automatically* because $\sum_i d_{i\cdot} = 0 = \sum_j d_{\cdot j}$.

As a result, we have only $(k + m - 1)$ linearly independent constraints, and the overall number of degrees of freedom in χ_{obs}^2 is

$$\mathrm{d.f.} = km - (k + m - 1) = (k - 1)(m - 1).$$

Chi-square test for independence	Test statistic $$\chi_{\mathrm{obs}}^2 = \sum_{i=1}^{k} \sum_{j=1}^{m} \frac{\left\{Obs(i,j) - \widehat{\mathrm{Exp}}(i,j)\right\}^2}{\widehat{\mathrm{Exp}}(i,j)},$$ where $Obs(i,j) = n_{ij}$ are observed counts, $\widehat{\mathrm{Exp}}(i,j) = \dfrac{(n_{i\cdot})(n_{\cdot j})}{n}$ are estimated expected counts, and χ_{obs}^2 has $(k - 1)(m - 1)$ d.f.

As always in this section, this test is one-sided and right-tail.

Example 10.4 (SPAM AND ATTACHMENTS). Modern e-mail servers and anti-spam filters attempt to identify spam e-mails and direct them to a junk folder. There are various ways to detect spam, and research still continues. In this regard, an information security officer tries to confirm that the chance for an e-mail to be spam depends on whether it contains images or not. The following data were collected on $n = 1000$ random e-mail messages,

$Obs(i,j) = n_{ij}$	With images	No images	$n_{i\cdot}$
Spam	160	240	400
No spam	140	460	600
$n_{\cdot j}$	300	700	1000

Testing H_0: "being spam and containing images are independent factors" vs H_A: "these factors are dependent", calculate the estimated expected counts,

$$\widehat{Exp}(1,1) = \frac{(300)(400)}{1000} = 120, \qquad \widehat{Exp}(1,2) = \frac{(700)(400)}{1000} = 280,$$

$$\widehat{Exp}(2,1) = \frac{(300)(600)}{1000} = 180, \qquad \widehat{Exp}(2,2) = \frac{(700)(600)}{1000} = 420.$$

$\widehat{Exp}(i,j) = \frac{(n_{i\cdot})(n_{\cdot j})}{n}$	With images	No images	$n_{i\cdot}$
Spam	120	280	400
No spam	180	420	600
$n_{\cdot j}$	300	700	1000

You can always check that all the row totals, the column totals, and the whole table total are the same for the observed and the expected counts (so, if there is a mistake, catch it here).

$$\chi^2_{obs} = \frac{(160 - 120)^2}{120} + \frac{(240 - 280)^2}{280} + \frac{(140 - 180)^2}{180} + \frac{(460 - 420)^2}{420} = 31.75.$$

From Table A6 with $(2 - 1)(2 - 1) = 1$ d.f., we find that the P-value $P < 0.001$. We have a significant evidence that an e-mail having an attachment is somehow related to being spam. Therefore, this piece of information can be used in anti-spam filters. $\diamond$

Example 10.5 (INTERNET SHOPPING ON DIFFERENT DAYS OF THE WEEK). A web designer suspects that the chance for an internet shopper to make a purchase through her web site varies depending on the day of the week. To test this claim, she collects data during one week, when the web site recorded 3758 hits.

Observed	Mon	Tue	Wed	Thu	Fri	Sat	Sun	Total
No purchase	399	261	284	263	393	531	502	2633
Single purchase	119	72	97	51	143	145	150	777
Multiple purchases	39	50	20	15	41	97	86	348
Total	557	383	401	329	577	773	738	3758

Testing independence (i.e., probability of making a purchase or multiple purchases is the same on any day of the week), we compute the estimated expected counts,

$$\widehat{Exp}(i,j) = \frac{(n_{i\cdot})(n_{\cdot j})}{n} \quad \text{for} \quad i = 1,\ldots,7, \ j = 1,2,3.$$

Expected	Mon	Tue	Wed	Thu	Fri	Sat	Sun	Total
No purchase	390.26	268.34	280.96	230.51	404.27	541.59	517.07	2633
Single purchase	115.16	79.19	82.91	68.02	119.30	159.82	152.59	777
Multiple purchases	51.58	35.47	37.13	30.47	53.43	71.58	68.34	348
Total	557	383	401	329	577	773	738	3758

Then, the test statistic is

$$\chi^2_{\text{obs}} = \frac{(399 - 390.26)^2}{390.26} + \ldots + \frac{(86 - 68.34)^2}{68.34} = 60.79,$$

and it has $(7-1)(3-1) = 12$ degrees of freedom. From Table A6, we find that the P-value is $P < 0.001$, so indeed, there is significant evidence that the probability of making a single purchase or multiple purchases varies during the week. ◊

R and MATLAB notes

The χ^2 test for independence is readily available in R and MATLAB. For example, here is a quick solution to Example 10.4 on page 323.

— R —————

```
counts = c(160,140,240,460)      # These counts are given in Example 10.4
T = matrix(counts,2,2)           # Create a contingency table T
chisq.test(T,correct=0)          # Chi-square test for independence
```

All components of this χ^2 test are saved in the data frame `chisq.test(T,correct=0)`, and you can see the list of them by typing `names(chisq.test(T,correct=0))`. For example, to get the expected counts, enter a line `chisq.test(T,correct=0)$expected`.

The option `correct=0` asks to perform the test precisely as in Example 10.4. Without this option, R uses a slightly different method that includes a continuity correction.

The same command can be applied to counts, as we just did, or to raw data. We can create a dataset with the same counts and then test the two factors for independence as follows.

— R —————

```
X1 = rbind(matrix("Spam",400,1),matrix("No spam",600,1))
X2 = rbind(matrix("Images",160,1),matrix("No images",240,1),
          matrix("Images",140,1),matrix("No images",460,1))
chisq.test(X1,X2,correct=0)
```

A similar tool for the χ^2 test exists in MATLAB.

— MATLAB —————

```
X1 = [ones(400,1); zeros(600,1)];
X2 = [ones(160,1); zeros(240,1); ones(140,1); zeros(460,1)];
[ observed, statistic, pvalue ] = crosstab(X1,X2)
```

This way, we define variables `observed`, `statistic`, and `pvalue` that will contain the observed counts, the χ^2 statistic, and the P-value of the test allowing us to decide about independence of images and spam in e-mails.

Without special commands for χ^2 tests offered by R and MATLAB, users of other software languages can write a code along the following lines.

```
X = [160 240; 140 460];                        % Matrix of observed counts
Row = sum(X')'; Col = sum(X); Tot = sum(Row);  % Row and column totals
k = length(Col); m = length(Row);              % Dimensions of the table
e = zeros(size(X));                            % Expected counts
for i=1:k; for j=1:m; e(i,j) = Row(i)*Col(j)/Tot; end; end;
chisq = (X-e).^2./e;                           % Chi-square terms
chistat = sum(sum(chisq));                     % Chi-square statistic
Pvalue = 1-chi2cdf(chistat,(k-1)*(m-1))        % P-value
```

This code is actually executable in MATLAB. It computes the expected counts, the test statistic, and the P-value of the test step by step.

10.2 Nonparametric statistics

Parametric statistical methods are always designed for a specific family of distributions such as Normal, Poisson, Gamma, etc. *Nonparametric statistics* does not assume any particular distribution. On one hand, nonparametric methods are less powerful because the less you assume about the data the less you can find out from it. On the other hand, having fewer requirements, they are applicable to wider applications. Here are three typical examples.

Example 10.6 (UNKNOWN DISTRIBUTION). Our test in Example 9.29 on p. 283 requires the data, battery lives, to follow Normal distribution. Certainly, this t-test is *parametric*. However, looking back at this example, we are not sure if this assumption could be made. Samples of 12 and 18 data measurements are probably too small to find an evidence for or against the Normal distribution. Can we test the same hypothesis ($\mu_X = \mu_Y$) without relying on this assumption? ◇

Example 10.7 (OUTLIERS). Sample sizes in Example 9.37 on p. 292 are large enough ($m = n = 50$), so we can refer to the Central Limit Theorem to justify the use of a Z-test there. However, these data on running times of some software may contain occasional outliers. Suppose that one computer accidentally got frozen during the experiment, and this added 40 minutes to its running time. Certainly, it is an outlier, one out of 100 observations, but it makes an impact on the result.

Indeed, if one observation Y_i increases by 40 minutes, then the new average of 50 running times after the upgrade becomes $\overline{Y} = 7.2 + 40/50 = 8.0$. The test statistic is now $Z = (8.5 - 8.0)/0.36 = 1.39$ instead of 3.61, and the P-value is $P = 0.0823$ instead of $P = 0.0002$. In Example 9.37, we concluded that the evidence of an efficient upgrade is overwhelming, but now we are not so sure!

We need a method that will not be so sensitive to a few outliers. ◇

Example 10.8 (ORDINAL DATA). A software company conducts regular surveys where customers rank their level of satisfaction on a standard scale "strongly agree, agree, neutral, disagree, strongly disagree". Then, statisticians are asked to conduct tests comparing the customer satisfaction of different products.

These data are not even numerical! Yes, they contain important information but how can we conduct tests without any numbers? Can we assign numerical values to customers' responses, say, "strongly agree"=5, "agree"=4, ..., "strongly disagree"=1? This approach seems to claim some additional information that is not given in the data. For example, it implies that the difference between "agree" and "strongly agree" is the same as between "agree" and "neutral". And further, the difference between "strongly agree" and "neutral" is the same as between "agree" and "disagree". We do have informative data, but this particular information is not given, so we cannot use it.

Are there methods in Statistics that allow us to work with the data that are not numbers but *ranks*, so they can be ordered from the lowest (like "strongly disagree") to the highest (like "strongly agree")?   ◇

The situation in Example 10.8 is rather typical. Besides satisfaction surveys, one may want to analyze and compare the levels of education, high school to Ph.D.; military officer ranks, from a Second Lieutenant to a General; or chess skills, from Class J to Grandmaster. Also, many surveys ask for an interval instead of precise quantities, for example, income brackets - "is your annual income less than $20,000?" "between $20,000 and $30,000?" etc.

DEFINITION 10.1

> Data that can be ordered from the lowest to the highest without any numeric values are called **ordinal data**.

Most nonparametric methods can deal with situations described in Examples 10.6–10.8. Here we discuss three very common nonparametric tests for one sample and two samples: the sign test, the signed rank test, and the rank sum test.

10.2.1   Sign test

Here is a simple test for one population. It refers to a population *median M* and tests

$$H_0 : M = m$$

against a one-sided or a two-sided alternative, $H_A : M < m$, $H_A : M > m$, or $H_A : M \neq m$. So, we are testing whether exactly a half of the population is below m and a half is above m.

To conduct the *sign test*, simply count how many observations are above m,

$$S_{\text{obs}} = S(X_1, \ldots, X_n) = \text{number of } \{i : X_i > m\}.$$

Typically, it is assumed that the underlying distribution is continuous, so $P\{X_i = m\} = 0$. In this case, if H_0 is true and m is the median, then each X_i is equally likely to be above m or below m, and S has *Binomial* distribution with parameters n and $p = 1/2$. Using Table

A2 of Binomial distribution for small n or Table A4 of Normal distribution for large n (with a proper standardization and continuity correction, of course), we compute the P-value and arrive at a conclusion.

The right-tail alternative will be supported by large values of S_{obs}, so the P-value should be $P = \boldsymbol{P}\{S \geq S_{\mathrm{obs}}\}$. Similarly, for the left-tail alternative, $P = \boldsymbol{P}\{S \leq S_{\mathrm{obs}}\}$, and for the two-sided test, $P = 2\min\left(\boldsymbol{P}\{S \leq S_{\mathrm{obs}}\}, \boldsymbol{P}\{S \geq S_{\mathrm{obs}}\}\right)$.

The sign test

Test of the median, $H_0 : M = m$

Test statistic $S = $ number of $X_i > m$

Null distribution $S \sim Binomial(n, 1/2)$
For large n, $S \approx Normal(n/2, \sqrt{n}/2)$
(if the distribution of X_i is continuous)

Example 10.9 (UNAUTHORIZED USE OF A COMPUTER ACCOUNT, CONTINUED).

Example 9.39 on p. 292 shows a very significant evidence that a computer account was used by an unauthorized person. This conclusion was based on the following times between keystrokes (data set `Keystrokes`),

.24, .22, .26, .34, .35, .32, .33, .29, .19, .36, .30, .15, .17, .28, .38, .40, .37, .27 seconds,

whereas the account owner usually makes 0.2 sec between keystrokes. Our solution was based on a T-test and we had to assume the Normal distribution of data. However, the histogram on Figure 10.1 does not confirm this assumption. Also, the sample size is too small to conduct a Chi-square goodness-of-fit test (Exercise 10.5).

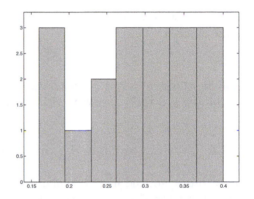

FIGURE 10.1: *The histogram of times between keystrokes does not support or deny a Normal distribution.*

Let's apply *the sign test* because it does not require the Normal distribution. The test statistic for $H_0 : M = 0.2$ vs $H_A : M \neq 0.2$ is $S_{\mathrm{obs}} = 15$ because 15 of 18 recorded times exceed 0.2.

Then, from Table A2 with $n = 18$ and $p = 0.5$, we find the P-value,

$$
\begin{aligned}
P &= 2\min\left(\boldsymbol{P}\{S \leq S_{\mathrm{obs}}\}, \boldsymbol{P}\{S \geq S_{\mathrm{obs}}\}\right) \\
&= 2\min(0.0038, 0.9993) = 0.0076.
\end{aligned}
$$

(A hint for using the table here: $\boldsymbol{P}\{S \geq 15\} = \boldsymbol{P}\{S \leq 3\}$, because the Binomial(0.5) distribution is symmetric.) So, the sign test rejects H_0 at any $\alpha > 0.0076$), which is an evidence that the account was used by an unauthorized person.

Here is a large-sample version of the sign test.

Example 10.10 (ARE THE CARS SPEEDING?). Eric suspects that the median speed of cars on his way to the university exceeds the speed limit, which is 30 mph. As an experiment, he drives with Alyssa precisely at the speed of 30 mph, and Alyssa counts the cars. At the end, Alyssa reports that 56 cars were driving faster than Eric, and 44 cars were driving slower. Does this confirm Eric's suspicion?

Solution. This is a one-sided right-tail test of

$$H_0 : M = 30 \text{ vs } H_A : M > 30,$$

because rejection of H_0 should imply that more than 50% of cars are speeding. The sign test statistic is $S = 56$, and the sample size is $n = 56 + 44 = 100$. The null distribution of S is approximately Normal with $\mu = n/2 = 50$ and $\sigma = \sqrt{n}/2 = 5$. Calculate the P-value (don't forget the continuity correction),

$$P = \boldsymbol{P}\{S \geq 56\} = \boldsymbol{P}\left\{Z \geq \frac{55.5 - 50}{5}\right\} = 1 - \Phi(1.1) = 1 - .8643 = 0.1357,$$

from Table A4. We can conclude that Alyssa and Eric *did not find a significant evidence that the median speed of cars is above the speed limit.*

Notice that we used *ordinal data.* Knowing the exact speed of each car was not necessary for the sign test.   ◊

Applying the *sign* test, statisticians used to mark the data above the tested value ($X_i > m$) with pluses and those below m with minuses... which fully explains the test's name.

10.2.2   Wilcoxon signed rank test

You certainly noticed how little information the sign test uses. All we had to know was how many observed values are above m and below m. But perhaps the unused information can be handy, and it should not be wasted. The next test takes magnitudes into account, in addition to signs.

To test the population *median M*, we again split the sample into two parts - below m and above m. But now we compare *how far* below m is the first group, collectively, and *how far* above m is the second group. This comparison is done in terms of statistics called *ranks*.

DEFINITION 10.2 ————————————————————————————

Rank of any unit of a sample is its position when the sample is arranged in the increasing order. In a sample of size n, the smallest observation has rank 1, the second smallest has rank 2, and so on, and the largest observation has rank n. If two or more observations are equal, their ranks are typically replaced by their average rank.

NOTATION ‖ R_i = rank of the i-th observation ‖

Having $R_i = r$ means that X_i is the r-th smallest observation in the sample.

Example 10.11. Consider a sample

$$3, 7, 5, 6, 5, 4.$$

The smallest observation, $X_1 = 3$, has rank 1. The second smallest is $X_6 = 4$; it has rank 2. The 3rd and 4th smallest are $X_3 = X_5 = 5$; their ranks 3 and 4 are averaged, and each gets rank 3.5. Then, $X_4 = 6$ and $X_2 = 7$ get ranks 5 and 6. So, we have the following ranks,

$$R_1 = 1, \ R_2 = 6, \ R_3 = 3.5, \ R_4 = 5, R_5 = 3.5, R_6 = 2.$$

$\Diamond$

Wilcoxon signed rank test of $H_0 : M = m$ is conducted as follows.

1. Consider the *distances* between observations and the tested value, $d_i = |X_i - m|$.

2. Order these distances and compute their *ranks* R_i. *Notice! These are ranks of d_i, not X_i.*

3. Take only the ranks corresponding to observations X_i greater than m. Their sum is the test statistic W. Namely,

$$W = \sum_{i: \, X_i > m} R_i$$

(see Figure 10.3 on p. 331 to visualize distances d_i and their ranks R_i).

4. Large values of W suggest rejection of H_0 in favor of $H_A : M > m$; small values support $H_A : M < m$; and both support a two-sided alternative $H_A : M \neq m$.

This test was proposed by an Ireland-born American statistician Frank Wilcoxon (1892-1965), and it is often called *Wilcoxon test*.

For convenience, Wilcoxon proposed to use *signed ranks*, giving a "+" sign to a rank R_i if $X_i > m$ and a "−" sign if $X_i < m$. His statistic W then equals the sum of positive signed ranks. (Some people take the sum of negative ranks instead, or its simple transformation. These statistics are one-to-one functions of each other, so each of them can be used for the Wilcoxon test.)

We assume that the distribution of X is *symmetric* about its median M, which means

$$\boldsymbol{P}\{X \leq M - a\} = \boldsymbol{P}\{X \geq M + a\}.$$

for any a (Figure 10.2). If the mean $\mu = \mathbf{E}(X)$ exists, it will also equal M.

Let's also assume that the distribution is continuous. So, there are no equal observations and no ranks need to be averaged.

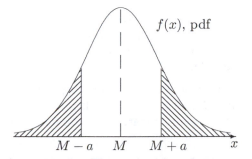

FIGURE 10.2: *Symmetric distribution. The shaded areas-probabilities are equal for any a.*

Under these assumptions, the null distribution of Wilcoxon test statistic W is derived later in this section on p. 333. For $n \leq 30$, the critical values are given in Table A8. For $n \geq 15$, one can use a Normal approximation, with a continuity correction because the distribution of W is discrete.

Remark about acceptance and rejection regions: When we construct acceptance and rejection regions for some test statistic T, we always look for a critical value T_α such that $P\{T \le T_\alpha\} = \alpha$ or $P\{T \ge T_\alpha\} = \alpha$.

But in the case of Wilcoxon test, statistic W takes only a finite number of possible values for any n. As a result, there may be no such value w that makes probability $P\{W \le w\}$ or $P\{W \ge w\}$ exactly equal α. One can achieve only inequalities $P\{W \le w\} \le \alpha$ and $P\{W \ge w\} \le \alpha$.

Critical values in Table A8 make these inequalities as tight as possible (that is, as close as possible to equalities), still making sure that

$$P\{\text{Type I error}\} \le \alpha.$$

Wilcoxon signed rank test	Test of the median, $H_0: M = m$. Test statistic $W = \sum\limits_{i:\, X_i > m} R_i$, where R_i is the rank of $d_i = \|X_i - m\|$. Null distribution: Table A8 or recursive formula (10.6). For $n \ge 15$, $W \approx Normal\left(\dfrac{n(n+1)}{4}, \sqrt{\dfrac{n(n+1)(2n+1)}{24}}\right)$ Assumptions: the distribution of X_i is continuous and symmetric

Example 10.12 (SUPPLY AND DEMAND). Having sufficient supply to match the demand always requires the help of statisticians.

Suppose that you are managing a student computer lab. In particular, your duty is to ensure that the printers don't run out of paper. During the first six days of your work, the lab consumed

$$7, 5.5, 9.5, 6, 3.5, \text{ and } 9 \text{ cartons of paper.}$$

Does this imply significant evidence, at the 5% level of significance, that the median daily consumption of paper is more than 5 cartons? It is fair to assume that the amounts of used paper are independent on different days, and these six days are as good as a simple random sample.

Let's test $H_0: M = 5$ vs $H_A: M > 5$. According to Table A8 for the right-tail test with $n = 6$ and $\alpha = 0.05$, we'll reject H_0 when the sum of positive ranks $T \ge 19$. To compute Wilcoxon test statistic T, compute distances $d_i = |X_i - 5|$ and rank them from the smallest to the largest.

i	X_i	$X_i - 5$	d_i	R_i	sign
1	7	2	2	4	+
2	5.5	0.5	0.5	1	+
3	9.5	4.5	4.5	6	+
4	6	1	1	2	+
5	3.5	−1.5	1.5	3	−
6	9	4	4	5	+

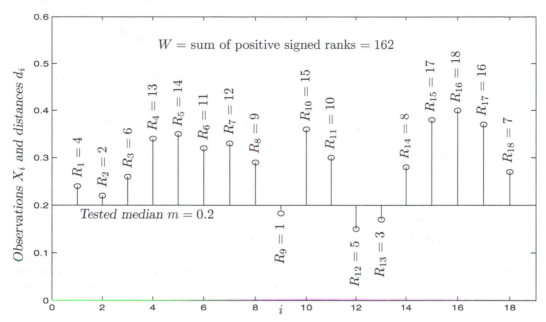

FIGURE 10.3: *Ranks of distances* $d_i = |X_i - m|$ *for the Wilcoxon signed rank test in Example 10.13.*

Then, compute T adding the "positive" ranks only, $T = 4 + 1 + 6 + 2 + 5 = 18$. It did not make it to the rejection region, and therefore, *at the 5% level, these data do not provide significance evidence that the median consumption of paper exceeds 5 cartons per day.* $\Diamond$

Example 10.13 (UNAUTHORIZED USE OF A COMPUTER ACCOUNT, CONTINUED). Applying Wilcoxon signed rank test to test $M = 0.2$ vs $M \neq 0.2$ in Example 10.9 (data set Keystrokes), we compute the distances $d_1 = |X_1 - m| = |0.24 - 0.2| = 0.04, \ldots,$ $d_{18} = |0.27 - 0.2| = 0.07$ and rank them; see the ranks on Figure 10.3. Here we notice that the 9-th, 12-th, and 13-th observations are below the tested value $m = 0.2$ while all the others are above $m = 0.2$. Computing the sum of only positive signed ranks, without R_9, R_{12}, and R_{13}, we find the test statistic

$$W = \sum_{i:\ X_i > m} R_i = 162.$$

Can you find a shortcut in this computation? Look, the sum of all ranks is $1 + \ldots + 18 = (18)(19)/2 = 171$. From this, we have to deduct the ranks corresponding to small X_i, so $W_{\text{obs}} = 171 - R_9 - R_{12} - R_{13} = 171 - 1 - 5 - 3 = 162$.

Compute a P-value. This is a two-sided test, therefore,

$$P = 2 \min \left(P\{W \leq 162\}, P\{S \geq 162\} \right) < 2 \cdot 0.001 = 0.002,$$

if we use Table A8 with $n = 18$.

For the sample size $n = 18$, we can also use the Normal approximation. The null distribution of W has

$$\mathbf{E}(W|H_0) = \frac{(18)(19)}{4} = 85.5 \quad \text{and} \quad \text{Std}(W|H_0) = \sqrt{\frac{(18)(19)(37)}{24}} = 23.0.$$

Making a *continuity correction* (because W is discrete), we find the P-value for this two-sided test,

$$P = 2\mathbf{P}\left\{ Z \geq \frac{161.5 - 85.5}{23.0} \right\} = 2(1 - \Phi(3.30)) = 2 \cdot 0.0005 = 0.001.$$

Wilcoxon test shows strong evidence that the account was used by an unauthorized person.

As always, for a one-sided alternative, the P-value is $\mathbf{P}\{W \geq W_{\text{obs}}\}$ or $\mathbf{P}\{W \leq W_{\text{obs}}\}$.

Comparing Examples 10.9 and 10.13, we can see that the evidence obtained by the Wilcoxon test is stronger. This should generally be the case. Wilcoxon signed rank test utilizes more information contained in the data, and therefore, it is more *powerful*, i.e., more sensitive to a violation of H_0. In other words, if H_0 is not true, the Wilcoxon test is more likely to show that.

R and MATLAB notes

R has a built-in command `wilcox.test` to conduct the Wilcoxon signed rank test. For example, here is how we can test $H_0 : M = 0.2$ vs $H_A : M \neq 0.2$ from Example 10.13 in R.

—— R ——————————
```
X<-c(.24,.22,.26,.34,.35,.32,.33,.29,.19,.36,.30,.15,.17,.28,.38,
 .40,.37,.27)
wilcox.test(X, mu=0.2, alternative="two.sided")
```

For the same test in MATLAB, use `signrank`. The following code will return the P-value, the Wilcoxon test statistic W, and the Z-score computed from it.

—— MATLAB ——————
```
X=[.24,.22,.26,.34,.35,.32,.33,.29,.19,.36,.30,.15,.17,.28,.38,.40,
 .37,.27];
[ pvalue, decision, stats ] = signrank(X, 0.2)
```

Users of a software that does not have built-in tools for the Wilcoxon signed rank test can write a code along the lines of a MATLAB program below.

```
x = [ ]; y=[ ];            % Split the data into X_i > m and X_i < m
for k = 1:length(X);
   if X(k)>m; x=[x X(k)];
   else y=[y X(k)]; end; end;
dx = abs(x-m); d = abs(X-m);  % Distances to the tested median
W=0; for k=1:length(x);       % Each rank is the number of distances d
   W=W+sum(d<dx(k))+1; end;    % not exceeding the given distance d_x(k)
W                             % W is the sum of positive ranks
```

Null distribution of Wilcoxon test statistic

Under the assumption of a symmetric and continuous distribution of $X_1, \ldots, X_n$, here we derive the null distribution of W. Elegant ideas are used in this derivation.

Exact distribution

The exact null distribution of test statistic W can be found *recursively*. That is, we start with the trivial case of a sample of size $n = 1$, then compute the distribution of W for $n = 2$ from it, use that for $n = 3$, and so on, similarly to the mathematical induction.

In a sample of size n, all ranks are different integers between 1 and n. Because of the symmetry of the distribution, each of them can be a "positive rank" or a "negative rank" with probabilities 0.5 and 0.5, if H_0 is true.

Let $p_n(w)$ denote the probability mass function of W for a sample of size n. Then, for $n = 1$, the distribution of W is

$$p_1(0) = \boldsymbol{P}\{W = 0 \mid n = 1\} = 0.5 \quad \text{and} \quad p_1(1) = \boldsymbol{P}\{W = 1 \mid n = 1\} = 0.5.$$

Now, make a transition from size $(n-1)$ to size n, under H_0. By symmetry, the most distant observation from M that produces the highest rank n can be above M or below M with probabilities 0.5 and 0.5. So, the sum of positive ranks W equals w in two cases:

(1) the sum of positive ranks without rank n equals w, and rank n is "negative"; or

(2) the sum of positive ranks without rank n equals $(w - n)$, and rank n is "positive", in which case it adds n to the sum making the sum equal w.

This gives us a recursive formula for calculating the pmf of Wilcoxon statistic W,

$$p_n(w) = 0.5 p_{n-1}(w) + 0.5 p_{n-1}(w - n). \tag{10.6}$$

Normal approximation

For large samples ($n \geq 15$ is a rule of thumb), the distribution of W is approximately Normal. Let us find its parameters $\boldsymbol{E}(W)$ and $\text{Var}(W)$.

Introduce Bernoulli(0.5) variables $Y_1, \ldots, Y_n$, where $Y_i = 1$ if the i-th signed rank is positive. The test statistic can then be written as

$$W = \sum_{i=1}^{n} i Y_i.$$

In this sum, negative signed ranks will be multiplied by 0, so only the positive ones will be counted. Recall that for Bernoulli(0.5), $\boldsymbol{E}(Y_i) = 0.5$ and $\text{Var}(Y_i) = 0.25$. Using independence of Y_i, compute

$$\boldsymbol{E}(W|H_0) = \sum_{i=1}^{n} i\,\boldsymbol{E}(Y_i) = \left(\sum_{i=1}^{n} i\right)(0.5) = \left(\frac{n(n+1)}{2}\right)(0.5) = \frac{n(n+1)}{4},$$

$$\text{Var}(W|H_0) = \sum_{i=1}^{n} i^2\,\text{Var}(Y_i) = \left(\frac{n(n+1)(2n+1)}{6}\right)\left(\frac{1}{4}\right) = \frac{n(n+1)(2n+1)}{24}.$$

Applying this Normal approximation, do not forget to make a *continuity correction* because the distribution of W is discrete.

The following MATLAB code can be used to calculate the probability mass function of W under H_0 for $n = 1, \ldots, 15$.

```
N=15; S=N*(N+1)/2;                         % max value of W
p=zeros(N,S+1);                            % pmf; columns 1..S+1 for w=0..S
p(1,1)=0.5; p(1,2)=0.5;                    % pmf for n=1
for n=2:N;                                 % Sample sizes until N
  for w=0:n-1;                             % Possible values of W
    p(n,w+1) = 0.5*p(n-1,w+1);
  end;
  for w=n:S;                               % Possible values of W
    p(n,w+1) = 0.5*p(n-1,w+1) + 0.5*p(n-1,w-n+1);
  end;
end;
```

An R code for the pmf of statistic W is quite similar. Then, taking partial sums $F(n,w)=sum(p(n,1:w))$, we can find the cdf. Table A8 is based on these calculations.

10.2.3 Mann–Whitney–Wilcoxon rank sum test

What about two-sample procedures? Suppose we have samples from two populations, and we need to compare their medians or just some location parameters to see how different they are.

Wilcoxon signed rank test can be extended to a two-sample problem as follows.

We are comparing two populations, the population of X and the population of Y. In terms of their cumulative distribution functions, we test

$$H_0 : F_X(t) = F_Y(t) \quad \text{for all } t.$$

Assume that under the alternative H_A, either Y is *stochastically larger* than X, and $F_X(t) > F_Y(t)$, or it is *stochastically smaller* than X, and $F_X(t) < F_Y(t)$. For example, it is the case when one distribution is obtained from the other by a simple shift, as on Figure 10.4.

Observed are two independent samples $X_1, \ldots, X_n$ and $Y_1, \ldots, Y_m$.

In the previous section, we compared observations with the fixed tested value m. Instead, now we'll compare one sample against the other! Here is how we conduct this test.

1. Combine all X_i and Y_j into one sample.

2. Rank observations in this combined sample. Ranks R_i are from 1 to $(n + m)$. Some of these ranks correspond to X-variables, others to Y-variables.

3. The test statistic U is the sum of all X-ranks.

When U is small, it means that X-variables have low ranks in the combined sample, so they are generally smaller than Y-variables. This implies that Y is stochastically larger than X, supporting the alternative $H_A : F_Y(t) < F_X(t)$ (Figure 10.4).

This test was proposed by Frank Wilcoxon for equal sample sizes $n = m$ and later generalized to any n and m by other statisticians, Henry Mann and Donald Whitney.

The null distribution of the Mann–Whitney–Wilcoxon test statistic U is in Table A9. For large sample sizes n and m (the rule of thumb is to have both $n > 10$ and $m > 10$), U is approximately Normal.

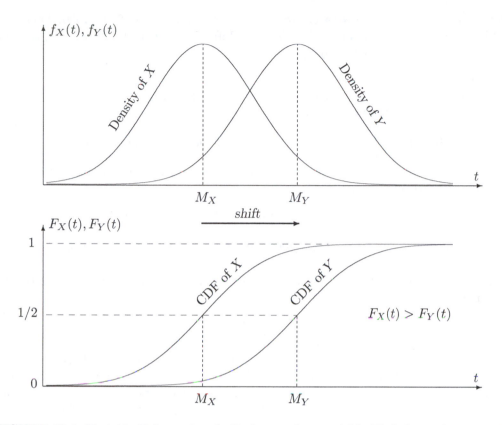

FIGURE 10.4: *Variable Y is stochastically larger than variable X. It has a larger median and a smaller cdf, $M_Y > M_X$ and $F_Y(t) < F_X(t)$.*

Mann–Whitney–Wilcoxon two-sample rank-sum test	Test of two populations, $H_0 : F_X = F_Y$. Test statistic $U = \sum_i R_i$, where R_i are ranks of X_i in the combined sample of X_i and Y_i. Null distribution: Table A9 or recursive formula (10.7). For $n, m \geq 10$, $U \approx \text{Normal}\left(\dfrac{n(n+m+1)}{2}, \sqrt{\dfrac{nm(n+m+1)}{12}} \right)$ Assumptions: the distributions of X_i and Y_i are continuous; $F_X(t) = F_Y(t)$ under H_0; $F_X(t) < F_Y(t)$ for all t or $F_X(t) > F_Y(t)$ for all t under H_A

Example 10.14 (ONLINE INCENTIVES). Managers of an internet shopping portal suspect that more customers participate in online shopping if they are offered some incentive, such as a discount or cash back. To verify this hypothesis, they chose 12 days at random, offered a 5% discount on 6 randomly selected days but did not offer any incentives on the other 6 days. The discounts were indicated on the links leading to this shopping portal.

With the discount, the portal received (rounded to 100s) 1200, 1700, 2600, 1500, 2400, and 2100 hits. Without the discount, 1400, 900, 1300, 1800, 700, and 1000 hits were registered. Does this support the managers' hypothesis?

Solution. Let F_X and F_Y be the cdf of the number of hits without the discount and with the discount, respectively. We need to test

$$H_0 : F_X = F_Y \quad \text{vs} \quad H_A : X \text{ is stochastically smaller than } Y.$$

Indeed, the discount is suspected to boost the number of hits to the shopping portal, so Y should be stochastically larger than X.

To compute the test statistic, combine all the observations and order them,

700, 900, 1000, 1200, 1300, 1400, 1500, 1700, 1800, 2100, 2400, 2600.

X-variables are underlined here. In the combined sample, their ranks are 1, 2, 3, 5, 6, and 9, and their sum is

$$U_{\text{obs}} = \sum_{i=1}^{6} R_i = 26.$$

From Table A9 with $n = m = 6$, we find that the one-sided left-tail P-value is $p \in (0.01, 0.025]$. Although it implies some evidence that discounts help increase the online shopping activity, this evidence is not overwhelming. Namely, we can conclude that the evidence supporting the managers' claim is significant at any significance level $\alpha > 0.025$.

◊

Example 10.15 (PINGS). Round-trip transit times (pings) at two locations are given in Example 8.19 on p. 235 and data set Pings. Arranged in the increasing order, they are

Location I: 0.0156, 0.0210, 0.0215, 0.0280, 0.0308, 0.0327, 0.0335, 0.0350, 0.0355, 0.0396, 0.0419, 0.0437, 0.0480, 0.0483, 0.0543 seconds

Location II: 0.0039, 0.0045, 0.0109, 0.0167, 0.0198, 0.0298, 0.0387, 0.0467, 0.0661, 0.0674, 0.0712, 0.0787 seconds

Is there evidence that the median ping depends on the location?

Let us apply the Mann–Whitney–Wilcoxon test for

$$H_0 : F_X = F_Y \quad \text{vs} \quad H_A : F_X \neq F_Y,$$

where X and Y are pings at the two locations. We observed samples of sizes $n = 15$ and $m = 12$. Among them, X-pings have ranks 4, 7, 8, 9, 11, 12, 13, 14, 15, 17, 18, 19, 21, 22, and 23, and their sum is

$$U_{\text{obs}} = \sum_{i=1}^{15} R_i = 213.$$

(It may be easier to calculate the Y-ranks and subtract their sum from $1 + \ldots + 27 = (27)(28)/2 = 378$.)

Preparing to use the Normal approximation, compute

$$\mathbf{E}(U \mid H_0) = \frac{n(n+m+1)}{2} = 210,$$

$$\mathrm{Var}(U \mid H_0) = \frac{nm(n+m+1)}{12} = 420.$$

Then compute the P-value for this two-sided test,

$$
\begin{aligned}
P &= 2\min\left(\mathbf{P}\left\{U \le 213\right\}, \mathbf{P}\left\{U \ge 213\right\}\right) = 2\min\left(\mathbf{P}\left\{U \le 213.5\right\}, \mathbf{P}\left\{U \ge 212.5\right\}\right) \\
&= 2\min\left(\mathbf{P}\left\{Z \le \frac{213.5-210}{\sqrt{420}}\right\}, \mathbf{P}\left\{Z \ge \frac{212.5-210}{\sqrt{420}}\right\}\right) \\
&= 2\mathbf{P}\left\{Z \ge \frac{212.5-210}{\sqrt{420}}\right\} = 2(1-\Phi(0.12)) = 2(0.4522) = \underline{0.9044}
\end{aligned}
$$

(from Table A4). There is no evidence that pings at the two locations have different distributions. $\diamond$

Example 10.16 (COMPETITION). Two computer manufacturers, A and B, compete for a profitable and prestigious contract. In their rivalry, each claims that their computers are faster. How should the customer choose between them?

It was decided to start execution of the same program simultaneously on seven computers of each company and see which ones finish earlier. As a result, two computers produced by A finished first, followed by three computers produced by B, then five computers produced by A, and finally, four computers produced by B. The actual times have never been recorded.

Use these data to test whether computers produced by A are stochastically faster.

Solution. In the absence of numerical data, we should use nonparametric tests. Apply the Mann–Whitney–Wilcoxon test for testing

$$H_0 : F_A = F_b \quad \text{vs} \quad H_A : X_A \text{ is stochastically smaller than } X_b,$$

where X_A and X_b are the program execution times spent by computers produced by A and by B, and F_A and F_b are there respective cdf.

From results of this competition,

$$A, A, B, B, B, A, A, A, A, A, B, B, B, B,$$

we see that A-computers have ranks 1, 2, 6, 7, 8, 9, and 10. Their sum is

$$U_{\mathrm{obs}} = \sum_{i=1}^{7} R_i = \underline{43}.$$

From Table A9, for the left-tail test with $n_1 = n_2 = 7$, we find that the P-value is between 0.1 and 0.2. There is no significant evidence that computers produced by company A are (stochastically) faster. $\diamond$

Mann–Whitney–Wilcoxon test in R and MATLAB

The same R command `wilcox.test` can perform the Mann–Whitney–Wilcoxon test if you enter two variables in it instead of one. Just do not forget to turn off the option `paired`; otherwise, R will understand it as a matched pairs test.

— R ———————

```
X <- c(1200,1700,2600,1500,2400,2100)
Y <- c(1400,900,1300,1800,700,1000)
wilcox.test(X, Y, alternative="greater", paired=0)
```

In MATLAB, the general command `ranksum(x,y)` returns a P-value for the two-sided test. To see the test statistic, write `[P,H,stat] = ranksum(x,y)`. Then `stats` will contain the test statistic U (MATLAB calculates the smaller of the sum of X-ranks and the sum of Y-ranks in the combined sample), and its standardized value $Z = (U - \mathbf{E}(U))/\operatorname{Std}(U)$. Also, P will be a two-sided P-value, and H will just equal 1 if H_0 is rejected at the 5% level and 0 otherwise.

The following code can be used for Example 10.15.

— MATLAB ———————

```
x = [ 0.0156, 0.0210, 0.0215, 0.0280, 0.0308, 0.0327, 0.0335, 0.0350, ...
0.0355, 0.0396, 0.0419, 0.0437, 0.0480, 0.0483, 0.0543 ];
y = [0.0039, 0.0045, 0.0109, 0.0167, 0.0198, 0.0298, 0.0387, 0.0467, ...
0.0661, 0.0674, 0.0712, 0.0787 ];
[p,h,stats] = ranksum(x,y);
```

Variable `stats` contains the sum of Y-ranks, 165, because it is smaller than the sum of X-ranks, and $Z = 0.1220$. Then, P is 0.9029, slightly different from our $P = 0.9044$ because we rounded the value of Z to 0.12.

Without these built-in tools, the Mann–Whitney–Wilcoxon test statistic can be computed by simple commands

```
U = 0; XY = [X Y];                          % Joint sample
for k = 1:length(X); U = U + sum( XY < X(k) ) + 1;
end; U                          % Statistic U is the sum of X-ranks
```

Null distribution of Mann–Whitney–Wilcoxon test statistic

A recursive formula for the exact distribution of test statistic U follows the main ideas of the derivation of (10.6).

If H_0 is true, the combined sample of X and Y is a sample of identically distributed random variables of size $(n + m)$. Therefore, all $\binom{n+m}{n}$ allocations of X-variables in the ordered combined sample are equally likely. Among them, $\binom{n+m-1}{n}$ allocations have a Y-variable as the largest observation in the combined sample. In this case, deleting this largest observation from the sample will not affect statistic U, which is the sum of X-ranks. The other $\binom{n+m-1}{n-1}$ allocations have an X-variable as the largest observation, and it has a rank $(n + m)$. In this case, deleting it will reduce the sum of X-ranks by $(n + m)$.

Now, let $N_{n,m}(u)$ be the number of allocations of X-variables and Y-variables that results in $U = u$. Under H_0, we have the situation of equally likely outcomes, so the probability of $U = u$ equals

$$p_{n,m}(u) = \boldsymbol{P}\{U = u\} = \frac{N_{n,m}(u)}{\dbinom{n+m}{n}}.$$

From the previous paragraph,

$$
\begin{aligned}
N_{n,m}(u) &= N_{n,m-1}(u) + N_{n-1,m}(u - n - m) \\
&= \dbinom{n+m-1}{n} p_{n,m-1}(u) + \dbinom{n+m-1}{n-1} N_{n,m}(u-n-m),
\end{aligned}
$$

and therefore, dividing all parts of this equation by $\dbinom{n+m}{n}$, we get

$$p_{n,m}(u) = \frac{m}{n+m} p_{n,m-1}(u) + \frac{n}{n+m} p_{n-1,m}(u - n - m). \qquad (10.7)$$

This is a recursive formula for the probability mass function of U. To start the recursions, let's notice that without any X-variables, $p_{0,m} = 0$, and without any Y variables, $p_{n,0} = \sum_{i=1}^{n} i = n(n+1)/2$. These formulas are used to construct Table A9.

Normal approximation

When both $n \geq 10$ and $m \geq 10$, statistic U has approximately Normal distribution. Some work is needed to derive its parameters $\boldsymbol{E}(U)$ and $\mathrm{Var}(U)$, but here are the main steps.

We notice that each X-rank equals

$$R_i = 1 + \#\{j : X_j < X_i\} + \#\{k : Y_k < X_i\}$$

(each # denotes the number of elements in the corresponding set). Therefore,

$$U = \sum_{i=1}^{n} R_i = n + \binom{n}{2} + \sum_{i=1}^{n} \xi_i,$$

where $\xi_i = \sum_{k=1}^{m} I\{Y_k < X_i\}$ is the number of Y-variables that are smaller than X_i.

Under the hypothesis H_0, all X_i and Y_k are identically distributed, so $\boldsymbol{P}\{Y_k < X_i\} = 1/2$. Then, $\boldsymbol{E}(\xi_i) = m/2$, and

$$\boldsymbol{E}(U) = n + \binom{n}{2} + \frac{nm}{2} = \frac{n(n+m+1)}{2}.$$

For $\mathrm{Var}(U) = \mathrm{Var}\sum_i \xi_i = \sum_i \sum_j \mathrm{Cov}(\xi_i, \xi_j)$, we have to derive $\mathrm{Var}(\xi_i) = (m^2 + 2m)/12$ and $\mathrm{Cov}(\xi_i, \xi_j) = m/12$. This is not trivial, but you can fill in all the steps, as a nice non-required exercise! If you decide to do that, notice that all X_i and Y_k have the same distribution, and therefore, for any $i \neq j$ and $k \neq l$,

$$\boldsymbol{P}\{Y_k < X_i \cap Y_l < X_i\} = \boldsymbol{P}\{X_i \text{ is the smallest of } X_i, Y_k, Y_l\} = \frac{1}{3},$$

$$\boldsymbol{P}\{Y_k < X_i \cap Y_l < X_j\} = \frac{6 \text{ favorable allocations of } X_i, X_j, Y_k, Y_l}{4! \text{ allocations of } X_i, X_j, Y_k, Y_l} = \frac{1}{4}.$$

Substituting into $\mathrm{Var}(U)$ and simplifying, we obtain

$$\mathrm{Var}(U) = \frac{nm(n+m+1)}{12}.$$

10.3   Bootstrap

Bootstrap is used to estimate population parameters by Monte Carlo simulations when it is too difficult to do it analytically. When computers became powerful enough to handle large-scale simulations, the bootstrap methods got very popular for their ability to evaluate properties of various estimates.

Consider, for example, the *standard errors*. From the previous chapters, we know the standard errors of a sample mean and a sample proportion,

$$s(\overline{X}) = \frac{\sigma}{\sqrt{n}} \ \text{ and } \ s(\widehat{p}) = \sqrt{\frac{p(1-p)}{n}},$$

and they are quite simple. Well, try to derive standard errors of other statistics - sample median, sample variance, sample interquartile range, and so on. Each of these will require a substantial amount of work.

Many complicated estimates are being used nowadays in modern Statistics. How can one evaluate their performance, estimate their standard error, bias, etc.?

The difficulty is that we observe an estimator $\widehat{\theta}$ only once. That is, we have one sample $\mathcal{S} = (X_1, \ldots, X_n)$ from the population $\mathcal{P}$, and from this sample we compute $\widehat{\theta}$. We would very much like to observe many $\widehat{\theta}$'s and then compute their sample variance, for example, but we don't have this luxury. From one sample, we observe only one $\widehat{\theta}$, and this is just not enough to compute its sample variance!

10.3.1   Bootstrap distribution and all bootstrap samples

In 1970s, an American mathematician *Bradley Efron*, Professor at Stanford University, proposed a rather simple approach. He called it **bootstrap** referring to the idiom *"to pull oneself up by one's bootstraps"*, which means to find a solution relying on your own sources and without any help from outside (a classical example is Baron Münchausen from the 18th century collection of tales by R. E. Raspe, who avoided drowning in a lake by pulling himself from the water by his own hair!). In our situation, even though we really need several samples to explore the properties of an estimator $\widehat{\theta}$, we'll manage to do it with what we have, which is just one sample.

For the invention of bootstrap, Professor Efron is awarded the 2019 International Prize in Statistics which is said to be an equivalent of the Nobel Prize in Statistics.

To start, let's notice that many commonly used statistics are constructed in the same way as the corresponding population parameters. We are used to estimating a population mean μ by a sample mean $\overline{X}$, a population variance σ^2 by a sample variance s^2, population quantiles q_p by sample quantiles $\widehat{q}_p$, and so on. To estimate a certain parameter θ, we collect a random sample $\mathcal{S}$ from $\mathcal{P}$ and essentially compute the estimator $\widehat{\theta}$ from this sample by the same mechanism as θ was computed from the whole population!

In other words, there is a function g that one can use to compute a parameter θ from the population $\mathcal{P}$ (Figure 10.5). Then

$$\theta = g(\mathcal{P}) \ \text{ and } \ \widehat{\theta} = g(\mathcal{S}).$$

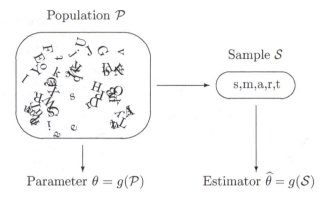

Population $\mathcal{P}$

Sample $\mathcal{S}$

s,m,a,r,t

Parameter $\theta = g(\mathcal{P})$

Estimator $\widehat{\theta} = g(\mathcal{S})$

FIGURE 10.5: *Parameter θ and its estimator $\widehat{\theta}$ computed by the same mechanism g applied to the population and to the sample.*

Example 10.17 (POPULATION MEAN AND SAMPLE MEAN). Imagine a strange calculator that can only do one operation g – averaging. Give it 4 and 6, and it returns their average $g(4, 6) = 5$. Give it 3, 7, and 8, and it returns $g(3, 7, 8) = 6$.

Give it the whole population $\mathcal{P}$, and it returns the parameter $\theta = g(\mathcal{P}) = \mu$. Give it a sample $\mathcal{S} = (X_1, \dots, X_n)$, and it returns the estimator $\widehat{\theta} = g(X_1, \dots, X_n) = \overline{X}$. ◇

You can imagine similar calculators for the variances, medians, and other parameters and their estimates. Essentially, each estimator $\widehat{\theta}$ is a mechanism that copies the way a parameter θ is obtained from the population and then applies it to the sample.

Bradley Efron proposed one further step. Suppose some estimator is difficult to figure out. For example, we are interested in the variance of a sample median, $\eta = \text{Var}(\widehat{M}) = h(\mathcal{P})$. This mechanism h consists of taking all possible samples from the population, taking their sample medians $\widehat{M}$, and then calculating their variance,

$$\eta = h(\mathcal{P}) = \mathbf{E}(\widehat{M} - \mathbf{E}\widehat{M})^2.$$

Of course, we typically cannot observe all possible samples; that's why we cannot compute $h(\mathcal{P})$, and parameter η is unknown. How can we estimate it based on just one sample $\mathcal{S}$?

Efron's **bootstrap approach** is to apply the same mechanism to the sample $\mathcal{S}$. That is, take all possible samples from $\mathcal{S}$, compute their medians, and then compute the sample variance of those, as on Figure 10.6. It may sound strange, but yes, we are proposing to create more samples from one sample $\mathcal{S}$ given to us. After all, exactly the same algorithm was used to compute $\eta = h(\mathcal{P})$. The advantage is that we know the observed sample $\mathcal{S}$, and therefore, we can perform all the bootstrap steps and estimate η.

Sampling from the observed sample $\mathcal{S}$ is a statistical technique called *resampling*. Bootstrap is one of the resampling methods. The obtained samples from $\mathcal{S}$ are called *bootstrap samples*. Each bootstrap sample $\mathcal{B}_j = (X_{1j}^*, \dots, X_{nj}^*)$ consists of values X_{ij}^* sampled from $\mathcal{S} = (X_1, \dots, X_n)$ independently and with equal probabilities. That is,

$$X_{ij}^* = \begin{cases} X_1 & \text{with probability } 1/n \\ X_2 & \text{with probability } 1/n \\ \dots & \dots \quad \dots \quad \dots \quad \dots \\ X_n & \text{with probability } 1/n \end{cases}$$

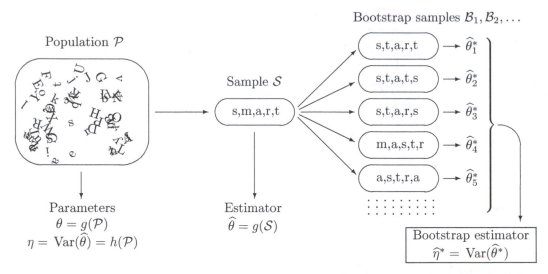

FIGURE 10.6: *Bootstrap procedure estimates* $\eta = \text{Var}(\widehat{\theta})$ *by the variance of* $\widehat{\theta}_i^*$ *'s, obtained from bootstrap samples.*

This is *sampling with replacement*, which means that the same observation X_i can be sampled more than once. An asterisk (*) is used to denote the contents of bootstrap samples.

DEFINITION 10.3 ————————————————————————————————

A **bootstrap sample** is a random sample drawn with replacement from the observed sample $\mathcal{S}$ of the same size as $\mathcal{S}$.

The distribution of a statistic across bootstrap samples is called a **bootstrap distribution**.

An estimator that is computed on basis of bootstrap samples is a **bootstrap estimator**.

Example 10.18 (VARIANCE OF A SAMPLE MEDIAN). Suppose that we observed a small sample $\mathcal{S} = (2, 5, 7)$ and estimated the population median M with the sample median $\widehat{M} = 5$. How can we estimate its variance $\text{Var}(\widehat{M})$?

Solution. Table 10.1 lists all $3^3 = 27$ equally likely bootstrap samples that can be drawn from $\mathcal{S}$. Among these, 7 samples have $\widehat{M}_i^* = 2$, 13 samples have $\widehat{M}_i^* = 5$, and 7 samples have $\widehat{M}_i^* = 7$. So, the *bootstrap distribution* of a sample median is

$$P^*(2) = 7/27, \quad P^*(5) = 13/27, \quad P^*(7) = 7/27. \tag{10.8}$$

i	$\mathcal{B}_i$	$\widehat{M}_i$	i	$\mathcal{B}_i$	$\widehat{M}_i$	i	$\mathcal{B}_i$	$\widehat{M}_i$
1	$(2,2,2)$	2	10	$(5,2,2)$	2	19	$(7,2,2)$	2
2	$(2,2,5)$	2	11	$(5,2,5)$	5	20	$(7,2,5)$	5
3	$(2,2,7)$	2	12	$(5,2,7)$	5	21	$(7,2,7)$	7
4	$(2,5,2)$	2	13	$(5,5,2)$	5	22	$(7,5,2)$	5
5	$(2,5,5)$	5	14	$(5,5,5)$	5	23	$(7,5,5)$	5
6	$(2,5,7)$	5	15	$(5,5,7)$	5	24	$(7,5,7)$	7
7	$(2,7,2)$	2	16	$(5,7,2)$	5	25	$(7,7,2)$	7
8	$(2,7,5)$	5	17	$(5,7,5)$	5	26	$(7,7,5)$	7
9	$(2,7,7)$	7	18	$(5,7,7)$	7	27	$(7,7,7)$	7

TABLE 10.1: *All bootstrap samples $\mathcal{B}_i$ drawn from $\mathcal{S}$ and the corresponding sample medians for Example 10.18.*

We use it to estimate $h(\mathcal{P}) = \mathrm{Var}(\widehat{M})$ with the *bootstrap estimator*

$$\widehat{\mathrm{Var}}^{*}(\widehat{M}) = h(\mathcal{S}) = \sum_x x^2 P^*(x) - \left(\sum_x x P^*(x)\right)^2$$

$$= (4)\left(\frac{7}{27}\right) + (25)\left(\frac{13}{27}\right) + (49)\left(\frac{7}{27}\right) - \left\{(2)\left(\frac{7}{27}\right) + (5)\left(\frac{13}{27}\right) + (7)\left(\frac{7}{27}\right)\right\}^2$$

$$= \underline{3.303}. \hspace{6cm} \diamond$$

Here is a summary of the bootstrap method that we applied in this example.

Bootstrap (all bootstrap samples)	To estimate parameter η of the distribution of $\widehat{\theta}$: 1. Consider all possible bootstrap samples drawn with replacement from the given sample $\mathcal{S}$ as well as statistics $\widehat{\theta}^*$ computed from them. 2. Derive the bootstrap distribution of $\widehat{\theta}^*$. 3. Compute the parameter of this bootstrap distribution that has the same meaning as η.

All right, one may say, this certainly works for a "toy" sample of size three. But how about bigger samples? We can list all n^n possible bootstrap samples for very small n. However, a rather modest sample of size $n = 60$ can produce $60^{60} \approx 4.9 \cdot 10^{106}$ different bootstrap samples, which is almost five million times larger than *the googol*!

Certainly, we are not going to list a googol of bootstrap samples. Instead, we'll discuss two alternative approaches. The first one proposes to compute the bootstrap distribution without listing all the bootstrap samples. This, however, is still feasible only in relatively simple situations. The second solution, by far the most popular one, uses Monte Carlo simulations to produce a large number b of bootstrap samples. Although this way we may not get *all* possible bootstrap samples, results will be very close for large b.

Bootstrap distribution

The only reason for considering all possible bootstrap samples in Example 10.18 was to find the distribution (10.8) and then to obtain the bootstrap estimator $\widehat{\eta}^* = \widehat{\text{Var}}(M^*)$ from it.

Sometimes it is possible to compute the bootstrap distribution without listing all bootstrap samples. Here is an example.

Example 10.19 (BIAS OF A SAMPLE MEDIAN). A sample median may be biased or unbiased for estimating the population median. It depends on the underlying distribution of data. Suppose we observed a sample

$$S = (3,\, 5,\, 8,\, 5,\, 5,\, 8,\, 5,\, 4,\, 2).$$

Find a bootstrap estimator of $\eta = \text{Bias}(\widehat{M})$, the bias of the sample median.

Solution. First, find the bootstrap distribution of a sample median $\widehat{M}^*$. Based on the given sample of size $n = 9$, the sample median of bootstrap samples can be equal to 2, 3, 4, 5, or 8. Let us compute the probability of each value.

Sampling from S, values $X_{ij}^* = 2$, 3, and 4 appear with probability 1/9 because only one of each of them appears in the given sample. Next, $X_{ij}^* = 5$ with probability 4/9, and $X_{ij}^* = 8$ with probability 2/9.

Now, the sample median $\widehat{M}_i^*$ in any bootstrap sample $\mathcal{B}_i$ is the central or the 5th smallest observation. Thus, it equals 2 if at least 5 of 9 values in $\mathcal{B}_i$ equal 2. The probability of that is

$$P^*(2) = \boldsymbol{P}(Y \geq 5) = \sum_{y=5}^{9} \binom{9}{y} \left(\frac{1}{9}\right)^y \left(\frac{8}{9}\right)^{9-y} = 0.0014$$

for a Binomial$(n = 9, p = 1/9)$ variable Y.

Similarly, $\widehat{M}_i^* \leq 3$ if at least 5 of 9 values in $\mathcal{B}_i$ do not exceed 3. The probability of that is

$$F^*(3) = \boldsymbol{P}(Y \geq 5) = \sum_{y=5}^{9} \binom{9}{y} \left(\frac{2}{9}\right)^y \left(\frac{7}{9}\right)^{9-y} = 0.0304$$

for a Binomial$(n = 9, p)$ variable Y, where $p = 2/9$ is a probability to sample either $X_{ij}^* = 2$ or $X_{ij}^* = 3$.

Proceeding in a similar fashion, we get

$$F^*(4) = \sum_{y=5}^{9} \binom{9}{y} \left(\frac{3}{9}\right)^y \left(\frac{6}{9}\right)^{9-y} = 0.1448,$$

$$F^*(5) = \sum_{y=5}^{9} \binom{9}{y} \left(\frac{7}{9}\right)^y \left(\frac{2}{9}\right)^{9-y} = 0.9696, \text{ and}$$

$$F^*(8) = 1.$$

From this cdf, we can find the bootstrap probability mass function of $\widehat{M}_i^*$,

$$P^*(2) = 0.0014, \quad P^*(3) = 0.0304 - 0.0014 = 0.0290, \quad P^*(4) = 0.1448 - 0.0304 = 0.1144,$$

$$P^*(5) = 0.9696 - 0.1448 = 0.8248, \quad P^*(8) = 1 - 0.9696 = 0.0304.$$

From this, the bootstrap estimator of $\mathbf{E}(\widehat{M})$ is the expected value of the bootstrap distribution,

$$\mathbf{E}^*(\widehat{M_i^*}) = (2)(0.0014) + (3)(0.0290) + (4)(0.1144) + (5)(0.8248) + (8)(0.0304) = 4.9146.$$

Last step! The bias of $\widehat{M}$ is defined as $h(\mathcal{P}) = \mathrm{Bias}(\widehat{M}) = \mathbf{E}(\widehat{M}) - M$. We have estimated the first term, $\mathbf{E}(\widehat{M})$. Following the bootstrap ideas, what should we use as an estimator of M, the second term of the bias? The answer is simple. We have agreed to estimate $g(\mathcal{P})$ with $g(\mathcal{S})$, so we just estimate the population median $g(\mathcal{P}) = M$ with the sample median $g(\mathcal{S}) = \widehat{M} = 5$ obtained from our original sample $\mathcal{S}$.

The bootstrap estimator of $\mathrm{Bias}(\widehat{M}) = \mathbf{E}(\widehat{M}) - M$ (based on all possible bootstrap samples) is

$$\eta(\mathcal{S}) = \widehat{\mathrm{Bias}}(\widehat{M}) = \mathbf{E}^*(\widehat{M}) - \widehat{M} = 4.9146 - 5 = \boxed{-0.0852}. \qquad \Diamond$$

Although the sample in Example 10.19 is still rather small, the method presented can be extended to a sample of any size. Manual computations may be rather tedious here, but one can write a suitable computer code.

10.3.2 Computer generated bootstrap samples

Modern Statistics makes use of many complicated estimates. As the earliest examples, Efron used bootstrap to explore properties of the sample correlation coefficient, the trimmed mean[1], and the excess error[2], but certainly, there are more complex situations. Typically, samples will be too large to list all the bootstrap samples, as in Table 10.1, and the statistics will be too complicated to figure out their bootstrap distributions, as in Example 10.19.

This is where *Monte Carlo simulations* kick in. Instead of listing all possible bootstrap samples, we use a computer to generate a large number b of them. The rest follows our general scheme on Figure 10.6, p. 342.

Bootstrap (generated bootstrap samples)	To estimate parameter η of the distribution of $\widehat{\theta}$: 1. Generate a large number b of bootstrap samples drawn with replacement from the given sample $\mathcal{S}$. 2. From each bootstrap sample $\mathcal{B}_i$, compute statistic $\widehat{\theta}_i^*$ the same way as $\widehat{\theta}$ is computed from the original sample $\mathcal{S}$. 3. Estimate parameter η from the obtained values of $\widehat{\theta}_1^*, \dots, \widehat{\theta}_b^*$.

This is a classical bootstrap method of evaluating properties of parameter estimates. Do you think it is less accurate than the one based on *all* the bootstrap samples? Notice that

---

[1] Trimmed mean is a version of a sample mean, where a certain portion of the smallest and the largest observations is dropped before computing the arithmetic average. Trimmed means are not sensitive to a few extreme observations, and they are used if extreme outliers may be suspected in the sample.

[2] This measures how well one can estimate the error of prediction in regression analysis. We study regression in the next chapter.

b, the number of generated bootstrap samples, can be very large. Increasing b gives more work to your computer, but it does not require a more advanced computer code or a larger original sample. And of course, as $b \to \infty$, our estimator of η becomes just as good as if we had a complete list of bootstrap samples. Typically, thousands or tens of thousands of bootstrap samples are being generated.

Software notes

The following MATLAB code generates b bootstrap samples from the given sample $\boldsymbol{X} = (X_1, \ldots, X_n)$.

—— MATLAB ————

```
n = length(X);             % Sample size
U = ceil(n*rand(b,n));     % A b × n matrix of random integers from 1 to n
B = X(U);                  % A matrix of bootstrap samples.
                           % The i-th bootstrap sample is in the i-th row.
```

For example, based on a sample $\boldsymbol{X} = (10, 20, 30, 40, 50)$ and generated matrix U of random indices, we obtain a matrix B of bootstrap samples,

$$
U = \begin{pmatrix}
1 & 4 & 5 & 5 & 1 \\
3 & 2 & 3 & 1 & 5 \\
3 & 4 & 5 & 2 & 3 \\
1 & 4 & 1 & 2 & 3 \\
2 & 4 & 3 & 5 & 1 \\
1 & 3 & 1 & 3 & 5 \\
4 & 1 & 5 & 5 & 4 \\
2 & 2 & 1 & 1 & 2 \\
3 & 5 & 4 & 2 & 3 \\
1 & 1 & 5 & 1 & 3
\end{pmatrix},
\quad
B = \begin{pmatrix}
B_1 \\
B_2 \\
B_3 \\
B_4 \\
B_5 \\
B_6 \\
B_7 \\
B_8 \\
B_9 \\
B_{10}
\end{pmatrix}
=
\begin{pmatrix}
10 & 40 & 50 & 50 & 10 \\
30 & 20 & 30 & 10 & 50 \\
30 & 40 & 50 & 20 & 30 \\
10 & 40 & 10 & 20 & 30 \\
20 & 40 & 30 & 50 & 10 \\
10 & 30 & 10 & 30 & 50 \\
40 & 10 & 50 & 50 & 40 \\
20 & 20 & 10 & 10 & 20 \\
30 & 50 & 40 & 20 & 30 \\
10 & 10 & 50 & 10 & 30
\end{pmatrix}
$$

If b and n are so large that storing the entire matrices U and b requires too many computer resources, we can generate bootstrap samples in a do-loop, one at a time, and keep the statistics $\hat{\theta}_i$ only.

In fact, MATLAB has a special command **bootstrp** for generating bootstrap samples and computing estimates from them. For example, the code

$$M = \texttt{bootstrp(50000,@median,S)};$$

takes the given sample $\mathcal{S}$, generates $b = 50,000$ bootstrap samples from it, computes medians from each of them, and stores these median in a vector M. After that, we can compute various sample statistics of M, the mean, standard deviation, etc., and use them to evaluate the properties of a sample median.

In R, generating b bootstrap samples and saving their medians is rather simple, making use of the command **sample**.

— R ——————

```
for (k in 1:B){             # Do-loop to produce B bootstrap samples
b <- sample( n, n, replace=1 )   # Random subsample with replacement
BootMedian[k] <- median(X[b]) }  # Compute medians of bootstrap samples
```

After this, `sd(BootMedian)` returns the bootstrap estimator of $\mathrm{Std}(\widehat{M})$, the standard deviation of the sample median; `quantile(BootMedian,0.10)` is the tenth percentile of the distribution of $\widehat{M}$, etc.

Also, R has a special package `boot` for bootstrap inference. To take advantage of it, we have to define our statistic of interest as a function of a sample.

— R ——————

```
median.fn <- function(X,subsample){ return(median(X[subsample])) }
# Now we invoke package "boot" and apply it to our new function median.fn
install.packages("boot"); library(boot);
boot( X, median.fn, R=10000 )
```

This will generate $b = 10,000$ bootstrap samples from sample X, compute their medians, and use them to calculate bootstrap estimates of the bias and standard error of the sample median. Other bootstrap statistics can be computed from the whole set of generated bootstrap medians $\widehat{\theta}_1^*, \ldots, \widehat{\theta}_b^*$. We obtain it as component `t` of the object resulting from the `boot` command. For example,

$$\texttt{BootMed <- boot(X, median.fn, R=10000)\$t}$$

Would you expect the same results when you run this code again? You should not. We realize that this estimation algorithm includes random number generation, and so, our results are random, and they will differ from each other. On the other hand, the differences between our bootstrap estimates will be small when the number of bootstrap samples b is large.

Example 10.20 (BIAS OF A SAMPLE MEDIAN (CONTINUED)). Based on the sample

$$\mathcal{S} = (3, 5, 8, 5, 5, 8, 5, 4, 2)$$

from Example 10.19, we estimated the population median θ with the sample median $\widehat{\theta} = 5$. Next, we investigate the properties of $\widehat{\theta}$. These computer codes will estimate the bias of $\widehat{\theta}$, the standard error, the first quartile, and the probability of $\widehat{\theta} > 4$.

— R ——————

```
 S  <-  c(3,5,8,5,5,8,5,4,2);
 b  <-  100000;
median.fn <- function(X,subsample){
      return(median(X[subsample])) }
install.packages("boot"); library(boot);
BootMed <- boot(X,median.fn,R=10000)$t
 biasM <- mean(M)-median(S)
 sterrM <- sd(M)
 q25 <- quantile(BootMed,0.25)
 prob4 <- mean(M > 4)
```

— MATLAB ——————

```
 S  =  [3 5 8 5 5 8 5 4 2];
 b  =  100000;
 M  =  bootstrp(b,@median,S);
biasM = mean(M)-median(S);
sterrM = std(M);
q25 = quantile(BootMed,0.25);
prob4 = mean(M > 4);
```

Vector M contains bootstrap statistics $\widehat{\theta}_1^*, \ldots, \widehat{\theta}_b^*$. Based on $b = 100,000$ bootstrap samples, we obtained

$$\widehat{\text{Bias}^*}(\widehat{\theta}) = \overline{M} - \widehat{\theta} = -0.0858,$$

$$s^*(\widehat{\theta}) = s(M) = 0.7062, \quad \text{and}$$

$$\widehat{P}^*(\widehat{\theta} > 4) = \frac{\#\left\{i : \widehat{\theta}_i^* > 4\right\}}{b} = 0.8558.$$

$\Diamond$

10.3.3 Bootstrap confidence intervals

In the previous chapters, we learned how to construct confidence intervals for the population mean, proportion, variance, and also, difference of means, difference of proportions, and ratio of variances. Normal, T, χ^2, and F distributions were used in these cases. These methods required either a Normal distribution of the observed data or sufficiently large samples.

There are many situations, however, where these conditions will not hold, or the distribution of the test statistic is too difficult to obtain. Then the problem becomes nonparametric, and we can use *bootstrap* to construct approximately $(1 - \alpha)100\%$ confidence intervals for the population parameters.

Two methods of bootstrap confidence intervals are rather popular.

Parametric method, based on the bootstrap estimation of the standard error

This method is used when we need a confidence interval for a parameter θ, and its estimator $\widehat{\theta}$ has approximately Normal distribution.

In this case, we compute $s^*(\widehat{\theta})$, the bootstrap estimator of the standard error $\sigma(\widehat{\theta})$, and use it to compute the approximately $(1 - \alpha)100\%$ confidence interval for θ,

$$\textbf{Parametric bootstrap} \quad \boxed{\widehat{\theta} \pm z_{\alpha/2} s^*(\widehat{\theta})} \qquad (10.9)$$
$$\textbf{confidence interval}$$

It is similar to our usual formula for the confidence interval in the case of Normal distribution (9.3), and $z_{\alpha/2} = q_{1-\alpha/2}$ in it, as always, denotes the $(1 - \alpha/2)$-quantile from the Standard Normal distribution.

Example 10.21 (CONFIDENCE INTERVAL FOR A CORRELATION COEFFICIENT). Example 8.20 on p. 237 and data set Antivirus contain the data on the number of times X the antivirus software was launched on 30 computers during 1 month and the number Y of detected worms,

X	30	30	30	30	30	30	30	30	30	30	30	15	15	15	10
Y	0	0	1	0	0	0	1	1	0	0	0	0	1	1	0

X	10	10	6	6	5	5	5	4	4	4	4	4	1	1	1
Y	0	2	0	4	1	2	0	2	1	0	1	0	6	3	1

Scatter plots on Figure 8.11 showed existence of a negative correlation between X and Y, which means that in general, the number of worms reduces if the antivirus software is used more often.

Next, the computer manager asks for a 90% confidence interval for the correlation coefficient ρ between X and Y.

The following MATLAB code can be used to solve this problem. This is a detailed step-by-step solution that our readers can translate line by line to other software languages.

```
—— MATLAB ————

alpha = 0.10;
X = [30 30 30 30 30 30 30 30 30 30 30 15 15 15 10 10 10 6 6 5 5 5 4 4 4 4 4 1 1 1];
Y = [0 0 1 0 0 0 1 1 0 0 0 0 1 1 0 0 2 0 4 1 2 0 2 1 0 1 0 6 3 1];
r = corrcoef(X,Y); % correlation coefficient from the given sample
r = r(2,1); % because corrcoef returns a matrix of correlation coefficients
b = 10000; n = length(X);
U = ceil(n*rand(b,n));
BootX = X(U); BootY = Y(U); % Values X and Y of generated bootstrap samples
BootR = zeros(b,1); % Initiate the vector of bootstrap corr. coefficients
for i=1:b;
  BR = corrcoef(BootX(i,:),BootY(i,:));
  BootR(i) = BR(2,1);
end;
s = std(BootR); % Bootstrap estimator of the standard error of r
CI = [ r + s*norminv(alpha/2,0,1), r + s*norminv(1-alpha/2,0,1) ];
disp(CI) % Bootstrap confidence interval
```

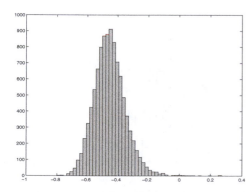

FIGURE 10.7: *The histogram of bootstrap correlation coefficients.*

As a result of this code, we get the sample correlation coefficient $r = -0.4533$, and also, $b = 10,000$ bootstrap correlation coefficients $r_1^*, \ldots, r_b^*$ obtained from b generated bootstrap samples.

Next, we notice that for the sample of size $n = 30$, r has approximately Normal distribution. For example, this can be confirmed by the histogram of bootstrap correlation coefficients on Figure 10.7.

Applying the parametric method, we compute $s^*(r) = 0.1028$, the standard error of $r_1^*, \ldots, r_b^*$, and use it to construct the 90% confidence interval

$$r \pm z_{\alpha/2} s^*(r) = -0.4533 \pm (1.645)(0.1028) = \underline{[\text{-0.6224, -0.2841}]}$$

This step-by-step MATLAB code can be translated to other software languages, including R. However, let's use the R package **boot** to see how we can handle the correlation coefficient which is a function of *two* variables X and Y instead of one.

We'll define a data frame D consisting of these two observed variables and then a function correl of it.

```R
— R —————————
X <- c(rep(30,11),15,15,15,10,10,10,6,6,5,5,5,4,4,4,4,4,1,1,1)
Y <- c(0,0,1,0,0,0,1,1,0,0,0,0,1,1,0,0,2,0,4,1,2,0,2,1,0,1,0,6,3,1)
D <- data.frame(X,Y)
correl <- function(D,subsample){
X <- D[subsample,1]; Y <- D[subsample,2];
return( cor(X,Y) )
}
install.packages("boot"); library(boot);
BootR <- boot(data=D, statistic=correl, R=10000)$t
alpha <- 0.10; r <- cor(X,Y); s <- sd(BootR);
CI <- c( r + s*qnorm(alpha/2), r + s*qnorm(1-alpha/2) )
print(CI)
```

◇

Nonparametric method, based on the bootstrap quantiles

Equation 10.9 simply estimates two quantiles $q_{\alpha/2}$ and $q_{1-\alpha/2}$ of the distribution of statistic $\widehat{\theta}$, so that

$$P\left\{ q_{\alpha/2} \leq \widehat{\theta} \leq q_{1-\alpha/2} \right\} = 1 - \alpha. \tag{10.10}$$

Since $\widehat{\theta}$ estimates parameter θ, this becomes an approximately $(1 - \alpha)100\%$ confidence interval for θ.

This method fails if the distribution of $\widehat{\theta}$ is not Normal. The coverage probability in (10.10) may be rather different from $(1-\alpha)$ in this case. However, the idea to construct a confidence interval based on the quantiles of $\widehat{\theta}$ is still valid.

The quantiles $q_{\alpha/2}$ and $q_{1-\alpha/2}$ of the distribution of $\widehat{\theta}$ will then be estimated from the bootstrap samples. To do this, we generate b bootstrap samples, compute statistic $\widehat{\theta}^*$ for each of them, and determine the sample quantiles $\widehat{q}^*_{\alpha/2}$ and $\widehat{q}^*_{1-\alpha/2}$ from them. These quantiles become the end points of the $(1 - \alpha)100\%$ confidence interval for θ.

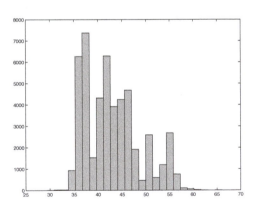

FIGURE 10.8: *The histogram of bootstrap medians.*

Nonparametric bootstrap confidence interval for parameter θ	$\left[\widehat{q}^*_{\alpha/2},\ \widehat{q}^*_{1-\alpha/2} \right],$ where $q^*_{\alpha/2}$ and $\widehat{q}^*_{1-\alpha/2}$ are quantiles of the distribution of $\widehat{\theta}$ estimated from bootstrap samples	(10.11)

Example 10.22 (CONFIDENCE INTERVAL FOR THE MEDIAN CPU TIME). The population median was estimated in Example 8.12 on p. 223, based on the following observed CPU times (data set CPU):

$$
\begin{array}{cccccccccc}
70 & 36 & 43 & 69 & 82 & 48 & 34 & 62 & 35 & 15 \\
59 & 139 & 46 & 37 & 42 & 30 & 55 & 56 & 36 & 82 \\
38 & 89 & 54 & 25 & 35 & 24 & 22 & 9 & 56 & 19
\end{array}
$$

Let us now compute the 95% bootstrap confidence interval for the median CPU time.

— R ————————

```
alpha <- 0.05;
X <- c( 70, 36, 43, 69, 82, 48, 34, 62, 35, 15, 59, 139, 46, 37, 42,
30, 55, 56, 36, 82, 38, 89, 54, 25, 35, 24, 22, 9, 56, 19)
median.fn <- function(X,subsample){ return(median(X[subsample])) }
install.packages("boot"); library(boot);
BootMed <- boot(data=X, statistic=median.fn, R=50000)$t
print( c( quantile(BootMed,alpha/2), quantile(BootMed,1-alpha/2) ) )
```

— MATLAB ————————

```
alpha = 0.05;
X = [ 70 36 43 69 82 48 34 62 35 15 59 139 46 37 42 ...
       30 55 56 36 82 38 89 54 25 35 24 22 9 56 19]';
b = 50000; n = length(X);
U = ceil(n*rand(b,n)); BootX = X(U); BootM = zeros(b,1);
for i=1:b; BootM(i) = median(BootX(i,:)); end;
CI = [ quantile(BootM,alpha/2), quantile(BootM,1-alpha/2) ]
```

These programs generate $b = 50,000$ bootstrap samples and compute their sample medians $\widehat{\theta}_1^*, \ldots, \widehat{\theta}_b^*$ (variable BootM). Based on these sample medians, the 0.025- and 0.975-quantiles are calculated. The 95% confidence interval CI then stretches between these quantiles, from $\widehat{q}_{0.025}^*$ to $\widehat{q}_{0.975}^*$.

This algorithm results in a confidence interval $[35.5, 55.5]$.

By the way, the histogram of bootstrap medians $\widehat{\theta}_1^*, \ldots, \widehat{\theta}_b^*$ on Figure 10.8 shows a rather non-Normal distribution. We were essentially forced to use the nonparametric approach.

$\diamond$

MATLAB has a special command bootci for the construction of bootstrap confidence intervals. The problem in Example 10.22 can be solved by just one command

```
bootci(50000,{@median,X},'alpha',0.05,'type','percentile')
```

where 0.05 denotes the α-level, and 'percentile' requests a confidence interval computed by the method of percentiles. This is precisely the nonparametric method that we have just discussed. Replace it with type 'normal' to obtain a parametric bootstrap confidence interval that is based on the Normal distribution (10.9).

10.4   Bayesian inference

Interesting results and many new statistical methods can be obtained when we take a rather different look at statistical problems.

The difference is in our treatment of *uncertainty*.

So far, random samples were the only source of uncertainty in all the discussed statistical problems. The only distributions, expectations, and variances considered so far were distributions, expectations, and variances of data and various statistics computed from data. Population parameters were considered fixed. Statistical procedures were based on the distribution of data given these parameters,

$$f(\boldsymbol{x} \mid \theta) = f(X_1, \ldots, X_n \mid \theta).$$

This is the **frequentist approach**. According to it, all probabilities refer to random samples of data and possible long-run frequencies, and so do such concepts as unbiasedness, consistency, confidence level, and significance level:

- an estimator $\widehat{\theta}$ is *unbiased* if in a long run of random samples, it averages to the parameter θ;

- a test has significance level α if in a long run of random samples, $(100\alpha)\%$ of times the true hypothesis is rejected;

- an interval has confidence level $(1-\alpha)$ if in a long run of random samples, $(1-\alpha)100\%$ of obtained confidence intervals contain the parameter, as shown in Figure 9.2, p. 255;

- and so on.

However, there is another approach: the **Bayesian approach**. According to it, uncertainty is also attributed to the unknown parameter θ. Some values of θ are more likely than others. Then, as long as we talk about the likelihood, we can define a whole distribution of values of θ. Let us call it a *prior distribution*. It reflects our ideas, beliefs, and past experiences about the parameter before we collect and use the data.

Example 10.23 (SALARIES). What do you think is the average starting annual salary of a Computer Science graduate? Is it $20,000 per year? Unlikely, that's too low. Perhaps, $200,000 per year? No, that's too high for a fresh graduate. Between $40,000 and $70,000 sounds like a reasonable range. We can certainly collect data on 100 recent graduates, compute their average salary, and use it as an estimate, but before that, we already have our beliefs on what the mean salary may be. We can express it as some distribution with the most likely range between $40,000 and $70,000 (Figure 10.9).   ◊

Collected data may force us to change our initial idea about the unknown parameter. Probabilities of different values of θ may change. Then we'll have a *posterior distribution* of θ.

One benefit of this approach is that we no longer have to explain our results in terms of a "long run." Often we collect just one sample for our analysis and don't experience any long runs of samples. Instead, with the Bayesian approach, we can state the result in terms of the posterior distribution of θ. For example, we can clearly state the *posterior probability* for a parameter to belong to the obtained confidence interval, or the *posterior probability* that the hypothesis is true.

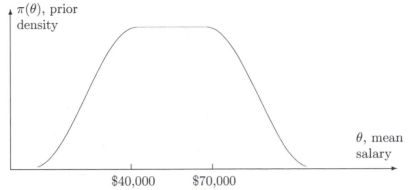

FIGURE 10.9: *Our prior distribution for the average starting salary.*

10.4.1   Prior and posterior

Now we have two sources of information to use in our Bayesian inference:

1. collected and observed data;

2. prior distribution of the parameter.

Here is how these two pieces are combined via the **Bayes formula** (see p. 29 and Figure 10.10).

Prior to the experiment, our knowledge about the parameter θ is expressed in terms of the **prior distribution** (prior pmf or pdf)

$$\pi(\theta).$$

The observed sample of data $\boldsymbol{X} = (X_1, \ldots, X_n)$ has distribution (pmf or pdf)

$$f(\boldsymbol{x}|\theta) = f(x_1, \ldots, x_n|\theta).$$

This distribution is conditional on θ. That is, different values of the parameter θ generate different distributions of data, and thus, conditional probabilities about $\boldsymbol{X}$ generally depend on the condition, θ.

Observed data add information about the parameter. The updated knowledge about θ can be expressed as the **posterior distribution**.

Posterior distribution
$$\pi(\theta|\boldsymbol{x}) = \pi(\theta|\boldsymbol{X} = \boldsymbol{x}) = \frac{f(\boldsymbol{x}|\theta)\pi(\theta)}{m(\boldsymbol{x})}. \tag{10.12}$$

Posterior distribution of the parameter θ is now conditioned on data $\boldsymbol{X} = \boldsymbol{x}$. Naturally, conditional distributions $f(\boldsymbol{x}|\theta)$ and $\pi(\theta|\boldsymbol{x})$ are related via the Bayes Rule (2.9).

According to the Bayes Rule, the denominator of (10.12), $m(\boldsymbol{x})$, represents the unconditional distribution of data $\boldsymbol{X}$. This is the **marginal distribution** (pmf or pdf) of the sample $\boldsymbol{X}$. Being unconditional means that it is constant for different values of the parameter θ. It can be computed by the *Law of Total Probability* (p. 31) or its continuous-case version below.

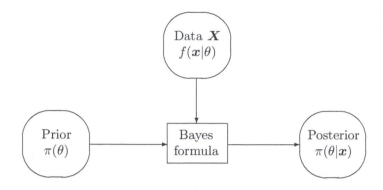

FIGURE 10.10: *Two sources of information about the parameter θ.*

| Marginal distribution of data | $$m(\boldsymbol{x}) = \sum_{\theta} f(x|\theta)\pi(\theta)$$
 for discrete prior distributions π

 $$m(\boldsymbol{x}) = \int_{\theta} f(x|\theta)\pi(\theta)d\theta$$
 for continuous prior distributions π | (10.13) |

Example 10.24 (QUALITY INSPECTION). A manufacturer claims that the shipment contains only 5% of defective items, but the inspector feels that in fact it is 10%. We have to decide whether to accept or to reject the shipment based on θ, the proportion of defective parts.

Before we see the real data, let's assign a 50-50 chance to both suggested values of θ, i.e.,

$$\pi(0.05) = \pi(0.10) = 0.5.$$

A random sample of 20 parts has 3 defective ones. Calculate the posterior distribution of θ.

Solution. Apply the Bayes formula (10.12). Given θ, the distribution of the number of defective parts X is Binomial$(n = 20, \theta)$. For $x = 3$, Table A2 gives

$$f(x \mid \theta = 0.05) = F(3 \mid \theta = 0.05) - F(2 \mid \theta = 0.05) = 0.9841 - 0.9245 = 0.0596$$

and

$$f(x \mid \theta = 0.10) = F(3 \mid \theta = 0.10) - F(2 \mid \theta = 0.10) = 0.8670 - 0.6769 = 0.1901.$$

The marginal distribution of X (for $x = 3$) is

$$
\begin{aligned}
m(3) &= f(x \mid 0.05)\pi(0.05) + f(x \mid 0.10)\pi(0.10) \\
&= (0.0596)(0.5) + (0.1901)(0.5) = 0.12485.
\end{aligned}
$$

Posterior probabilities of $\theta = 0.05$ and $\theta = 0.10$ are now computed as

$$\pi(0.05 \mid X = 3) \quad = \quad \frac{f(x \mid 0.05)\pi(0.05)}{m(3)} = \frac{(0.0596)(0.5)}{0.1248} = 0.2387;$$

$$\pi(0.10 \mid X = 3) \quad = \quad \frac{f(x \mid 0.10)\pi(0.10)}{m(3)} = \frac{(0.1901)(0.5)}{0.1248} = 0.7613.$$

Conclusion. In the beginning, we had no preference between the two suggested values of θ. Then we observed a rather high proportion of defective parts, 3/20=15%. Taking this into account, $\theta = 0.10$ is now about three times as likely than $\theta = 0.05$. $\diamond$

NOTATION	$\pi(\theta)$	$=$	prior distribution
	$\pi(\theta \mid \boldsymbol{x})$	$=$	posterior distribution
	$f(x\|\theta)$	$=$	distribution of data (model)
	$m(\boldsymbol{x})$	$=$	marginal distribution of data
	$\boldsymbol{X}$	$=$	$(X_1, \ldots, X_n)$, sample of data
	$\boldsymbol{x}$	$=$	$(x_1, \ldots, x_n)$, observed values of $X_1, \ldots, X_n$.

Conjugate distribution families

A suitably chosen prior distribution of θ may lead to a very tractable form of the posterior.

DEFINITION 10.4

A family of prior distributions π is **conjugate** to the model $f(\boldsymbol{x}|\theta)$ if the posterior distribution belongs to the same family.

Three classical examples of conjugate families are given below.

Gamma family is conjugate to the Poisson model

Let $(X_1, \ldots, X_n)$ be a sample from Poisson(θ) distribution with a Gamma(α, λ) prior distribution of θ.

Then

$$f(\boldsymbol{x}|\theta) = \prod_{i=1}^{n} f(x_i|\theta) = \prod_{i=1}^{n} \frac{e^{-\theta}\theta^{x_i}}{x_i!} \sim e^{-n\theta}\theta^{\sum x_i}. \tag{10.14}$$

Remark about dropping constant coefficients. In the end of (10.14), we dropped $(x_i!)$ and wrote that the result is "proportional" ($\sim$) to $e^{-n\theta}\theta^{\sum x_i}$. Dropping terms that don't contain θ often simplifies the computation. The form of the posterior distribution can be obtained without the constant term, and if needed, we can eventually evaluate the normalizing constant in the end, making $\pi(\theta|\boldsymbol{x})$ a fine distribution with the total probability 1, for example, as we did in Example 4.1 on p. 77. In particular, the marginal distribution

$m(\boldsymbol{x})$ can be dropped because it is θ-free. But keep in mind that in this case we obtain the posterior distribution "up to a constant coefficient."

The Gamma prior distribution of θ has density

$$\pi(\theta) \sim \theta^{\alpha-1}e^{-\lambda\theta}.$$

As a function of θ, this prior density has the same form as the model $f(\boldsymbol{x}|\theta)$ – a power of θ multiplied by an exponential function. This is the general idea behind conjugate families.

Then, the posterior distribution of θ given $\boldsymbol{X} = \boldsymbol{x}$ is

$$\begin{aligned}\pi(\theta|\boldsymbol{x}) \quad &\sim \quad f(\boldsymbol{x}|\theta)\pi(\theta) \\ &\sim \quad \left(e^{-n\theta}\theta^{\sum x_i}\right)\left(\theta^{\alpha-1}e^{-\lambda\theta}\right) \\ &\sim \quad \theta^{\alpha+\sum x_i - 1}e^{-(\lambda+n)\theta}.\end{aligned}$$

Comparing with the general form of a Gamma density (say, (4.7) on p. 85), we see that $\pi(\theta|\boldsymbol{x})$ is the Gamma distribution with new parameters,

$$\alpha_x = \alpha + \sum_{i=1}^{n} x_i \text{ and } \lambda_x = \lambda + n.$$

We can conclude that:

1. Gamma family of prior distributions is conjugate to Poisson models.

2. Having observed a Poisson sample $\boldsymbol{X} = \boldsymbol{x}$, we update the Gamma$(\alpha, \lambda)$ prior distribution of θ to the Gamma$(\alpha + \sum x_i, \lambda + n)$ posterior.

Gamma distribution family is rather rich; it has two parameters. There is often a good chance to find a member of this large family that suitably reflects our knowledge about θ.

Example 10.25 (NETWORK BLACKOUTS). The number of network blackouts each week has Poisson(θ) distribution. The weekly rate of blackouts θ is not known exactly, but according to the past experience with similar networks, it averages 4 blackouts with a standard deviation of 2.

There exists a Gamma distribution with the given mean $\mu = \alpha/\lambda = 4$ and standard deviation $\sigma = \sqrt{\alpha}/\lambda = 2$. Its parameters α and λ can be obtained by solving the system,

$$\begin{cases} \alpha/\lambda &= 4 \\ \sqrt{\alpha}/\lambda &= 2 \end{cases} \Rightarrow \begin{cases} \alpha &= (4/2)^2 &= 4 \\ \lambda &= 2^2/4 &= 1 \end{cases}$$

Hence, we can assume the Gamma$(\alpha = 4, \lambda = 1)$ prior distribution θ. It is convenient to have a conjugate prior because the posterior will then belong to the Gamma family too.

Suppose there were $X_1 = 2$ blackouts this week. Given that, the posterior distribution of θ is Gamma with parameters

$$\alpha_x = \alpha + 2 = 6, \quad \lambda_x = \lambda + 1 = 2.$$

If no blackouts occur during the next week, the updated posterior parameters become

$$\alpha_x = \alpha + 2 + 0 = 6, \quad \lambda_x = \lambda + 2 = 3.$$

This posterior distribution has the average weekly rate of $6/3 = 2$ blackouts per week. Two weeks with very few blackouts reduced our estimate of the average rate from 4 to 2. $\Diamond$

Beta family is conjugate to the Binomial model

A sample from Binomial(k, θ) distribution (assume k is known) has the probability mass function

$$f(\boldsymbol{x} \mid \theta) = \prod_{i=1}^{n} \binom{k}{x_i} \theta^{x_i} (1-\theta)^{k-x_i} \sim \theta^{\sum x_i} (1-\theta)^{nk - \sum x_i}.$$

Density of Beta(α, β) prior distribution has the same form, as a function of θ,

$$\pi(\theta) \sim \theta^{\alpha-1} (1-\theta)^{\beta-1} \quad \text{for } 0 < \theta < 1$$

(see Section A.2.2 in the Appendix). Then, the posterior density of θ is

$$\pi(\theta \mid \boldsymbol{x}) \sim f(\boldsymbol{x} \mid \theta) \pi(\theta) \sim \theta^{\alpha + \sum_{i=1}^{n} x_i - 1} (1-\theta)^{\beta + nk - \sum_{i=1}^{n} x_i - 1},$$

and we recognize the Beta density with new parameters

$$\alpha_x = \alpha + \sum_{i=1}^{n} x_i \quad \text{and} \quad \beta_x = \beta + nk - \sum_{i=1}^{n} x_i.$$

Hence,

1. Beta family of prior distributions is conjugate to the Binomial model.

2. Posterior parameters are $\alpha_x = \alpha + \sum x_i$ and $\beta_x = \beta + nk - \sum x_i$.

Normal family is conjugate to the Normal model

Consider now a sample from Normal distribution with an unknown mean θ and a known variance σ^2:

$$
\begin{aligned}
f(\boldsymbol{x} \mid \theta) &= \prod_{i=1}^{n} \frac{1}{\sqrt{2\pi}} \exp\left\{ -\frac{(x_i - \theta)^2}{2\sigma^2} \right\} \sim \exp\left\{ -\sum_{i=1}^{n} \frac{(x_i - \theta)^2}{2\sigma^2} \right\} \\
&\sim \exp\left\{ \theta \frac{\sum x_i}{\sigma^2} - \theta^2 \frac{n}{2\sigma^2} \right\} = \exp\left\{ \left(\theta \overline{X} - \frac{\theta^2}{2} \right) \frac{n}{\sigma^2} \right\}.
\end{aligned}
$$

If the prior distribution of θ is also Normal, with prior mean μ and prior variance τ^2, then

$$\pi(\theta) \sim \exp\left\{ -\frac{(\theta - \mu)^2}{2\tau^2} \right\} \sim \exp\left\{ \left(\theta\mu - \frac{\theta^2}{2} \right) \frac{1}{\tau^2} \right\},$$

and again, it has a similar form as $f(\boldsymbol{x}|\theta)$.

The posterior density of θ equals

$$
\begin{aligned}
\pi(\theta \mid \boldsymbol{x}) &\sim f(\boldsymbol{x} \mid \theta)\pi(\theta) \sim \exp\left\{ \theta\left(\frac{n\overline{X}}{\sigma^2} + \frac{\mu}{\tau^2} \right) - \frac{\theta^2}{2} \left(\frac{n}{\sigma^2} + \frac{1}{\tau^2} \right) \right\} \\
&\sim \exp\left\{ -\frac{(\theta - \mu_x)^2}{2\tau_x^2} \right\},
\end{aligned}
$$

where

$$\mu_x = \frac{n\overline{X}/\sigma^2 + \mu/\tau^2}{n/\sigma^2 + 1/\tau^2} \quad \text{and} \quad \tau_x^2 = \frac{1}{n/\sigma^2 + 1/\tau^2}. \tag{10.15}$$

This posterior distribution is certainly Normal with parameters μ_x and τ_x.

We can conclude that:

1. Normal family of prior distributions is conjugate to the Normal model with unknown mean;

2. Posterior parameters are given by (10.15).

We see that the posterior mean μ_x is a weighted average of the prior mean μ and the sample mean $\overline{X}$. This is how the prior information and the observed data are combined in case of Normal distributions.

How will the posterior mean behave when it is computed from a large sample? As the sample size n increases, we get more information from the data, and as a result, the frequentist estimator will dominate. According to (10.15), the posterior mean converges to the sample mean $\overline{X}$ as $n \to \infty$.

Posterior mean will also converge to $\overline{X}$ when $\tau \to \infty$. Large τ means a lot of uncertainty in the prior distribution of θ; thus, naturally, we should rather use observed data as a more reliable source of information in this case.

On the other hand, large σ indicates a lot of uncertainty in the observed sample. If that is the case, the prior distribution is more reliable, and as we see in (10.15), $\mu_x \approx \mu$ for large σ.

Results of this section are summarized in Table 10.2. You will find more examples of conjugate prior distributions among the exercises.

Model $f(\boldsymbol{x}\|\theta)$	Prior $\pi(\theta)$	Posterior $\pi(\theta\|\boldsymbol{x})$
Poisson(θ)	Gamma(α, λ)	Gamma$(\alpha + n\overline{X}, \lambda + n)$
Binomial(k, θ)	Beta(α, β)	Beta$(\alpha + n\overline{X}, \beta + n(k - \overline{X}))$
Normal(θ, σ)	Normal(μ, τ)	Normal$\left(\dfrac{n\overline{X}/\sigma^2 + \mu/\tau^2}{n/\sigma^2 + 1/\tau^2}, \dfrac{1}{\sqrt{n/\sigma^2 + 1/\tau^2}} \right)$

TABLE 10.2: *Three classical conjugate families.*

10.4.2 Bayesian estimation

We have already completed the most important step in Bayesian inference. We obtained the posterior distribution. All the knowledge about the unknown parameter is now included in the posterior, and that is what we'll use for further statistical analysis (Figure 10.11).

To estimate θ, we simply compute the **posterior mean**,

$$\widehat{\theta}_{\mathrm{B}} = \mathbf{E}\left\{\theta | \boldsymbol{X} = \boldsymbol{x}\right\} = \begin{cases} \displaystyle\sum_{\theta} \theta \pi(\theta|\boldsymbol{x}) &= \dfrac{\sum \theta f(\boldsymbol{x}|\theta)\pi(\theta)}{\sum f(\boldsymbol{x}|\theta)\pi(\theta)} & \text{if } \theta \text{ is discrete} \\[2em] \displaystyle\int_{\theta} \theta \pi(\theta|\boldsymbol{x})d\theta &= \dfrac{\int \theta f(\boldsymbol{x}|\theta)\pi(\theta)d\theta}{\int f(\boldsymbol{x}|\theta)\pi(\theta)d\theta} & \text{if } \theta \text{ is continuous} \end{cases}$$

The result is a conditional expectation of θ given data $\boldsymbol{X}$. In abstract terms, the **Bayes estimator** $\widehat{\theta}_{\mathrm{B}}$ is what we "expect" θ to be, after we observed a sample.

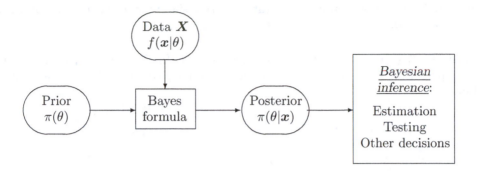

FIGURE 10.11: *Posterior distribution is the basis for Bayesian inference.*

How accurate is this estimator? Among all estimators, $\widehat{\theta}_{\mathrm{B}} = \mathbf{E}\{\theta|\boldsymbol{x}\}$ has the lowest squared-error **posterior risk**

$$\rho(\widehat{\theta}) = \mathbf{E}\left\{(\widehat{\theta} - \theta)^2 \mid \boldsymbol{X} = \boldsymbol{x}\right\}.$$

For the Bayes estimator $\widehat{\theta}_{\mathrm{B}} = \mathbf{E}\{\theta \mid \boldsymbol{x}\}$, posterior risk equals **posterior variance**,

$$\rho(\widehat{\theta}) = \mathbf{E}\left\{(\mathbf{E}\{\theta|\boldsymbol{x}\} - \theta)^2 \mid \boldsymbol{x}\right\} = \mathbf{E}\left\{(\theta - \mathbf{E}\{\theta|\boldsymbol{x}\})^2 \mid \boldsymbol{x}\right\} = \mathrm{Var}\{\theta|\boldsymbol{x}\},$$

which measures variability of θ around $\widehat{\theta}_{\mathrm{B}}$, according to the posterior distribution of θ.

Example 10.26 (NORMAL CASE). The Bayes estimator of the mean θ of Normal(θ, σ) distribution with a Normal(μ, τ) prior is

$$\widehat{\theta}_{\mathrm{B}} = \mu_x = \frac{n\overline{X}/\sigma^2 + \mu/\tau^2}{n/\sigma^2 + 1/\tau^2},$$

and its posterior risk is

$$\rho(\widehat{\theta}_{\mathrm{B}}) = \tau_x^2 = \frac{1}{n/\sigma^2 + 1/\tau^2}$$

(Table 10.2). As we expect, this risk decreases to 0 as the sample size grows to infinity. $\lozenge$

Example 10.27 (NETWORK BLACKOUTS, CONTINUED). After two weeks of data, the weekly rate of network blackouts, according to Example 10.25 on p. 356, has Gamma posterior distribution with parameters $\alpha_x = 6$ and $\lambda_x = 3$.

The Bayes estimator of the weekly rate θ is

$$\widehat{\theta}_{\mathrm{B}} = \mathbf{E}\{\theta|\boldsymbol{x}\} = \frac{\alpha_x}{\lambda_x} = 2 \text{ (blackouts per week)}$$

with a posterior risk

$$\rho(\widehat{\theta}_{\mathrm{B}}) = \mathrm{Var}\{\theta|\boldsymbol{x}\} = \frac{\alpha_x}{\lambda_x^2} = \frac{2}{3}.$$

$\lozenge$

Although conjugate priors simplify our statistics, Bayesian inference can certainly be done for other priors too.

Example 10.28 (QUALITY INSPECTION, CONTINUED). In Example 10.24 on p. 354, we computed posterior distribution of the proportion of defective parts θ. This was a discrete distribution,

$$\pi(0.05 \mid \boldsymbol{x}) = 0.2387; \qquad \pi(0.10 \mid \boldsymbol{x}) = 0.7613.$$

Now, the Bayes estimator of θ is

$$\widehat{\theta}_{\mathrm{B}} = \sum_{\theta} \theta \pi(\theta \mid \boldsymbol{x}) = (0.05)(0.2387) + (0.10)(0.7613) = 0.0881.$$

It does not agree with the manufacturer (who claims $\theta = 0.05$) or with the quality inspector (who feels that $\theta = 0.10$) but its value is much closer to the inspector's estimate.

The posterior risk of $\widehat{\theta}_{\mathrm{B}}$ is

$$
\begin{aligned}
\mathrm{Var}\{\theta|\boldsymbol{x}\} &= \mathbf{E}\{\theta^2|\boldsymbol{x}\} - \mathbf{E}^2\{\theta|\boldsymbol{x}\} \\
&= (0.05)^2(0.2387) + (0.10)^2(0.7613) - (0.0881)^2 = 0.0004,
\end{aligned}
$$

which means a rather low posterior standard deviation of 0.02. $\diamond$

10.4.3 Bayesian credible sets

Confidence intervals have a totally different meaning in Bayesian analysis. Having a posterior distribution of θ, we no longer have to explain the confidence level $(1 - \alpha)$ in terms of a long run of samples. Instead, we can give an interval $[a, b]$ or a set C that has a posterior probability $(1-\alpha)$ and state that *the parameter θ belongs to this set with probability $(1-\alpha)$*. Such a statement was impossible before we considered prior and posterior distributions. This set is called a $(1 - \alpha)100\%$ *credible set*.

DEFINITION 10.5 —————

> Set C is a $(1 - \alpha)100\%$ **credible set** for the parameter θ if the posterior probability for θ to belong to C equals $(1 - \alpha)$. That is,
>
> $$P\{\theta \in C \mid \boldsymbol{X} = \boldsymbol{x}\} = \int_C \pi(\theta|\boldsymbol{x})d\theta = 1 - \alpha.$$

Such a set is not unique. Recall that for two-sided, left-tail, and right-tail hypothesis testing, we took different portions of the area under the Normal curve, all equal $(1 - \alpha)$.

Minimizing the length of set C among all $(1-\alpha)100\%$ credible sets, we just have to include all the points θ with a high posterior density $\pi(\theta|\boldsymbol{x})$,

$$C = \{\theta : \ \pi(\theta|\boldsymbol{x}) \geq c\}$$

(see Figure 10.12). Such a set is called the **highest posterior density credible set**, or just the **HPD set**.

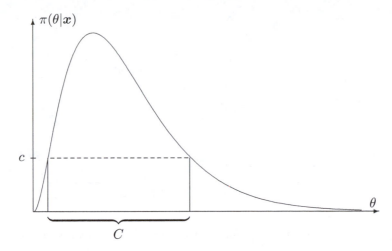

FIGURE 10.12: *The* $(1-\alpha)100\%$ *highest posterior density credible set.*

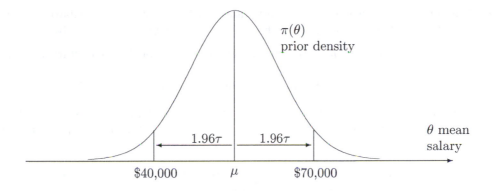

FIGURE 10.13: *Normal prior distribution and the 95% HPD credible set for the mean starting salary of Computer Science graduates (Example 10.29).*

For the Normal(μ_x, τ_x) posterior distribution of θ, the $(1-\alpha)100\%$ HPD set is

$$\mu_x \pm z_{\alpha/2}\tau_x = [\mu_x - z_{\alpha/2}\tau_x, \mu_x + z_{\alpha/2}\tau_x].$$

Example 10.29 (SALARIES, CONTINUED). In Example 10.23 on p. 352, we "decided" that the most likely range for the mean starting salary θ of Computer Science graduates is between \$40,000 and \$70,000. Expressing this in a form of a prior distribution, we let the prior mean be $\mu = (40,000 + 70,000)/2 = 55,000$. Further, if we feel that the range [40,000; 70,000] is 95% likely, and we accept a Normal prior distribution for θ, then this range should be equal

$$[40,000; 70,000] = \mu \pm z_{0.025/2}\tau = \mu \pm 1.96\tau,$$

where τ is the prior standard deviation (Figure 10.13). We can now evaluate the prior standard deviation parameter τ from this information,

$$\tau = \frac{70,000 - 40,000}{2(1.96)} = 7,653.$$

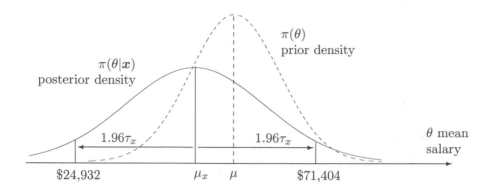

FIGURE 10.14: *Normal prior and posterior distributions for the mean starting salary (Example 10.29).*

This is the advantage of using a rich (two-parameter) family of prior distributions: we are likely to find a member of this family that reflects our prior beliefs adequately.

Then, *prior to collecting any data*, the 95% HPD credible set of the mean starting salary θ is

$$\mu \pm z_{0.025}\tau = [40{,}000; \ 70{,}000].$$

Suppose a random sample of 100 graduates has the mean starting salary $\overline{X} = 48{,}000$ with a sample standard deviation $s = 12{,}000$. From Table 10.2, we determine the posterior mean and standard deviation,

$$
\begin{aligned}
\mu_x &= \frac{n\overline{X}/\sigma^2 + \mu/\tau^2}{n/\sigma^2 + 1/\tau^2} = \frac{(100)(48{,}000)/(12{,}000)^2 + (55{,}000)/(7{,}653)^2}{100/(12{,}000)^2 + 1/(7653)^2} \\
&= 48{,}168; \\
\tau_x &= \frac{1}{\sqrt{n/\sigma^2 + 1/\tau^2}} = \frac{1}{\sqrt{100/(12{,}000)^2 + 1/(7653)^2}} = 11{,}855.
\end{aligned}
$$

We used the sample standard deviation s in place of the population standard deviation σ assuming that a sample of size 100 estimates the latter rather accurately. Alternatively, we could put a prior distribution on unknown σ too and estimate it by Bayesian methods. Since the observed sample mean is smaller than our prior mean, the resulting posterior distribution is shifted to the left of the prior (Figure 10.14).

Conclusion. After seeing the data, the Bayes estimator for the mean starting salary of CS graduates is

$$\widehat{\theta}_{\mathrm{B}} = \mu_x = 48{,}168 \text{ dollars},$$

and the 95% HPD credible set for this mean salary is

$$\mu_x \pm z_{0.025}\tau_x = 48{,}168 \pm (1.96)(11{,}855) = 48{,}168 \pm 23{,}236 = \underline{[24{,}932; \ 71{,}404]}$$

Lower observed salaries than the ones predicted a priori extended the lower end of our credible interval.   ◊

Example 10.30 (TELEPHONE COMPANY). A new telephone company predicts to handle an average of 1000 calls per hour. During 10 randomly selected hours of operation, it handled a total of 7265 calls.

How should it update the initial estimate of the frequency of telephone calls? Construct a 95% HPD credible set. Telephone calls are placed according to a Poisson process. The hourly rate of calls has an Exponential prior distribution.

Solution. We need a Bayesian estimator of the frequency θ of telephone calls. The number of calls during 1 hour has Poisson(θ) distribution, where θ is unknown, with

$$\text{Exponential}(\lambda) = \text{Gamma}(1, \lambda)$$

prior distribution that has an expectation of

$$\mathbf{E}(\theta) = \frac{1}{\lambda} = 1000 \text{ calls.}$$

Hence, $\lambda = 0.001$. We observe a sample of size $n = 10$, totaling

$$\sum_{i=1}^{n} X_i = n\overline{X} = 7265 \text{ calls.}$$

As we know (see Table 10.2 on p. 358), the posterior distribution in this case is Gamma(α_x, λ_x) with

$$\begin{aligned} \alpha_x &= \alpha + n\overline{X} = 7266, \\ \lambda_x &= \lambda + n = 10.001. \end{aligned}$$

This distribution has mean

$$\mu_x = \alpha_x / \lambda_x = 726.53$$

and standard deviation

$$\tau_x = \sqrt{\alpha_x} / \lambda_x = 8.52.$$

The Bayes estimator of θ is

$$\mathbf{E}(\theta | \mathbf{X}) = \mu_x = \underline{726.53 \text{ calls per hour.}}$$

It almost coincides with the sample mean $\overline{X}$ showing that the sample was informative enough to dominate over the prior information.

For the credible set, we notice that α_x is sufficiently large to make the Gamma posterior distribution approximately equal the Normal distribution with parameters μ_x and τ_x. The 95% HPD credible set is then

$$\mu_x \pm z_{0.05/2} \tau_x = 726.53 \pm (1.96)(8.52) = 726.53 \pm 16.70 = \underline{[709.83, 743.23]}$$

$\Diamond$

10.4.4 Bayesian hypothesis testing

Bayesian hypothesis testing is very easy to interpret. We can compute prior and posterior probabilities for the hypothesis H_0 and alternative H_A to be true and decide from there which one to accept or to reject.

Computing such probabilities was not possible without prior and posterior distributions of the parameter θ. In non-Bayesian statistics, θ was not random, thus H_0 and H_A were either true (with probability 1) or false (with probability 1).

For Bayesian tests, in order for H_0 to have a meaningful, non-zero probability, it often represents a set of parameter values instead of just one θ_0, and we are testing

$$H_0 : \theta \in \Theta_0 \quad \text{vs} \quad H_A : \theta \in \Theta_1.$$

This actually makes sense because exact equality $\theta = \theta_0$ is unlikely to hold anyway, and in practice it is understood as $\theta \approx \theta_0$.

Comparing posterior probabilities of H_0 and H_A,

$$\boldsymbol{P}\{\Theta_0 \mid \boldsymbol{X} = \boldsymbol{x}\} \quad \text{and} \quad \boldsymbol{P}\{\Theta_1 \mid \boldsymbol{X} = \boldsymbol{x}\},$$

we decide whether $\boldsymbol{P}\{\Theta_1 \mid \boldsymbol{X} = \boldsymbol{x}\}$ is large enough to present significant evidence and to reject the null hypothesis. One can again compare it with $(1 - \alpha)$ such as 0.90, 0.95, 0.99, or state the result in terms of likelihood, "the null hypothesis is this much likely to be true."

Example 10.31 (Telephone company, continued). Let us test whether the telephone company in Example 10.30 can actually face a call rate of 1000 calls *or more* per hour. We are testing

$$H_0 : \theta \geq 1000 \quad \text{vs} \quad H_A : \theta < 1000,$$

where θ is the hourly rate of telephone calls.

According to the Gamma(α_x, λ_x) posterior distribution of θ and its Normal $(\mu_x = 726.53, \tau_x = 72.65)$ approximation,

$$\boldsymbol{P}\{H_0 \mid \boldsymbol{X} = \boldsymbol{x}\} = \boldsymbol{P}\left\{\frac{\theta - \mu_x}{\tau_x} \geq \frac{1000 - \mu_x}{\tau_x}\right\} = 1 - \Phi(3.76) = 0.0001.$$

By the complement rule, $\boldsymbol{P}\{H_A \mid \boldsymbol{X} = \boldsymbol{x}\} = 0.9999$, and this presents sufficient evidence against H_0.

We conclude that it's extremely unlikely for this company to face a frequency of 1000+ calls per hour. $\diamond$

Loss and risk

Often one can anticipate the consequences of Type I and Type II errors in hypothesis testing and assign a **loss** $L(\theta, a)$ associated with each possible error. Here θ is the parameter, and a is our action, the decision on whether we accept or reject the null hypothesis.

Each decision then has its **posterior risk** $\rho(a)$, defined as the expected loss computed under the posterior distribution. The action with the lower posterior risk is our **Bayes action**.

Suppose that the Type I error causes the loss

$$w_0 = Loss(\text{Type I error}) = L(\Theta_0, \text{reject } H_0),$$

and the Type II error causes the loss

$$w_1 = Loss(\text{Type II error}) = L(\Theta_1, \text{accept } H_0).$$

Posterior risks of each possible action are then computed as

$$\begin{aligned}
\rho(\text{reject } H_0) &= w_0 \pi(\Theta_0 \mid \boldsymbol{x}), \\
\rho(\text{accept } H_0) &= w_1 \pi(\Theta_1 \mid \boldsymbol{x}).
\end{aligned}$$

Now we can determine the Bayes action. If $w_0 \pi(\Theta_0 \mid \boldsymbol{x}) \leq w_1 \pi(\Theta_1 \mid \boldsymbol{x})$, the Bayes action is to accept H_0. If $w_0 \pi(\Theta_0 \mid \boldsymbol{x}) \geq w_1 \pi(\Theta_1 \mid \boldsymbol{x})$, the Bayes action is to reject H_0.

Example 10.32 (QUALITY INSPECTION, CONTINUED). In Example 10.24 on p. 354, we are testing

$$H_0 : \theta = 0.05 \quad \text{vs} \quad H_A : \theta = 0.10$$

for the proportion θ of defective parts. Suppose that the Type I error is three times as costly here as the Type II error. What is the Bayes action in this case?

Example 10.28 gives posterior probabilities

$$\pi(\Theta_0 \mid \boldsymbol{X} = \boldsymbol{x}) = 0.2387 \text{ and } \pi(\Theta_1 \mid \boldsymbol{X} = \boldsymbol{x}) = 0.7613.$$

Since $w_0 = 3w_1$, the posterior risks are

$$\begin{aligned}
\rho(\text{reject } H_0) &= w_0 \pi(\Theta_0 \mid \boldsymbol{x}) = 3w_1(0.2387) = 0.7161 w_1, \\
\rho(\text{accept } H_0) &= w_1 \pi(\Theta_1 \mid \boldsymbol{x}) = 0.7613 w_1.
\end{aligned}$$

Thus, rejecting H_0 has a lower posterior risk, and therefore, it is the Bayes action. Reject H_0. $\diamond$

Summary and conclusions

A number of popular methods of Statistical Inference are presented in this chapter.

Chi-square tests represent a general technique based on counts. Comparing the observed counts with the counts expected under the null hypothesis through the Chi-square statistic, one can test for the goodness of fit and for the independence of two factors. Contingency tables are widely used for the detection of significant relations between categorical variables.

Nonparametric statistical methods are not based on any particular distribution of data. So, they are often used when the distribution is unknown or complicated. They are also very handy when the sample may contain some outliers, and even when the data are not numerical.

A sign test and Wilcoxon signed rank test for one sample and a two-sample Mann–Whitney–Wilcoxon rank sum test are introduced in this chapter for testing and comparing distributions and their medians. Combinatorics helps to find the exact null distribution of each considered test statistic. For large samples, this distribution is approximately Normal.

Bootstrap is a popular modern resampling technique that is widely used nowadays for studying properties and evaluating performance of various statistics. A simple idea behind the bootstrap allows us to analyze complicated statistics using only the power of our computer. This chapter showed the most basic applications of bootstrap to the estimation of standard errors and biases of parameter estimates and construction of parametric and nonparametric confidence intervals.

Bayesian inference combines the information contained in the data and in the prior distribution of the unknown parameter. It is based on the posterior distribution, which is the conditional distribution of the unknown parameter, given the data.

The most commonly used Bayesian parameter estimator is the posterior mean. It minimizes the squared-error posterior risk among all the estimates.

Bayesian $(1-\alpha)100\%$ *credible sets* also contain the parameter θ with probability $(1-\alpha)$, but this time the probability refers to the distribution of θ. Explaining a $(1-\alpha)100\%$ credible set, we can say that given the observed data, θ belongs to the obtained set with probability $(1-\alpha)$.

For *Bayesian hypothesis testing*, we compute posterior probabilities of H_0 and H_A and decide if the former is sufficiently smaller than the latter to suggest rejection of H_0. We can also take a Bayesian action that minimizes the posterior risk.

Exercises

10.1. Does the number of unsolicited (spam) e-mails follow a Poisson distribution? Here is the record of the number of spam e-mails received during 60 consecutive days (data set Spam).

12	6	4	0	13	5	1	3	10	1	29	12	4	4	22
2	2	27	7	27	9	34	10	10	2	28	7	0	9	4
32	4	5	9	1	13	10	20	5	5	0	6	9	20	28
22	10	8	11	15	1	14	0	9	9	1	9	0	7	13

Choose suitable bins and conduct a goodness-of-fit test at the 1% level of significance.

10.2. Applying the theory of M/M/1 queuing systems, we assume that the service times follow Exponential distribution. The following service times, in minutes, have been observed during 24 hours of operation (data set `ServiceTimes`):

10.5	1.2	6.3	3.7	0.9	7.1	3.3	4.0	1.7	11.6	5.1	2.8	4.8	2.0	8.0	4.6
3.1	10.2	5.9	12.6	4.5	8.8	7.2	7.5	4.3	8.0	0.2	4.4	3.5	9.6	5.5	0.3
2.7	4.9	6.8	8.6	0.8	2.2	2.1	0.5	2.3	2.9	11.7	0.6	6.9	11.4	3.8	3.2
2.6	1.9	1.0	4.1	2.4	13.6	15.2	6.4	5.3	5.4	1.4	5.0	3.9	1.8	4.7	0.7

Is the assumption of Exponentiality supported by these data?

10.3. The following sample, data set `RandomNumbers` is collected to verify the accuracy of a new random number generator (it is already ordered for your convenience).

```
-2.434  -2.336  -2.192  -2.010  -1.967  -1.707  -1.678  -1.563  -1.476  -1.388
-1.331  -1.269  -1.229  -1.227  -1.174  -1.136  -1.127  -1.124  -1.120  -1.073
-1.052  -1.051  -1.032  -0.938  -0.884  -0.847  -0.846  -0.716  -0.644  -0.625
-0.588  -0.584  -0.496  -0.489  -0.473  -0.453  -0.427  -0.395  -0.386  -0.386
-0.373  -0.344  -0.280  -0.246  -0.239  -0.211  -0.188  -0.155  -0.149  -0.112
-0.103  -0.101  -0.033  -0.011   0.033   0.110   0.139   0.143   0.218   0.218
 0.251   0.261   0.308   0.343   0.357   0.463   0.477   0.482   0.489   0.545
 0.590   0.638   0.652   0.656   0.673   0.772   0.775   0.776   0.787   0.969
 0.978   1.005   1.013   1.039   1.072   1.168   1.185   1.263   1.269   1.297
 1.360   1.370   1.681   1.721   1.735   1.779   1.792   1.881   1.903   2.009
```

(a) Apply the χ^2 goodness-of-fit test to check if this sample comes from the Standard Normal distribution.

(b) Test if this sample comes from the Uniform(-3,3) distribution.

(c) Is it theoretically possible to accept both null hypotheses in (a) and (b) although they are contradicting to each other? Why does it make sense?

10.4. In Example 10.3 on p. 319, we tested whether the number of concurrent users is approximately Normal. How does the result of the Chi-square test depend on our choice of bins? For the same data, test the assumption of a Normal distribution using a different set of bins $B_1, \ldots, B_N$.

10.5. Show that the sample size is too small in Example 10.9 on p. 327 to conduct a χ^2 goodness-of-fit test of Normal distribution that involves estimation of its both parameters.

10.6. Two computer makers, A and B, compete for a certain market. Their users rank the quality of computers on a 4-point scale as "Not satisfied", "Satisfied", "Good quality", and "Excellent quality, will recommend to others". The following counts were observed,

Computer maker	"Not satisfied"	"Satisfied"	"Good quality"	"Excellent quality"
A	20	40	70	20
B	10	30	40	20

Is there a significant difference in customer satisfaction of the computers produced by A and by B?

10.7. An AP test has been given in two schools. In the first school, 162 girls and 567 boys passed it whereas 69 girls and 378 boys failed. In the second school, 462 girls and 57 boys passed the test whereas 693 girls and 132 boys failed it.

(a) In the first school, are the results significantly different for girls and boys?

(b) In the second school, are the results significantly different for girls and boys?

(c) In both schools together, are the results significantly different for girls and boys?

For each school, construct a contingency table and apply the Chi-square test.

Remark: This data set is an example of a strange phenomenon known as **Simpson's paradox**. Look, the girls performed better than the boys in *each* school; however, in both schools together, the boys did better!!!

Check for yourself... In the first school, 70% of girls and only 60% of boys passed the test. In the second school, 40% of girls and only 30% of boys passed. But in both schools together, 55% of boys and only 45% of girls passed the test. Wow!

10.8. A computer manager decides to install the new antivirus software on all the company's computers. Three competing antivirus solutions (X, Y, and Z) are offered to her for a free 30-day trial. She installs each solution on 50 computers and records infections during the following 30 days. Results of her study are in the table.

Antivirus software	X	Y	Z
Computers not infected	36	28	32
Computers infected once	12	16	14
Computers infected more than once	2	6	4

Does the computer manager have significant evidence that the three antivirus solutions are *not* of the same quality?

10.9. The Probability and Statistics course has three sections - S01, S02, and S03. Among 120 students in section S01, 40 got an A in the course, 50 got a B, 20 got a C, 2 got a D, and 8 got an F. Among 100 students in section S02, 20 got an A, 40 got a B, 25 got a C, 5 got a D, and 10 got an F. Finally, among 60 students in section S03, 20 got an A, 20 got a B, 15 got a C, 2 got a D, and 3 got an F. Do the three sections differ in their students' performance?

10.10. Among 25 jobs sent to the printer at random times, 6 jobs were printed in less than 20 seconds each, and 19 jobs took more than 20 seconds each. Is there evidence that the median response time for this printer exceeds 20 sec? Apply the sign test.

10.11. At a computer factory, the median of certain measurements of some computer parts should equal m. If it is found to be less than m or greater than m, the process is recalibrated. Every day, a quality control technician takes measurements from a sample of 20 parts. According to a 5% level sign test, how many measurements on either side of m justify recalibration? (In other words, what is the rejection region?)

10.12. When a computer chip is manufactured, its certain crucial layer should have the median thickness of 45 nm (nanometers; one nanometer is one billionth of a metre). Measurements are made on a sample of 60 produced chips (data set `Chips`), and the measured thickness is recorded as

34.9	35.9	38.9	39.4	39.9	41.3	41.5	41.7	42.0	42.1	42.5	43.5	43.7	43.9	44.2
44.4	44.6	45.3	45.7	45.9	46.0	46.2	46.4	46.6	46.8	47.2	47.6	47.7	47.8	48.8
49.1	49.2	49.4	49.5	49.8	49.9	50.0	50.2	50.5	50.7	50.9	51.0	51.3	51.4	51.5
51.6	51.8	52.0	52.5	52.6	52.8	52.9	53.1	53.7	53.8	54.3	56.8	57.1	57.8	58.9

(This data set is already ordered, for your convenience.)

Will a 1% level sign test conclude that the median thickness slipped from 45 nm?

10.13. Refer to Exercise 10.12. It is also important to verify that the first quartile Q_1 of the layer thickness does not exceed 43 nm. Apply the idea of the sign test to the first quartile instead of the median and test

$$H_0 : Q_1 = 43 \quad \text{vs} \quad H_A : Q_1 > 43.$$

The test statistic will be the number of measurements that exceed 43 nm. Find the null distribution of this test statistic, compute the P-value of the test, and state the conclusion about the first quartile.

10.14. The median of certain measurements on the produced computer parts should never exceed 5 inches. To verify that the process is conforming, engineers decided to measure 100 parts, one by one. To everyone's surprise, after 63 of the first 75 measurements exceeded 5 in, one engineer suggests to halt measurements and fix the process. She claims that whatever the remaining 25 measurements are, the median will be inevitably found significantly greater than 5 inches after all 100 measurements are made.

Do you agree with this statement? Why or why not? Certainly, if the remaining 25 measurements make no impact on the test then it should be a good resource-saving decision to stop testing early and fix the process.

10.15. The teacher states that the median score on the last test was 84. Masha asks 12 of her classmates and records their scores as

$$76, 96, 74, 88, 79, 95, 75, 82, 90, 60, 77, 56.$$

Assuming that she picks classmates at random, can she treat these data as evidence that the class median was less than 84? Can she get stronger evidence by using the sign test or the Wilcoxon signed rank test?

10.16. The starting salaries of eleven software developers are

$$47, 52, 68, 72, 55, 44, 58, 63, 54, 59, 77 \text{ thousand of dollars.}$$

Does the 5% level Wilcoxon signed rank test provide significant evidence that the median starting salary of software developers is above \$50,000?

10.17. Refer to Exercise 10.12. Does the Wilcoxon signed rank test confirm that the median thickness no longer equals 45 nm?

10.18. Refer to Exercise 10.2. Do these data provide significant evidence that the median service time is less than 5 min 40 sec? Conduct the Wilcoxon signed rank test at the 5% level of significance. What assumption of this test may not be fully satisfied by these data?

10.19. Use the recursive formula (10.6) to calculate the null distribution of Wilcoxon test statistic W for sample sizes $n = 2$, $n = 3$, and $n = 4$.

10.20. Apply the Mann–Whitney–Wilcoxon test to the quiz grades in Exercise 9.23 on p. 312 to see if Anthony's median grade is significantly higher than Eric's. What is the P-value?

10.21. Two internet service providers claim that they offer the fastest internet in the area. A local company requires the download speed of at least 20 Megabytes per second (Mbps), for its normal operation. It decides to conduct a fair contest by sending 10 packets of equal size through each network and recording their download speed.

For the 1st internet service provider, the download speed is recorded as 26.7, 19.0, 26.5, 29.1, 26.2, 27.6, 26.8, 24.2, 25.7, 23.0 Mbps. For the 2nd provider, the download speed is recorded as 19.3, 22.1, 23.4, 24.8, 25.9, 22.2, 18.3, 20.1, 19.2, 27.9 Mbps.

(a) According to the sign test, is there significant evidence that the median download speed for the 1st provider is at least 20 Mbps? What about the 2nd provider? Calculate each P-value.

(b) Repeat (a) using the Wilcoxon signed rank test. Do the P-values show that this test is more sensitive than the sign test?

(c) At the 1% level, is there significant evidence that the median download speed for the 1st provider exceeds the median download speed for the 2nd provider? Use the suitable test.

These data are also available in data set `Internet`.

10.22. Fifteen e-mail attachments were classified as benign and malicious. Seven benign attachments were 0.4, 2.1, 3.6, 0.6, 0.8, 2.4, and 4.0 Mbytes in size. Eight malicious attachments had sizes 1.2, 0.2, 0.3, 3.3, 2.0, 0.9, 1.1, and 1.5 Mbytes. Does the Mann–Whitney–Wilcoxon test detect a significant difference in the distribution of sizes of benign and malicious attachments? (If so, the size could help classify e-mail attachments and warn about possible malicious codes.)

10.23. During freshman year, Eric's textbooks cost $89, $99, $119, $139, $189, $199, and $229. During his senior year, he had to pay $109, $159, $179, $209, $219, $259, $279, $299, and $309 for his textbooks. Is this significant evidence that the median cost of textbooks is rising, according to the Mann–Whitney–Wilcoxon test?

10.24. Two teams, six students each, competed at a programming contest. The judges gave the overall 1st, 3rd, 6th, 7th, 9th, and 10th places to members of Team A. Now the captain of Team A claims the overall victory over Team B, according to a one-sided Mann–Whitney–Wilcoxon test? Do you concur with his conclusion? What hypothesis are you testing?

10.25. Refer to Exercise 10.2 and data set `ServiceTimes`. After the first 32 service times were recorded (the first two rows of data), the server was substantially modified. Conduct the Mann–Whitney–Wilcoxon test at the 10% level to see if this modification led to a reduction of the median service time.

10.26. On five days of the week, Natasha spends 2, 2, 3, 3, and 5 hours doing her homework.

(a) List all possible bootstrap samples and find the probability of each of them.

(b) Use your list to find the bootstrap distribution of the sample median.

(c) Use this bootstrap distribution to estimate the standard error and the bias of a sample median.

10.27. In Exercise 10.16, we tested the median starting salary of software developers. We can actually estimate this starting salary by the sample median, which for these data equals $\widehat{M} = \$58,000$.

(a) How many different bootstrap samples can be generated? Find the number of all possible ordered and unordered samples.

(b) Find the bootstrap distribution of the sample median. Do not list all the bootstrap samples!

(c) Use this distribution to estimate the standard error of $\widehat{M}$.

(d) Construct an 88% bootstrap confidence interval for the population median salary M.

(e) Use the bootstrap distribution to estimate the probability that 11 randomly selected software developers have their median starting salary above \$50,000.

Exercises 10.28–10.30 require the use of a computer.

10.28. Refer to Exercise 10.15. Eric estimates the class median score by the sample median, which is any number between 77 and 79 for these data (Eric takes the middle value 78). Generate 10,000 bootstrap samples and use them for the following inference.

(a) Estimate the standard error of the sample median used by Eric.

(b) Is this sample median a biased estimator for the population median? Estimate its bias.

(c) Construct a 95% bootstrap confidence interval for the population median. Can it be used to test whether the class median score is 84?

10.29. Refer to data set `Chips` and Exercise 10.13 where the first quartile of layer thickness is being tested. This quartile is estimated by the first sample quartile $\widehat{Q}_1$ (in case of a whole interval of sample quartiles, its middle is taken). Use the bootstrap method to estimate the standard error and the bias of this estimator.

10.30. Is there a correlation between the Eric's and Anthony's grades in Exercise 9.23 on p. 312? Construct an 80% nonparametric bootstrap confidence interval for the correlation coefficient between their quiz grades (a small sample size does not allow to assume the Normal distribution of the sample correlation coefficient).

10.31. A new section of a highway is opened, and $X = 4$ accidents occurred there during one month. The number of accidents has Poisson(θ) distribution, where θ is the expected number of accidents during one month. Experience from the other sections of this highway suggests that the prior distribution of θ is Gamma(5,1). Find the Bayes estimator of θ under the squared-error loss and find its posterior risk.

10.32. The data set consists of a sample $X = (2,3,5,8,2)$ from the Geometric distribution with unknown parameter θ.

(a) Show that the Beta family of prior distributions is conjugate.

(b) Taking the Beta(3,3) distribution as the prior, compute the Bayes estimator of θ.

10.33. Service times of a queuing system, in minutes, follow Exponential distribution with an unknown parameter θ. The prior distribution of θ is Gamma(3,1). The service times of five random jobs are 4 min, 3 min, 2 min, 5 min, and 5 min.

(a) Show that the Gamma family of prior distributions is conjugate.

(b) Compute the Bayes estimator of parameter θ and its posterior risk.

(c) Compute the posterior probability that $\theta \geq 0.5$, i.e., at least one job every two minutes can be served.

(d) If the Type I error and the Type II error cause approximately the same loss, test the hypothesis $H_0 : \theta \geq 0.5$ versus $H_A : \theta < 0.5$.

10.34. An internet service provider studies the distribution of the number of concurrent users of the network. This number has Normal distribution with mean θ and standard deviation 4,000 people. The prior distribution of θ is Normal with mean 14,000 and standard deviation 2,000.

The data on the number of concurrent users are collected; see data set `ConcurrentUsers` and Exercise 8.2 on p. 240.

(a) Give the Bayes estimator for the mean number of concurrent users θ.

(b) Construct the highest posterior density 90% credible set for θ and interpret it.

(c) Is there significant evidence that the mean number of concurrent users exceeds 16,000?

10.35. Continue Exercise 10.34. Another statistician conducts a non-Bayesian analysis of the data in Exercise 8.2 on p. 240 about concurrent users.

(a) Give the non-Bayesian estimator for the mean number of concurrent users θ.

(b) Construct a 90% confidence interval for θ and interpret it.

(c) Is there significant evidence that the mean number of concurrent users exceeds 16,000?

(d) How do your results differ from the previous exercise?

10.36. In Example 9.13 on p. 258, we constructed a confidence interval for the population mean μ based on the observed Normally distributed measurements. Suppose that prior to the experiment we thought this mean should be between 5.0 and 6.0 with probability 0.95.

(a) Find a conjugate prior distribution that fully reflects your prior beliefs.

(b) Derive the posterior distribution and find the Bayes estimator of μ. Compute its posterior risk.

(c) Compute a 95% HPD credible set for μ. Is it different from the 95% confidence interval? What causes the differences?

10.37. If ten coin tosses result in ten straight heads, can this coin still be fair and unbiased?

By looking at a coin, you believe that it is fair (a 50-50 chance of each side) with probability 0.99. This is your prior probability. With probability 0.01, you allow the coin to be biased, one way or another, so its probability of heads is Uniformly distributed between 0 and 1. Then you toss the coin ten times, and each time it turns up heads. Compute the posterior probability that it is a fair coin.

10.38. Observed is a sample from Uniform$(0, \theta)$ distribution.

(a) Find a conjugate family of prior distributions (you can find it in our inventory in Section A.2).

(b) Assuming a prior distribution from this family, derive a form of the Bayes estimator and its posterior risk.

10.39. A sample $X_1, \ldots, X_n$ is observed from *Beta* $(\theta, 1)$ distribution. Derive the general form of a Bayes estimator of θ under the Gamma(α, λ) prior distribution.

10.40. Anton played five chess games and won all of them. Let us estimate θ, his probability of winning the next game. Suppose that parameter θ has Beta(4,1) prior distribution and that the game results are independent of each other.

(a) Compute the Bayes estimator of θ and its posterior risk.

(b) Construct a 90% HPD credible set for θ.

(c) Is there significant evidence that Anton's winning probability is more than 0.7? To answer this question, find the posterior probabilities of $\theta \leq 0.7$ and $\theta > 0.7$ and use them to test $H_0 : \theta \leq 0.7$ vs $H_A : \theta > 0.7$.

Notice that the standard frequentist estimator of θ is the sample proportion $\widehat{p} = 1$, which is rather unrealistic because it gives Anton no chance to lose!

Chapter 11

Regression

In Chapters 8, 9, and 10, we were concerned about the distribution of *one random variable*, its parameters, expectation, variance, median, symmetry, skewness, etc. In this chapter, we study *relations* among variables.

Many variables observed in real life are related. The type of their relation can often be expressed in a mathematical form called *regression*. Establishing and testing such a relation enables us:

– to understand interactions, causes, and effects among variables;

– to predict unobserved variables based on the observed ones;

– to determine which variables significantly affect the variable of interest.

11.1 Least squares estimation

Regression models relate a *response variable* to one or several predictors. Having observed predictors, we can forecast the response by computing its *conditional expectation*, given all the available predictors.

DEFINITION 11.1 —————————

> **Response** or *dependent variable* Y is a variable of interest that we predict based on one or several predictors.
>
> **Predictors** or *independent variables* $X^{(1)}, \ldots, X^{(k)}$ are used to predict the values and behavior of the response variable Y.
>
> **Regression** of Y on $X^{(1)}, \ldots, X^{(k)}$ is the conditional expectation,
>
> $$G(x^{(1)}, \ldots, x^{(k)}) = \mathbf{E}\left\{ Y \mid X^{(1)} = x^{(1)}, \ldots, X^{(k)} = x^{(k)} \right\}.$$
>
> It is a function of $x^{(1)}, \ldots, x^{(k)}$ whose form can be estimated from data.

11.1.1 Examples

Consider several situations when we can predict a *dependent* variable of interest from *independent* predictors.

Example 11.1 (WORLD POPULATION). According to the International Data Base of the *U.S. Census Bureau*, population of the world grows according to Table 11.1 and data set `PopulationWorld`. How can we use these data to predict the world population in years 2020 and 2030?

Figure 11.1 shows that the population (response) is tightly related to the year (predictor),

$$\text{population} \approx G(\text{year}).$$

Population increases every year, and its growth is almost linear. If we estimate the *regression function* G (the dotted line on Figure 11.1) relating the response and the predictor and

Year	Population mln. people	Year	Population mln. people	Year	Population mln. people	Year	Population mln. people
1950	2557	1970	3708	1990	5273	2010	6835
1955	2781	1975	4084	1995	5682	2015	7226
1960	3041	1980	4447	2000	6072	2020	?
1965	3347	1985	4844	2005	6449	2030	?

TABLE 11.1: *Population of the world, 1950–2030.*

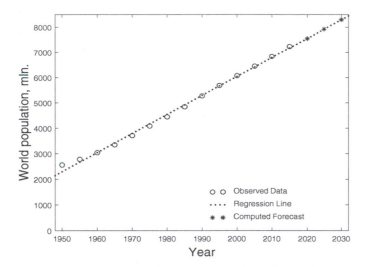

FIGURE 11.1: *World population in 1950–2019 and its regression forecast until 2030.*

extend its graph to the year 2030, the forecast will be ready. We can simply compute $G(2020)$ and $G(2030)$.

A straight line that fits the observed data for years 1950–2015 predicts the population of 7.54 billion in 2020, 7.92 billion in 2025, and 8.29 billion in 2030. It also shows that between 2025 and 2030, around the year 2026, the world population reaches the historical mark of 8 billion. ◊

How accurate is the forecast obtained in this example? The observed population during 1950–2019 appears to grow rather closely to the estimated regression line in Figure 11.1. It is reasonable to hope that it will continue to do so through 2030.

The situation is different in the next example.

Example 11.2 (HOUSE PRICES). Seventy house sale prices in a certain county are depicted in Figure 11.2 along with the house area.

First, we see a clear relation between these two variables, and in general, bigger houses are more expensive. However, the trend no longer seems linear.

Second, there is a large amount of variability around this trend. Indeed, area is not the only factor determining the house price. Houses with the same area may still be priced differently.

Then, how can we estimate the price of a 3200-square-foot house? We can estimate the general trend (the dotted line in Figure 11.2) and plug 3200 into the resulting formula, but due to obviously high variability, our estimation will not be as accurate as in Example 11.1. ◊

To improve our estimation in the last example, we may take other factors into account: the number of bedrooms and bathrooms, the backyard area, the average income of the

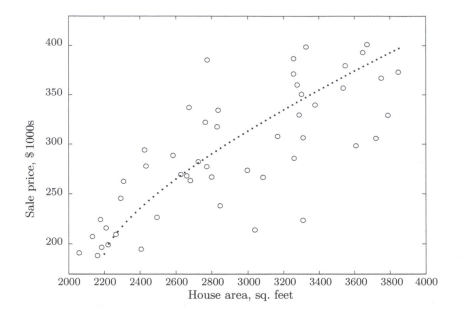

FIGURE 11.2: *House sale prices and their footage.*

neighborhood, etc. If all the added variables are relevant for pricing a house, our model will have a closer fit and will provide more accurate predictions. Regression models with multiple predictors are studied in Section 11.3.

11.1.2   Method of least squares

Our immediate goal is to estimate the **regression function** G that connects response variable Y with predictors $X^{(1)}, \ldots, X^{(k)}$. First we focus on *univariate regression* predicting response Y based on one predictor X. The method will be extended to k predictors in Section 11.3.

In univariate regression, we observe *pairs* $(x_1, y_1), \ldots, (x_n, y_n)$, shown in Figure 11.3a.

For accurate forecasting, we are looking for the function $\widehat{G}(x)$ that passes as close as possible to the observed data points. This is achieved by minimizing distances between observed data points

$$y_1, \ldots, y_n$$

and the corresponding points on the fitted regression line,

$$\widehat{y}_1 = \widehat{G}(x_1), \ldots, \widehat{y}_n = \widehat{G}(x_n)$$

(see Figure 11.3b). Method of least squares minimizes the sum of squared distances.

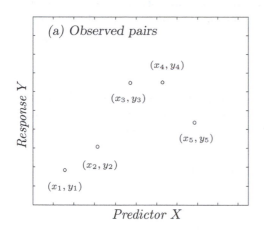
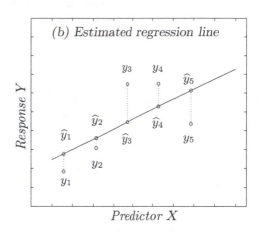

FIGURE 11.3: *Least squares estimation of the regression line.*

DEFINITION 11.2

Residuals
$$e_i = y_i - \widehat{y}_i$$

are differences between observed responses y_i and their **fitted values** $\widehat{y}_i = \widehat{G}(x_i)$.

Method of least squares finds a regression function $\widehat{G}(x)$ that minimizes the sum of squared residuals

$$\sum_{i=1}^{n} e_i^2 = \sum_{i=1}^{n} (y_i - \widehat{y}_i)^2. \tag{11.1}$$

Function $\widehat{G}$ is usually sought in a suitable form: linear, quadratic, logarithmic, etc. The simplest form is linear.

11.1.3 Linear regression

Linear regression model assumes that the conditional expectation

$$G(x) = \mathbf{E}\{Y \mid X = x\} = \beta_0 + \beta_1 x$$

is a *linear function* of x. As any linear function, it has an intercept β_0 and a slope β_1.

The **intercept**
$$\beta_0 = G(0)$$

equals the value of the regression function for $x = 0$. Sometimes it has no physical meaning. For example, nobody will try to predict the value of a computer with 0 random access memory (RAM), and nobody will consider the Federal reserve rate in year 0. In other cases, intercept is quite important. For example, according to the *Ohm's Law* ($V = R\,I$)

the voltage across an *ideal* conductor is proportional to the current. A non-zero intercept $(V = V_0 + R\,I)$ would show that the circuit is not ideal, and there is an external loss of voltage.

The **slope**

$$\beta_1 = G(x+1) - G(x)$$

is the predicted change in the response variable when predictor changes by 1. This is a very important parameter that shows how fast we can change the expected response by varying the predictor. For example, customer satisfaction will increase by $\beta_1(\Delta x)$ when the quality of produced computers increases by (Δx).

A zero slope means absence of a linear relationship between X and Y. In this case, Y is expected to stay constant when X changes.

Estimation in linear regression

Let us estimate the slope and intercept by **method of least squares**. Following (11.1), we minimize the sum of squared residuals

$$Q = \sum_{i=1}^{n}(y_i - \widehat{y}_i)^2 = \sum_{i=1}^{n}\left(y_i - \widehat{G}(x_i)\right)^2 = \sum_{i=1}^{n}(y_i - \beta_0 - \beta_1\,x_i)^2\,.$$

We can do it by taking partial derivatives of Q, equating them to 0, and solving the resulting equations for β_0 and β_1.

The partial derivatives are

$$\frac{\partial Q}{\partial \beta_0} = -2\sum_{i=1}^{n}(y_i - \beta_0 - \beta_1\,x_i)\,;$$

$$\frac{\partial Q}{\partial \beta_1} = -2\sum_{i=1}^{n}(y_i - \beta_0 - \beta_1\,x_i)\,x_i.$$

Equating them to 0, we obtain so-called *normal equations*,

$$\begin{cases} \displaystyle\sum_{i=1}^{n}(y_i - \beta_0 - \beta_1\,x_i) & = & 0 \\[2mm] \displaystyle\sum_{i=1}^{n}x_i\,(y_i - \beta_0 - \beta_1\,x_i) & = & 0 \end{cases}$$

From the first normal equation,

$$\beta_0 = \frac{\sum y_i - \beta_1 \sum x_i}{n} = \overline{y} - \beta_1 \overline{x}. \tag{11.2}$$

Substituting this into the second normal equation, we get

$$\sum_{i=1}^{n}x_i\,(y_i - \beta_0 - \beta_1\,x_i) = \sum_{i=1}^{n}x_i\,((y_i - \overline{y}) - \beta_1(x_i - \overline{x})) = S_{xy} - \beta_1 S_{xx} = 0, \tag{11.3}$$

where

$$S_{xx} = \sum_{i=1}^{n} x_i(x_i - \overline{x}) = \sum_{i=1}^{n} (x_i - \overline{x})^2 \qquad (11.4)$$

and

$$S_{xy} = \sum_{i=1}^{n} x_i(y_i - \overline{y}) = \sum_{i=1}^{n} (x_i - \overline{x})(y_i - \overline{y}) \qquad (11.5)$$

are *sums of squares and cross-products*. Notice that it is all right to subtract $\overline{x}$ from x_i in the right-hand sides of (11.4) and (11.5) because $\sum(x_i - \overline{x}) = 0$ and $\sum(y_i - \overline{y}) = 0$.

Finally, we obtain the **least squares estimates** of intercept β_0 and slope β_1 from (11.2) and (11.3).

Regression estimates

$$
\begin{aligned}
b_0 &= \widehat{\beta}_0 = \overline{y} - b_1 \overline{x} \\[2mm]
b_1 &= \widehat{\beta}_1 = S_{xy}/S_{xx} \\[2mm]
&\text{where} \\[1mm]
S_{xx} &= \sum_{i=1}^{n} (x_i - \overline{x})^2 \\
S_{xy} &= \sum_{i=1}^{n} (x_i - \overline{x})(y_i - \overline{y})
\end{aligned}
$$

(11.6)

Example 11.3 (WORLD POPULATION). In Example 11.1, x_i is the year, and y_i is the world population during that year. To estimate the regression line in Figure 11.1, we compute

$$\overline{x} = 1984; \quad \overline{y} = 4843;$$

$$
\begin{aligned}
S_{xx} &= (1950 - \overline{x})^2 + \ldots + (2019 - \overline{x})^2 = 27370; \\
S_{xy} &= (1950 - \overline{x})(2558 - \overline{y}) + \ldots + (2010 - \overline{x})(6864 - \overline{y}) = 2053529.
\end{aligned}
$$

Then

$$
\begin{aligned}
b_1 &= S_{xy}/S_{xx} = 75 \\
b_0 &= \overline{y} - b_1 \overline{x} = -144013.
\end{aligned}
$$

The estimated regression line is

$$\widehat{G}(x) = b_0 + b_1 x = \underline{-144013 + 75\,x}.$$

We conclude that the world population grows at the average rate of 75 million every year.

We can use the obtained equation to predict the future growth of the world population. Regression predictions for years 2020 and 2030 are

$$
\begin{aligned}
\widehat{G}(2020) &= b_0 + 2020\,b_1 = \underline{7544 \text{ million people}} \\
\widehat{G}(2030) &= b_0 + 2030\,b_1 = \underline{8295 \text{ million people}}
\end{aligned}
$$

$\Diamond$

11.1.4 Regression and correlation

Recall from Section 3.3.5 that **covariance**

$$\mathrm{Cov}(X, Y) = \mathbf{E}(X - \mathbf{E}(X))(Y - \mathbf{E}(Y))$$

and **correlation coefficient**

$$\rho = \frac{\mathrm{Cov}(X, Y)}{(\mathrm{Std} X)(\mathrm{Std} Y)}$$

measure the direction and strength of a linear relationship between variables X and Y. From observed data, we estimate $\mathrm{Cov}(X, Y)$ and ρ by the **sample covariance**

$$s_{xy} = \frac{\sum\limits_{i=1}^{n} (x_i - \overline{x})(y_i - \overline{y})}{n - 1}$$

(it is unbiased for the population covariance) and the **sample correlation coefficient**

$$r = \frac{s_{xy}}{s_x s_y}, \tag{11.7}$$

where

$$s_x = \sqrt{\frac{\sum (x_i - \overline{x})^2}{n - 1}} \quad \text{and} \quad s_y = \sqrt{\frac{\sum (y_i - \overline{y})^2}{n - 1}}$$

are sample standard deviations of X and Y.

Comparing (11.3) and (11.7), we see that the estimated slope b_1 and the sample regression coefficient r are proportional to each other. Now we have two new formulas for the regression slope.

Estimated regression slope

$$b_1 = \frac{S_{xy}}{S_{xx}} = \frac{s_{xy}}{s_x^2} = r \left(\frac{s_y}{s_x} \right)$$

Like the correlation coefficient, regression slope is positive for positively correlated X and Y and negative for negatively correlated X and Y. The difference is that r is dimensionless whereas the slope is measured in units of Y per units of X. Thus, its value by itself does not indicate whether the dependence is weak or strong. It depends on the units, the scale of X and Y. We test significance of the regression slope in Section 11.2.

11.1.5 Overfitting a model

Among all possible straight lines, the method of least squares chooses one line that is closest to the observed data. Still, as we see in Figure 11.3b, we did have some residuals $e_i = (y_i - \widehat{y}_i)$ and some positive sum of squared residuals. The straight line has not accounted for all 100% of variation among y_i.

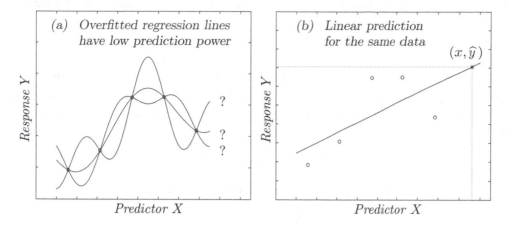

FIGURE 11.4: *Regression-based prediction.*

Why, one might ask, have we considered only linear models? As long as all x_i are different, we can always find a regression function $\widehat{G}(x)$ that passes through all the observed points without any error. Then, the sum $\sum e_i^2 = 0$ will truly be minimized!

Trying to fit the data perfectly is a rather dangerous habit. Although we can achieve an excellent fit to the observed data, it never guarantees a good prediction. The model will be *overfitted*, too much attached to the given data. Using it to predict unobserved responses is very questionable (see Figure 11.4a,b).

11.2 Analysis of variance, prediction, and further inference

In this section, we

- evaluate the *goodness of fit* of the chosen regression model to the observed data,

- estimate the response variance,

- test significance of regression parameters,

- construct confidence and prediction intervals.

11.2.1 ANOVA and R-square

Analysis of variance (ANOVA) explores variation among the observed responses. A portion of this variation can be explained by predictors. The rest is attributed to "error."

For example, there exists some variation among the house sale prices on Figure 11.2. Why are the houses priced differently? Well, the price depends on the house area, and bigger houses tend to be more expensive. So, to some extent, variation among prices is explained by variation among house areas. However, two houses with the same area may still have different prices. These differences cannot be explained by the area.

The total variation among observed responses is measured by the **total sum of squares**

$$SS_{\text{TOT}} = \sum_{i=1}^{n}(y_i - \overline{y})^2 = (n-1)s_y^2.$$

This is the variation of y_i about their sample mean *regardless* of our regression model.

A portion of this total variation is attributed to predictor X and the regression model connecting predictor and response. This portion is measured by the **regression sum of squares**

$$SS_{\text{REG}} = \sum_{i=1}^{n}(\widehat{y}_i - \overline{y})^2.$$

This is the portion of total variation *explained by the model*. It is often computed as

$$
\begin{aligned}
SS_{\text{REG}} &= \sum_{i=1}^{n}(b_0 + b_1 x_i - \overline{y})^2 \\
&= \sum_{i=1}^{n}(\overline{y} - b_1\overline{x} + b_1 x_i - \overline{y})^2 \\
&= \sum_{i=1}^{n}b_1^2(x_i - \overline{x})^2 \\
&= b_1^2 S_{xx} \text{ or } (n-1)b_1^2 s_x^2.
\end{aligned}
$$

The rest of total variation is attributed to "error." It is measured by the **error sum of squares**

$$SS_{\text{ERR}} = \sum_{i=1}^{n}(y_i - \widehat{y}_i)^2 = \sum_{i=1}^{n}e_i^2.$$

This is the portion of total variation *not explained by the model*. It equals the sum of squared residuals that the method of least squares minimizes. Thus, applying this method, we minimize the *error sum of squares*.

Regression and error sums of squares partition SS_{TOT} into two parts (Exercise 11.6),

$$SS_{\text{TOT}} = SS_{\text{REG}} + SS_{\text{ERR}}.$$

The *goodness of fit*, appropriateness of the predictor and the chosen regression model can be judged by the proportion of SS_{TOT} that the model can explain.

DEFINITION 11.3 ————————————

> **R-square**, or **coefficient of determination** is the proportion of the total variation explained by the model,
>
> $$R^2 = \frac{SS_{\text{REG}}}{SS_{\text{TOT}}}.$$
>
> It is always between 0 and 1, with high values generally suggesting a good fit.

In univariate regression, R-square also equals the squared sample correlation coefficient (Exercise 11.7),

$$R^2 = r^2.$$

Example 11.4 (WORLD POPULATION, CONTINUED). Continuing Example 11.3, we find

$$
\begin{aligned}
SS_{\text{TOT}} &= (n-1)s_y^2 = (12)(2.093 \cdot 10^6) = 1.545 \cdot 10^8, \\
SS_{\text{REG}} &= b_1^2 S_{xx} = (74.1)^2(4550) = 1.541 \cdot 10^8, \\
SS_{\text{ERR}} &= SS_{\text{TOT}} - SS_{\text{REG}} = 4.498 \cdot 10^5.
\end{aligned}
$$

A linear model for the growth of the world population has a very high R-square of

$$
R^2 = \frac{SS_{\text{REG}}}{SS_{\text{TOT}}} = \underline{0.997} \text{ or } \underline{99.7\%}.
$$

This is a very good fit although some portion of the remaining 0.3% of total variation can still be explained by adding non-linear terms into the model. $\Diamond$

11.2.2 Tests and confidence intervals

Methods of estimating a regression line and partitioning the total variation do not rely on any distribution; thus, we can apply them to virtually any data.

For further analysis, we introduce **standard regression assumptions**. We will assume that observed responses y_i are independent Normal random variables with mean

$$
\mathbf{E}(Y_i) = \beta_0 + \beta_1 \, x_i
$$

and constant variance σ^2. Predictors x_i are considered *non-random*.

As a consequence, regression estimates b_0 and b_1 have Normal distribution. After we estimate the variance σ^2, they can be studied by T-tests and T-intervals.

Degrees of freedom and variance estimation

According to the standard assumptions, responses $Y_1, \ldots, Y_n$ have different means but the same variance. This variance equals the mean squared deviation of responses from their respective expectations. Let us estimate it.

First, we estimate each expectation $\mathbf{E}(Y_i) = G(x_i)$ by

$$
\widehat{G}(x_i) = b_0 + b_1 x_i = \widehat{y}_i.
$$

Then, we consider deviations $e_i = y_i - \widehat{y}_i$, square them, and add. We obtain the *error sum of squares*

$$
SS_{\text{ERR}} = \sum_{i=1}^{n} e_i^2.
$$

It remains to divide this sum by its number of degrees of freedom (this is how we estimated variances in Section 8.2.4).

Let us compute degrees of freedom for all three SS in the regression ANOVA.

The total sum of squares $SS_{\text{TOT}} = (n-1)s_y^2$ has $\text{df}_{\text{TOT}} = n-1$ degrees of freedom because it is computed directly from the sample variance s_y^2.

Out of them, the regression sum of squares SS_{REG} has $\text{df}_{\text{REG}} = 1$ degree of freedom. Recall (from Section 9.3.4, p. 268) that the number of degrees of is the dimension of the corresponding space. Regression line, which is just a straight line, has dimension 1.

This leaves $\text{df}_{\text{ERR}} = n - 2$ degrees of freedom for the error sum of squares, so that

$$\text{df}_{\text{TOT}} = \text{df}_{\text{REG}} + \text{df}_{\text{ERR}}.$$

The error degrees of freedom also follow from formula (9.10),

$$\text{df}_{\text{ERR}} = \text{sample size} - \frac{\text{number of estimated}}{\text{location parameters}} = n - 2,$$

with 2 degrees of freedom deducted for 2 estimated parameters, β_0 and β_1.

Equipped with this, we now estimate the variance.

Regression variance	$s^2 = \dfrac{SS_{\text{ERR}}}{n - 2}$

It estimates $\sigma^2 = \text{Var}(Y)$ unbiasedly.

Remark: Notice that the usual sample variance

$$s_y^2 = \frac{SS_{\text{TOT}}}{n - 1} = \frac{\sum (y_i - \bar{y})^2}{n - 1}$$

is biased because $\bar{y}$ no longer estimates the expectation of Y_i.

A standard way to present analysis of variance is the *ANOVA table*.

	Source	Sum of squares	Degrees of freedom	Mean squares	F
Univariate ANOVA	Model	SS_{REG} $= \sum (\widehat{y}_i - \bar{y})^2$	1	MS_{REG} $= SS_{\text{REG}}$	$\dfrac{MS_{\text{REG}}}{MS_{\text{ERR}}}$
	Error	SS_{ERR} $= \sum (y_i - \widehat{y}_i)^2$	$n - 2$	MS_{ERR} $= \dfrac{SS_{\text{ERR}}}{n - 2}$	
	Total	SS_{TOT} $= \sum (y_i - \bar{y})^2$	$n - 1$		

Mean squares MS_{REG} and MS_{ERR} are obtained from the corresponding sums of squares dividing them by their degrees of freedom. We see that the sample regression variance is the mean squared error,

$$s^2 = MS_{\text{ERR}}.$$

The estimated standard deviation s is usually called *root mean squared error* or *RMSE*.

The *F-ratio*

$$F = \frac{MS_{\text{REG}}}{MS_{\text{ERR}}}$$

is used to test significance of the entire regression model.

Inference about the regression slope

Having estimated the regression variance σ^2, we can proceed with tests and confidence intervals for the regression slope β_1. As usually, we start with the estimator of β_1 and its sampling distribution.

The slope is estimated by

$$b_1 = \frac{S_{xy}}{S_{xx}} = \frac{\sum(x_i - \bar{x})(y_i - \bar{y})}{S_{xx}} = \frac{\sum(x_i - \bar{x})y_i}{S_{xx}}$$

(we can drop $\bar{y}$ because it is multiplied by $\sum(x_i - \bar{x}) = 0$).

According to *standard regression assumptions*, y_i are Normal random variables and x_i are non-random. Being a linear function of y_i, the estimated slope b_1 is also Normal with the expectation

$$\mathbf{E}(b_1) = \frac{\sum(x_i - \bar{x})\,\mathbf{E}(y_i)}{S_{xx}} = \frac{\sum(x_i - \bar{x})(\beta_0 + \beta_1 x_i)}{S_{xx}} = \frac{\sum(x_i - \bar{x})^2(\beta_1)}{\sum(x_i - \bar{x})^2} = \beta_1,$$

(which shows that b_1 is an *unbiased estimator* of β_1), and the variance

$$\text{Var}(b_1) = \frac{\sum(x_i - \bar{x})^2\,\text{Var}(y_i)}{S_{xx}^2} = \frac{\sum(x_i - \bar{x})^2\sigma^2}{S_{xx}^2} = \frac{\sigma^2}{S_{xx}}.$$

Summarizing the results,

Sampling distribution of a regression slope

b_1 is Normal(μ_b, σ_b),

where

$$\mu_b = \mathbf{E}(b_1) = \beta_1$$

$$\sigma_b = \text{Std}(b_1) = \frac{\sigma}{\sqrt{S_{xx}}}$$

We estimate the standard error of b_1 by

$$s(b_1) = \frac{s}{\sqrt{S_{xx}}},$$

and therefore, use T-intervals and T-tests.

Following the general principles, a $(1 - \alpha)100\%$ **confidence interval** for the slope is

$$\text{Estimator} \pm t_{\alpha/2} \left(\begin{array}{c} \text{estimated} \\ \text{st. deviation} \\ \text{of the estimator} \end{array} \right) = b_1 \pm t_{\alpha/2} \frac{s}{\sqrt{S_{xx}}}.$$

Testing hypotheses $H_0 : \beta_1 = B$ about the regression slope, use the T-statistic

$$t = \frac{b_1 - B}{s(b_1)} = \frac{b_1 - B}{s/\sqrt{S_{xx}}}.$$

P-values, acceptance and rejection regions are computed from Table A5 in the Appendix, T-distribution with $(n - 2)$ degrees of freedom. These are degrees of freedom used in the estimation of σ^2.

As always, the form of the alternative hypothesis determines whether it is a two-sided, right-tail, or left-tail test.

A non-zero slope indicates significance of the model, relevance of predictor X in the inference about response Y, and existence of a linear relation among them. It means that a change in X causes changes in Y. In the absence of such relation, $\mathbf{E}(Y) = \beta_0$ remains constant.

To see if X is significant for the prediction of Y, test the null hypothesis

$$H_0 : \beta_1 = 0 \quad \text{vs} \quad H_a : \beta_1 \neq 0.$$

ANOVA F-test

A more universal, and therefore, more popular method of testing significance of a model is the **ANOVA F-test**. It compares the portion of variation explained by regression with the portion that remains unexplained. Significant models explain a relatively large portion.

Each portion of the total variation is measured by the corresponding *sum of squares*, SS_{REG} for the explained portion and SS_{ERR} for the unexplained portion (error). Dividing each SS by the number of degrees of freedom, we obtain **mean squares**,

$$MS_{\text{REG}} = \frac{SS_{\text{REG}}}{\text{df}_{\text{REG}}} = \frac{SS_{\text{REG}}}{1} = SS_{\text{REG}}$$

and

$$MS_{\text{ERR}} = \frac{SS_{\text{ERR}}}{\text{df}_{\text{ERR}}} = \frac{SS_{\text{ERR}}}{n - 2} = s^2.$$

Under the null hypothesis

$$H_0 : \beta_1 = 0,$$

both mean squares, MS_{REG} and MS_{ERR} are independent, and their ratio

$$F = \frac{MSR}{MSE} = \frac{SSR}{s^2}$$

has **F-distribution** with $\text{df}_{\text{REG}} = 1$ and $\text{df}_{\text{ERR}} = n - 2$ degrees of freedom (d.f.).

As we discovered in Section 9.5.4, this F-distribution has two parameters, numerator d.f. and denominator d.f., and it is very popular for testing ratios of variances and significance of models. Its critical values for the most popular significance levels between $\alpha = 0.001$ and $\alpha = 0.25$ are tabulated in Table A7.

ANOVA F-test is always *one-sided* and *right-tail* because only large values of the F-statistic show a large portion of explained variation and the overall significance of the model.

F-test and T-test

We now have two tests for the model significance, a T-test for the regression slope and the ANOVA F-test. For the univariate regression, they are absolutely equivalent. In fact, the F-statistic equals the squared T-statistic for testing $H_0 : \beta_1 = 0$ because

$$t^2 = \frac{b_1^2}{s^2/S_{xx}} = \frac{(S_{xy}/S_{xx})^2}{s^2/S_{xx}} = \frac{S_{xy}^2}{S_{xx}S_{yy}} \frac{S_{yy}}{s^2} = \frac{r^2 SS_{\text{TOT}}}{s^2} = \frac{SS_{\text{REG}}}{s^2} = F.$$

Hence, both tests give us the same result.

Example 11.5 (ASSUMPTION OF INDEPENDENCE). Can we apply the introduced methods to Examples 11.1–11.3? For the world population data in Example 11.1, the sample correlation coefficient between residuals e_i and e_{i-1} is 0.9915, which is very high. Hence, we cannot assume independence of y_i, and one of the standard assumptions is violated.

Our least squares regression line is still correct; however, in order to proceed with tests and confidence intervals, we need more advanced *time series* methods accounting not only for the variance but also for covariances among the observed responses.

For the house prices in Example 11.2, there is no evidence of any dependence. These 70 houses are sampled at random, and they are likely to be priced independently of each other.

$\Diamond$

Remark: Notice that we used residuals $e_i = y_i - \widehat{y}_i$ for the correlation study. Indeed, according to our regression model, responses y_i have different expected values, so their sample mean $\overline{y}$ does not estimate the population mean of any of them; therefore, the sample correlation coefficient based on that mean is misleading. On the other hand, if the linear regression model is correct, all *residuals* have the same mean $\mathbf{E}(e_i) = 0$. In the population, the difference between y_i and ε_i is non-random, $y_i - \varepsilon_i = G(x_i)$; therefore, the population correlation coefficients between y_i and y_j and between ε_i and ε_j are the same.

Example 11.6 (EFFICIENCY OF COMPUTER PROGRAMS). A computer manager needs to know how efficiency of her new computer program depends on the size of incoming data. Efficiency will be measured by the number of processed requests per hour. Applying the program to data sets of different sizes, she gets the following results,

Data size (gigabytes), x	6	7	7	8	10	10	15
Processed requests, y	40	55	50	41	17	26	16

In general, larger data sets require more computer time, and therefore, fewer requests are processed within 1 hour. The response variable here is the number of processed requests (y), and we attempt to predict it from the size of a data set (x).

(a) ESTIMATION OF THE REGRESSION LINE. We can start by computing

$$n = 7, \ \overline{x} = 9, \ \overline{y} = 35, \ S_{xx} = 56, \ S_{xy} = -232, \ S_{yy} = 1452.$$

Estimate regression slope and intercept by

$$b_1 = \frac{S_{xy}}{S_{xx}} = -4.14 \quad \text{and} \quad b_0 = \overline{y} - b_1\overline{x} = 72.3.$$

Then, the estimated regression line has an equation

$$y = 72.3 - 4.14x.$$

Notice the negative slope. It means that *increasing* incoming data sets by 1 gigabyte, we expect to process 4.14 *fewer* requests per hour.

(b) ANOVA TABLE AND VARIANCE ESTIMATION. Let us compute all components of the ANOVA. We have

$$SS_{\text{TOT}} = S_{yy} = 1452$$

partitioned into

$$SS_{\text{REG}} = b_1^2 S_{xx} = 961 \quad \text{and} \quad SS_{\text{ERR}} = SS_{\text{TOT}} - SS_{\text{REG}} = 491.$$

Simultaneously, $n - 1 = 6$ degrees of freedom of SS_{TOT} are partitioned into $df_{\text{REG}} = 1$ and $df_{\text{ERR}} = 5$ degrees of freedom.

Fill the rest of the ANOVA table,

Source	Sum of squares	Degrees of freedom	Mean squares	F
Model	961	1	961	9.79
Error	491	5	98.2	
Total	1452	6		

REGRESSION VARIANCE σ^2 is estimated by

$$s^2 = MS_{\text{ERR}} = 98.2.$$

R-SQUARE is

$$R^2 = \frac{SS_{\text{REG}}}{SS_{\text{TOT}}} = \frac{961}{1452} = 0.662.$$

That is, 66.2% of the total variation of the number of processed requests is explained by sizes of data sets only.

(c) INFERENCE ABOUT THE SLOPE. Is the slope statistically significant? Does the number of processed requests really depend on the size of data sets? To test the null hypothesis $H_0 : \beta_1 = 0$, compute the T-statistic

$$t = \frac{b_1}{\sqrt{s^2/S_{xx}}} = \frac{-4.14}{\sqrt{98.2/56}} = -3.13.$$

Checking the T-distribution table (Table A5) with 5 d.f., we find that the P-value for the *two-sided* test is between 0.02 and 0.04. We conclude that the slope is *moderately significant*. Precisely, it is significant at any level $\alpha \geq 0.04$ and not significant at any $\alpha \leq 0.02$.

(d) ANOVA F-TEST. A similar result is suggested by the F-test. From Table A7, the F-statistic of 9.79 is not significant at the 0.025 level, but significant at the 0.05 level.

$\Diamond$

11.2.3 Prediction

One of the main applications of regression analysis is making forecasts, predictions of the response variable Y based on the known or controlled predictors X.

Let x_* be the value of the predictor X. The corresponding value of the response Y is computed by evaluating the estimated regression line at x_*,

$$\widehat{y}_* = \widehat{G}(x_*) = b_0 + b_1 x_*.$$

This is how we predicted the world population in years 2020 and 2030 in Example 11.3. As it happens with any forecast, our predicted values are understood as the most intelligent guesses, and not as guaranteed exact sizes of the population during these years.

How reliable are regression predictions, and how close are they to the real true values? As a good answer, we can construct

- a $(1 - \alpha)100\%$ **confidence interval** for the expectation

$$\mu_* = \mathbf{E}(Y \mid X = x_*)$$

and

- a $(1 - \alpha)100\%$ **prediction interval** for the actual value of $Y = y_*$ when $X = x_*$.

Confidence interval for the mean of responses

The expectation
$$\mu_* = \mathbf{E}(Y \mid X = x_*) = G(x_*) = \beta_0 + \beta_1 x_*$$

is a population parameter. This is the mean response for the entire subpopulation of units where the independent variable X equals x_*. For example, it corresponds to the average price of all houses with the area $x_* = 2500$ square feet.

First, we estimate μ_* by

$$
\begin{aligned}
\widehat{y}_* &= b_0 + b_1 x_* \\
&= \bar{y} - b_1 \bar{x} + b_1 x_* \\
&= \bar{y} + b_1 (x_* - \bar{x}) \\
&= \frac{1}{n} \sum y_i + \frac{\sum (x_i - \bar{x}) y_i}{S_{xx}} (x_* - \bar{x}) \\
&= \sum_{i=1}^{n} \left(\frac{1}{n} + \frac{\sum (x_i - \bar{x})}{S_{xx}} (x_* - \bar{x}) \right) y_i.
\end{aligned}
$$

We see again that the estimator is a linear function of responses y_i. Then, under standard regression assumptions, $\widehat{y}_*$ is Normal, with expectation

$$\mathbf{E}(\widehat{y}_*) = \mathbf{E}b_0 + \mathbf{E}b_1 x_* = \beta_0 + \beta_1 x_* = \mu_*$$

(it is unbiased), and variance

$$
\begin{aligned}
\text{Var}(\widehat{y}_*) &= \sum \left(\frac{1}{n} + \frac{\sum(x_i - \overline{x})}{S_{xx}}(x_* - \overline{x}) \right)^2 \text{Var}(y_i) \\
&= \sigma^2 \left(\sum_{i=1}^{n} \frac{1}{n^2} + 2\sum_{i=1}^{n}(x_i - \overline{x})\frac{x_* - \overline{x}}{S_{xx}} + \frac{S_{xx}(x_* - \overline{x})^2}{S_{xx}^2} \right) \\
&= \sigma^2 \left(\frac{1}{n} + \frac{(x_* - \overline{x})^2}{S_{xx}} \right)
\end{aligned}
$$ (11.8)

(because $\sum(x_i - \overline{x}) = 0$).

Then, we estimate the regression variance σ^2 by s^2 and obtain the following confidence interval.

$(1 - \alpha)100\%$ **confidence interval for the mean** $\mu_* = \mathbf{E}(Y \mid X = x_*)$ **of all responses with** $X = x_*$

$$
b_0 + b_1 x_* \pm t_{\alpha/2} \, s \sqrt{\frac{1}{n} + \frac{(x_* - \overline{x})^2}{S_{xx}}}
$$

Prediction interval for the individual response

Often we are more interested in predicting the actual response rather than the mean of all possible responses. For example, we may be interested in the price of one particular house that we are planning to buy, not in the average price of all similar houses.

Instead of estimating a *population parameter*, we are now predicting the *actual value* of a random variable.

DEFINITION 11.4 ⎯⎯⎯⎯⎯⎯⎯

> An interval $[a, b]$ is a $(1 - \alpha)100\%$ **prediction interval** for the individual response Y corresponding to predictor $X = x_*$ if it contains the value of Y with probability $(1 - \alpha)$,
>
> $$ P\{a \leq Y \leq b \mid X = x_*\} = 1 - \alpha. $$

This time, all three quantities, Y, a, and b, are random variables. Predicting Y by $\widehat{y}_*$, estimating the standard deviation

$$
\text{Std}(Y - \widehat{y}_*) = \sqrt{\text{Var}(Y) + \text{Var}(\widehat{y}_*)} = \sigma\sqrt{1 + \frac{1}{n} + \frac{(x_* - \overline{x})^2}{S_{xx}}}
$$ (11.9)

by

$$
\widehat{\text{Std}}(Y - \widehat{y}_*) = s\sqrt{1 + \frac{1}{n} + \frac{(x_* - \overline{x})^2}{S_{xx}}},
$$

and standardizing all three parts of the inequality

$$a \leq Y \leq b,$$

we realize that the $(1 - \alpha)100\%$ prediction interval for Y has to satisfy the equation

$$P\left\{ \frac{a - \widehat{y}_*}{\widehat{\text{Std}}(Y - \widehat{y}_*)} \leq \frac{Y - \widehat{y}_*}{\widehat{\text{Std}}(Y - \widehat{y}_*)} \leq \frac{b - \widehat{y}_*}{\widehat{\text{Std}}(Y - \widehat{y}_*)} \right\} = 1 - \alpha.$$

At the same time, the properly standardized $(Y - \widehat{y}_*)$ has T-distribution, and

$$P\left\{ -t_{\alpha/2} \leq \frac{Y - \widehat{y}_*}{\widehat{\text{Std}}(Y - \widehat{y}_*)} \leq t_{\alpha/2} \right\} = 1 - \alpha.$$

A prediction interval is now computed by solving equations

$$\frac{a - \widehat{y}_*}{\widehat{\text{Std}}(Y - \widehat{y}_*)} = -t_{\alpha/2} \quad \text{and} \quad \frac{b - \widehat{y}_*}{\widehat{\text{Std}}(Y - \widehat{y}_*)} = t_{\alpha/2}$$

in terms of a and b.

$(1 - \alpha)100\%$ prediction interval for the individual response Y when $X = x_*$	$b_0 + b_1 x_* \pm t_{\alpha/2}\, s \sqrt{1 + \dfrac{1}{n} + \dfrac{(x_* - \overline{x})^2}{S_{xx}}}$	(11.10)

Several conclusions are apparent from this.

First, compare the standard deviations in (11.8) and (11.9). Response Y that we are predicting made its contribution into the variance. This is the difference between a confidence interval for the mean of all responses and a prediction interval for the individual response. Predicting the individual value is a more difficult task; therefore, the prediction interval is always wider than the confidence interval for the mean response. More uncertainty is involved, and as a result, the margin of a prediction interval is larger than the margin of a confidence interval.

Second, we get more accurate estimates and more accurate predictions from large samples. When the sample size n (and therefore, typically, S_{xx}), tends to ∞, the margin of the confidence interval converges to 0.

On the other hand, the margin of a prediction interval converges to $(t_{\alpha/2}\sigma)$. As we collect more and more observations, our estimates of b_0 and b_1 become more accurate; however, uncertainty about the individual response Y will never vanish.

Third, we see that regression estimation and prediction are most accurate when x_* is close to $\overline{x}$ so that

$$(x_* - \overline{x})^2 \approx 0.$$

The margin increases as the independent variable x_* drifts away from $\overline{x}$. We conclude that it is easiest to make forecasts under normal and "standard" conditions, and it is hardest to predict anomalies. And this agrees with our common sense.

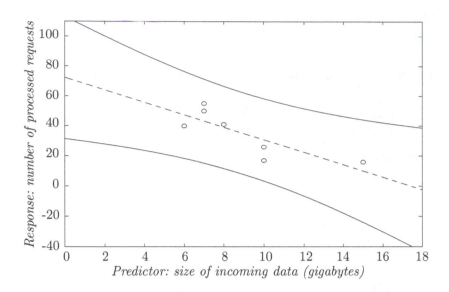

FIGURE 11.5: *Regression prediction of program efficiency.*

Example 11.7 (PREDICTING THE PROGRAM EFFICIENCY). Suppose we need to start processing requests that refer to $x_* = 16$ gigabytes of data. Based on our regression analysis of the program efficiency in Example 11.6, we predict

$$y_* = b_0 + b_1 x_* = 72.3 - 4.14(16) = 6$$

requests processed within 1 hour. A 95% prediction interval for the number of processed requests is

$$
\begin{aligned}
y_* \pm t_{0.025}\, s \sqrt{1 + \frac{1}{n} + \frac{(x_* - \bar{x})^2}{S_{xx}}} \;&=\; 6 \pm (2.571)\sqrt{98.2}\sqrt{1 + \frac{1}{7} + \frac{(16-9)^2}{56}} \\
&=\; 6 \pm 36.2 = [0; 42].
\end{aligned}
$$

(using Table A5 with 5 d.f.). We rounded both ends of the prediction interval knowing that there cannot be a negative or fractional number of requests. $\Diamond$

Prediction bands

For all possible values of a predictor x_*, we can prepare a graph of $(1 - \alpha)$ **prediction bands** given by (11.10). Then, for each value of x_*, one can draw a vertical line and obtain a $100(1 - \alpha)\%$ prediction interval between these bands.

Figure 11.5 shows the 95% prediction bands for the number of processed requests in Example 11.7. These are two curves on each side of the fitted regression line. As we have already noticed, prediction is most accurate when x_* is near the sample mean $\bar{x}$. Prediction intervals get wider when we move away from $\bar{x}$.

11.3 Multivariate regression

In the previous two sections, we learned how to predict a response variable Y from a predictor variable X. We hoped in several examples that including more information and using multiple predictors instead of one will enhance our prediction.

Now we introduce **multiple linear regression** that will connect a response Y with several predictors $X^{(1)}$, $X^{(2)}$, ..., $X^{(k)}$.

11.3.1 Introduction and examples

Example 11.8 (ADDITIONAL INFORMATION). In Example 11.2, we discussed predicting price of a house based on its area. We decided that perhaps this prediction is not very accurate due to a high variability among house prices.

What is the source of this variability? Why are houses of the same size priced differently?

Certainly, area is not the only important parameter of a house. Prices are different due to different design, location, number of rooms and bathrooms, presence of a basement, a garage, a swimming pool, different size of a backyard, etc. When we take all this information into account, we'll have a rather accurate description of a house and hopefully, a rather accurate prediction of its price. ◊

Example 11.9 (U.S. POPULATION AND NONLINEAR TERMS). One can often reduce variability around the trend and do more accurate analysis by adding nonlinear terms into the regression model. In Example 11.3, we predicted the world population for years 2020–2030 based on the *linear model*

$$\mathbf{E}(\text{population}) = \beta_0 + \beta_1(\text{year}).$$

We showed in Example 11.4 that this model has a pretty good fit.

However, a linear model does a poor prediction of the U.S. population between 1790 and 2010 (see Figure 11.6a). The population growth over a longer period of time is clearly nonlinear.

On the other hand, a *quadratic model* in Figure 11.6b gives an amazingly excellent fit! It seems to account for everything except a temporary decrease in the rate of growth during the World War II (1939–1945).

For this model, we assume

$$\mathbf{E}(\text{population}) = \beta_0 + \beta_1(\text{year}) + \beta_2(\text{year})^2,$$

or in a more convenient but equivalent form,

$$\mathbf{E}(\text{population}) = \beta_0 + \beta_1(\text{year-1800}) + \beta_2(\text{year-1800})^2.$$

◊

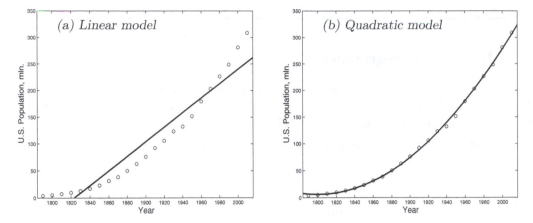

FIGURE 11.6: *U.S. population in 1790–2010 (million people).*

A **multivariate linear regression model** assumes that the conditional expectation of a response

$$\mathbf{E}\left\{Y \mid X^{(1)} = x^{(1)}, \ldots, X^{(k)} = x^{(k)}\right\} = \beta_0 + \beta_1 x^{(1)} + \ldots + \beta_k x^{(k)} \qquad (11.11)$$

is a linear function of predictors $x^{(1)}, \ldots, x^{(k)}$.

This regression model has one intercept and a total of k slopes, and therefore, it defines a k-dimensional *regression plane* in a $(k+1)$-dimensional space of $(X^{(1)}, \ldots, X^{(k)}, Y)$.

The **intercept** β_0 is the expected response when all predictors equal zero.

Each **regression slope** β_j is the expected change of the response Y when the corresponding predictor $X^{(j)}$ changes by 1 *while all the other predictors remain constant.*

In order to estimate all the parameters of model (11.11), we collect a sample of n *multivariate observations*

$$\begin{cases} \boldsymbol{X}_1 &= \left(X_1^{(1)}, X_1^{(2)}, \ldots, X_1^{(k)}\right) \\ \boldsymbol{X}_2 &= \left(X_2^{(1)}, X_2^{(2)}, \ldots, X_2^{(k)}\right) \\ \vdots \quad \vdots & \qquad\quad \vdots \\ \boldsymbol{X}_n &= \left(X_n^{(1)}, X_n^{(2)}, \ldots, X_n^{(k)}\right) \end{cases}.$$

Essentially, we collect a sample of n units (say, houses) and measure all k predictors on each unit (area, number of rooms, etc.). Also, we measure responses, $Y_1, \ldots, Y_n$. We then estimate $\beta_0, \beta_1, \ldots, \beta_k$ by the method of least squares, generalizing it from the univariate case of Section 11.1 to multivariate regression.

11.3.2   Matrix approach and least squares estimation

According to the *method of least squares*, we find such slopes $\beta_1, \ldots, \beta_k$ and such an intercept β_0 that will minimize the sum of squared "errors"

$$Q = \sum_{i=1}^{n} (y_i - \widehat{y}_i)^2 = \sum_{i=1}^{n} \left(y_i - \beta_0 - \beta_1 x_i^{(1)} - \ldots - \beta_k x_i^{(k)}\right)^2 .$$

Minimizing Q, we can again take partial derivatives of Q with respect to all the unknown parameters and solve the resulting system of equations. It can be conveniently written in a *matrix form* (which requires basic knowledge of linear algebra; if needed, refer to Appendix, Section A.5).

Matrix approach to multivariate linear regression

We start with the data. Observed are an $n \times 1$ response vector $\boldsymbol{Y}$ and an $n \times (k+1)$ predictor matrix $\boldsymbol{X}$,

$$\boldsymbol{Y} = \begin{pmatrix} Y_1 \\ \vdots \\ Y_n \end{pmatrix} \quad \text{and} \quad \boldsymbol{X} = \begin{pmatrix} 1 & \boldsymbol{X}_1 \\ \vdots & \vdots \\ 1 & \boldsymbol{X}_n \end{pmatrix} = \begin{pmatrix} 1 & X_1^{(1)} & \cdots & X_1^{(k)} \\ \vdots & \vdots & \vdots & \vdots \\ 1 & X_n^{(1)} & \cdots & X_n^{(k)} \end{pmatrix}.$$

It is convenient to augment the predictor matrix with a column of 1's because now the multivariate regression model (11.11) can be written as

$$\mathbf{E} \begin{pmatrix} Y_1 \\ \vdots \\ Y_n \end{pmatrix} = \begin{pmatrix} 1 & X_1^{(1)} & \cdots & X_1^{(k)} \\ \vdots & \vdots & \vdots & \vdots \\ 1 & X_n^{(1)} & \cdots & X_n^{(k)} \end{pmatrix} \begin{pmatrix} \beta_0 \\ \beta_1 \\ \vdots \\ \beta_k \end{pmatrix},$$

or simply

$$\mathbf{E}(\boldsymbol{Y}) = \boldsymbol{X}\boldsymbol{\beta}.$$

Now the multidimensional parameter

$$\boldsymbol{\beta} = \begin{pmatrix} \beta_0 \\ \beta_1 \\ \vdots \\ \beta_k \end{pmatrix} \in \mathbb{R}^{k+1}$$

includes the intercept and all the slopes. In fact, the intercept β_0 can also be treated as one of the slopes that corresponds to the added column of 1's.

Our goal is to estimate $\boldsymbol{\beta}$ with a vector of **sample regression slopes**

$$\boldsymbol{b} = \begin{pmatrix} b_0 \\ b_1 \\ \vdots \\ b_k \end{pmatrix}.$$

Fitted values will then be computed as

$$\widehat{\boldsymbol{y}} = \begin{pmatrix} \widehat{y}_1 \\ \vdots \\ \widehat{y}_n \end{pmatrix} = \boldsymbol{X}\boldsymbol{b}.$$

Thus, the least squares problem reduces to minimizing

$$\begin{aligned} Q(\boldsymbol{b}) &= \sum_{i=1}^{n} (y_i - \widehat{y}_i)^2 = (\boldsymbol{y} - \widehat{\boldsymbol{y}})^T (\boldsymbol{y} - \widehat{\boldsymbol{y}}) \\ &= (\boldsymbol{y} - \boldsymbol{X}\boldsymbol{b})^T (\boldsymbol{y} - \boldsymbol{X}\boldsymbol{b}). \end{aligned} \tag{11.12}$$

with T denoting a transposed vector.

Least squares estimates

In the matrix form, the minimum of the sum of squares

$$Q(b) = (y - Xb)^T(y - Xb) = b^T(X^TX)b - 2y^TXb + y^Ty$$

is attained by

Estimated slopes in multivariate regression	$b = (X^TX)^{-1}X^Ty$

As we can see from this formula, all the estimated slopes are

– *linear* functions of observed responses $(y_1, \ldots, y_n)$,

– *unbiased* for the regression slopes because

$$\mathbf{E}(b) = (X^TX)^{-1}X^T\,\mathbf{E}(y) = (X^TX)^{-1}X^TX\beta = \beta,$$

– *Normal* if the response variable Y is Normal.

This is a multivariate analogue of $b = S_{xy}/S_{xx}$ that we derived for the univariate case.

11.3.3   Analysis of variance, tests, and prediction

We can again partition the *total sum of squares* measuring the total variation of responses into the *regression sum of squares* and the *error sum of squares*.

The **total sum of squares** is still

$$SS_{\text{TOT}} = \sum_{i=1}^{n}(y_i - \overline{y})^2 = (y - \overline{y})^T(y - \overline{y}),$$

with $\text{df}_{\text{TOT}} = (n - 1)$ degrees of freedom, where

$$\overline{y} = \begin{pmatrix} \overline{y} \\ \vdots \\ \overline{y} \end{pmatrix} = \overline{y}\begin{pmatrix} 1 \\ \vdots \\ 1 \end{pmatrix}.$$

Again, $SS_{\text{TOT}} = SS_{\text{REG}} + SS_{\text{ERR}}$, where

$$SS_{\text{REG}} = \sum_{i=1}^{n}(\widehat{y}_i - \overline{y})^2 = (\widehat{y} - \overline{y})^T(\widehat{y} - \overline{y})$$

is the **regression sum of squares**, and

$$SS_{\text{ERR}} = \sum_{i=1}^{n}(y_i - \widehat{y}_i)^2 = (y - \widehat{y})^T(y - \widehat{y}) = e^Te$$

is the **error sum of squares**, the quantity that we minimized when we applied the method of least squares.

The multivariate regression model (11.11) defines a k-dimensional regression plane where the fitted values belong to. Therefore, the regression sum of squares has

$$\text{df}_{\text{REG}} = k$$

degrees of freedom, whereas by subtraction,

$$\text{df}_{\text{ERR}} = \text{df}_{\text{TOT}} - \text{df}_{\text{REG}} = n - k - 1$$

degrees of freedom are left for SS_{ERR}. This is again the sample size n minus k estimated slopes and 1 estimated intercept.

We can then write the ANOVA table,

Multivariate ANOVA

Source	Sum of squares	Degrees of freedom	Mean squares	F
Model	SS_{REG} $= (\widehat{\boldsymbol{y}} - \overline{\boldsymbol{y}})^T (\widehat{\boldsymbol{y}} - \overline{\boldsymbol{y}})$	k	MS_{REG} $= \dfrac{SS_{\text{REG}}}{k}$	$\dfrac{MS_{\text{REG}}}{MS_{\text{ERR}}}$
Error	SS_{ERR} $= (\boldsymbol{y} - \widehat{\boldsymbol{y}})^T (\boldsymbol{y} - \widehat{\boldsymbol{y}})$	$n-k-1$	MS_{ERR} $= \dfrac{SS_{\text{ERR}}}{n-k-1}$	
Total	SS_{TOT} $= (\boldsymbol{y} - \overline{\boldsymbol{y}})^T (\boldsymbol{y} - \overline{\boldsymbol{y}})$	$n-1$		

The **coefficient of determination**

$$R^2 = \frac{SS_{\text{REG}}}{SS_{\text{TOT}}}$$

again measures the proportion of the total variation explained by regression, just like it did in the univariate case. When we add new predictors to our model, we explain additional portions of SS_{TOT}; therefore, R^2 can only go up. Thus, we should expect to increase R^2 and generally, get a better fit by going from univariate to multivariate regression.

Testing significance of the entire model

Further inference requires **standard multivariate regression assumptions** of Y_i being independent Normal random variables with means

$$\mathbf{E}(Y_i) = \beta_0 + \beta_1 X_i^{(1)} + \ldots + \beta_k X_i^{(k)}$$

and constant variance σ^2 while all predictors $X_i^{(j)}$ are non-random.

ANOVA F-test in multivariate regression tests significance of the entire model. The model is significant as long as at least one slope is not zero. Thus, we are testing

$$H_0: \ \beta_1 = \ldots = \beta_k = 0 \quad \text{vs} \quad H_A: \ \text{not } H_0; \text{ at least one } \beta_j \neq 0.$$

We compute the F-statistic

$$F = \frac{MS_{\text{REG}}}{MS_{\text{ERR}}} = \frac{SS_{\text{REG}}/k}{SS_{\text{ERR}}/(n-k-1)}$$

and check it against the F-distribution with k and $(n-k-1)$ degrees of freedom in Table A7.

This is always a one-sided right-tail test. Only large values of F correspond to large SS_{REG} indicating that fitted values $\hat{y}_i$ are far from the overall mean $\bar{y}$, and therefore, the expected response really changes along the regression plane according to predictors.

Variance estimator

Regression variance $\sigma^2 = \text{Var}(Y)$ is then estimated by the mean squared error

$$s^2 = MS_{\text{ERR}} = \frac{SS_{\text{ERR}}}{n-k-1}.$$

It is an unbiased estimator of σ^2 that can be used in further inference.

Testing individual slopes

For the inference about **individual regression slopes** β_j, we compute all the variances $\text{Var}(\beta_j)$. Matrix

$$\text{VAR}(\boldsymbol{b}) = \begin{pmatrix} \text{Var}(b_1) & \text{Cov}(b_1, b_2) & \cdots & \text{Cov}(b_1, b_k) \\ \text{Cov}(b_2, b_1) & \text{Var}(b_2) & \cdots & \text{Cov}(b_2, b_k) \\ \vdots & \vdots & \vdots & \vdots \\ \text{Cov}(b_k, b_1) & \text{Cov}(b_k, b_2) & \cdots & \text{Var}(b_k) \end{pmatrix}$$

is called a **variance-covariance matrix** of a vector $\boldsymbol{b}$. It equals

$$\begin{aligned} \text{VAR}(\boldsymbol{b}) &= \text{VAR}\left((\boldsymbol{X}^T\boldsymbol{X})^{-1}\boldsymbol{X}^T\boldsymbol{y}\right) \\ &= (\boldsymbol{X}^T\boldsymbol{X})^{-1}\boldsymbol{X}^T\,\text{VAR}(\boldsymbol{y})\boldsymbol{X}(\boldsymbol{X}^T\boldsymbol{X})^{-1} \\ &= \sigma^2(\boldsymbol{X}^T\boldsymbol{X})^{-1}\boldsymbol{X}^T\boldsymbol{X}(\boldsymbol{X}^T\boldsymbol{X})^{-1} \\ &= \sigma^2(\boldsymbol{X}^T\boldsymbol{X})^{-1}. \end{aligned}$$

Diagonal elements of this $k \times k$ matrix are variances of individual regression slopes,

$$\sigma^2(b_1) = \sigma^2(\boldsymbol{X}^T\boldsymbol{X})^{-1}_{11}, \ldots, \ \sigma^2(b_k) = \sigma^2(\boldsymbol{X}^T\boldsymbol{X})^{-1}_{kk}.$$

We estimate them by sample variances,

$$s^2(b_1) = s^2(\boldsymbol{X}^T\boldsymbol{X})^{-1}_{11}, \ldots, \; s^2(b_k) = s^2(\boldsymbol{X}^T\boldsymbol{X})^{-1}_{kk}.$$

Now we are ready for the inference about individual slopes. Hypothesis

$$H_0: \; \beta_j = B$$

can be tested with a T-statistic

$$t = \frac{b_j - B}{s(b_j)}.$$

Compare this T-statistic against the T-distribution with $\mathrm{df}_{\mathrm{ERR}} = n - k - 1$ degrees of freedom, Table A5. This test may be two-sided or one-sided, depending on the alternative.

A test of

$$H_0: \; \beta_j = 0 \quad \text{vs} \;\; H_A: \; \beta_j \neq 0$$

shows whether predictor $X^{(j)}$ is relevant for the prediction of Y. If the alternative is true, the expected response

$$\mathbf{E}(Y) = \beta_0 + \beta_1 X^{(1)} + \ldots + \beta_j X^{(j)} + \ldots + \beta_k X^{(k)}$$

changes depending on $X^{(j)}$ even if all the other predictors remain constant.

Prediction

For the given vector of predictors $\boldsymbol{X}_* = (X^{(1)}_* = x^{(1)}_*, \ldots, X^{(k)}_* = x^{(k)}_*)$, we estimate the expected response by

$$\widehat{y}_* = \widehat{\mathbf{E}}\{Y \mid \boldsymbol{X}_* = \boldsymbol{x}_*\} = \boldsymbol{x}_* \boldsymbol{b}$$

and predict the individual response by the same statistic.

To produce confidence and prediction intervals, we compute the variance,

$$\mathrm{Var}(\widehat{y}_*) = \mathrm{Var}(\boldsymbol{x}_*\boldsymbol{b}) = \boldsymbol{x}_*^T \, \mathrm{Var}(\boldsymbol{b})\boldsymbol{x}_* = \sigma^2 \boldsymbol{x}_*^T(\boldsymbol{X}^T\boldsymbol{X})^{-1}\boldsymbol{x}_*,$$

where $\boldsymbol{X}$ is the matrix of predictors used to estimate the regression slope $\boldsymbol{\beta}$.

Estimating σ^2 by s^2, we obtain a $(1-\alpha)100\%$ **confidence interval** for $\mu_* = \mathbf{E}(Y)$.

$(1-\alpha)100\%$ confidence interval for the mean $\mu_* = \mathbf{E}(Y \mid \boldsymbol{X}_* = \boldsymbol{x}_*)$ **of all responses with** $X_* = x_*$	$\boldsymbol{x}_*\boldsymbol{b} \pm t_{\alpha/2}\, s\, \sqrt{\boldsymbol{x}_*^T(\boldsymbol{X}^T\boldsymbol{X})^{-1}\boldsymbol{x}_*}$

Accounting for the additional variation of the individual response y_*, we get a $(1-\alpha)100\%$ **prediction interval** for y_*.

$(1-\alpha)100\%$ prediction interval for the individual response Y when $\boldsymbol{X}_* = \boldsymbol{x}_*$	$\boldsymbol{x}_* \boldsymbol{b} \pm t_{\alpha/2}\, s\, \sqrt{1 + \boldsymbol{x}_*^T (\boldsymbol{X}^T \boldsymbol{X})^{-1} \boldsymbol{x}_*}$

In both expressions, $t_{\alpha/2}$ refers to the T-distribution with $(n - k - 1)$ degrees of freedom.

Example 11.10 (DATABASE STRUCTURE). The computer manager in Examples 11.6 and 11.7 tries to improve the model by adding another predictor. She decides that in addition to the size of data sets, efficiency of the program may depend on the database structure. In particular, it may be important to know how many tables were used to arrange each data set. Putting all this information together, we have

Data size (gigabytes), x_1	6	7	7	8	10	10	15
Number of tables, x_2	4	20	20	10	10	2	1
Processed requests, y	40	55	50	41	17	26	16

(a) LEAST SQUARES ESTIMATION. The *predictor matrix* and the *response vector* are

$$
\boldsymbol{X} = \begin{pmatrix} 1 & 6 & 4 \\ 1 & 7 & 20 \\ 1 & 7 & 20 \\ 1 & 8 & 10 \\ 1 & 10 & 10 \\ 1 & 10 & 2 \\ 1 & 15 & 1 \end{pmatrix}, \qquad \boldsymbol{Y} = \begin{pmatrix} 40 \\ 55 \\ 50 \\ 41 \\ 17 \\ 26 \\ 16 \end{pmatrix}.
$$

We then compute

$$
\boldsymbol{X}^T \boldsymbol{X} = \begin{pmatrix} 7 & 63 & 67 \\ 63 & 623 & 519 \\ 67 & 519 & 1021 \end{pmatrix} \quad \text{and} \quad \boldsymbol{X}^T \boldsymbol{Y} = \begin{pmatrix} 245 \\ 1973 \\ 2908 \end{pmatrix},
$$

to obtain the estimated *vector of slopes*

$$
\boldsymbol{b} = (\boldsymbol{X}^T \boldsymbol{X})^{-1} (\boldsymbol{X}^T \boldsymbol{Y}) = \begin{pmatrix} 52.7 \\ -2.87 \\ 0.85 \end{pmatrix}.
$$

Thus, the regression equation is

$$
y = 52.7 - 2.87 x_1 + 0.85 x_2,
$$

or

$$
\begin{pmatrix} \text{number of} \\ \text{requests} \end{pmatrix} = 52.7 - 2.87 \begin{pmatrix} \text{size of} \\ \text{data} \end{pmatrix} + 0.85 \begin{pmatrix} \text{number of} \\ \text{tables} \end{pmatrix}.
$$

(b) ANOVA AND F-TEST. The total sum of squares is still $SS_{\text{TOT}} = S_{yy} = 1452$. It is the same for all the models with this response.

Having figured a vector of *fitted values*

$$\widehat{\boldsymbol{y}} = \boldsymbol{X}\boldsymbol{b} = \begin{pmatrix} 38.9 \\ 49.6 \\ 49.6 \\ 38.2 \\ 32.5 \\ 25.7 \\ 10.5 \end{pmatrix},$$

we can immediately compute

$$SS_{\text{REG}} = (\widehat{\boldsymbol{y}} - \overline{\boldsymbol{y}})^T (\widehat{\boldsymbol{y}} - \overline{\boldsymbol{y}}) = 1143.3 \quad \text{and} \quad SS_{\text{ERR}} = (\boldsymbol{y} - \widehat{\boldsymbol{y}})^T (\boldsymbol{y} - \widehat{\boldsymbol{y}}) = 308.7.$$

The ANOVA table is then completed as

Source	Sum of squares	Degrees of freedom	Mean squares	F
Model	1143.3	2	571.7	7.41
Error	308.7	4	77.2	
Total	1452	6		

Notice 2 degrees of freedom for the model because we now use two predictor variables.

R-SQUARE is now $R^2 = SS_{\text{REG}}/SS_{\text{TOT}} = 0.787$, which is 12.5% higher than in Example 11.6. These additional 12.5% of the total variation are explained by the new predictor x_2 that is used in the model in addition to x_1. R-square can only increase when new variables are added.

ANOVA F-TEST statistic of 7.41 with 2 and 4 d.f. shows that the model is significant at the level of 0.05 but not at the level of 0.025.

REGRESSION VARIANCE σ^2 is estimated by $s^2 = 77.2$.

(c) INFERENCE ABOUT THE NEW SLOPE. Is the new predictor variable x_2 significant? It is, as long as the corresponding slope β_2 is proved to be non-zero. Let us test $H_0 : \beta_2 = 0$.

The vector of slopes $\boldsymbol{b}$ has an estimated variance-covariance matrix

$$\widehat{\text{VAR}}(\boldsymbol{b}) = s^2 (\boldsymbol{X}^T \boldsymbol{X})^{-1} = \begin{pmatrix} 284.7 & -22.9 & -7.02 \\ -22.9 & 2.06 & 0.46 \\ -7.02 & 0.46 & 0.30 \end{pmatrix}.$$

From this, $s(b_2) = \sqrt{0.30} = 0.55$. The T-statistic is then

$$t = \frac{b_2}{s(b_2)} = \frac{0.85}{0.55} = 1.54,$$

and for a two-sided test this is not significant at any level up to 0.10. This suggests that adding the data structure into the model does not bring a significant improvement. ◇

11.4 Model building

Multivariate regression opens an almost unlimited opportunity for us to improve prediction by adding more and more X-variables into our model. On the other hand, we saw in Section 11.1.5 that overfitting a model leads to a low prediction power. Moreover, it will often result in large variances $\sigma^2(b_j)$ and therefore, unstable regression estimates.

Then, how can we build a model with the right, optimal set of predictors $X^{(j)}$ that will give us a good, accurate fit?

Two methods of variable selection are introduced here. One is based on the *adjusted R-square* criterion, the other is derived from the *extra sum of squares principle*.

11.4.1 Adjusted R-square

It is shown mathematically that R^2, the coefficient of determination, can only increase when we add predictors to the regression model. No matter how irrelevant it is for the response Y, any new predictor can only increase the proportion of explained variation.

Therefore, R^2 is not a fair criterion when we compare models with different numbers of predictors (k). Including irrelevant predictors should be penalized whereas R^2 can only reward for this.

A fair measure of goodness-of-fit is the *adjusted R-square*.

DEFINITION 11.5

Adjusted R-square

$$R^2_{\text{adj}} = 1 - \frac{SS_{\text{ERR}}/(n-k-1)}{SS_{\text{TOT}}/(n-1)} = 1 - \frac{SS_{\text{ERR}}/df_{\text{ERR}}}{SS_{\text{TOT}}/df_{\text{TOT}}}$$

is a criterion of variable selection. It rewards for adding a predictor only if it considerably reduces the error sum of squares.

Comparing with

$$R^2 = \frac{SS_{\text{REG}}}{SS_{\text{TOT}}} = \frac{SS_{\text{TOT}} - SS_{\text{ERR}}}{SS_{\text{TOT}}} = 1 - \frac{SS_{\text{ERR}}}{SS_{\text{TOT}}},$$

adjusted R-square includes degrees of freedom into this formula. This adjustment may result in a penalty when a useless X-variable is added to the regression mode.

Indeed, imagine adding a non-significant predictor. The number of estimated slopes k increments by 1. However, if this variable is not able to explain any variation of the response, the sums of squares, SS_{REG} and SS_{ERR}, will remain the same. Then, $SS_{\text{ERR}}/(n-k-1)$ will increase and R^2_{adj} will decrease, penalizing us for including such a poor predictor.

Adjusted R-square criterion: *choose a model with the highest adjusted R-square.*

11.4.2   Extra sum of squares, partial F-tests, and variable selection

Suppose we have K predictors available for predicting a response. Technically, to select a subset that maximizes adjusted R-square, we need to fit all 2^K models and choose the one with the highest R^2_{adj}. This is possible for rather moderate K, and such schemes are built in some statistical software.

Fitting all models is not feasible when the total number of predictors is large. Instead, we consider a *sequential* scheme that will follow a reasonable path through possible regression models and consider only a few of them. At every step, it will compare some set of predictors

$$\boldsymbol{X}(Full) = \left(X^{(1)},\dots,X^{(k)},X^{(k+1)},\dots,X^{(m)}\right)$$

and the corresponding *full model*

$$\mathbf{E}(Y \mid \boldsymbol{X}=\boldsymbol{x}) = \beta_0 + \beta_1 x^{(1)} + \dots + \beta_k x^{(k)} + \beta_{k+1} x^{(k+1)} + \dots + \beta_m x^{(m)}$$

with a subset

$$\boldsymbol{X}(Reduced) = \left(X^{(1)},\dots,X^{(k)}\right)$$

and the corresponding *reduced model*

$$\mathbf{E}(Y \mid \boldsymbol{X}=\boldsymbol{x}) = \beta_0 + \beta_1 x^{(1)} + \dots + \beta_k x^{(k)}.$$

If the full model is significantly better, expanding the set of predictors is justified. If it is just as good as the reduced model, we should keep the smaller number of predictors in order to attain lower variances of the estimated regression slopes, more accurate predictions, and a lower adjusted R-square.

DEFINITION 11.6

A model with a larger set of predictors is called a **full model**.

Including only a subset of predictors, we obtain a **reduced model**.

The difference in the variation explained by the two models is the **extra sum of squares**,

$$\begin{aligned} SS_{\text{EX}} &= SS_{\text{REG}}(Full) - SS_{\text{REG}}(Reduced) \\ &= SS_{\text{ERR}}(Reduced) - SS_{\text{ERR}}(Full). \end{aligned}$$

Extra sum of squares measures the *additional* amount of variation explained by additional predictors $X^{(k+1)},\dots,X^{(m)}$. By subtraction, it has

$$\text{df}_{\text{EX}} = \text{df}_{\text{REG}}(Full) - \text{df}_{\text{REG}}(Reduced) = m - k$$

degrees of freedom.

Significance of the additional explained variation (measured by SS_{EX}) is tested by a **partial F-test statistic**

$$F = \frac{SS_{\text{EX}}/\text{df}_{\text{EX}}}{MS_{\text{ERR}}(Full)} = \frac{SS_{\text{ERR}}(Reduced) - SS_{\text{ERR}}(Full)}{SS_{\text{ERR}}(Full)}\left(\frac{n-m-1}{m-k}\right).$$

As a set, $X^{(k+1)}, \ldots, X^{(m)}$ affect the response Y if at least one of the slopes $\beta_{k+1}, \ldots, \beta_m$ is not zero in the full model. The partial F-test is a test of

$$H_0 : \ \beta_{k+1} = \ldots = \beta_m = 0 \quad \text{vs} \quad H_A : \ \text{not } H_0.$$

If the null hypothesis is true, the partial F-statistic has the *F-distribution* with

$$\text{df}_{\text{EX}} = m - k \quad \text{and} \quad \text{df}_{\text{ERR}}(Full) = n - m - 1$$

degrees of freedom, Table A7.

The *partial F-test* is used for sequential selection of predictors in multivariate regression. Let us look at two algorithms that are based on the partial F-test: *stepwise selection* and *backward elimination*.

Stepwise (forward) selection

The **stepwise selection** *algorithm* starts with the simplest model that excludes all the predictors,

$$G(\boldsymbol{x}) = \beta_0.$$

Then, predictors enter the model sequentially, one by one. Every new predictor should make the most significant contribution, among all the predictors that have not been included yet.

According to this rule, the first predictor $X^{(s)}$ to enter the model is the one that has the most significant univariate ANOVA F-statistic

$$F_1 = \frac{MS_{\text{REG}}(X^{(s)})}{MS_{\text{ERR}}(X^{(s)})}.$$

All F-tests considered at this step refer to the same F-distribution with 1 and $(n - 2)$ d.f. Therefore, the largest F-statistic implies the lowest P-value and the most significant slope β_s

The model is now

$$G(\boldsymbol{x}) = \beta_0 + \beta_s x^{(s)}.$$

The next predictor $X^{(t)}$ to be selected is the one that makes the most significant contribution, in addition to $X^{(s)}$. Among all the remaining predictors, it should maximize the partial F-statistic

$$F_2 = \frac{SS_{\text{ERR}}(Reduced) - SS_{\text{ERR}}(Full)}{MS_{\text{ERR}}(Full)}$$

designed to test significance of the slope β_t when the first predictor $X^{(s)}$ is already included. At this step, we compare the "full model" $G(\boldsymbol{x}) = \beta_0 + \beta_s x^{(s)} + \beta_t x^{(t)}$ against the "reduced model" $G(\boldsymbol{x}) = \beta_0 + \beta_s x^{(s)}$. Such a partial F-statistic is also called **F-to-enter**.

All F-statistics at this step are compared against the same F-distribution with 1 and $(n - 3)$ d.f., and again, the largest F-statistic points to the most significant slope β_t.

If the second predictor is included, the model becomes

$$G(\boldsymbol{x}) = \beta_0 + \beta_s x^{(s)} + \beta_t x^{(t)}.$$

The algorithm continues until the F-to-enter statistic is not significant for all the remaining predictors, according to a pre-selected significance level α. The final model will have all predictors significant at this level.

Backward elimination

The **backward elimination algorithm** works in the direction opposite to stepwise selection.

It starts with the full model that contains all possible predictors,

$$G(\boldsymbol{x}) = \beta_0 + \beta_1 x^{(1)} + \ldots + \beta_m x^{(m)}.$$

Predictors are *removed* from the model sequentially, one by one, starting with the *least significant* predictor, until all the remaining predictors are statistically significant.

Significance is again determined by a partial F-test. In this scheme, it is called **F-to-remove**.

The first predictor to be removed is the one that *minimizes* the F-to-remove statistic

$$F_{-1} = \frac{SS_{\text{ERR}}(Reduced) - SS_{\text{ERR}}(Full)}{MS_{\text{ERR}}(Full)}.$$

Again, the test with the lowest value of F_{-1} has the highest P-value indicating the least significance.

Suppose the slope β_u is found the least significant. Predictor $X^{(u)}$ is removed, and the model becomes

$$G(\boldsymbol{x}) = \beta_0 + \beta_1 x^{(1)} + \ldots + \beta_{u-1} x^{(u-1)} + \beta_{u+1} x^{(u+1)} + \ldots + \beta_m x^{(m)}.$$

Then we choose the next predictor to be removed by comparing all F_{-2} statistics, then go to F_{-3}, etc. The algorithm stops at the stage when all F-to-remove tests reject the corresponding null hypotheses. It means that in the final resulting model, all the remaining slopes are significant.

Both sequential model selection schemes, stepwise and backward elimination, involve fitting at most K models. This requires much less computing power than the adjusted R^2 method, where all 2^K models are considered.

Modern statistical computing packages (SAS, Splus, SPSS, JMP, and others) are equipped with all three considered model selection procedures.

Example 11.11 (PROGRAM EFFICIENCY: CHOICE OF A MODEL). How should we predict the program efficiency in Examples 11.6, 11.7, and 11.10 after all? Should we use the size of data sets x_1 alone, or the data structure x_2 alone, or both variables?

(a) ADJUSTED R-SQUARE CRITERION. For the *full model*,

$$R^2_{\text{adj}} = 1 - \frac{SS_{\text{ERR}}/df_{\text{ERR}}}{SS_{\text{TOT}}/df_{\text{TOT}}} = 1 - \frac{308.7/4}{1452/6} = 0.681.$$

Reduced model with only one predictor x_1 (Example 11.6) has

$$R^2_{\text{adj}} = 1 - \frac{491/5}{1452/6} = 0.594,$$

and another *reduced model* with only x_2 has $R^2_{\text{adj}} = 0.490$ (Exercise 11.9).

How do we interpret these R^2_{adj}? The price paid for including both predictors x_1 and x_2 is the division by 4 d.f. instead of 5 when we computed R^2_{adj} for the full model. Nevertheless, the full model explains such a large portion of the total variation that fully compensates for this penalty and makes the full model preferred to reduced ones. According to the *adjusted R-square criterion, the full model is best.*

(b) PARTIAL F-TEST. How significant was addition of a new variable x_2 into our model? Comparing the *full model* in Example 11.10 with the *reduced model* in Example 11.6, we find the *extra sum of squares*

$$SS_{\text{EX}} = SS_{\text{REG}}(Full) - SS_{\text{REG}}(Reduced) = 1143 - 961 = 182.$$

This is the additional amount of the total variation of response explained by x_2 when x_1 is already in the model. It has 1 d.f. because we added only 1 variable. The *partial F-test statistic* is

$$F = \frac{SS_{\text{EX}}/\text{df}_{\text{EX}}}{MS_{\text{ERR}}(Full)} = \frac{182/1}{309} = 0.59.$$

From Table A7 with 1 and 4 d.f., we see that this F-statistic is *not significant* at the 0.25 level. It means that a relatively small additional variation of 182 that the second predictor can explain does not justify its inclusion into the model.

(c) SEQUENTIAL MODEL SELECTION. What models should be selected by stepwise and backward elimination routines?

Stepwise model selection starts by including the first predictor x_1. It is significant at the 5% level, as we know from Example 11.6, hence we keep it in the model. Next, we include x_2. As we have just seen, it fails to result in a significant gain, $F_2 = 0.59$, and thus, we do not keep it in the model. The resulting model predicts the program efficiency y based on the size of data sets x_1 only.

Backward elimination scheme starts with the full model and looks for ways to reduce it. Among the two reduced models, the model with x_1 has a higher regression sum of squares SS_{REG}, hence the other variable x_2 is the first one to be removed. The remaining variable x_1 is significant at the 5% level; therefore, we again arrive to the reduced model predicting y based on x_1.

Two different model selection criteria, adjusted R-square and partial F-tests, lead us to two different models. Each of them is best in a different sense. Not a surprise.   ◊

11.4.3   Categorical predictors and dummy variables

Careful model selection is one of the most important steps in practical statistics. In regression, only a wisely chosen subset of predictors delivers accurate estimates and good prediction.

At the same time, any useful information should be incorporated into our model. We conclude this chapter with a note on using *categorical* (that is, non-numerical) predictors in regression modeling.

Often a good portion of the variation of response Y can be explained by *attributes* rather than numbers. Examples are

– computer manufacturer (Dell, IBM, Hewlett Packard, etc.);
– operating system (Unix, Windows, DOS, etc.);
– major (Statistics, Computer Science, Electrical Engineering, etc.);
– gender (female, male);
– color (white, blue, green, etc.).

Unlike numerical predictors, attributes have no particular order. For example, it is totally *wrong* to code operating systems with numbers (1 = Unix, 2 = Windows, 3 = DOS), create a new predictor $X^{(k+1)}$, and include it into the regression model

$$G(\boldsymbol{x}) = \beta_0 + \beta_1 x^{(1)} + \ldots + \beta_k x^{(k)} + \beta_{k+1} x^{(k+1)}.$$

If we do so, it puts Windows right in the middle between Unix and DOS and tells that changing an operating system from Unix to Windows has exactly the same effect on the response Y as changing it from Windows to DOS!

However, performance of a computer really depends on the operating system, manufacturer, type of the processor, and other categorical variables. How can we use them in our regression model?

We need to create so-called **dummy variables**. A dummy variable is binary, taking values 0 or 1,

$$Z_i^{(j)} = \begin{cases} 1 & \text{if unit } i \text{ in the sample has category } j \\ 0 & \text{otherwise} \end{cases}$$

For a categorical variable with C categories, we create $(C-1)$ dummy predictors, $\boldsymbol{Z}^{(1)}, \ldots, \boldsymbol{Z}^{(C-1)}$. They carry the entire information about the attribute. Sampled items from category C will be marked by all $(C-1)$ dummies equal to 0.

Example 11.12 (DUMMY VARIABLES FOR THE OPERATING SYSTEM). In addition to numerical variables, we would like to include the operating system into the regression model. Suppose that each sampled computer has one of three operating systems: Unix, Windows, or DOS. In order to use this information for the regression modeling and more accurate forecasting, we create *two* dummy variables,

$$Z_i^{(1)} = \begin{cases} 1 & \text{if computer } i \text{ has Unix} \\ 0 & \text{otherwise} \end{cases}$$

$$Z_i^{(2)} = \begin{cases} 1 & \text{if computer } i \text{ has Windows} \\ 0 & \text{otherwise} \end{cases}$$

Together with numerical predictors $\boldsymbol{X}^{(1)}, \ldots, \boldsymbol{X}^{(k)}$, the regression model will be

$$G(\boldsymbol{x}, \boldsymbol{z}) = \beta_0 + \beta_1 x^{(1)} + \ldots + \beta_k x^{(k)} + \gamma_1 z^{(1)} + \gamma_2 z^{(2)}.$$

$\Diamond$

Fitting the model, all dummy variables are included into the *predictor matrix* $\boldsymbol{X}$ as columns.

Avoid singularity by creating only $(C-1)$ dummies

Notice that if we make a mistake and create C dummies for an attribute with C categories, one dummy per category, this would cause a linear relation

$$\boldsymbol{Z}^{(1)} + \ldots + \boldsymbol{Z}^{(C)} = 1.$$

A column of 1's is already included into the predictor matrix $\boldsymbol{X}$, and therefore, such a linear relation will cause singularity of $(\boldsymbol{X}^T\boldsymbol{X})$ when we compute the least squares estimates $\boldsymbol{b} = (\boldsymbol{X}^T\boldsymbol{X})^{-1}\boldsymbol{X}^T\boldsymbol{y}$. Thus, it is necessary and sufficient to have only $(C-1)$ dummy variables.

Interpretation of slopes for dummy variables

Each slope γ_j for a dummy predictor $Z^{(j)}$ is the expected change in the response caused by incrementing $Z^{(j)}$ by 1 while keeping all other predictors constant. Such an increment occurs when we compare the last category C with category j.

Thus, the slope γ_j is the difference in the expected response comparing category C with category j. The difference of two slopes $(\gamma_j - \gamma_C)$ compares category j with category C.

To test significance of a categorical variable, we test all the corresponding slopes γ_j simultaneously. This is done by a partial F-test.

R notes

For linear regression, R has a simple command `lm` (linear model).

$$\texttt{lm(Y} \sim \texttt{X1 + X2 + X3)}$$

Just typing this will only produce regression coefficients, an intercept and slopes. In fact, `lm` creates the whole object with a lot of information in it.

Let's give this object a name such as, for example, `reg <- lm( Y ~ X1 + X2 + X3 )`. Then

- `summary(reg)` will give us all the inference about regression coefficients, with their standard errors and significance testing, along with the standard error s, ANOVA F-test, R^2, and adjusted $R^@$.

- `anova(reg)` will show the full analysis of variance table with SS, MS, partial F-statistics, and their corresponding p-values.

This can also be used for variable selection because `summary(reg)` reports p-values of T-tests for testing significance of each X-variable when all the other variables are already in the model whereas `anova(reg)` tests each variable with a partial F-test measuring its contribution to those variables written ahead of it in the regression equation.

In fact, R has several powerful tools for the regression model building. We can do the forward and backward selection with `step`. Let us define the *full* model using all independent variables, and on the other extreme, the *null* model with no variables in it.

```
— R ——————————
full = lm(Y ~ X1 + X2 + X3 + X4 + X5)
null = lm(Y ~ 1)
step( null, scope=list( lower=null, upper=full ), dorection="forward" )
```

The whole forward selection algorithm will run, showing all the details of it and the resulting final model.

For backward elimination, we need to change the direction and start from the full model,

```
step( full, scope=list( lower=null, upper=full ), direction="backward" )
```

For a partial F-test deciding between some reduced model `reduced` and a full model `full`, use `anova` with two arguments, listing the reduced model first (because we do not want to see a negative extra sum of squares!),

```
anova(reduced, full)
```

Confidence intervals for the slopes can be obtained by `confint(reg)` or `confint(reg, level=0.90)` (the default confidence level is 95%).

R will recognize categorical variables and create dummies automatically... as long as they are not represented by numbers. Sometimes categories of a categorical predictor are coded with numbers such as, for example, 1 = Alabama, 2 = Alaska, 3 = Arizona, ... In this case, we have to tell R to treat this variable as categorical by replacing X with `as.factor(X)`.

MATLAB notes

MATLAB (MATrix LABoratory) is great for matrix computations, so all the regression analysis can be done by writing the matrix formulas, `b=inv(X'T*X)*(X*Y)` for the regression slope, `Yhat=X*b` for the fitted values, `e=Y-Yhat` for residuals, etc., given a vector of responses Y and a matrix of predictors X. For more instructions on that, see the last paragraph of Section A.5.

Also, MATLAB's Statistics Toolbox has special tools for regression. The general command `regress(Y,X)` returns a sample regression slope with a response Y and predictor X. Notice that if you'd like to fit regression with an *intercept*, then the vector of ones has to be included into matrix X. For example,

```
— MATLAB ——————
X0 = ones(size(Y));
regress(Y,[X0,X1,X2]);
```

will create a vector of ones of the same size as the vector of responses and use it to fit a regression model $Y = \beta_0 + \beta_1 X_1 + \beta_2 X_2 + \varepsilon$. To get $(1 - \alpha)100\%$ confidence intervals for all the regression slopes, write `[b bint] = regress(Y,[X0,X1,X2],alpha)`.

Many components of regression analysis are included in the MATLAB command `regstats(Y,X)`. A list of options appears where you can mark which statistics you'd like to obtain – mean squared error s^2, ANOVA F-statistic, R^2, adjusted R^2, etc.

You can also use `stepwise(X,Y)` to select variables for the multivariate regression. A window opens allowing you to run the stepwise variable selection algorithm step by step and reporting the main regression statistics at each step.

Summary and conclusions

This chapter provides methods of estimating mathematical relations between one or several predictor variables and a response variable. Results are used to explain behavior of the response and to predict its value for any new set of predictors.

Method of least squares is used to estimate regression parameters. Coefficient of determination R^2 shows the portion of the total variation that the included predictors can explain. The unexplained portion is considered as "error."

Analysis of variance (ANOVA) partitions the total variation into explained and unexplained parts and estimates regression variance by the mean squared error. This allows further statistical inference, testing slopes, constructing confidence intervals for mean responses and prediction intervals for individual responses. ANOVA F-test is used to test significance of the entire model.

For accurate estimation and efficient prediction, it is important to select the right subset of predictors. Sequential model selection algorithms are based on partial F-tests comparing full and reduced models at each step.

Categorical predictors are included into regression modeling by creating dummy variables.

Exercises

Suppose that the standard regression assumptions, univariate or multivariate, hold in Exercises 11.2, 11.3, 11.4, 11.5, 11.9, 11.14, and 11.15.

11.1. The time it takes to transmit a file always depends on the file size. Suppose you transmitted 30 files, with the average size of 126 Kbytes and the standard deviation of 35 Kbytes. The average transmittance time was 0.04 seconds with the standard deviation of 0.01 seconds. The correlation coefficient between the time and the size was 0.86.

Based on this data, fit a linear regression model and predict the time it will take to transmit a 400 Kbyte file.

11.2. The following statistics were obtained from a sample of size $n = 75$:
– the predictor variable X has mean 32.2, variance 6.4;
– the response variable Y has mean 8.4, variance 2.8; and
– the sample covariance between X and Y is 3.6.

(a) Estimate the linear regression equation predicting Y based on X.

(b) Complete the ANOVA table. What portion of the total variation of Y is explained by variable X?

(c) Construct a 99% confidence interval for the regression slope. Is the slope significant?

11.3. At a gas station, 180 drivers were asked to record the mileage of their cars and the number of miles per gallon. The results are summarized in the table.

	Sample mean	Standard deviation
Mileage	24,598	14,634
Miles per gallon	23.8	3.4

The sample correlation coefficient is $r = -0.17$.

(a) Compute the least squares regression line which describes how the number of miles per gallon depends on the mileage. What do the obtained slope and intercept mean in this situation?

(b) Use R^2 to evaluate its goodness of fit. Is this a good model?

(c) You purchase a used car with 35,000 miles on it. Predict the number of miles per gallon. Give a 95% prediction interval for your car and a 95% confidence interval for the average number of miles per gallon of all cars with such a mileage.

11.4. The data below represent investments, in $1000s, in the development of new software by some computer company over an 11-year period,

Year, X	2008	2009	2010	2011	2012	2013	2014	2015	2016	2017	2018
Investment, Y	17	23	31	29	33	39	39	40	41	44	47

(a) In the regression model with Y as a dependent variable, estimate the variance of Y.

(b) Test whether the investment increases by *more* than $ 1,800 every year, on the average.

(c) Give a 95% prediction interval for the investment in new-product development in the year 2022.

(d) Interpret this interval (explain the meaning of 95%) and state all assumptions used in this procedure.

Subtracting 2000 from each year will certainly simplify the calculations.
These data are also available in data set `Investments`.

11.5. In the previous problem, a market analyst notices that the investment amount may depend on whether the company shows profit in its financial reports. A categorical variable Z contains the additional information.

Year, X	2008	2009	2010	2011	2012	2013	2014	2015	2016	2017	2018
Reporting profit, Z	no	no	yes	no	yes	yes	yes	yes	no	yes	yes

(a) Add a dummy variable to the model considered in Exercise 11.4 and estimate the intercept and slopes of the new multivariate regression model.

(b) If the company will report a profit during year 2015, predict the investment amount.

(c) How would your prediction change if the company reports a loss during year 2015?

(d) Complete a multivariate ANOVA table and test significance of the entire model.

(e) Does the new variable Z explain a significant portion of the total variation, in addition to the time trend?

11.6. For a univariate linear regression, show that

$$SS_{\text{TOT}} = SS_{\text{REG}} + SS_{\text{ERR}}.$$

Hint: Write $SS_{\text{TOT}} = \sum_{i=1}^n (y_i - \bar{y})^2 = \sum_{i=1}^n ((y_i - \hat{y}_i) + (\hat{y}_i - \bar{y}))^2$.

11.7. For a univariate linear regression, show that R-square is the squared sample correlation coefficient,

$$R^2 = r^2.$$

Hint: Write the regression sum of squares as

$$SS_{\text{REG}} = \sum_{i=1}^n (b_0 + b_1 x_i - \bar{y})^2$$

and substitute our derived expressions for the regression intercept b_0 and slope b_1.

11.8. Anton wants to know if there is a relation between the number of hours he spends preparing for his weekly quiz and the grade he receives on it. He keeps records for 10 weeks.

It turns out that on the average, he spends 3.6 hours a week preparing for the quiz, with the standard deviation of 0.5 hours. His average grade is 82 (out of 100), with the standard deviation of 14. The correlation between the two variables is $r = 0.62$.

(a) Find the equation of the regression line predicting the quiz grade based on the time spent on preparation.

(b) This week, Anton studied for 4 hours. Predict his grade.

(c) Does this linear model explain most of the variation? Is it a good fit? Why?

11.9. For the data in Example 11.10 and data set `Efficiency`, fit a linear regression model predicting the program efficiency (number of processed requests) based on the database structure x_2 only. Complete the ANOVA table, compute R-square and adjusted R-square. Is this reduced model significant?

11.10. Refer to Exercise 8.5 on p. 241 and data set `PopulationUSA`.

(a) Fit a linear regression model estimating the time trend of the U.S. population. For simplicity, subtract 1800 from each year and let $x = \text{year} - 1800$ serve as a predictor.

(b) Complete the ANOVA table and compute R-square.

(c) According to the linear model, what population would you predict for years 2020 and 2025? Is this a reasonable prediction? Comment on the prediction ability of this linear model.

11.11. Here we improve the model of the previous exercise.

(a) Add the quadratic component and fit a multivariate regression model

$$\mathbf{E}(\text{population}) = \beta_0 + \beta_1(\text{year-1800}) + \beta_2(\text{year-1800})^2.$$

Estimate the regression slopes and intercept.

(b) What population does this model predict for years 2020 and 2025?

(c) Complete the ANOVA table and compute R-square. Comparing with the previous exercise, how much of the total variation does the quadratic term explain?

(d) Now we have two competing models for the U.S. population. A linear (reduced) model is studied in Exercise 11.10, and a quadratic (full) model is considered here. Also, consider a model with the quadratic term only (without the linear term). Which model is preferred, according to the adjusted R-square criterion?

(e) Do your findings agree with Figure 11.6 on p. 396? Comment.

11.12. Masha weighed 7.5 lbs when she was born. Then her weight increased according to the table.

Age (months)	0	2	4	6	8	10	12	14	16	18	20	22
Weight (lbs)	7.5	10.3	12.7	14.9	16.8	18.5	19.9	21.3	22.5	23.6	24.5	25.2

(a) Construct a time plot of Masha's weight. What regression model seems appropriate for these data?

(b) Fit a quadratic model $y = \beta_0 + \beta_1 x + \beta_2 x^2 + \varepsilon$. Why does the estimate of β_2 appear negative?

(c) What additional portion of the total variation is explained by the quadratic term? Would you keep it in the model?

(d) Predict Masha's weight at 24 months using the quadratic model and using the linear model. Which prediction is more reasonable?

(e) Can one test significance of the quadratic term by an F-test? Why or why not?

11.13. Refer to Exercise 8.6 on p. 241 and data set `PopulationUSA`. Does a linear regression model provide an adequate fit? Estimate regression parameters, make a plot of 10-year increments in the U.S. population, along with your estimated regression line. What can you infer about the U.S. population growth?

11.14. Refer to Exercise 8.7 on p. 241 and data set `PopulationUSA`.

(a) Fit a linear regression model to the 10-year relative change of the U.S. population. Estimate the regression parameters.

(b) Complete the ANOVA table and compute R-square.

(c) Conduct ANOVA F-test and comment on the significance of the fitted model.

(d) Compute a 95% confidence interval for the regression slope.

(e) Compute 95% prediction intervals for the relative change of the population between years 2010 and 2020, and for the relative change between years 2020 and 2030.

(f) Construct a histogram of regression residuals. Does it support our assumption of the normal distribution?

11.15. Consider the program efficiency study in Examples 11.6–11.7 and 11.10–11.11 and data set **Efficiency**. The computer manager makes another attempt to improve the prediction power. This time she would like to consider the fact that the first four times the program worked under the operational system A and then switched to the operational system B.

Data size (gigabytes), x_1	6	7	7	8	10	10	15
Number of tables, x_2	4	20	20	10	10	2	1
Operational system, x_3	A	A	A	A	B	B	B
Processed requests, y	40	55	50	41	17	26	16

(a) Introduce a dummy variable responsible for the operational system and include it into the regression analysis. Estimate the new regression equation.

(b) Does the new variable improve the goodness of fit? What is the new R-square?

(c) Is the new variable significant?

(d) What is the final regression equation that you would recommend to the computer manager any time when she needs to predict the number of processed requests given the size of data sets, the number of tables, and the operational system? Select the best regression equation using different model selection criteria.

Appendix

A.1   Data sets

All data sets are given in the ASCII (.txt, plain text, tab separated) and comma-separated values (.csv) formats. The first row of each data file is reserved for variable names. These data sets are available on the web site http://fs2.american.edu/baron/www/Book/. Also, each data set is printed in this book at least once, typically on the page where it is mentioned for the first time.

Reading data into R and MATLAB

To start working with any of these data sets, you can either save on your computer and read into your program, or you can read it directly from the web site. Here is how you can do it in R and MATLAB.

```
— R ——————————
setwd("C:\\Documents\\Data")          # Set the working directory
read.table("datafile.txt", header=1)  # Read text data
read.csv("datafile.csv")              # Read CSV data
```

In practice, you will put your real folder name instead of "C:
Documents
Data" and will replace `datafile` with the actual data file name, such as `CPU.txt` or `Pings.csv`. The `header` option tells R to read variable names from the first row. This option is not necessary when reading csv files, since R expects to see variable names in the top row by default.

Appendix

R can also read the data directly from the web site, for example,

`read.csv(url("http://fs2.american.edu/baron/www/Book/Data/Antivirus.csv"))`

To use the data, you should save it as a data frame, attach it, and... it is ready for all the nice tools that we've learned. For example:

```
— R ———————————
A <- read.table("CPU.txt", header=1)   # Read data into data frame A
attach(A)                              # Attach data A
names(A)                               # A list of variables in A
summary(A)                             # Descriptive statistics of variables
head(A)                                # Top few rows of A
mean(CPUtime)                          # Sample mean of variable CPUtime
```

In MATLAB, similarly, we can set the directory to a folder where our data sets are and read them into our code. Then each variable is denoted as `data.name` such as, for example, `A.CPUtime`.

```
— MATLAB ———————
cd 'C:\\Documents\\Data';    % Set the working directory
A = readtable("CPU.txt");     % Read text data
A = readtable("CPU.csv");     % Read CSV data with the same function
mean(A.CPUtime)               % Sample mean of variable CPUtime in data A
```

Data inventory

Data set	Variables	Pages
Accounts	N	242
Antivirus	X, Y	237, 238, 348
Attachments	Size, Classification	370
Books	Cost, Year	370
Chips	Thickness	368, 369, 371
ConcurrentUsers	N	240, 372
CPU	CPUtime	217, 219, 223, 224, 226, 229, 230, 231, 234, 236
Efficiency	Size, Tables, OS, Requests	389, 394, 402, 407, 414, 416
Incentive	Hits, Discount	335
Internet	Speed, Provider	370
Intrusions	Attempts, Firewall	240
Investments	Year, Investment, Profit	413
Keystrokes	Time	267, 282, 287, 292, 327, 367
Pings	Ping, Location	235, 336
PopulationUSA	Year, Population	241, 395, 414, 414, 415, 415
PopulationWorld	Year, Population	376, 381, 385, 389
RandomNumbers	U	367
ServiceTimes	Time	366, 369, 370
Spam	N	366
Symmetry	N, Set	241
TestScores	X	369
Weight	Age, Weight	415

A.2   Inventory of distributions

Here we list the most commonly used families of discrete and continuous distributions along with their main characteristics.

For all these distributions, R and MATLAB have special built-in functions that can be used to compute their (1) cumulative distribution functions, (2) probability mass functions or probability density functions, (3) quantiles, and (4) to generate random variables from the given distribution. These functions have a similar structure. Say, for a (made-up) distribution family "fam", R and MATLAB have the following commands.

	R	MATLAB
Cumulative distribution function (cdf)	pfam	famcdf
Probability mass function (pmf) or probability density function (pdf)	dfam	fampdf
Quantile	qfam	faminv
Random number generator	rfam	famrnd

TABLE A.1: *R and MATLAB built-in commands for common distributions.*

In each command, you will specify parameters of the distribution, or the software will use standard default parameters, if they exist. Use ? in R and help in MATLAB to see what parameters must be entered.

Random number generators can usually produce a whole vector or even a matrix of independent random variables from the given distribution, a user just specifies how many of them to generate.

Example A.1 (NORMAL DISTRIBUTION). The following R and MATLAB commands are associated with the Normal distribution.

pnorm(1,2,3) in R and normcdf(1,2,3) in MATLAB return the cdf $F(1) = P\{X \le 1\}$ for a Normal variable X with mean 2 and standard deviation 3.

dnorm(1,2,3) in R and normpdf(1,2,3) in MATLAB return the density $f(1)$ for the same Normal(2,3) distribution.

qnorm(0.95,2,3) in R and normpdf(0.95,2,3) in MATLAB return the 0.95-quantile or the 95th percentile from the Normal(2,3) distribution. It is the value $x = 6.9346 = F^{-1}(0.95)$ that solves the equation $F(x) = P\{X \le x\} = 0.95$.

pnorm(1), dnorm(1), and qnorm(0.95) in R as well as normcdf(1), normpdf(1), and norminv(0.95) in MATLAB repeat these commands for the Standard Normal distribution. For example, qnorm(0.95) in R returns $z_{0.05} = 1.645$.

rnorm(10,2,3) in R and normrnd(2,3,10,1) in MATLAB generate a vector of 10 independent Normal(2,3) random variables. Also in MATLAB, normrnd(2,3,10) produces a 10×10 square matrix of random variables, and normrnd(2,3,10,5) returns a 10×5 matrix. ◇

A.2.1   Discrete families

Bernoulli(p)

Generic description:	0 or 1, success or failure, result of one Bernoulli trial
Range of values:	$x = 0, 1$
Parameter:	$p \in (0, 1)$, probability of success
Probability mass function:	$P(x) = p^x (1 - p)^{1-x}$
Expectation:	$\mu = p$
Variance:	$\sigma^2 = p(1 - p)$
Relations:	Special case of *Binomial*(n, p) when $n = 1$
	A sum of n independent Bernoulli(p) variables is *Binomial*(n, p)
R and MATLAB:	Use Binomial distribution with $n = 1$

Binomial(n, p)

Generic description:	Number of successes in n independent Bernoulli trials
Range of values:	$x = 0, 1, 2, \ldots, n$
Parameters:	$n = 1, 2, 3, \ldots$, number of Bernoulli trials
	$p \in (0, 1)$, probability of success
Probability mass function:	$P(x) = \binom{n}{x} p^x (1 - p)^{n-x}$
Cdf:	Table A2
Expectation:	$\mu = np$
Variance:	$\sigma^2 = np(1 - p)$
Relations:	Binomial(n, p) is a sum of n independent *Bernoulli*(p) variables
	Binomial$(1, p) = Bernoulli(p)$
Table:	Appendix, Table A2
R:	`pbinom, dbinom, qbinom, rbinom`
MATLAB:	`binocdf, binopdf, binoinv, binornd`

Geometric(p)

Generic description:	Number of Bernoulli trials until the first success
Range of values:	$x = 1, 2, 3, \ldots$
Parameter:	$p \in (0, 1)$, probability of success
Probability mass function:	$P(x) = p(1 - p)^{x-1}$
Cdf:	$1 - (1 - p)^x$
Expectation:	$\mu = \dfrac{1}{p}$
Variance:	$\sigma^2 = \dfrac{1 - p}{p^2}$
Relations:	Special case of *Negative Binomial*(k, p) when $k = 1$

A sum of n independent Geometric(p) variables is
Negative Binomial(n, p)

R: pgeom, dgeom, qgeom, rgeom

MATLAB: geocdf, geopdf, geoinv, geornd

R and MATLAB use zero-based Geometric distributions, $x = 0, 1, \ldots$.

To calculate $F(x)$, use pgeom(x-1,p) and geocdf(x-1,p).

Negative Binomial(k, p)

Generic description: Number of Bernoulli trials until the kth success

Range of values: $x = k, k+1, k+2, \ldots$

Parameters: $k = 1, 2, 3, \ldots$, number of successes

$p \in (0, 1)$, probability of success

Probability mass function: $P(x) = \binom{x-1}{k-1}(1-p)^{x-k}p^k$

Expectation: $\mu = \dfrac{k}{p}$

Variance: $\sigma^2 = \dfrac{k(1-p)}{p^2}$

Relations: Negative Binomial(k, p) is a sum of n independent Geometric(p) variables

Negative Binomial($1, p$) = Geometric(p)

R: pnbinom, dnbinom, qnbinom, rnbinom

MATLAB: nbincdf, nbinpdf, nbininv, nbinrnd

These commands use zero-based distributions with $x = 0, 1, 2, \ldots$

To calculate $F(x)$, use pnbinom(x-k,k,p) and nbincdf(x-k,k,p).

Poisson(λ)

Generic description: Number of "rare events" during a fixed time interval

Range of values: $x = 0, 1, 2, \ldots$

Parameter: $\lambda \in (0, \infty)$, frequency of "rare events"

Probability mass function: $P(x) = e^{-\lambda}\dfrac{\lambda^x}{x!}$

Cdf: Table A3

Expectation: $\mu = \lambda$

Variance: $\sigma^2 = \lambda$

Relations: Limiting case of *Binomial*(n, p) when $n \to \infty$, $p \to 0$, $np \to \lambda$

Table: Appendix, Table A3

R: ppois, dpois, qpois, rpois

MATLAB: poisscdf, poisspdf, poissinv, poissrnd

A.2.2 Continuous families

Beta(α, β)

Generic description:	A random variable taking values in $[0, 1]$; distribution of R^2 under the standard regression assumptions.
Range of values:	$0 < x < 1$
Parameters:	$\alpha, \beta \in (0, \infty)$

Density:
$$f(x) = \frac{\Gamma(\alpha + \beta)}{\Gamma(\alpha)\Gamma(\beta)} \, x^{\alpha-1}(1 - x)^{\beta-1}$$

Expectation: $\mu = \alpha/(\alpha + \beta)$

Variance: $\sigma^2 = \alpha\beta/(\alpha + \beta)^2(\alpha + \beta + 1)$

Relations:	For independent $X \sim Gamma(\alpha, \lambda)$ and $Y \sim Gamma(\beta, \lambda)$, the ratio $\frac{X}{X+Y}$ is Beta(α, β). As a prior distribution, Beta family is conjugate to the *Binomial* model. Beta(1,1) distribution is *Uniform(0,1)*. In a sample of *Uniform(0,1)* variables of size n, the kth smallest observation is Beta$(k, n + 1 - k)$.
R:	`pbeta, dbeta, qbeta, rbeta`
MATLAB:	`betacdf, betapdf, betainv, betarnd`

Chi-square $\chi^2(\nu)$

Generic description:	Sum of squares of ν iid Normal variables
Range of values:	$x > 0$
Parameter:	$\nu \in (0, \infty)$, degrees of freedom

Density:
$$f(x) = \frac{1}{2^{\nu/2}\Gamma(\nu/2)} \, x^{\nu/2-1} e^{-x/2}$$

Expectation: $\mu = \nu$

Variance: $\sigma^2 = 2\nu$

Relations:	Special case of $Gamma(\alpha, \lambda)$ when $\alpha = \nu/2$ and $\lambda = 1/2$. Ratio of two independent χ^2 variables, divided by their respective degrees of freedom, has *F-distribution*.
R:	`pchisq, dchisq, qchisq, rchisq`
MATLAB:	`chi2cdf, chi2pdf, chi2inv, chi2rnd`

Exponential(λ)

Generic description:	In a Poisson process, time between consecutive events
Range of values:	$x > 0$
Parameter:	$\lambda \in (0, \infty)$, frequency of events, inverse scale parameter

Density: $f(x) = \lambda e^{-\lambda x}$

Cdf: $F(x) = 1 - e^{-\lambda x}$

Expectation: $\mu = 1/\lambda$

Variance:	$\sigma^2 = 1/\lambda^2$
Relations:	Special case of $Gamma(\alpha, \lambda)$ when $\alpha = 1$ and $Chi\text{-}square$ with $\nu = 2$ degrees of freedom. A sum of α independent $Exponential(\lambda)$ variables is $Gamma(\alpha, \lambda)$
R:	`pexp, dexp, qexp, rexp`
MATLAB:	`expcdf, exppdf, expinv, exprnd`
	MATLAB uses the *scale* parameter $\sigma = 1/\lambda$ instead of frequency λ. For $F(x)$, use `expcdf(x,1/lambda)`. R uses the *frequency* parameter λ.

Fisher-Snedecor $\mathbf{F}(\nu_1, \nu_2)$

Generic description:	Ratio of independent mean squares
Range of values:	$x > 0$
Parameters:	$\nu_1 > 0$, numerator degrees of freedom
	$\nu_2 > 0$, denominator degrees of freedom
Density function:	$f(x) = \dfrac{\Gamma\left(\frac{\nu_1+\nu_2}{2}\right)}{x\Gamma\left(\frac{\nu_1}{2}\right)\Gamma\left(\frac{\nu_2}{2}\right)} \sqrt{\dfrac{(\nu_1 x)^{\nu_1}\nu_2^{\nu_2}}{(\nu_1 x + \nu_2)^{\nu_1+\nu_2}}}$
Expectation:	$\dfrac{\nu_2}{\nu_2 - 2}$ for $\nu_2 > 2$
Variance:	$\dfrac{2\nu^2(\nu_1 + \nu_2 - 2)}{\nu_1(\nu_2 - 2)^2(\nu_2 - 4)}$ for $\nu_2 > 4$
Relations:	For independent $\chi^2(\nu_1)$ and $\chi^2(\nu_2)$, the ratio $\frac{\chi^2(\nu_1)/\nu_1}{\chi^2(\nu_2)/\nu_2}$ is $F(\nu_1, \nu_2)$
	Special case: a square of $t(\nu)$ variable is $F(1, \nu)$
Table:	Appendix, Table A7
R:	`pf, df, qf, rf`
MATLAB:	`fcdf, fpdf, finv, frnd`

Gamma(α, λ)

Generic description:	In a Poisson process, the time of the α-th event
Range of values:	$x > 0$
Parameters:	$\alpha \in (0, \infty)$, shape parameter
	$\lambda \in (0, \infty)$, frequency of events, inverse scale parameter
Density function:	$f(x) = \dfrac{\lambda^\alpha}{\Gamma(\alpha)} x^{\alpha-1} e^{-\lambda x}$
Expectation:	$\mu = \alpha/\lambda$
Variance:	$\sigma^2 = \alpha/\lambda^2$
Relations:	For integer α, Gamma(α, λ) is a sum of α independent $Exponential(\lambda)$ variables
	Gamma$(1, \lambda) = Exp(\lambda)$; Gamma$(\nu/2, 1/2) = Chi\text{-}square(\nu)$
	As a prior distribution, Gamma family is conjugate to the $Poisson(\theta)$ model and the $Exponential(\theta)$ model.

R: pgamma, dgamma, qgamma, rgamma
MATLAB: gamcdf, gampdf, gaminv, gamrnd
MATLAB uses the *scale* parameter $\sigma = 1/\lambda$ instead of frequency λ.
To find $F(x)$, use gamcdf(x,alpha,1/lambda). R uses parameter λ.

Normal(μ, σ)

Generic description: Often used as distribution of errors, measurements, sums, averages, etc.

Range of values: $-\infty < x < +\infty$

Parameters: $\mu \in (-\infty, \infty)$, expectation, location parameter
$\sigma \in (0, \infty)$, standard deviation, scale parameter

Density function:
$$f(x) = \frac{1}{\sigma\sqrt{2\pi}} \exp\left\{ \frac{-(x-\mu)^2}{2\sigma^2} \right\}$$

Cdf: $F(x) = \Phi\left(\dfrac{x-\mu}{\sigma}\right)$, use Table A4 for $\Phi(z)$

Expectation: μ

Variance: σ^2

Relations: Limiting case of standardized sums of random variables, including *Binomial(n,p)*, *Neg. Binomial(k,p)*, and *Gamma(α, λ)*, as $n \to \infty$, $k \to \infty$, $\alpha \to \infty$

Table: Appendix, Table A4

R: pnorm, dnorm, qnorm, rnorm

MATLAB: normcdf, normpdf, norminv, normrnd, or just randn

Pareto(θ, σ)

Generic description: A heavy-tail distribution often used to model amount of internet traffic, various financial and sociological data

Range of values: $x > \sigma$

Parameters: $\theta \in (0, \infty)$, shape parameter
$\sigma \in (0, \infty)$, scale parameter

Density function: $f(x) = \theta\sigma^\theta x^{-\theta-1}$

Cdf: $F(x) = 1 - \left(\dfrac{x}{\sigma}\right)^{-\theta}$

Expectation: $\dfrac{\theta\sigma}{\theta-1}$ for $\theta > 1$; does not exist for $\theta \leq 1$

Variance: $\dfrac{\theta\sigma^2}{(\theta-1)^2(\theta-2)}$ for $\theta > 2$; does not exist for $\theta \leq 2$

Relations: If X is *Exponential(θ)* then $Y = \sigma e^X$ is *Pareto(θ, σ)*

R: ppareto, dpareto, qpareto, rpareto (but see Remark 1 below)

MATLAB: gpcdf, gppdf, gpinv, gprnd (but see Remark 2 below)

Student's T(ν)

Generic description: Standardized statistic with an estimated standard deviation

Range of values: $-\infty < x < +\infty$

Parameter: $\nu > 0$, number of degrees of freedom

Density function:	$f(x) = \dfrac{\Gamma\left(\frac{\nu+1}{2}\right)}{\sqrt{\nu\pi}\,\Gamma\left(\frac{\nu}{2}\right)}\left(1 + \dfrac{x^2}{\nu}\right)^{-\frac{\nu+1}{2}}$
Expectation:	0
Variance:	$\dfrac{\nu}{\nu-2}$ for $\nu > 2$
Relations:	Converges to the Standard Normal distribution as $\nu \to \infty$
Table:	Appendix, Table A5
R:	pt, dt, qt, rt
MATLAB:	tcdf, tpdf, tinv, trnd

Uniform(a, b)

Generic description:	A number selected "at random" from a given interval
Range of values:	$a < x < b$
Parameters:	$-\infty < a < b < +\infty$, ends of the interval
Density function:	$f(x) = \dfrac{1}{b-a}$
Expectation:	$\mu = \dfrac{a+b}{2}$
Variance:	$\sigma^2 = \dfrac{(b-a)^2}{12}$
Relations:	Uniform(0,1) distribution is *Beta(1,1)*
R:	punif, dunif, qunif, runif
MATLAB:	unifdf, unifpdf, unifinv, unifrnd, or just rand

Remark 1: These functions come in R package "actuar" that can be invoked by entering install.packages("actuar"); library(actuar); Also, these commands refer to a *zero-based* Pareto distribution, where $X > 0$ instead of $X > \sigma$, so we need to replace x with $x - \sigma$ while using these tools. For example, to calculate $F(x)$, type ppareto(x-sigma,theta,sigma).

Remark 2: These commands are for the Generalized Pareto distribution. For Pareto(θ, σ), use gpcdf(x,1/theta,sigma/theta,sigma), gppdf(x,1/theta,sigma/theta,sigma), etc.

A.3 Distribution tables

Table A1. Uniform(0,1) random numbers

	1	2	3	4	5	6	7	8	9	10
1	.9501	.8381	.7948	.4154	.6085	.4398	.2974	.7165	.7327	.8121
2	.2311	.0196	.9568	.3050	.0158	.3400	.0492	.5113	.4222	.6101
3	.6068	.6813	.5226	.8744	.0164	.3142	.6932	.7764	.9614	.7015
4	.4860	.3795	.8801	.0150	.1901	.3651	.6501	.4893	.0721	.0922
5	.8913	.8318	.1730	.7680	.5869	.3932	.9830	.1859	.5534	.4249
6	.7621	.5028	.9797	.9708	.0576	.5915	.5527	.7006	.2920	.3756
7	.4565	.7095	.2714	.9901	.3676	.1197	.4001	.9827	.8580	.1662
8	.0185	.4289	.2523	.7889	.6315	.0381	.1988	.8066	.3358	.8332
9	.8214	.3046	.8757	.4387	.7176	.4586	.6252	.7036	.6802	.8386
10	.4447	.1897	.7373	.4983	.6927	.8699	.7334	.4850	.0534	.4516
11	.6154	.1934	.1365	.2140	.0841	.9342	.3759	.1146	.3567	.9566
12	.7919	.6822	.0118	.6435	.4544	.2644	.0099	.6649	.4983	.1472
13	.9218	.3028	.8939	.3200	.4418	.1603	.4199	.3654	.4344	.8699
14	.7382	.5417	.1991	.9601	.3533	.8729	.7537	.1400	.5625	.7694
15	.1763	.1509	.2987	.7266	.1536	.2379	.7939	.5668	.6166	.4442
16	.4057	.6979	.6614	.4120	.6756	.6458	.9200	.8230	.1133	.6206
17	.9355	.3784	.2844	.7446	.6992	.9669	.8447	.6739	.8983	.9517
18	.9169	.8600	.4692	.2679	.7275	.6649	.3678	.9994	.7546	.6400
19	.4103	.8537	.0648	.4399	.4784	.8704	.6208	.9616	.7911	.2473
20	.8936	.5936	.9883	.9334	.5548	.0099	.7313	.0589	.8150	.3527
21	.0579	.4966	.5828	.6833	.1210	.1370	.1939	.3603	.6700	.1879
22	.3529	.8998	.4235	.2126	.4508	.8188	.9048	.5485	.2009	.4906
23	.8132	.8216	.5155	.8392	.7159	.4302	.5692	.2618	.2731	.4093
24	.0099	.6449	.3340	.6288	.8928	.8903	.6318	.5973	.6262	.4635
25	.1389	.8180	.4329	.1338	.2731	.7349	.2344	.0493	.5369	.6109
26	.2028	.6602	.2259	.2071	.2548	.6873	.5488	.5711	.0595	.0712
27	.1987	.3420	.5798	.6072	.8656	.3461	.9316	.7009	.0890	.3143
28	.6038	.2897	.7604	.6299	.2324	.1660	.3352	.9623	.2713	.6084
29	.2722	.3412	.5298	.3705	.8049	.1556	.6555	.7505	.4091	.1750
30	.1988	.5341	.6405	.5751	.9084	.1911	.3919	.7400	.4740	.6210
31	.0153	.7271	.2091	.4514	.2319	.4225	.6273	.4319	.9090	.2460
32	.7468	.3093	.3798	.0439	.2393	.8560	.6991	.6343	.5962	.5874
33	.4451	.8385	.7833	.0272	.0498	.4902	.3972	.8030	.3290	.5061
34	.9318	.5681	.6808	.3127	.0784	.8159	.4136	.0839	.4782	.4648
35	.4660	.3704	.4611	.0129	.6408	.4608	.6552	.9455	.5972	.5414
36	.4186	.7027	.5678	.3840	.1909	.4574	.8376	.9159	.1614	.9423
37	.8462	.5466	.7942	.6831	.8439	.4507	.3716	.6020	.8295	.3418
38	.5252	.4449	.0592	.0928	.1739	.4122	.4253	.2536	.9561	.4018
39	.2026	.6946	.6029	.0353	.1708	.9016	.5947	.8735	.5955	.3077
40	.6721	.6213	.0503	.6124	.9943	.0056	.5657	.5134	.0287	.4116

Table A2. Binomial distribution

$$F(x) = P\{X \le x\} = \sum_{k=0}^{x} \binom{n}{k} p^k (1-p)^{n-k}$$

n	x	.050	.100	.150	.200	.250	.300	.350	.400	.450	.500	.550	.600	.650	.700	.750	.800	.850	.900	.950
1	0	.950	.900	.850	.800	.750	.700	.650	.600	.550	.500	.450	.400	.350	.300	.250	.200	.150	.100	.050
2	0	.903	.810	.723	.640	.563	.490	.423	.360	.303	.250	.203	.160	.123	.090	.063	.040	.023	.010	.003
	1	.998	.990	.978	.960	.938	.910	.878	.840	.798	.750	.698	.640	.578	.510	.438	.360	.278	.190	.098
3	0	.857	.729	.614	.512	.422	.343	.275	.216	.166	.125	.091	.064	.043	.027	.016	.008	.003	.001	.000
	1	.993	.972	.939	.896	.844	.784	.718	.648	.575	.500	.425	.352	.282	.216	.156	.104	.061	.028	.007
	2	1.0	.999	.997	.992	.984	.973	.957	.936	.909	.875	.834	.784	.725	.657	.578	.488	.386	.271	.143
4	0	.815	.656	.522	.410	.316	.240	.179	.130	.092	.063	.041	.026	.015	.008	.004	.002	.001	.000	.000
	1	.986	.948	.890	.819	.738	.652	.563	.475	.391	.313	.241	.179	.126	.084	.051	.027	.012	.004	.000
	2	1.0	.996	.988	.973	.949	.916	.874	.821	.759	.688	.609	.525	.437	.348	.262	.181	.110	.052	.014
	3	1.0	1.0	.999	.998	.996	.992	.985	.974	.959	.938	.908	.870	.821	.760	.684	.590	.478	.344	.185
5	0	.774	.590	.444	.328	.237	.168	.116	.078	.050	.031	.018	.010	.005	.002	.001	.000	.000	.000	.000
	1	.977	.919	.835	.737	.633	.528	.428	.337	.256	.188	.131	.087	.056	.031	.016	.007	.002	.000	.000
	2	.999	.991	.973	.942	.896	.837	.765	.683	.593	.500	.407	.317	.235	.163	.104	.058	.027	.009	.001
	3	1.0	1.0	.998	.993	.984	.969	.946	.913	.869	.813	.744	.663	.572	.472	.367	.263	.165	.081	.023
	4	1.0	1.0	1.0	1.0	.999	.998	.995	.990	.982	.969	.950	.922	.884	.832	.763	.672	.556	.410	.226
6	0	.735	.531	.377	.262	.178	.118	.075	.047	.028	.016	.008	.004	.002	.001	.000	.000	.000	.000	.000
	1	.967	.886	.776	.655	.534	.420	.319	.233	.164	.109	.069	.041	.022	.011	.005	.002	.000	.000	.000
	2	.998	.984	.953	.901	.831	.744	.647	.544	.442	.344	.255	.179	.117	.070	.038	.017	.006	.001	.000
	3	1.0	.999	.994	.983	.962	.930	.883	.821	.745	.656	.558	.456	.353	.256	.169	.099	.047	.016	.002
	4	1.0	1.0	1.0	.998	.995	.989	.978	.959	.931	.891	.836	.767	.681	.580	.466	.345	.224	.114	.033
	5	1.0	1.0	1.0	1.0	1.0	.999	.998	.996	.992	.984	.972	.953	.925	.882	.822	.738	.623	.469	.265
7	0	.698	.478	.321	.210	.133	.082	.049	.028	.015	.008	.004	.002	.001	.000	.000	.000	.000	.000	.000
	1	.956	.850	.717	.577	.445	.329	.234	.159	.102	.063	.036	.019	.009	.004	.001	.000	.000	.000	.000
	2	.996	.974	.926	.852	.756	.647	.532	.420	.316	.227	.153	.096	.056	.029	.013	.005	.001	.000	.000
	3	1.0	.997	.988	.967	.929	.874	.800	.710	.608	.500	.392	.290	.200	.126	.071	.033	.012	.003	.000
	4	1.0	1.0	.999	.995	.987	.971	.944	.904	.847	.773	.684	.580	.468	.353	.244	.148	.074	.026	.004
	5	1.0	1.0	1.0	1.0	.999	.996	.991	.981	.964	.938	.898	.841	.766	.671	.555	.423	.283	.150	.044
	6	1.0	1.0	1.0	1.0	1.0	1.0	.999	.998	.996	.992	.985	.972	.951	.918	.867	.790	.679	.522	.302
8	0	.663	.430	.272	.168	.100	.058	.032	.017	.008	.004	.002	.001	.000	.000	.000	.000	.000	.000	.000
	1	.943	.813	.657	.503	.367	.255	.169	.106	.063	.035	.018	.009	.004	.001	.000	.000	.000	.000	.000
	2	.994	.962	.895	.797	.679	.552	.428	.315	.220	.145	.088	.050	.025	.011	.004	.001	.000	.000	.000
	3	1.0	.995	.979	.944	.886	.806	.706	.594	.477	.363	.260	.174	.106	.058	.027	.010	.003	.000	.000
	4	1.0	1.0	.997	.990	.973	.942	.894	.826	.740	.637	.523	.406	.294	.194	.114	.056	.021	.005	.000
	5	1.0	1.0	1.0	.999	.996	.989	.975	.950	.912	.855	.780	.685	.572	.448	.321	.203	.105	.038	.006
	6	1.0	1.0	1.0	1.0	1.0	.999	.996	.991	.982	.965	.937	.894	.831	.745	.633	.497	.343	.187	.057
	7	1.0	1.0	1.0	1.0	1.0	1.0	1.0	.999	.998	.996	.992	.983	.968	.942	.900	.832	.728	.570	.337
9	0	.630	.387	.232	.134	.075	.040	.021	.010	.005	.002	.001	.000	.000	.000	.000	.000	.000	.000	.000
	1	.929	.775	.599	.436	.300	.196	.121	.071	.039	.020	.009	.004	.001	.000	.000	.000	.000	.000	.000
	2	.992	.947	.859	.738	.601	.463	.337	.232	.150	.090	.050	.025	.011	.004	.001	.000	.000	.000	.000
	3	.999	.992	.966	.914	.834	.730	.609	.483	.361	.254	.166	.099	.054	.025	.010	.003	.001	.000	.000
	4	1.0	.999	.994	.980	.951	.901	.828	.733	.621	.500	.379	.267	.172	.099	.049	.020	.006	.001	.000
	5	1.0	1.0	.999	.997	.990	.975	.946	.901	.834	.746	.639	.517	.391	.270	.166	.086	.034	.008	.001
	6	1.0	1.0	1.0	1.0	.999	.996	.989	.975	.950	.910	.850	.768	.663	.537	.399	.262	.141	.053	.008
	7	1.0	1.0	1.0	1.0	1.0	1.0	.999	.996	.991	.980	.961	.929	.879	.804	.700	.564	.401	.225	.071
	8	1.0	1.0	1.0	1.0	1.0	1.0	1.0	1.0	.999	.998	.995	.990	.979	.960	.925	.866	.768	.613	.370
10	0	.599	.349	.197	.107	.056	.028	.013	.006	.003	.001	.000	.000	.000	.000	.000	.000	.000	.000	.000
	1	.914	.736	.544	.376	.244	.149	.086	.046	.023	.011	.005	.002	.001	.000	.000	.000	.000	.000	.000
	2	.988	.930	.820	.678	.526	.383	.262	.167	.100	.055	.027	.012	.005	.002	.000	.000	.000	.000	.000
	3	.999	.987	.950	.879	.776	.650	.514	.382	.266	.172	.102	.055	.026	.011	.004	.001	.000	.000	.000
	4	1.0	.998	.990	.967	.922	.850	.751	.633	.504	.377	.262	.166	.095	.047	.020	.006	.001	.000	.000
	5	1.0	1.0	.999	.994	.980	.953	.905	.834	.738	.623	.496	.367	.249	.150	.078	.033	.010	.002	.000
	6	1.0	1.0	1.0	.999	.996	.989	.974	.945	.898	.828	.734	.618	.486	.350	.224	.121	.050	.013	.001
	7	1.0	1.0	1.0	1.0	1.0	.998	.995	.988	.973	.945	.900	.833	.738	.617	.474	.322	.180	.070	.012
	8	1.0	1.0	1.0	1.0	1.0	1.0	.999	.998	.995	.989	.977	.954	.914	.851	.756	.624	.456	.264	.086
	9	1.0	1.0	1.0	1.0	1.0	1.0	1.0	1.0	1.0	.999	.997	.994	.987	.972	.944	.893	.803	.651	.401

Table A2, continued. Binomial distribution

n	x	.050	.100	.150	.200	.250	.300	.350	.400	.450	.500	.550	.600	.650	.700	.750	.800	.850	.900	.950
11	0	.569	.314	.167	.086	.042	.020	.009	.004	.001	.000	.000	.000	.000	.000	.000	.000	.000	.000	.000
	1	.898	.697	.492	.322	.197	.113	.061	.030	.014	.006	.002	.001	.000	.000	.000	.000	.000	.000	.000
	2	.985	.910	.779	.617	.455	.313	.200	.119	.065	.033	.015	.006	.002	.001	.000	.000	.000	.000	.000
	3	.998	.981	.931	.839	.713	.570	.426	.296	.191	.113	.061	.029	.012	.004	.001	.000	.000	.000	.000
	4	1.0	.997	.984	.950	.885	.790	.668	.533	.397	.274	.174	.099	.050	.022	.008	.002	.000	.000	.000
	5	1.0	1.0	.997	.988	.966	.922	.851	.753	.633	.500	.367	.247	.149	.078	.034	.012	.003	.000	.000
	6	1.0	1.0	1.0	.998	.992	.978	.950	.901	.826	.726	.603	.467	.332	.210	.115	.050	.016	.003	.000
	7	1.0	1.0	1.0	1.0	.999	.996	.988	.971	.939	.887	.809	.704	.574	.430	.287	.161	.069	.019	.002
	8	1.0	1.0	1.0	1.0	1.0	.999	.998	.994	.985	.967	.935	.881	.800	.687	.545	.383	.221	.090	.015
	9	1.0	1.0	1.0	1.0	1.0	1.0	1.0	.999	.998	.994	.986	.970	.939	.887	.803	.678	.508	.303	.102
	10	1.0	1.0	1.0	1.0	1.0	1.0	1.0	1.0	1.0	1.0	.999	.996	.991	.980	.958	.914	.833	.686	.431
12	0	.540	.282	.142	.069	.032	.014	.006	.002	.001	.000	.000	.000	.000	.000	.000	.000	.000	.000	.000
	1	.882	.659	.443	.275	.158	.085	.042	.020	.008	.003	.001	.000	.000	.000	.000	.000	.000	.000	.000
	2	.980	.889	.736	.558	.391	.253	.151	.083	.042	.019	.008	.003	.001	.000	.000	.000	.000	.000	.000
	3	.998	.974	.908	.795	.649	.493	.347	.225	.134	.073	.036	.015	.006	.002	.000	.000	.000	.000	.000
	4	1.0	.996	.976	.927	.842	.724	.583	.438	.304	.194	.112	.057	.026	.009	.003	.001	.000	.000	.000
	5	1.0	.999	.995	.981	.946	.882	.787	.665	.527	.387	.261	.158	.085	.039	.014	.004	.001	.000	.000
	6	1.0	1.0	.999	.996	.986	.961	.915	.842	.739	.613	.473	.335	.213	.118	.054	.019	.005	.001	.000
	7	1.0	1.0	1.0	.999	.997	.991	.974	.943	.888	.806	.696	.562	.417	.276	.158	.073	.024	.004	.000
	8	1.0	1.0	1.0	1.0	1.0	.998	.994	.985	.964	.927	.866	.775	.653	.507	.351	.205	.092	.026	.002
	9	1.0	1.0	1.0	1.0	1.0	1.0	.999	.997	.992	.981	.958	.917	.849	.747	.609	.442	.264	.111	.020
	10	1.0	1.0	1.0	1.0	1.0	1.0	1.0	1.0	.999	.997	.992	.980	.958	.915	.842	.725	.557	.341	.118
	11	1.0	1.0	1.0	1.0	1.0	1.0	1.0	1.0	1.0	1.0	.999	.998	.994	.986	.968	.931	.858	.718	.460
13	0	.513	.254	.121	.055	.024	.010	.004	.001	.000	.000	.000	.000	.000	.000	.000	.000	.000	.000	.000
	1	.865	.621	.398	.234	.127	.064	.030	.013	.005	.002	.001	.000	.000	.000	.000	.000	.000	.000	.000
	2	.975	.866	.692	.502	.333	.202	.113	.058	.027	.011	.004	.001	.000	.000	.000	.000	.000	.000	.000
	3	.997	.966	.882	.747	.584	.421	.278	.169	.093	.046	.020	.008	.003	.001	.000	.000	.000	.000	.000
	4	1.0	.994	.966	.901	.794	.654	.501	.353	.228	.133	.070	.032	.013	.004	.001	.000	.000	.000	.000
	5	1.0	.999	.992	.970	.920	.835	.716	.574	.427	.291	.179	.098	.046	.018	.006	.001	.000	.000	.000
	6	1.0	1.0	.999	.993	.976	.938	.871	.771	.644	.500	.356	.229	.129	.062	.024	.007	.001	.000	.000
	7	1.0	1.0	1.0	.999	.994	.982	.954	.902	.821	.709	.573	.426	.284	.165	.080	.030	.008	.001	.000
	8	1.0	1.0	1.0	1.0	.999	.996	.987	.968	.930	.867	.772	.647	.499	.346	.206	.099	.034	.006	.000
	9	1.0	1.0	1.0	1.0	1.0	.999	.997	.992	.980	.954	.907	.831	.722	.579	.416	.253	.118	.034	.003
	10	1.0	1.0	1.0	1.0	1.0	1.0	1.0	.999	.996	.989	.973	.942	.887	.798	.667	.498	.308	.134	.025
	11	1.0	1.0	1.0	1.0	1.0	1.0	1.0	1.0	.999	.998	.995	.987	.970	.936	.873	.766	.602	.379	.135
	12	1.0	1.0	1.0	1.0	1.0	1.0	1.0	1.0	1.0	1.0	1.0	.999	.996	.990	.976	.945	.879	.746	.487
14	0	.488	.229	.103	.044	.018	.007	.002	.001	.000	.000	.000	.000	.000	.000	.000	.000	.000	.000	.000
	1	.847	.585	.357	.198	.101	.047	.021	.008	.003	.001	.000	.000	.000	.000	.000	.000	.000	.000	.000
	2	.970	.842	.648	.448	.281	.161	.084	.040	.017	.006	.002	.001	.000	.000	.000	.000	.000	.000	.000
	3	.996	.956	.853	.698	.521	.355	.220	.124	.063	.029	.011	.004	.001	.000	.000	.000	.000	.000	.000
	4	1.0	.991	.953	.870	.742	.584	.423	.279	.167	.090	.043	.018	.006	.002	.000	.000	.000	.000	.000
	5	1.0	.999	.988	.956	.888	.781	.641	.486	.337	.212	.119	.058	.024	.008	.002	.000	.000	.000	.000
	6	1.0	1.0	.998	.988	.962	.907	.816	.692	.546	.395	.259	.150	.075	.031	.010	.002	.000	.000	.000
	7	1.0	1.0	1.0	.998	.990	.969	.925	.850	.741	.605	.454	.308	.184	.093	.038	.012	.002	.000	.000
	8	1.0	1.0	1.0	1.0	.998	.992	.976	.942	.881	.788	.663	.514	.359	.219	.112	.044	.012	.001	.000
	9	1.0	1.0	1.0	1.0	1.0	.998	.994	.982	.957	.910	.833	.721	.577	.416	.258	.130	.047	.009	.000
	10	1.0	1.0	1.0	1.0	1.0	1.0	.999	.996	.989	.971	.937	.876	.780	.645	.479	.302	.147	.044	.004
	11	1.0	1.0	1.0	1.0	1.0	1.0	1.0	.999	.998	.994	.983	.960	.916	.839	.719	.552	.352	.158	.030
	12	1.0	1.0	1.0	1.0	1.0	1.0	1.0	1.0	1.0	.999	.997	.992	.979	.953	.899	.802	.643	.415	.153
	13	1.0	1.0	1.0	1.0	1.0	1.0	1.0	1.0	1.0	1.0	1.0	.999	.998	.993	.982	.956	.897	.771	.512
15	0	.463	.206	.087	.035	.013	.005	.002	.000	.000	.000	.000	.000	.000	.000	.000	.000	.000	.000	.000
	1	.829	.549	.319	.167	.080	.035	.014	.005	.002	.000	.000	.000	.000	.000	.000	.000	.000	.000	.000
	2	.964	.816	.604	.398	.236	.127	.062	.027	.011	.004	.001	.000	.000	.000	.000	.000	.000	.000	.000
	3	.995	.944	.823	.648	.461	.297	.173	.091	.042	.018	.006	.002	.000	.000	.000	.000	.000	.000	.000
	4	.999	.987	.938	.836	.686	.515	.352	.217	.120	.059	.025	.009	.003	.001	.000	.000	.000	.000	.000
	5	1.0	.998	.983	.939	.852	.722	.564	.403	.261	.151	.077	.034	.012	.004	.001	.000	.000	.000	.000
	6	1.0	1.0	.996	.982	.943	.869	.755	.610	.452	.304	.182	.095	.042	.015	.004	.001	.000	.000	.000
	7	1.0	1.0	.999	.996	.983	.950	.887	.787	.654	.500	.346	.213	.113	.050	.017	.004	.001	.000	.000
	8	1.0	1.0	1.0	.999	.996	.985	.958	.905	.818	.696	.548	.390	.245	.131	.057	.018	.004	.000	.000
	9	1.0	1.0	1.0	1.0	.999	.996	.988	.966	.923	.849	.739	.597	.436	.278	.148	.061	.017	.002	.000
	10	1.0	1.0	1.0	1.0	1.0	.999	.997	.991	.975	.941	.880	.783	.648	.485	.314	.164	.062	.013	.001
	11	1.0	1.0	1.0	1.0	1.0	1.0	1.0	.998	.994	.982	.958	.909	.827	.703	.539	.352	.177	.056	.005
	12	1.0	1.0	1.0	1.0	1.0	1.0	1.0	1.0	.999	.996	.989	.973	.938	.873	.764	.602	.396	.184	.036
	13	1.0	1.0	1.0	1.0	1.0	1.0	1.0	1.0	1.0	1.0	.998	.995	.986	.965	.920	.833	.681	.451	.171
	14	1.0	1.0	1.0	1.0	1.0	1.0	1.0	1.0	1.0	1.0	1.0	1.0	.998	.995	.987	.965	.913	.794	.537

Table A2, continued. Binomial distribution

n	x	.050	.100	.150	.200	.250	.300	.350	.400	.450	.500	.550	.600	.650	.700	.750	.800	.850	.900	.950
16	1	.811	.515	.284	.141	.063	.026	.010	.003	.001	.000	.000	.000	.000	.000	.000	.000	.000	.000	.000
	2	.957	.789	.561	.352	.197	.099	.045	.018	.007	.002	.001	.000	.000	.000	.000	.000	.000	.000	.000
	3	.993	.932	.790	.598	.405	.246	.134	.065	.028	.011	.003	.001	.000	.000	.000	.000	.000	.000	.000
	4	.999	.983	.921	.798	.630	.450	.289	.167	.085	.038	.015	.005	.001	.000	.000	.000	.000	.000	.000
	5	1.0	.997	.976	.918	.810	.660	.490	.329	.198	.105	.049	.019	.006	.002	.000	.000	.000	.000	.000
	6	1.0	.999	.994	.973	.920	.825	.688	.527	.366	.227	.124	.058	.023	.007	.002	.000	.000	.000	.000
	7	1.0	1.0	.999	.993	.973	.926	.841	.716	.563	.402	.256	.142	.067	.026	.007	.001	.000	.000	.000
	8	1.0	1.0	1.0	.999	.993	.974	.933	.858	.744	.598	.437	.284	.159	.074	.027	.007	.001	.000	.000
	9	1.0	1.0	1.0	1.0	.998	.993	.977	.942	.876	.773	.634	.473	.312	.175	.080	.027	.006	.001	.000
	10	1.0	1.0	1.0	1.0	1.0	.998	.994	.981	.951	.895	.802	.671	.510	.340	.190	.082	.024	.003	.000
	11	1.0	1.0	1.0	1.0	1.0	1.0	.999	.995	.985	.962	.915	.833	.711	.550	.370	.202	.079	.017	.001
	12	1.0	1.0	1.0	1.0	1.0	1.0	1.0	.999	.997	.989	.972	.935	.866	.754	.595	.402	.210	.068	.007
	13	1.0	1.0	1.0	1.0	1.0	1.0	1.0	1.0	.999	.998	.993	.982	.955	.901	.803	.648	.439	.211	.043
	14	1.0	1.0	1.0	1.0	1.0	1.0	1.0	1.0	1.0	1.0	.999	.997	.990	.974	.937	.859	.716	.485	.189
	15	1.0	1.0	1.0	1.0	1.0	1.0	1.0	1.0	1.0	1.0	1.0	1.0	.999	.997	.990	.972	.926	.815	.560
18	1	.774	.450	.224	.099	.039	.014	.005	.001	.000	.000	.000	.000	.000	.000	.000	.000	.000	.000	.000
	2	.942	.734	.480	.271	.135	.060	.024	.008	.003	.001	.000	.000	.000	.000	.000	.000	.000	.000	.000
	3	.989	.902	.720	.501	.306	.165	.078	.033	.012	.004	.001	.000	.000	.000	.000	.000	.000	.000	.000
	4	.998	.972	.879	.716	.519	.333	.189	.094	.041	.015	.005	.001	.000	.000	.000	.000	.000	.000	.000
	5	1.0	.994	.958	.867	.717	.534	.355	.209	.108	.048	.018	.006	.001	.000	.000	.000	.000	.000	.000
	6	1.0	.999	.988	.949	.861	.722	.549	.374	.226	.119	.054	.020	.006	.001	.000	.000	.000	.000	.000
	7	1.0	1.0	.997	.984	.943	.859	.728	.563	.391	.240	.128	.058	.021	.006	.001	.000	.000	.000	.000
	8	1.0	1.0	.999	.996	.981	.940	.861	.737	.578	.407	.253	.135	.060	.021	.005	.001	.000	.000	.000
	9	1.0	1.0	1.0	.999	.995	.979	.940	.865	.747	.593	.422	.263	.139	.060	.019	.004	.001	.000	.000
	10	1.0	1.0	1.0	1.0	.999	.994	.979	.942	.872	.760	.609	.437	.272	.141	.057	.016	.003	.000	.000
	11	1.0	1.0	1.0	1.0	1.0	.999	.994	.980	.946	.881	.774	.626	.451	.278	.139	.051	.012	.001	.000
	12	1.0	1.0	1.0	1.0	1.0	1.0	.999	.994	.982	.952	.892	.791	.645	.466	.283	.133	.042	.006	.000
	13	1.0	1.0	1.0	1.0	1.0	1.0	1.0	.999	.995	.985	.959	.906	.811	.667	.481	.284	.121	.028	.002
	14	1.0	1.0	1.0	1.0	1.0	1.0	1.0	1.0	.999	.996	.988	.967	.922	.835	.694	.499	.280	.098	.011
	15	1.0	1.0	1.0	1.0	1.0	1.0	1.0	1.0	1.0	.999	.997	.992	.976	.940	.865	.729	.520	.266	.058
	16	1.0	1.0	1.0	1.0	1.0	1.0	1.0	1.0	1.0	1.0	1.0	.999	.995	.986	.961	.901	.776	.550	.226
20	1	.736	.392	.176	.069	.024	.008	.002	.001	.000	.000	.000	.000	.000	.000	.000	.000	.000	.000	.000
	2	.925	.677	.405	.206	.091	.035	.012	.004	.001	.000	.000	.000	.000	.000	.000	.000	.000	.000	.000
	3	.984	.867	.648	.411	.225	.107	.044	.016	.005	.001	.000	.000	.000	.000	.000	.000	.000	.000	.000
	4	.997	.957	.830	.630	.415	.238	.118	.051	.019	.006	.002	.000	.000	.000	.000	.000	.000	.000	.000
	5	1.0	.989	.933	.804	.617	.416	.245	.126	.055	.021	.006	.002	.000	.000	.000	.000	.000	.000	.000
	6	1.0	.998	.978	.913	.786	.608	.417	.250	.130	.058	.021	.006	.002	.000	.000	.000	.000	.000	.000
	7	1.0	1.0	.994	.968	.898	.772	.601	.416	.252	.132	.058	.021	.006	.001	.000	.000	.000	.000	.000
	8	1.0	1.0	.999	.990	.959	.887	.762	.596	.414	.252	.131	.057	.020	.005	.001	.000	.000	.000	.000
	9	1.0	1.0	1.0	.997	.986	.952	.878	.755	.591	.412	.249	.128	.053	.017	.004	.001	.000	.000	.000
	10	1.0	1.0	1.0	.999	.996	.983	.947	.872	.751	.588	.409	.245	.122	.048	.014	.003	.000	.000	.000
	11	1.0	1.0	1.0	1.0	.999	.995	.980	.943	.869	.748	.586	.404	.238	.113	.041	.010	.001	.000	.000
	12	1.0	1.0	1.0	1.0	1.0	.999	.994	.979	.942	.868	.748	.584	.399	.228	.102	.032	.006	.000	.000
	13	1.0	1.0	1.0	1.0	1.0	1.0	.998	.994	.979	.942	.870	.750	.583	.392	.214	.087	.022	.002	.000
	14	1.0	1.0	1.0	1.0	1.0	1.0	1.0	.998	.994	.979	.945	.874	.755	.584	.383	.196	.067	.011	.000
	15	1.0	1.0	1.0	1.0	1.0	1.0	1.0	1.0	.998	.994	.981	.949	.882	.762	.585	.370	.170	.043	.003
	16	1.0	1.0	1.0	1.0	1.0	1.0	1.0	1.0	1.0	.999	.995	.984	.956	.893	.775	.589	.352	.133	.016
25	2	.873	.537	.254	.098	.032	.009	.002	.000	.000	.000	.000	.000	.000	.000	.000	.000	.000	.000	.000
	3	.966	.764	.471	.234	.096	.033	.010	.002	.000	.000	.000	.000	.000	.000	.000	.000	.000	.000	.000
	4	.993	.902	.682	.421	.214	.090	.032	.009	.002	.000	.000	.000	.000	.000	.000	.000	.000	.000	.000
	5	.999	.967	.838	.617	.378	.193	.083	.029	.009	.002	.000	.000	.000	.000	.000	.000	.000	.000	.000
	6	1.0	.991	.930	.780	.561	.341	.173	.074	.026	.007	.002	.000	.000	.000	.000	.000	.000	.000	.000
	7	1.0	.998	.975	.891	.727	.512	.306	.154	.064	.022	.006	.001	.000	.000	.000	.000	.000	.000	.000
	8	1.0	1.0	.992	.953	.851	.677	.467	.274	.134	.054	.017	.004	.001	.000	.000	.000	.000	.000	.000
	9	1.0	1.0	.998	.983	.929	.811	.630	.425	.242	.115	.044	.013	.003	.000	.000	.000	.000	.000	.000
	10	1.0	1.0	1.0	.994	.970	.902	.771	.586	.384	.212	.096	.034	.009	.002	.000	.000	.000	.000	.000
	11	1.0	1.0	1.0	.998	.989	.956	.875	.732	.543	.345	.183	.078	.025	.006	.001	.000	.000	.000	.000
	12	1.0	1.0	1.0	1.0	.997	.983	.940	.846	.694	.500	.306	.154	.060	.017	.003	.000	.000	.000	.000
	13	1.0	1.0	1.0	1.0	.999	.994	.975	.922	.817	.655	.457	.268	.125	.044	.011	.002	.000	.000	.000
	14	1.0	1.0	1.0	1.0	1.0	.998	.991	.966	.904	.788	.616	.414	.229	.098	.030	.006	.000	.000	.000
	15	1.0	1.0	1.0	1.0	1.0	1.0	.997	.987	.956	.885	.758	.575	.370	.189	.071	.017	.002	.000	.000
	16	1.0	1.0	1.0	1.0	1.0	1.0	.999	.996	.983	.946	.866	.726	.533	.323	.149	.047	.008	.000	.000
	17	1.0	1.0	1.0	1.0	1.0	1.0	1.0	.999	.994	.978	.936	.846	.694	.488	.273	.109	.025	.002	.000
	18	1.0	1.0	1.0	1.0	1.0	1.0	1.0	1.0	.998	.993	.974	.926	.827	.659	.439	.220	.070	.009	.000
	19	1.0	1.0	1.0	1.0	1.0	1.0	1.0	1.0	1.0	.998	.991	.971	.917	.807	.622	.383	.162	.033	.001
	20	1.0	1.0	1.0	1.0	1.0	1.0	1.0	1.0	1.0	1.0	.998	.991	.968	.910	.786	.579	.318	.098	.007

Table A3. Poisson distribution

$$F(x) = P\{X \le x\} = \sum_{k=0}^{x} \frac{e^{-\lambda}\lambda^k}{k!}$$

x	λ														
	0.1	0.2	0.3	0.4	0.5	0.6	0.7	0.8	0.9	1.0	1.1	1.2	1.3	1.4	1.5
0	.905	.819	.741	.670	.607	.549	.497	.449	.407	.368	.333	.301	.273	.247	.223
1	.995	.982	.963	.938	.910	.878	.844	.809	.772	.736	.699	.663	.627	.592	.558
2	1.00	.999	.996	.992	.986	.977	.966	.953	.937	.920	.900	.879	.857	.833	.809
3	1.00	1.00	1.00	.999	.998	.997	.994	.991	.987	.981	.974	.966	.957	.946	.934
4	1.00	1.00	1.00	1.00	1.00	1.00	.999	.999	.998	.996	.995	.992	.989	.986	.981
5	1.00	1.00	1.00	1.00	1.00	1.00	1.00	1.00	1.00	.999	.999	.998	.998	.997	.996
6	1.00	1.00	1.00	1.00	1.00	1.00	1.00	1.00	1.00	1.00	1.00	1.00	1.00	.999	.999
7	1.00	1.00	1.00	1.00	1.00	1.00	1.00	1.00	1.00	1.00	1.00	1.00	1.00	1.00	1.00

x	λ														
	1.6	1.7	1.8	1.9	2.0	2.1	2.2	2.3	2.4	2.5	2.6	2.7	2.8	2.9	3.0
0	.202	.183	.165	.150	.135	.122	.111	.100	.091	.082	.074	.067	.061	.055	.050
1	.525	.493	.463	.434	.406	.380	.355	.331	.308	.287	.267	.249	.231	.215	.199
2	.783	.757	.731	.704	.677	.650	.623	.596	.570	.544	.518	.494	.469	.446	.423
3	.921	.907	.891	.875	.857	.839	.819	.799	.779	.758	.736	.714	.692	.670	.647
4	.976	.970	.964	.956	.947	.938	.928	.916	.904	.891	.877	.863	.848	.832	.815
5	.994	.992	.990	.987	.983	.980	.975	.970	.964	.958	.951	.943	.935	.926	.916
6	.999	.998	.997	.997	.995	.994	.993	.991	.988	.986	.983	.979	.976	.971	.966
7	1.00	1.00	.999	.999	.999	.999	.998	.997	.997	.996	.995	.993	.992	.990	.988
8	1.00	1.00	1.00	1.00	1.00	1.00	1.00	.999	.999	.999	.999	.998	.998	.997	.996
9	1.00	1.00	1.00	1.00	1.00	1.00	1.00	1.00	1.00	1.00	1.00	.999	.999	.999	.999
10	1.00	1.00	1.00	1.00	1.00	1.00	1.00	1.00	1.00	1.00	1.00	1.00	1.00	1.00	1.00

x	λ														
	3.5	4.0	4.5	5.0	5.5	6.0	6.5	7.0	7.5	8.0	8.5	9.0	9.5	10.0	10.5
0	.030	.018	.011	.007	.004	.002	.002	.001	.001	.000	.000	.000	.000	.000	.000
1	.136	.092	.061	.040	.027	.017	.011	.007	.005	.003	.002	.001	.001	.000	.000
2	.321	.238	.174	.125	.088	.062	.043	.030	.020	.014	.009	.006	.004	.003	.002
3	.537	.433	.342	.265	.202	.151	.112	.082	.059	.042	.030	.021	.015	.010	.007
4	.725	.629	.532	.440	.358	.285	.224	.173	.132	.100	.074	.055	.040	.029	.021
5	.858	.785	.703	.616	.529	.446	.369	.301	.241	.191	.150	.116	.089	.067	.050
6	.935	.889	.831	.762	.686	.606	.527	.450	.378	.313	.256	.207	.165	.130	.102
7	.973	.949	.913	.867	.809	.744	.673	.599	.525	.453	.386	.324	.269	.220	.179
8	.990	.979	.960	.932	.894	.847	.792	.729	.662	.593	.523	.456	.392	.333	.279
9	.997	.992	.983	.968	.946	.916	.877	.830	.776	.717	.653	.587	.522	.458	.397
10	.999	.997	.993	.986	.975	.957	.933	.901	.862	.816	.763	.706	.645	.583	.521
11	1.00	.999	.998	.995	.989	.980	.966	.947	.921	.888	.849	.803	.752	.697	.639
12	1.00	1.00	.999	.998	.996	.991	.984	.973	.957	.936	.909	.876	.836	.792	.742
13	1.00	1.00	1.00	.999	.998	.996	.993	.987	.978	.966	.949	.926	.898	.864	.825
14	1.00	1.00	1.00	1.00	.999	.999	.997	.994	.990	.983	.973	.959	.940	.917	.888
15	1.00	1.00	1.00	1.00	1.00	.999	.999	.998	.995	.992	.986	.978	.967	.951	.932
16	1.00	1.00	1.00	1.00	1.00	1.00	1.00	.999	.998	.996	.993	.989	.982	.973	.960
17	1.00	1.00	1.00	1.00	1.00	1.00	1.00	1.00	.999	.998	.997	.995	.991	.986	.978
18	1.00	1.00	1.00	1.00	1.00	1.00	1.00	1.00	1.00	.999	.999	.998	.996	.993	.988
19	1.00	1.00	1.00	1.00	1.00	1.00	1.00	1.00	1.00	1.00	.999	.999	.998	.997	.994
20	1.00	1.00	1.00	1.00	1.00	1.00	1.00	1.00	1.00	1.00	1.00	1.00	.999	.998	.997

Table A3, continued. Poisson distribution

x	11	12	13	14	15	16	17	18	19	20	22	24	26	28	30
0	.000	.000	.000	.000	.000	.000	.000	.000	.000	.000	.000	.000	.000	.000	.000
1	.000	.000	.000	.000	.000	.000	.000	.000	.000	.000	.000	.000	.000	.000	.000
2	.001	.001	.000	.000	.000	.000	.000	.000	.000	.000	.000	.000	.000	.000	.000
3	.005	.002	.001	.000	.000	.000	.000	.000	.000	.000	.000	.000	.000	.000	.000
4	.015	.008	.004	.002	.001	.000	.000	.000	.000	.000	.000	.000	.000	.000	.000
5	.038	.020	.011	.006	.003	.001	.001	.000	.000	.000	.000	.000	.000	.000	.000
6	.079	.046	.026	.014	.008	.004	.002	.001	.001	.000	.000	.000	.000	.000	.000
7	.143	.090	.054	.032	.018	.010	.005	.003	.002	.001	.000	.000	.000	.000	.000
8	.232	.155	.100	.062	.037	.022	.013	.007	.004	.002	.001	.000	.000	.000	.000
9	.341	.242	.166	.109	.070	.043	.026	.015	.009	.005	.002	.000	.000	.000	.000
10	.460	.347	.252	.176	.118	.077	.049	.030	.018	.011	.004	.001	.000	.000	.000
11	.579	.462	.353	.260	.185	.127	.085	.055	.035	.021	.008	.003	.001	.000	.000
12	.689	.576	.463	.358	.268	.193	.135	.092	.061	.039	.015	.005	.002	.001	.000
13	.781	.682	.573	.464	.363	.275	.201	.143	.098	.066	.028	.011	.004	.001	.000
14	.854	.772	.675	.570	.466	.368	.281	.208	.150	.105	.048	.020	.008	.003	.001
15	.907	.844	.764	.669	.568	.467	.371	.287	.215	.157	.077	.034	.014	.005	.002
16	.944	.899	.835	.756	.664	.566	.468	.375	.292	.221	.117	.056	.025	.010	.004
17	.968	.937	.890	.827	.749	.659	.564	.469	.378	.297	.169	.087	.041	.018	.007
18	.982	.963	.930	.883	.819	.742	.655	.562	.469	.381	.232	.128	.065	.030	.013
19	.991	.979	.957	.923	.875	.812	.736	.651	.561	.470	.306	.180	.097	.048	.022
20	.995	.988	.975	.952	.917	.868	.805	.731	.647	.559	.387	.243	.139	.073	.035
21	.998	.994	.986	.971	.947	.911	.861	.799	.725	.644	.472	.314	.190	.106	.054
22	.999	.997	.992	.983	.967	.942	.905	.855	.793	.721	.556	.392	.252	.148	.081
23	1.00	.999	.996	.991	.981	.963	.937	.899	.849	.787	.637	.473	.321	.200	.115
24	1.00	.999	.998	.995	.989	.978	.959	.932	.893	.843	.712	.554	.396	.260	.157
25	1.00	1.00	.999	.997	.994	.987	.975	.955	.927	.888	.777	.632	.474	.327	.208
26	1.00	1.00	1.00	.999	.997	.993	.985	.972	.951	.922	.832	.704	.552	.400	.267
27	1.00	1.00	1.00	.999	.998	.996	.991	.983	.969	.948	.877	.768	.627	.475	.333
28	1.00	1.00	1.00	1.00	.999	.998	.995	.990	.980	.966	.913	.823	.697	.550	.403
29	1.00	1.00	1.00	1.00	1.00	.999	.997	.994	.988	.978	.940	.868	.759	.623	.476
30	1.00	1.00	1.00	1.00	1.00	.999	.999	.997	.993	.987	.959	.904	.813	.690	.548
31	1.00	1.00	1.00	1.00	1.00	1.00	.999	.998	.996	.992	.973	.932	.859	.752	.619
32	1.00	1.00	1.00	1.00	1.00	1.00	1.00	.999	.998	.995	.983	.953	.896	.805	.685
33	1.00	1.00	1.00	1.00	1.00	1.00	1.00	1.00	.999	.997	.989	.969	.925	.850	.744
34	1.00	1.00	1.00	1.00	1.00	1.00	1.00	1.00	.999	.999	.994	.979	.947	.888	.797
35	1.00	1.00	1.00	1.00	1.00	1.00	1.00	1.00	1.00	.999	.996	.987	.964	.918	.843
36	1.00	1.00	1.00	1.00	1.00	1.00	1.00	1.00	1.00	1.00	.998	.992	.976	.941	.880
37	1.00	1.00	1.00	1.00	1.00	1.00	1.00	1.00	1.00	1.00	.999	.995	.984	.959	.911
38	1.00	1.00	1.00	1.00	1.00	1.00	1.00	1.00	1.00	1.00	.999	.997	.990	.972	.935
39	1.00	1.00	1.00	1.00	1.00	1.00	1.00	1.00	1.00	1.00	1.00	.998	.994	.981	.954
40	1.00	1.00	1.00	1.00	1.00	1.00	1.00	1.00	1.00	1.00	1.00	.999	.996	.988	.968
41	1.00	1.00	1.00	1.00	1.00	1.00	1.00	1.00	1.00	1.00	1.00	.999	.998	.992	.978
42	1.00	1.00	1.00	1.00	1.00	1.00	1.00	1.00	1.00	1.00	1.00	1.00	.999	.995	.985
43	1.00	1.00	1.00	1.00	1.00	1.00	1.00	1.00	1.00	1.00	1.00	1.00	.999	.997	.990
44	1.00	1.00	1.00	1.00	1.00	1.00	1.00	1.00	1.00	1.00	1.00	1.00	1.00	.998	.994
45	1.00	1.00	1.00	1.00	1.00	1.00	1.00	1.00	1.00	1.00	1.00	1.00	1.00	.999	.996
46	1.00	1.00	1.00	1.00	1.00	1.00	1.00	1.00	1.00	1.00	1.00	1.00	1.00	.999	.998
47	1.00	1.00	1.00	1.00	1.00	1.00	1.00	1.00	1.00	1.00	1.00	1.00	1.00	1.00	.999
48	1.00	1.00	1.00	1.00	1.00	1.00	1.00	1.00	1.00	1.00	1.00	1.00	1.00	1.00	.999
49	1.00	1.00	1.00	1.00	1.00	1.00	1.00	1.00	1.00	1.00	1.00	1.00	1.00	1.00	.999
50	1.00	1.00	1.00	1.00	1.00	1.00	1.00	1.00	1.00	1.00	1.00	1.00	1.00	1.00	1.00

Table A4. Standard Normal distribution

$$\Phi(z) = P\left\{ Z \le z \right\} = \frac{1}{\sqrt{2\pi}} \int_{-\infty}^{z} e^{-x^2/2} dx$$

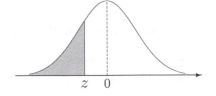

z	-0.09	-0.08	-0.07	-0.06	-0.05	-0.04	-0.03	-0.02	-0.01	-0.00
-(3.9+)	.0000	.0000	.0000	.0000	.0000	.0000	.0000	.0000	.0000	.0000
-3.8	.0001	.0001	.0001	.0001	.0001	.0001	.0001	.0001	.0001	.0001
-3.7	.0001	.0001	.0001	.0001	.0001	.0001	.0001	.0001	.0001	.0001
-3.6	.0001	.0001	.0001	.0001	.0001	.0001	.0001	.0001	.0002	.0002
-3.5	.0002	.0002	.0002	.0002	.0002	.0002	.0002	.0002	.0002	.0002
-3.4	.0002	.0003	.0003	.0003	.0003	.0003	.0003	.0003	.0003	.0003
-3.3	.0003	.0004	.0004	.0004	.0004	.0004	.0004	.0005	.0005	.0005
-3.2	.0005	.0005	.0005	.0006	.0006	.0006	.0006	.0006	.0007	.0007
-3.1	.0007	.0007	.0008	.0008	.0008	.0008	.0009	.0009	.0009	.0010
-3.0	.0010	.0010	.0011	.0011	.0011	.0012	.0012	.0013	.0013	.0013
-2.9	.0014	.0014	.0015	.0015	.0016	.0016	.0017	.0018	.0018	.0019
-2.8	.0019	.0020	.0021	.0021	.0022	.0023	.0023	.0024	.0025	.0026
-2.7	.0026	.0027	.0028	.0029	.0030	.0031	.0032	.0033	.0034	.0035
-2.6	.0036	.0037	.0038	.0039	.0040	.0041	.0043	.0044	.0045	.0047
-2.5	.0048	.0049	.0051	.0052	.0054	.0055	.0057	.0059	.0060	.0062
-2.4	.0064	.0066	.0068	.0069	.0071	.0073	.0075	.0078	.0080	.0082
-2.3	.0084	.0087	.0089	.0091	.0094	.0096	.0099	.0102	.0104	.0107
-2.2	.0110	.0113	.0116	.0119	.0122	.0125	.0129	.0132	.0136	.0139
-2.1	.0143	.0146	.0150	.0154	.0158	.0162	.0166	.0170	.0174	.0179
-2.0	.0183	.0188	.0192	.0197	.0202	.0207	.0212	.0217	.0222	.0228
-1.9	.0233	.0239	.0244	.0250	.0256	.0262	.0268	.0274	.0281	.0287
-1.8	.0294	.0301	.0307	.0314	.0322	.0329	.0336	.0344	.0351	.0359
-1.7	.0367	.0375	.0384	.0392	.0401	.0409	.0418	.0427	.0436	.0446
-1.6	.0455	.0465	.0475	.0485	.0495	.0505	.0516	.0526	.0537	.0548
-1.5	.0559	.0571	.0582	.0594	.0606	.0618	.0630	.0643	.0655	.0668
-1.4	.0681	.0694	.0708	.0721	.0735	.0749	.0764	.0778	.0793	.0808
-1.3	.0823	.0838	.0853	.0869	.0885	.0901	.0918	.0934	.0951	.0968
-1.2	.0985	.1003	.1020	.1038	.1056	.1075	.1093	.1112	.1131	.1151
-1.1	.1170	.1190	.1210	.1230	.1251	.1271	.1292	.1314	.1335	.1357
-1.0	.1379	.1401	.1423	.1446	.1469	.1492	.1515	.1539	.1562	.1587
-0.9	.1611	.1635	.1660	.1685	.1711	.1736	.1762	.1788	.1814	.1841
-0.8	.1867	.1894	.1922	.1949	.1977	.2005	.2033	.2061	.2090	.2119
-0.7	.2148	.2177	.2206	.2236	.2266	.2296	.2327	.2358	.2389	.2420
-0.6	.2451	.2483	.2514	.2546	.2578	.2611	.2643	.2676	.2709	.2743
-0.5	.2776	.2810	.2843	.2877	.2912	.2946	.2981	.3015	.3050	.3085
-0.4	.3121	.3156	.3192	.3228	.3264	.3300	.3336	.3372	.3409	.3446
-0.3	.3483	.3520	.3557	.3594	.3632	.3669	.3707	.3745	.3783	.3821
-0.2	.3859	.3897	.3936	.3974	.4013	.4052	.4090	.4129	.4168	.4207
-0.1	.4247	.4286	.4325	.4364	.4404	.4443	.4483	.4522	.4562	.4602
-0.0	.4641	.4681	.4721	.4761	.4801	.4840	.4880	.4920	.4960	.5000

Table A4, continued.

Standard Normal distribution

$$\Phi(z) = \boldsymbol{P}\{Z \le z\} = \frac{1}{\sqrt{2\pi}} \int_{-\infty}^{z} e^{-x^2/2}dx$$

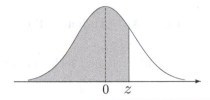

z	0.00	0.01	0.02	0.03	0.04	0.05	0.06	0.07	0.08	0.09
0.0	.5000	.5040	.5080	.5120	.5160	.5199	.5239	.5279	.5319	.5359
0.1	.5398	.5438	.5478	.5517	.5557	.5596	.5636	.5675	.5714	.5753
0.2	.5793	.5832	.5871	.5910	.5948	.5987	.6026	.6064	.6103	.6141
0.3	.6179	.6217	.6255	.6293	.6331	.6368	.6406	.6443	.6480	.6517
0.4	.6554	.6591	.6628	.6664	.6700	.6736	.6772	.6808	.6844	.6879
0.5	.6915	.6950	.6985	.7019	.7054	.7088	.7123	.7157	.7190	.7224
0.6	.7257	.7291	.7324	.7357	.7389	.7422	.7454	.7486	.7517	.7549
0.7	.7580	.7611	.7642	.7673	.7704	.7734	.7764	.7794	.7823	.7852
0.8	.7881	.7910	.7939	.7967	.7995	.8023	.8051	.8078	.8106	.8133
0.9	.8159	.8186	.8212	.8238	.8264	.8289	.8315	.8340	.8365	.8389
1.0	.8413	.8438	.8461	.8485	.8508	.8531	.8554	.8577	.8599	.8621
1.1	.8643	.8665	.8686	.8708	.8729	.8749	.8770	.8790	.8810	.8830
1.2	.8849	.8869	.8888	.8907	.8925	.8944	.8962	.8980	.8997	.9015
1.3	.9032	.9049	.9066	.9082	.9099	.9115	.9131	.9147	.9162	.9177
1.4	.9192	.9207	.9222	.9236	.9251	.9265	.9279	.9292	.9306	.9319
1.5	.9332	.9345	.9357	.9370	.9382	.9394	.9406	.9418	.9429	.9441
1.6	.9452	.9463	.9474	.9484	.9495	.9505	.9515	.9525	.9535	.9545
1.7	.9554	.9564	.9573	.9582	.9591	.9599	.9608	.9616	.9625	.9633
1.8	.9641	.9649	.9656	.9664	.9671	.9678	.9686	.9693	.9699	.9706
1.9	.9713	.9719	.9726	.9732	.9738	.9744	.9750	.9756	.9761	.9767
2.0	.9772	.9778	.9783	.9788	.9793	.9798	.9803	.9808	.9812	.9817
2.1	.9821	.9826	.9830	.9834	.9838	.9842	.9846	.9850	.9854	.9857
2.2	.9861	.9864	.9868	.9871	.9875	.9878	.9881	.9884	.9887	.9890
2.3	.9893	.9896	.9898	.9901	.9904	.9906	.9909	.9911	.9913	.9916
2.4	.9918	.9920	.9922	.9925	.9927	.9929	.9931	.9932	.9934	.9936
2.5	.9938	.9940	.9941	.9943	.9945	.9946	.9948	.9949	.9951	.9952
2.6	.9953	.9955	.9956	.9957	.9959	.9960	.9961	.9962	.9963	.9964
2.7	.9965	.9966	.9967	.9968	.9969	.9970	.9971	.9972	.9973	.9974
2.8	.9974	.9975	.9976	.9977	.9977	.9978	.9979	.9979	.9980	.9981
2.9	.9981	.9982	.9982	.9983	.9984	.9984	.9985	.9985	.9986	.9986
3.0	.9987	.9987	.9987	.9988	.9988	.9989	.9989	.9989	.9990	.9990
3.1	.9990	.9991	.9991	.9991	.9992	.9992	.9992	.9992	.9993	.9993
3.2	.9993	.9993	.9994	.9994	.9994	.9994	.9994	.9995	.9995	.9995
3.3	.9995	.9995	.9995	.9996	.9996	.9996	.9996	.9996	.9996	.9997
3.4	.9997	.9997	.9997	.9997	.9997	.9997	.9997	.9997	.9997	.9998
3.5	.9998	.9998	.9998	.9998	.9998	.9998	.9998	.9998	.9998	.9998
3.6	.9998	.9998	.9999	.9999	.9999	.9999	.9999	.9999	.9999	.9999
3.7	.9999	.9999	.9999	.9999	.9999	.9999	.9999	.9999	.9999	.9999
3.8	.9999	.9999	.9999	.9999	.9999	.9999	.9999	.9999	.9999	.9999
3.9+	1.00	1.00	1.00	1.00	1.00	1.00	1.00	1.00	1.00	1.00

Table A5. Student's T-distribution

t_α; critical values, such that $P\{t > t_\alpha\} = \alpha$

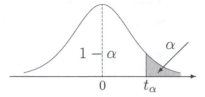

ν	α, the right-tail probability									
(d.f.)	.10	.05	.025	.02	.01	.005	.0025	.001	.0005	.0001
1	3.078	6.314	12.706	15.89	31.82	63.66	127.3	318.3	636.6	3185
2	1.886	2.920	4.303	4.849	6.965	9.925	14.09	22.33	31.60	70.71
3	1.638	2.353	3.182	3.482	4.541	5.841	7.453	10.21	12.92	22.20
4	1.533	2.132	2.776	2.999	3.747	4.604	5.598	7.173	8.610	13.04
5	1.476	2.015	2.571	2.757	3.365	4.032	4.773	5.894	6.869	9.676
6	1.440	1.943	2.447	2.612	3.143	3.707	4.317	5.208	5.959	8.023
7	1.415	1.895	2.365	2.517	2.998	3.499	4.029	4.785	5.408	7.064
8	1.397	1.860	2.306	2.449	2.896	3.355	3.833	4.501	5.041	6.442
9	1.383	1.833	2.262	2.398	2.821	3.250	3.690	4.297	4.781	6.009
10	1.372	1.812	2.228	2.359	2.764	3.169	3.581	4.144	4.587	5.694
11	1.363	1.796	2.201	2.328	2.718	3.106	3.497	4.025	4.437	5.453
12	1.356	1.782	2.179	2.303	2.681	3.055	3.428	3.930	4.318	5.263
13	1.350	1.771	2.160	2.282	2.650	3.012	3.372	3.852	4.221	5.111
14	1.345	1.761	2.145	2.264	2.624	2.977	3.326	3.787	4.140	4.985
15	1.341	1.753	2.131	2.249	2.602	2.947	3.286	3.733	4.073	4.880
16	1.337	1.746	2.120	2.235	2.583	2.921	3.252	3.686	4.015	4.790
17	1.333	1.740	2.110	2.224	2.567	2.898	3.222	3.646	3.965	4.715
18	1.330	1.734	2.101	2.214	2.552	2.878	3.197	3.610	3.922	4.648
19	1.328	1.729	2.093	2.205	2.539	2.861	3.174	3.579	3.883	4.590
20	1.325	1.725	2.086	2.197	2.528	2.845	3.153	3.552	3.850	4.539
21	1.323	1.721	2.080	2.189	2.518	2.831	3.135	3.527	3.819	4.492
22	1.321	1.717	2.074	2.183	2.508	2.819	3.119	3.505	3.792	4.452
23	1.319	1.714	2.069	2.177	2.500	2.807	3.104	3.485	3.768	4.416
24	1.318	1.711	2.064	2.172	2.492	2.797	3.091	3.467	3.745	4.382
25	1.316	1.708	2.060	2.167	2.485	2.787	3.078	3.450	3.725	4.352
26	1.315	1.706	2.056	2.162	2.479	2.779	3.067	3.435	3.707	4.324
27	1.314	1.703	2.052	2.158	2.473	2.771	3.057	3.421	3.689	4.299
28	1.313	1.701	2.048	2.154	2.467	2.763	3.047	3.408	3.674	4.276
29	1.311	1.699	2.045	2.150	2.462	2.756	3.038	3.396	3.660	4.254
30	1.310	1.697	2.042	2.147	2.457	2.750	3.030	3.385	3.646	4.234
32	1.309	1.694	2.037	2.141	2.449	2.738	3.015	3.365	3.622	4.198
34	1.307	1.691	2.032	2.136	2.441	2.728	3.002	3.348	3.601	4.168
36	1.306	1.688	2.028	2.131	2.434	2.719	2.990	3.333	3.582	4.140
38	1.304	1.686	2.024	2.127	2.429	2.712	2.980	3.319	3.566	4.115
40	1.303	1.684	2.021	2.123	2.423	2.704	2.971	3.307	3.551	4.094
45	1.301	1.679	2.014	2.115	2.412	2.690	2.952	3.281	3.520	4.049
50	1.299	1.676	2.009	2.109	2.403	2.678	2.937	3.261	3.496	4.014
55	1.297	1.673	2.004	2.104	2.396	2.668	2.925	3.245	3.476	3.985
60	1.296	1.671	2.000	2.099	2.390	2.660	2.915	3.232	3.460	3.962
70	1.294	1.667	1.994	2.093	2.381	2.648	2.899	3.211	3.435	3.926
80	1.292	1.664	1.990	2.088	2.374	2.639	2.887	3.195	3.416	3.899
90	1.291	1.662	1.987	2.084	2.368	2.632	2.878	3.183	3.402	3.878
100	1.290	1.660	1.984	2.081	2.364	2.626	2.871	3.174	3.390	3.861
200	1.286	1.653	1.972	2.067	2.345	2.601	2.838	3.131	3.340	3.789
∞	1.282	1.645	1.960	2.054	2.326	2.576	2.807	3.090	3.290	3.719

Table A6. Chi-Square Distribution

χ_α^2; critical values, such that $P\{\chi^2 > \chi_\alpha^2\} = \alpha$

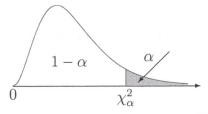

ν (d.f.)	.999	.995	.99	.975	.95	.90	.80	.20	.10	.05	.025	.01	.005	.001
							α, the right-tail probability							
1	0.00	0.00	0.00	0.00	0.00	0.02	0.06	1.64	2.71	3.84	5.02	6.63	7.88	10.8
2	0.00	0.01	0.02	0.05	0.10	0.21	0.45	3.22	4.61	5.99	7.38	9.21	10.6	13.8
3	0.02	0.07	0.11	0.22	0.35	0.58	1.01	4.64	6.25	7.81	9.35	11.3	12.8	16.3
4	0.09	0.21	0.30	0.48	0.71	1.06	1.65	5.99	7.78	9.49	11.1	13.3	14.9	18.5
5	0.21	0.41	0.55	0.83	1.15	1.61	2.34	7.29	9.24	11.1	12.8	15.1	16.7	20.5
6	0.38	0.68	0.87	1.24	1.64	2.20	3.07	8.56	10.6	12.6	14.4	16.8	18.5	22.5
7	0.60	0.99	1.24	1.69	2.17	2.83	3.82	9.80	12.0	14.1	16.0	18.5	20.3	24.3
8	0.86	1.34	1.65	2.18	2.73	3.49	4.59	11.0	13.4	15.5	17.5	20.1	22.0	26.1
9	1.15	1.73	2.09	2.70	3.33	4.17	5.38	12.2	14.7	16.9	19.0	21.7	23.6	27.9
10	1.48	2.16	2.56	3.25	3.94	4.87	6.18	13.4	16.0	18.3	20.5	23.2	25.2	29.6
11	1.83	2.60	3.05	3.82	4.57	5.58	6.99	14.6	17.3	19.7	21.9	24.7	26.8	31.3
12	2.21	3.07	3.57	4.40	5.23	6.30	7.81	15.8	18.5	21.0	23.3	26.2	28.3	32.9
13	2.62	3.57	4.11	5.01	5.89	7.04	8.63	17.0	19.8	22.4	24.7	27.7	29.8	34.5
14	3.04	4.07	4.66	5.63	6.57	7.79	9.47	18.2	21.1	23.7	26.1	29.1	31.3	36.1
15	3.48	4.60	5.23	6.26	7.26	8.55	10.3	19.3	22.3	25.0	27.5	30.6	32.8	37.7
16	3.94	5.14	5.81	6.91	7.96	9.31	11.1	20.5	23.5	26.3	28.8	32.0	34.3	39.3
17	4.42	5.70	6.41	7.56	8.67	10.1	12.0	21.6	24.8	27.6	30.2	33.4	35.7	40.8
18	4.90	6.26	7.01	8.23	9.39	10.9	12.9	22.8	26.0	28.9	31.5	34.8	37.2	42.3
19	5.41	6.84	7.63	8.91	10.1	11.7	13.7	23.9	27.2	30.1	32.9	36.2	38.6	43.8
20	5.92	7.43	8.26	9.59	10.9	12.4	14.6	25.0	28.4	31.4	34.2	37.6	40.0	45.3
21	6.45	8.03	8.90	10.3	11.6	13.2	15.4	26.2	29.6	32.7	35.5	38.9	41.4	46.8
22	6.98	8.64	9.54	11.0	12.3	14.0	16.3	27.3	30.8	33.9	36.8	40.3	42.8	48.3
23	7.53	9.26	10.2	11.7	13.1	14.8	17.2	28.4	32.0	35.2	38.1	41.6	44.2	49.7
24	8.08	9.89	10.9	12.4	13.8	15.7	18.1	29.6	33.2	36.4	39.4	43.0	45.6	51.2
25	8.65	10.5	11.5	13.1	14.6	16.5	18.9	30.7	34.4	37.7	40.6	44.3	46.9	52.6
26	9.22	11.2	12.2	13.8	15.4	17.3	19.8	31.8	35.6	38.9	41.9	45.6	48.3	54.1
27	9.80	11.8	12.9	14.6	16.2	18.1	20.7	32.9	36.7	40.1	43.2	47.0	49.6	55.5
28	10.4	12.5	13.6	15.3	16.9	18.9	21.6	34.0	37.9	41.3	44.5	48.3	51.0	56.9
29	11.0	13.1	14.3	16.0	17.7	19.8	22.5	35.1	39.1	42.6	45.7	49.6	52.3	58.3
30	11.6	13.8	15.0	16.8	18.5	20.6	23.4	36.3	40.3	43.8	47.0	50.9	53.7	59.7
31	12.2	14.5	15.7	17.5	19.3	21.4	24.3	37.4	41.4	45.0	48.2	52.2	55.0	61.1
32	12.8	15.1	16.4	18.3	20.1	22.3	25.1	38.5	42.6	46.2	49.5	53.5	56.3	62.5
33	13.4	15.8	17.1	19.0	20.9	23.1	26.0	39.6	43.7	47.4	50.7	54.8	57.6	63.9
34	14.1	16.5	17.8	19.8	21.7	24.0	26.9	40.7	44.9	48.6	52.0	56.1	59.0	65.2
35	14.7	17.2	18.5	20.6	22.5	24.8	27.8	41.8	46.1	49.8	53.2	57.3	60.3	66.6
36	15.3	17.9	19.2	21.3	23.3	25.6	28.7	42.9	47.2	51.0	54.4	58.6	61.6	68
37	16.0	18.6	20.0	22.1	24.1	26.5	29.6	44.0	48.4	52.2	55.7	59.9	62.9	69.3
38	16.6	19.3	20.7	22.9	24.9	27.3	30.5	45.1	49.5	53.4	56.9	61.2	64.2	70.7
39	17.3	20.0	21.4	23.7	25.7	28.2	31.4	46.2	50.7	54.6	58.1	62.4	65.5	72.1
40	17.9	20.7	22.2	24.4	26.5	29.1	32.3	47.3	51.8	55.8	59.3	63.7	66.8	73.4
41	18.6	21.4	22.9	25.2	27.3	29.9	33.3	48.4	52.9	56.9	60.6	65.0	68.1	74.7
42	19.2	22.1	23.7	26.0	28.1	30.8	34.2	49.5	54.1	58.1	61.8	66.2	69.3	76.1
43	19.9	22.9	24.4	26.8	29.0	31.6	35.1	50.5	55.2	59.3	63.0	67.5	70.6	77.4
44	20.6	23.6	25.1	27.6	29.8	32.5	36.0	51.6	56.4	60.5	64.2	68.7	71.9	78.7
45	21.3	24.3	25.9	28.4	30.6	33.4	36.9	52.7	57.5	61.7	65.4	70.0	73.2	80.1
46	21.9	25.0	26.7	29.2	31.4	34.2	37.8	53.8	58.6	62.8	66.6	71.2	74.4	81.4
47	22.6	25.8	27.4	30.0	32.3	35.1	38.7	54.9	59.8	64.0	67.8	72.4	75.7	82.7
48	23.3	26.5	28.2	30.8	33.1	35.9	39.6	56.0	60.9	65.2	69.0	73.7	77.0	84.0
49	24.0	27.2	28.9	31.6	33.9	36.8	40.5	57.1	62.0	66.3	70.2	74.9	78.2	85.4
50	24.7	28.0	29.7	32.4	34.8	37.7	41.4	58.2	63.2	67.5	71.4	76.2	79.5	86.7

Table A7. F-distribution

F_α; critical values such that $P\{F > F_\alpha\} = \alpha$

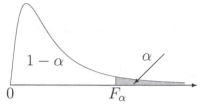

ν_2, denom. d.f.	α	ν_1, numerator degrees of freedom									
		1	2	3	4	5	6	7	8	9	10
1	0.25	5.83	7.5	8.2	8.58	8.82	8.98	9.1	9.19	9.26	9.32
	0.1	39.9	49.5	53.6	55.8	57.2	58.2	58.9	59.4	59.9	60.2
	0.05	161	199	216	225	230	234	237	239	241	242
	0.025	648	799	864	900	922	937	948	957	963	969
	0.01	4052	4999	5403	5625	5764	5859	5928	5981	6022	6056
	0.005	16211	19999	21615	22500	23056	23437	23715	23925	24091	24224
	0.001	405284	499999	540379	562500	576405	585937	592873	598144	602284	605621
2	0.25	2.57	3	3.15	3.23	3.28	3.31	3.34	3.35	3.37	3.38
	0.1	8.53	9	9.16	9.24	9.29	9.33	9.35	9.37	9.38	9.39
	0.05	18.5	19	19.2	19.2	19.3	19.3	19.4	19.4	19.4	19.4
	0.025	38.5	39	39.2	39.2	39.3	39.3	39.4	39.4	39.4	39.4
	0.01	98.5	99	99.2	99.2	99.3	99.3	99.4	99.4	99.4	99.4
	0.005	199	199	199	199	199	199	199	199	199	199
	0.001	999	999	999	999	999	999	999	999	999	999
3	0.25	2.02	2.28	2.36	2.39	2.41	2.42	2.43	2.44	2.44	2.44
	0.1	5.54	5.46	5.39	5.34	5.31	5.28	5.27	5.25	5.24	5.23
	0.05	10.1	9.55	9.28	9.12	9.01	8.94	8.89	8.85	8.81	8.79
	0.025	17.4	16	15.4	15.1	14.9	14.7	14.6	14.5	14.5	14.4
	0.01	34.1	30.8	29.5	28.7	28.2	27.9	27.7	27.5	27.3	27.2
	0.005	55.6	49.8	47.5	46.2	45.4	44.8	44.4	44.1	43.9	43.7
	0.001	167	149	141	137	135	133	132	131	130	129
4	0.25	1.81	2	2.05	2.06	2.07	2.08	2.08	2.08	2.08	2.08
	0.1	4.54	4.32	4.19	4.11	4.05	4.01	3.98	3.95	3.94	3.92
	0.05	7.71	6.94	6.59	6.39	6.26	6.16	6.09	6.04	6	5.96
	0.025	12.2	10.6	9.98	9.6	9.36	9.2	9.07	8.98	8.9	8.84
	0.01	21.2	18	16.7	16	15.5	15.2	15	14.8	14.7	14.5
	0.005	31.3	26.3	24.3	23.2	22.5	22	21.6	21.4	21.1	21
	0.001	74.1	61.2	56.2	53.4	51.7	50.5	49.7	49	48.5	48.1
5	0.25	1.69	1.85	1.88	1.89	1.89	1.89	1.89	1.89	1.89	1.89
	0.1	4.06	3.78	3.62	3.52	3.45	3.4	3.37	3.34	3.32	3.3
	0.05	6.61	5.79	5.41	5.19	5.05	4.95	4.88	4.82	4.77	4.74
	0.025	10	8.43	7.76	7.39	7.15	6.98	6.85	6.76	6.68	6.62
	0.01	16.3	13.3	12.1	11.4	11	10.7	10.5	10.3	10.2	10.1
	0.005	22.8	18.3	16.5	15.6	14.9	14.5	14.2	14	13.8	13.6
	0.001	47.2	37.1	33.2	31.1	29.8	28.8	28.2	27.6	27.2	26.9
6	0.25	1.62	1.76	1.78	1.79	1.79	1.78	1.78	1.78	1.77	1.77
	0.1	3.78	3.46	3.29	3.18	3.11	3.05	3.01	2.98	2.96	2.94
	0.05	5.99	5.14	4.76	4.53	4.39	4.28	4.21	4.15	4.1	4.06
	0.025	8.81	7.26	6.6	6.23	5.99	5.82	5.7	5.6	5.52	5.46
	0.01	13.7	10.9	9.78	9.15	8.75	8.47	8.26	8.1	7.98	7.87
	0.005	18.6	14.5	12.9	12	11.5	11.1	10.8	10.6	10.4	10.3
	0.001	35.5	27	23.7	21.9	20.8	20	19.5	19	18.7	18.4
8	0.25	1.54	1.66	1.67	1.66	1.66	1.65	1.64	1.64	1.63	1.63
	0.1	3.46	3.11	2.92	2.81	2.73	2.67	2.62	2.59	2.56	2.54
	0.05	5.32	4.46	4.07	3.84	3.69	3.58	3.5	3.44	3.39	3.35
	0.025	7.57	6.06	5.42	5.05	4.82	4.65	4.53	4.43	4.36	4.3
	0.01	11.3	8.65	7.59	7.01	6.63	6.37	6.18	6.03	5.91	5.81
	0.005	14.7	11	9.6	8.81	8.3	7.95	7.69	7.5	7.34	7.21
	0.001	25.4	18.5	15.8	14.4	13.5	12.9	12.4	12	11.8	11.5
10	0.25	1.49	1.6	1.6	1.59	1.59	1.58	1.57	1.56	1.56	1.55
	0.1	3.29	2.92	2.73	2.61	2.52	2.46	2.41	2.38	2.35	2.32
	0.05	4.96	4.1	3.71	3.48	3.33	3.22	3.14	3.07	3.02	2.98
	0.025	6.94	5.46	4.83	4.47	4.24	4.07	3.95	3.85	3.78	3.72
	0.01	10	7.56	6.55	5.99	5.64	5.39	5.2	5.06	4.94	4.85
	0.005	12.8	9.43	8.08	7.34	6.87	6.54	6.3	6.12	5.97	5.85
	0.001	21	14.9	12.6	11.3	10.5	9.93	9.52	9.2	8.96	8.75

Table A7, continued. F-distribution

ν_2, denom. d.f.	α	\multicolumn{10}{c}{ν_1, numerator degrees of freedom}
		15	20	25	30	40	50	100	200	500	∞
1	0.25	9.49	9.58	9.63	9.67	9.71	9.74	9.8	9.82	9.84	9.85
	0.1	61.2	61.7	62.1	62.3	62.5	62.7	63	63.2	63.3	63.3
	0.05	246	248	249	250	251	252	253	254	254	254
	0.025	985	993	998	1001	1006	1008	1013	1016	1017	1018
	0.01	6157	6209	6240	6261	6287	6303	6334	6350	6360	6366
	0.005	24630	24836	24960	25044	25148	25211	25337	25401	25439	25464
	0.001	615764	620908	624017	626099	628712	630285	633444	635030	635983	636619
2	0.25	3.41	3.43	3.44	3.44	3.45	3.46	3.47	3.47	3.47	3.48
	0.1	9.42	9.44	9.45	9.46	9.47	9.47	9.48	9.49	9.49	9.49
	0.05	19.4	19.4	19.5	19.5	19.5	19.5	19.5	19.5	19.5	19.5
	0.025	39.4	39.4	39.5	39.5	39.5	39.5	39.5	39.5	39.5	39.5
	0.01	99.4	99.4	99.5	99.5	99.5	99.5	99.5	99.5	99.5	99.5
	0.005	199	199	199	199	199	199	199	199	199	199
	0.001	999	999	999	999	999	999	999	999	999	999
3	0.25	2.46	2.46	2.46	2.47	2.47	2.47	2.47	2.47	2.47	2.47
	0.1	5.2	5.18	5.17	5.17	5.16	5.15	5.14	5.14	5.14	5.13
	0.05	8.7	8.66	8.63	8.62	8.59	8.58	8.55	8.54	8.53	8.53
	0.025	14.3	14.2	14.1	14.1	14	14	14	13.9	13.9	13.9
	0.01	26.9	26.7	26.6	26.5	26.4	26.4	26.2	26.2	26.1	26.1
	0.005	43.1	42.8	42.6	42.5	42.3	42.2	42	41.9	41.9	41.8
	0.001	127	126	126	125	125	125	124	124	124	123
4	0.25	2.08	2.08	2.08	2.08	2.08	2.08	2.08	2.08	2.08	2.08
	0.1	3.87	3.84	3.83	3.82	3.8	3.8	3.78	3.77	3.76	3.76
	0.05	5.86	5.8	5.77	5.75	5.72	5.7	5.66	5.65	5.64	5.63
	0.025	8.66	8.56	8.5	8.46	8.41	8.38	8.32	8.29	8.27	8.26
	0.01	14.2	14	13.9	13.8	13.7	13.7	13.6	13.5	13.5	13.5
	0.005	20.4	20.2	20	19.9	19.8	19.7	19.5	19.4	19.4	19.3
	0.001	46.8	46.1	45.7	45.4	45.1	44.9	44.5	44.3	44.1	44.1
5	0.25	1.89	1.88	1.88	1.88	1.88	1.88	1.87	1.87	1.87	1.87
	0.1	3.24	3.21	3.19	3.17	3.16	3.15	3.13	3.12	3.11	3.1
	0.05	4.62	4.56	4.52	4.5	4.46	4.44	4.41	4.39	4.37	4.36
	0.025	6.43	6.33	6.27	6.23	6.18	6.14	6.08	6.05	6.03	6.02
	0.01	9.72	9.55	9.45	9.38	9.29	9.24	9.13	9.08	9.04	9.02
	0.005	13.1	12.9	12.8	12.7	12.5	12.5	12.3	12.2	12.2	12.1
	0.001	25.9	25.4	25.1	24.9	24.6	24.4	24.1	24	23.9	23.8
6	0.25	1.76	1.76	1.75	1.75	1.75	1.75	1.74	1.74	1.74	1.74
	0.1	2.87	2.84	2.81	2.8	2.78	2.77	2.75	2.73	2.73	2.72
	0.05	3.94	3.87	3.83	3.81	3.77	3.75	3.71	3.69	3.68	3.67
	0.025	5.27	5.17	5.11	5.07	5.01	4.98	4.92	4.88	4.86	4.85
	0.01	7.56	7.4	7.3	7.23	7.14	7.09	6.99	6.93	6.9	6.88
	0.005	9.81	9.59	9.45	9.36	9.24	9.17	9.03	8.95	8.91	8.88
	0.001	17.6	17.1	16.9	16.7	16.4	16.3	16	15.9	15.8	15.7
8	0.25	1.62	1.61	1.6	1.6	1.59	1.59	1.58	1.58	1.58	1.58
	0.1	2.46	2.42	2.4	2.38	2.36	2.35	2.32	2.31	2.3	2.29
	0.05	3.22	3.15	3.11	3.08	3.04	3.02	2.97	2.95	2.94	2.93
	0.025	4.1	4	3.94	3.89	3.84	3.81	3.74	3.7	3.68	3.67
	0.01	5.52	5.36	5.26	5.2	5.12	5.07	4.96	4.91	4.88	4.86
	0.005	6.81	6.61	6.48	6.4	6.29	6.22	6.09	6.02	5.98	5.95
	0.001	10.8	10.5	10.3	10.1	9.92	9.8	9.57	9.45	9.38	9.33
10	0.25	1.53	1.52	1.52	1.51	1.51	1.5	1.49	1.49	1.49	1.48
	0.1	2.24	2.2	2.17	2.16	2.13	2.12	2.09	2.07	2.06	2.06
	0.05	2.85	2.77	2.73	2.7	2.66	2.64	2.59	2.56	2.55	2.54
	0.025	3.52	3.42	3.35	3.31	3.26	3.22	3.15	3.12	3.09	3.08
	0.01	4.56	4.41	4.31	4.25	4.17	4.12	4.01	3.96	3.93	3.91
	0.005	5.47	5.27	5.15	5.07	4.97	4.9	4.77	4.71	4.67	4.64
	0.001	8.13	7.8	7.6	7.47	7.3	7.19	6.98	6.87	6.81	6.76

Table A7, continued. F-distribution

ν_2, denom. d.f.	α	\multicolumn{10}{c}{ν_1, numerator degrees of freedom}									
		1	2	3	4	5	6	7	8	9	10
15	0.25	1.43	1.52	1.52	1.51	1.49	1.48	1.47	1.46	1.46	1.45
	0.1	3.07	2.7	2.49	2.36	2.27	2.21	2.16	2.12	2.09	2.06
	0.05	4.54	3.68	3.29	3.06	2.9	2.79	2.71	2.64	2.59	2.54
	0.025	6.2	4.77	4.15	3.8	3.58	3.41	3.29	3.2	3.12	3.06
	0.01	8.68	6.36	5.42	4.89	4.56	4.32	4.14	4	3.89	3.8
	0.005	10.8	7.7	6.48	5.8	5.37	5.07	4.85	4.67	4.54	4.42
	0.001	16.6	11.3	9.34	8.25	7.57	7.09	6.74	6.47	6.26	6.08
20	0.25	1.4	1.49	1.48	1.47	1.45	1.44	1.43	1.42	1.41	1.4
	0.1	2.97	2.59	2.38	2.25	2.16	2.09	2.04	2	1.96	1.94
	0.05	4.35	3.49	3.1	2.87	2.71	2.6	2.51	2.45	2.39	2.35
	0.025	5.87	4.46	3.86	3.51	3.29	3.13	3.01	2.91	2.84	2.77
	0.01	8.1	5.85	4.94	4.43	4.1	3.87	3.7	3.56	3.46	3.37
	0.005	9.94	6.99	5.82	5.17	4.76	4.47	4.26	4.09	3.96	3.85
	0.001	14.8	9.95	8.1	7.1	6.46	6.02	5.69	5.44	5.24	5.08
25	0.25	1.39	1.47	1.46	1.44	1.42	1.41	1.4	1.39	1.38	1.37
	0.1	2.92	2.53	2.32	2.18	2.09	2.02	1.97	1.93	1.89	1.87
	0.05	4.24	3.39	2.99	2.76	2.6	2.49	2.4	2.34	2.28	2.24
	0.025	5.69	4.29	3.69	3.35	3.13	2.97	2.85	2.75	2.68	2.61
	0.01	7.77	5.57	4.68	4.18	3.85	3.63	3.46	3.32	3.22	3.13
	0.005	9.48	6.6	5.46	4.84	4.43	4.15	3.94	3.78	3.64	3.54
	0.001	13.9	9.22	7.45	6.49	5.89	5.46	5.15	4.91	4.71	4.56
30	0.25	1.38	1.45	1.44	1.42	1.41	1.39	1.38	1.37	1.36	1.35
	0.1	2.88	2.49	2.28	2.14	2.05	1.98	1.93	1.88	1.85	1.82
	0.05	4.17	3.32	2.92	2.69	2.53	2.42	2.33	2.27	2.21	2.16
	0.025	5.57	4.18	3.59	3.25	3.03	2.87	2.75	2.65	2.57	2.51
	0.01	7.56	5.39	4.51	4.02	3.7	3.47	3.3	3.17	3.07	2.98
	0.005	9.18	6.35	5.24	4.62	4.23	3.95	3.74	3.58	3.45	3.34
	0.001	13.3	8.77	7.05	6.12	5.53	5.12	4.82	4.58	4.39	4.24
40	0.25	1.36	1.44	1.42	1.4	1.39	1.37	1.36	1.35	1.34	1.33
	0.1	2.84	2.44	2.23	2.09	2	1.93	1.87	1.83	1.79	1.76
	0.05	4.08	3.23	2.84	2.61	2.45	2.34	2.25	2.18	2.12	2.08
	0.025	5.42	4.05	3.46	3.13	2.9	2.74	2.62	2.53	2.45	2.39
	0.01	7.31	5.18	4.31	3.83	3.51	3.29	3.12	2.99	2.89	2.8
	0.005	8.83	6.07	4.98	4.37	3.99	3.71	3.51	3.35	3.22	3.12
	0.001	12.6	8.25	6.59	5.7	5.13	4.73	4.44	4.21	4.02	3.87
50	0.25	1.35	1.43	1.41	1.39	1.37	1.36	1.34	1.33	1.32	1.31
	0.1	2.81	2.41	2.2	2.06	1.97	1.9	1.84	1.8	1.76	1.73
	0.05	4.03	3.18	2.79	2.56	2.4	2.29	2.2	2.13	2.07	2.03
	0.025	5.34	3.97	3.39	3.05	2.83	2.67	2.55	2.46	2.38	2.32
	0.01	7.17	5.06	4.2	3.72	3.41	3.19	3.02	2.89	2.78	2.7
	0.005	8.63	5.9	4.83	4.23	3.85	3.58	3.38	3.22	3.09	2.99
	0.001	12.2	7.96	6.34	5.46	4.9	4.51	4.22	4	3.82	3.67
100	0.25	1.34	1.41	1.39	1.37	1.35	1.33	1.32	1.3	1.29	1.28
	0.1	2.76	2.36	2.14	2	1.91	1.83	1.78	1.73	1.69	1.66
	0.05	3.94	3.09	2.7	2.46	2.31	2.19	2.1	2.03	1.97	1.93
	0.025	5.18	3.83	3.25	2.92	2.7	2.54	2.42	2.32	2.24	2.18
	0.01	6.9	4.82	3.98	3.51	3.21	2.99	2.82	2.69	2.59	2.5
	0.005	8.24	5.59	4.54	3.96	3.59	3.33	3.13	2.97	2.85	2.74
	0.001	11.5	7.41	5.86	5.02	4.48	4.11	3.83	3.61	3.44	3.3
200	0.25	1.33	1.4	1.38	1.36	1.34	1.32	1.3	1.29	1.28	1.27
	0.1	2.73	2.33	2.11	1.97	1.88	1.8	1.75	1.7	1.66	1.63
	0.05	3.89	3.04	2.65	2.42	2.26	2.14	2.06	1.98	1.93	1.88
	0.025	5.1	3.76	3.18	2.85	2.63	2.47	2.35	2.26	2.18	2.11
	0.01	6.76	4.71	3.88	3.41	3.11	2.89	2.73	2.6	2.5	2.41
	0.005	8.06	5.44	4.41	3.84	3.47	3.21	3.01	2.86	2.73	2.63
	0.001	11.2	7.15	5.63	4.81	4.29	3.92	3.65	3.43	3.26	3.12
∞	0.25	1.32	1.39	1.37	1.35	1.33	1.31	1.29	1.28	1.27	1.25
	0.1	2.71	2.3	2.08	1.94	1.85	1.77	1.72	1.67	1.63	1.6
	0.05	3.84	3	2.6	2.37	2.21	2.1	2.01	1.94	1.88	1.83
	0.025	5.02	3.69	3.12	2.79	2.57	2.41	2.29	2.19	2.11	2.05
	0.01	6.63	4.61	3.78	3.32	3.02	2.8	2.64	2.51	2.41	2.32
	0.005	7.88	5.3	4.28	3.72	3.35	3.09	2.9	2.74	2.62	2.52
	0.001	10.8	6.91	5.42	4.62	4.1	3.74	3.47	3.27	3.1	2.96

Table A7, continued. F-distribution

ν_2, denom. d.f.	α	15	20	25	30	40	50	100	200	500	∞
15	0.25	1.43	1.41	1.4	1.4	1.39	1.38	1.37	1.37	1.36	1.36
	0.1	1.97	1.92	1.89	1.87	1.85	1.83	1.79	1.77	1.76	1.76
	0.05	2.4	2.33	2.28	2.25	2.2	2.18	2.12	2.1	2.08	2.07
	0.025	2.86	2.76	2.69	2.64	2.59	2.55	2.47	2.44	2.41	2.4
	0.01	3.52	3.37	3.28	3.21	3.13	3.08	2.98	2.92	2.89	2.87
	0.005	4.07	3.88	3.77	3.69	3.58	3.52	3.39	3.33	3.29	3.26
	0.001	5.54	5.25	5.07	4.95	4.8	4.7	4.51	4.41	4.35	4.31
20	0.25	1.37	1.36	1.35	1.34	1.33	1.32	1.31	1.3	1.3	1.29
	0.1	1.84	1.79	1.76	1.74	1.71	1.69	1.65	1.63	1.62	1.61
	0.05	2.2	2.12	2.07	2.04	1.99	1.97	1.91	1.88	1.86	1.84
	0.025	2.57	2.46	2.4	2.35	2.29	2.25	2.17	2.13	2.1	2.09
	0.01	3.09	2.94	2.84	2.78	2.69	2.64	2.54	2.48	2.44	2.42
	0.005	3.5	3.32	3.2	3.12	3.02	2.96	2.83	2.76	2.72	2.69
	0.001	4.56	4.29	4.12	4	3.86	3.77	3.58	3.48	3.42	3.38
25	0.25	1.34	1.33	1.31	1.31	1.29	1.29	1.27	1.26	1.26	1.25
	0.1	1.77	1.72	1.68	1.66	1.63	1.61	1.56	1.54	1.53	1.52
	0.05	2.09	2.01	1.96	1.92	1.87	1.84	1.78	1.75	1.73	1.71
	0.025	2.41	2.3	2.23	2.18	2.12	2.08	2	1.95	1.92	1.91
	0.01	2.85	2.7	2.6	2.54	2.45	2.4	2.29	2.23	2.19	2.17
	0.005	3.2	3.01	2.9	2.82	2.72	2.65	2.52	2.45	2.41	2.38
	0.001	4.06	3.79	3.63	3.52	3.37	3.28	3.09	2.99	2.93	2.89
30	0.25	1.32	1.3	1.29	1.28	1.27	1.26	1.25	1.24	1.23	1.23
	0.1	1.72	1.67	1.63	1.61	1.57	1.55	1.51	1.48	1.47	1.46
	0.05	2.01	1.93	1.88	1.84	1.79	1.76	1.7	1.66	1.64	1.62
	0.025	2.31	2.2	2.12	2.07	2.01	1.97	1.88	1.84	1.81	1.79
	0.01	2.7	2.55	2.45	2.39	2.3	2.25	2.13	2.07	2.03	2.01
	0.005	3.01	2.82	2.71	2.63	2.52	2.46	2.32	2.25	2.21	2.18
	0.001	3.75	3.49	3.33	3.22	3.07	2.98	2.79	2.69	2.63	2.59
40	0.25	1.3	1.28	1.26	1.25	1.24	1.23	1.21	1.2	1.19	1.19
	0.1	1.66	1.61	1.57	1.54	1.51	1.48	1.43	1.41	1.39	1.38
	0.05	1.92	1.84	1.78	1.74	1.69	1.66	1.59	1.55	1.53	1.51
	0.025	2.18	2.07	1.99	1.94	1.88	1.83	1.74	1.69	1.66	1.64
	0.01	2.52	2.37	2.27	2.2	2.11	2.06	1.94	1.87	1.83	1.8
	0.005	2.78	2.6	2.48	2.4	2.3	2.23	2.09	2.01	1.96	1.93
	0.001	3.4	3.14	2.98	2.87	2.73	2.64	2.44	2.34	2.28	2.23
50	0.25	1.28	1.26	1.25	1.23	1.22	1.21	1.19	1.18	1.17	1.16
	0.1	1.63	1.57	1.53	1.5	1.46	1.44	1.39	1.36	1.34	1.33
	0.05	1.87	1.78	1.73	1.69	1.63	1.6	1.52	1.48	1.46	1.44
	0.025	2.11	1.99	1.92	1.87	1.8	1.75	1.66	1.6	1.57	1.55
	0.01	2.42	2.27	2.17	2.1	2.01	1.95	1.82	1.76	1.71	1.68
	0.005	2.65	2.47	2.35	2.27	2.16	2.1	1.95	1.87	1.82	1.79
	0.001	3.2	2.95	2.79	2.68	2.53	2.44	2.25	2.14	2.07	2.03
100	0.25	1.25	1.23	1.21	1.2	1.18	1.17	1.14	1.13	1.12	1.11
	0.1	1.56	1.49	1.45	1.42	1.38	1.35	1.29	1.26	1.23	1.21
	0.05	1.77	1.68	1.62	1.57	1.52	1.48	1.39	1.34	1.31	1.28
	0.025	1.97	1.85	1.77	1.71	1.64	1.59	1.48	1.42	1.38	1.35
	0.01	2.22	2.07	1.97	1.89	1.8	1.74	1.6	1.52	1.47	1.43
	0.005	2.41	2.23	2.11	2.02	1.91	1.84	1.68	1.59	1.53	1.49
	0.001	2.84	2.59	2.43	2.32	2.17	2.08	1.87	1.75	1.67	1.62
200	0.25	1.23	1.21	1.19	1.18	1.16	1.15	1.12	1.1	1.09	1.07
	0.1	1.52	1.46	1.41	1.38	1.34	1.31	1.24	1.2	1.17	1.14
	0.05	1.72	1.62	1.56	1.52	1.46	1.41	1.32	1.26	1.22	1.19
	0.025	1.9	1.78	1.7	1.64	1.56	1.51	1.39	1.32	1.27	1.23
	0.01	2.13	1.97	1.87	1.79	1.69	1.63	1.48	1.39	1.33	1.28
	0.005	2.3	2.11	1.99	1.91	1.79	1.71	1.54	1.44	1.37	1.31
	0.001	2.67	2.42	2.26	2.15	2	1.9	1.68	1.55	1.46	1.39
∞	0.25	1.22	1.19	1.17	1.16	1.14	1.13	1.09	1.07	1.04	
	0.1	1.49	1.42	1.38	1.34	1.3	1.26	1.18	1.13	1.08	
	0.05	1.67	1.57	1.51	1.46	1.39	1.35	1.24	1.17	1.11	Undefined
	0.025	1.83	1.71	1.63	1.57	1.48	1.43	1.3	1.21	1.13	
	0.01	2.04	1.88	1.77	1.7	1.59	1.52	1.36	1.25	1.15	
	0.005	2.19	2	1.88	1.79	1.67	1.59	1.4	1.28	1.17	
	0.001	2.51	2.27	2.1	1.99	1.84	1.73	1.49	1.34	1.21	

Table A8. Critical Values for the Wilcoxon Signed Rank Test

For the left-tail test, the table gives the largest integer w such that $P\{W \leq w \mid H_0\} \leq \alpha$.
For the right-tail test, the table gives the smallest integer w such that $P\{W \geq w \mid H_0\} \leq \alpha$.
A missing table entry means that such an integer does not exist among possible values of W.

n	α, left-tail probability for the left-tail test							α, right-tail probability for the right-tail test						
	0.001	0.005	0.010	0.025	0.050	0.100	0.200	0.200	0.100	0.050	0.025	0.010	0.005	0.001
1	—	—	—	—	—	—	—	—	—	—	—	—	—	—
2	—	—	—	—	—	—	—	—	—	—	—	—	—	—
3	—	—	—	—	—	—	0	6	—	—	—	—	—	—
4	—	—	—	—	—	0	2	8	10	—	—	—	—	—
5	—	—	—	—	0	2	3	12	13	15	—	—	—	—
6	—	—	—	0	2	3	5	16	18	19	21	—	—	—
7	—	—	0	2	3	5	8	20	23	25	26	28	—	—
8	—	0	1	3	5	8	11	25	28	31	33	35	36	—
9	—	1	3	5	8	10	14	31	35	37	40	42	44	—
10	0	3	5	8	10	14	18	37	41	45	47	50	52	55
11	1	5	7	10	13	17	22	44	49	53	56	59	61	65
12	2	7	9	13	17	21	27	51	57	61	65	69	71	76
13	4	9	12	17	21	26	32	59	65	70	74	79	82	87
14	6	12	15	21	25	31	38	67	74	80	84	90	93	99
15	8	15	19	25	30	36	44	76	84	90	95	101	105	112
16	11	19	23	29	35	42	50	86	94	101	107	113	117	125
17	14	23	27	34	41	48	57	96	105	112	119	126	130	139
18	18	27	32	40	47	55	65	106	116	124	131	139	144	153
19	21	32	37	46	53	62	73	117	128	137	144	153	158	169
20	26	37	43	52	60	69	81	129	141	150	158	167	173	184
21	30	42	49	58	67	77	90	141	154	164	173	182	189	201
22	35	48	55	65	75	86	99	154	167	178	188	198	205	218
23	40	54	62	73	83	94	109	167	182	193	203	214	222	236
24	45	61	69	81	91	104	119	181	196	209	219	231	239	255
25	51	68	76	89	100	113	130	195	212	225	236	249	257	274
26	58	75	84	98	110	124	141	210	227	241	253	267	276	293
27	64	83	92	107	119	134	153	225	244	259	271	286	295	314
28	71	91	101	116	130	145	165	241	261	276	290	305	315	335
29	79	100	110	126	140	157	177	258	278	295	309	325	335	356
30	86	109	120	137	151	169	190	275	296	314	328	345	356	379

Table A9. Critical Values for the Mann–Whitney–Wilcoxon Rank-Sum Test

For the left-tail test, the table gives the largest integer u such that $P\{U \le u \mid H_0\} \le \alpha$.
For the right-tail test, the table gives the smallest integer u such that $P\{U \ge u \mid H_0\} \le \alpha$.
A missing table entry means that such an integer does not exist among possible values of U.

n_1	n_2	α, left-tail probability for the left-tail test $H_A : X$ is stochastically smaller than Y							α, right-tail probability for the right-tail test $H_A : X$ is stochastically larger than Y						
		0.001	0.005	0.010	0.025	0.050	0.100	0.200	0.200	0.100	0.050	0.025	0.010	0.005	0.001
3	2	—	—	—	—	—	6	7	11	12	—	—	—	—	—
3	3	—	—	—	—	6	7	8	13	14	15	—	—	—	—
3	4	—	—	—	—	6	7	9	15	17	18	—	—	—	—
3	5	—	—	—	6	7	8	10	17	19	20	21	—	—	—
3	6	—	—	—	7	8	9	11	19	21	22	23	—	—	—
3	7	—	—	6	7	8	10	12	21	23	25	26	27	—	—
3	8	—	—	6	8	9	11	13	23	25	27	28	30	—	—
3	9	—	6	7	8	10	11	14	25	28	29	31	32	33	—
3	10	—	6	7	9	10	12	15	27	30	32	33	35	36	—
3	11	—	6	7	9	11	13	16	29	32	34	36	38	39	—
3	12	—	7	8	10	11	14	17	31	34	37	38	40	41	—
4	2	—	—	—	—	—	10	11	17	18	—	—	—	—	—
4	3	—	—	—	—	10	11	13	19	21	22	—	—	—	—
4	4	—	—	—	10	11	13	14	22	23	25	26	—	—	—
4	5	—	—	10	11	12	14	15	25	26	28	29	30	—	—
4	6	—	10	11	12	13	15	17	27	29	31	32	33	34	—
4	7	—	10	11	13	14	16	18	30	32	34	35	37	38	—
4	8	—	11	12	14	15	17	20	32	35	37	38	40	41	—
4	9	—	11	13	14	16	19	21	35	37	40	42	43	45	—
4	10	10	12	13	15	17	20	23	37	40	43	45	47	48	50
4	11	10	12	14	16	18	21	24	40	43	46	48	50	52	54
4	12	10	13	15	17	19	22	26	42	46	49	51	53	55	58
5	2	—	—	—	—	15	16	17	23	24	25	—	—	—	—
5	3	—	—	—	15	16	17	19	26	28	29	30	—	—	—
5	4	—	—	15	16	17	19	20	30	31	33	34	35	—	—
5	5	—	15	16	17	19	20	22	33	35	36	38	39	40	—
5	6	—	16	17	18	20	22	24	36	38	40	42	43	44	—
5	7	—	16	18	20	21	23	26	39	42	44	45	47	49	—
5	8	15	17	19	21	23	25	28	42	45	47	49	51	53	55
5	9	16	18	20	22	24	27	30	45	48	51	53	55	57	59
5	10	16	19	21	23	26	28	32	48	52	54	57	59	61	64
5	11	17	20	22	24	27	30	34	51	55	58	61	63	65	68
5	12	17	21	23	26	28	32	36	54	58	62	64	67	69	73
6	2	—	—	—	—	21	22	23	31	32	33	—	—	—	—
6	3	—	—	—	22	23	24	26	34	36	37	38	—	—	—
6	4	—	21	22	23	24	26	28	38	40	42	43	44	45	—
6	5	—	22	23	24	26	28	30	42	44	46	48	49	50	—
6	6	—	23	24	26	28	30	33	45	48	50	52	54	55	—
6	7	21	24	25	27	29	32	35	49	52	55	57	59	60	63
6	8	22	25	27	29	31	34	37	53	56	59	61	63	65	68
6	9	23	26	28	31	33	36	40	56	60	63	65	68	70	73
6	10	24	27	29	32	35	38	42	60	64	67	70	73	75	78
6	11	25	28	30	34	37	40	44	64	68	71	74	78	80	83
6	12	25	30	32	35	38	42	47	67	72	76	79	82	84	89
7	2	—	—	—	—	28	29	31	39	41	42	—	—	—	—
7	3	—	—	28	29	30	32	34	43	45	47	48	49	—	—
7	4	—	28	29	31	32	34	36	48	50	52	53	55	56	—
7	5	—	29	31	33	34	36	39	52	55	57	58	60	62	—
7	6	28	31	32	34	36	39	42	56	59	62	64	66	67	70
7	7	29	32	34	36	39	41	45	60	64	66	69	71	73	76
7	8	30	34	35	38	41	44	48	64	68	71	74	77	78	82
7	9	31	35	37	40	43	46	50	69	73	76	79	82	84	88
7	10	33	37	39	42	45	49	53	73	77	81	84	87	89	93
7	11	34	38	40	44	47	51	56	77	82	86	89	93	95	99
7	12	35	40	42	46	49	54	59	81	86	91	94	98	100	105

Table A9, continued.

Critical Values for the Mann–Whitney–Wilcoxon Rank-Sum Test

For the left-tail test, the table gives the largest integer u such that $P\{U \le u \mid H_0\} \le \alpha$.
For the right-tail test, the table gives the smallest integer u such that $P\{U \ge u \mid H_0\} \le \alpha$.
A missing table entry means that such an integer does not exist among possible values of U.

n_1	n_2														

n_1	n_2	α, left-tail probability for the left-tail test H_A: X is stochastically smaller than Y							α, right-tail probability for the right-tail test H_A: X is stochastically larger than Y						
		0.001	0.005	0.010	0.025	0.050	0.100	0.200	0.200	0.100	0.050	0.025	0.010	0.005	0.001
8	2	—	—	—	36	37	38	40	48	50	51	52	—	—	—
8	3	—	—	36	38	39	41	43	53	55	57	58	60	—	—
8	4	—	37	38	40	41	43	46	58	61	63	64	66	67	—
8	5	36	38	40	42	44	46	49	63	66	68	70	72	74	76
8	6	37	40	42	44	46	49	52	68	71	74	76	78	80	83
8	7	38	42	43	46	49	52	56	72	76	79	82	85	86	90
8	8	40	43	45	49	51	55	59	77	81	85	87	91	93	96
8	9	41	45	47	51	54	58	62	82	86	90	93	97	99	103
8	10	42	47	49	53	56	60	65	87	92	96	99	103	105	110
8	11	44	49	51	55	59	63	69	91	97	101	105	109	111	116
8	12	45	51	53	58	62	66	72	96	102	106	110	115	117	123
9	2	—	—	—	45	46	47	49	59	61	62	63	—	—	—
9	3	—	45	46	47	49	50	53	64	67	68	70	71	72	—
9	4	—	46	48	49	51	54	56	70	72	75	77	78	80	—
9	5	46	48	50	52	54	57	60	75	78	81	83	85	87	89
9	6	47	50	52	55	57	60	64	80	84	87	89	92	94	97
9	7	48	52	54	57	60	63	67	86	90	93	96	99	101	105
9	8	50	54	56	60	63	67	71	91	95	99	102	106	108	112
9	9	52	56	59	62	66	70	75	96	101	105	109	112	115	119
9	10	53	58	61	65	69	73	78	102	107	111	115	119	122	127
9	11	55	61	63	68	72	76	82	107	113	117	121	126	128	134
9	12	57	63	66	71	75	80	86	112	118	123	127	132	135	141
10	2	—	—	—	55	56	58	60	70	72	74	75	—	—	—
10	3	—	55	56	58	59	61	64	76	79	81	82	84	85	—
10	4	55	57	58	60	62	65	68	82	85	88	90	92	93	95
10	5	56	59	61	63	66	68	72	88	92	94	97	99	101	104
10	6	58	61	63	66	69	72	76	94	98	101	104	107	109	112
10	7	60	64	66	69	72	76	80	100	104	108	111	114	116	120
10	8	61	66	68	72	75	79	84	106	111	115	118	122	124	129
10	9	63	68	71	75	79	83	88	112	117	121	125	129	132	137
10	10	65	71	74	78	82	87	93	117	123	128	132	136	139	145
10	11	67	73	77	81	86	91	97	123	129	134	139	143	147	153
10	12	69	76	79	84	89	94	101	129	136	141	146	151	154	161
11	2	—	—	—	66	67	69	71	83	85	87	88	—	—	—
11	3	—	66	67	69	71	73	76	89	92	94	96	98	99	—
11	4	66	68	70	72	74	77	80	96	99	102	104	106	108	110
11	5	68	71	73	75	78	81	85	102	106	109	112	114	116	119
11	6	70	73	75	79	82	85	89	109	113	116	119	123	125	128
11	7	72	76	78	82	85	89	94	115	120	124	127	131	133	137
11	8	74	79	81	85	89	93	99	121	127	131	135	139	141	146
11	9	76	82	84	89	93	97	103	128	134	138	142	147	149	155
11	10	78	84	88	92	97	102	108	134	140	145	150	154	158	164
11	11	81	87	91	96	100	106	112	141	147	153	157	162	166	172
11	12	83	90	94	99	104	110	117	147	154	160	165	170	174	181
12	2	—	—	—	79	80	82	84	96	98	100	101	—	—	—
12	3	—	79	80	82	83	86	89	103	106	109	110	112	113	—
12	4	78	81	83	85	87	90	94	110	114	117	119	121	123	126
12	5	80	84	86	89	91	95	99	117	121	125	127	130	132	136
12	6	82	87	89	92	95	99	104	124	129	133	136	139	141	146
12	7	85	90	92	96	99	104	109	131	136	141	144	148	150	155
12	8	87	93	95	100	104	108	114	138	144	148	152	157	159	165
12	9	90	96	99	104	108	113	119	145	151	156	160	165	168	174
12	10	92	99	102	107	112	117	124	152	159	164	169	174	177	184
12	11	95	102	106	111	116	122	129	159	166	172	177	182	186	193
12	12	98	105	109	115	120	127	134	166	173	180	185	191	195	202

A.4 Calculus review

This section is a very brief summary of Calculus skills required for reading this book.

A.4.1 Inverse function

Function g is the **inverse function** of function f if

$$g(f(x)) = x \quad \text{and} \quad f(g(y)) = y$$

for all x and y where $f(x)$ and $g(y)$ exist.

$$\underline{\text{NOTATION}} \; \| \text{ Inverse function} \quad g = f^{-1} \, \|$$

(Don't confuse the inverse $f^{-1}(x)$ with $1/f(x)$. These are different functions!)

To find the inverse function, solve the equation

$$f(x) = y.$$

The solution $g(y)$ is the inverse of f.

For example, to find the inverse of $f(x) = 3 + 1/x$, we solve the equation

$$3 + 1/x = y \quad \Rightarrow \quad 1/x = y - 3 \quad \Rightarrow \quad x = \frac{1}{y-3}.$$

The inverse function of f is $g(y) = 1/(y-3)$.

A.4.2 Limits and continuity

A function $f(x)$ has a **limit** L at a point x_0 if $f(x)$ approaches L when x approaches x_0. To say it more rigorously, for any ε there exists such δ that $f(x)$ is ε-close to L when x is δ-close to x_0. That is,

$$\text{if } |x - x_0| < \delta \text{ then } |f(x) - L| < \varepsilon.$$

A function $f(x)$ has a **limit** L at $+\infty$ if $f(x)$ approaches L when x goes to $+\infty$. Rigorously, for any ε there exists such N that $f(x)$ is ε-close to L when x gets beyond N, i.e.,

$$\text{if } x > N \text{ then } |f(x) - L| < \varepsilon.$$

Similarly, $f(x)$ has a **limit** L at $-\infty$ if for any ε there exists such N that $f(x)$ is ε-close to L when x gets below $(-N)$, i.e.,

$$\text{if } x < -N \text{ then } |f(x) - L| < \varepsilon.$$

$$\underline{\text{NOTATION}} \; \left\| \begin{array}{ll} \lim_{x \to x_0} f(x) = L, & \text{or} \quad f(x) \to L \text{ as } x \to x_0 \\ \lim_{x \to \infty} f(x) = L, & \text{or} \quad f(x) \to L \text{ as } x \to \infty \\ \lim_{x \to -\infty} f(x) = L, & \text{or} \quad f(x) \to L \text{ as } x \to -\infty \end{array} \right\|$$

Function f is **continuous** at a point x_0 if

$$\lim_{x \to x_0} f(x) = f(x_0).$$

Function f is **continuous** if it is continuous at every point.

A.4.3 Sequences and series

A **sequence** is a function of a positive integer argument,

$$f(n) \text{ where } n = 1, 2, 3, \ldots.$$

Sequence $f(n)$ **converges** to L if

$$\lim_{n \to \infty} f(n) = L$$

and **diverges** to infinity if for any M there exists N such that

$$f(n) > M \text{ when } n > N.$$

A **series** is a sequence of partial sums,

$$f(n) = \sum_{k=1}^{n} a_k = a_1 + a_2 + \ldots + a_n.$$

Geometric series is defined by

$$a_n = Cr^n,$$

where r is called the **ratio** of the series. In general,

$$\sum_{k=m}^{n} Cr^n = f(n) - f(m-1) = C \frac{r^m - r^{n+1}}{1 - r}.$$

For $m = 0$, we get

$$\sum_{k=0}^{n} Cr^n = C \frac{1 - r^{n+1}}{1 - r}.$$

A geometric series converges if and only if $|r| < 1$. In this case,

$$\lim_{n \to \infty} \sum_{k=m}^{\infty} Cr^n = \frac{Cr^m}{1 - r} \quad \text{and} \quad \sum_{k=0}^{\infty} Cr^n = \frac{C}{1 - r}.$$

A geometric series diverges to ∞ if $r \geq 1$.

A.4.4 Derivatives, minimum, and maximum

Derivative of a function f at a point x is the limit

$$f'(x) = \lim_{y \to x} \frac{f(y) - f(x)}{y - x}$$

provided that this limit exists. Taking derivative is called **differentiation**. A function that has derivatives is called **differentiable**.

$$\underline{\text{NOTATION}} \; \left\| \; f'(x) \quad \text{or} \quad \frac{d}{dx} f(x) \; \right\|$$

Differentiating a function of several variables, we take **partial derivatives** denoted as

$$\frac{\partial}{\partial x_1} f(x_1, x_2, \ldots), \quad \frac{\partial}{\partial x_2} f(x_1, x_2, \ldots), \quad \text{etc..}$$

The most important derivatives are:

Derivatives

$$
\begin{aligned}
(x^m)' &= m\,x^{m-1} \\
(e^x)' &= e^x \\
(\ln x)' &= 1/x \\
C' &= 0 \\
(f + g)'(x) &= f'(x) + g'(x) \\
(Cf)'(x) &= Cf'(x) \\
(f(x)g(x))' &= f'(x)g(x) + f(x)g'(x) \\
\left(\frac{f(x)}{g(x)}\right)' &= \frac{f'(x)g(x) - f(x)g'(x)}{g^2(x)}
\end{aligned}
$$

for any functions f and g and any number C

To differentiate a composite function

$$f(x) = g(h(x)),$$

we use a chain rule,

Chain rule

$$\frac{d}{dx} g(h(x)) = g'(h(x))h'(x)$$

For example,

$$\frac{d}{dx} \ln^3(x) = 3\ln^2(x)\frac{1}{x}.$$

Geometrically, derivative $f'(x)$ equals the slope of a tangent line at point x; see Figure A.1.

Computing maxima and minima

At the points where a differentiable function reaches its minimum or maximum, the tangent line is always flat; see points x_2 and x_3 on Figure A.1. The slope of a horizontal line is 0,

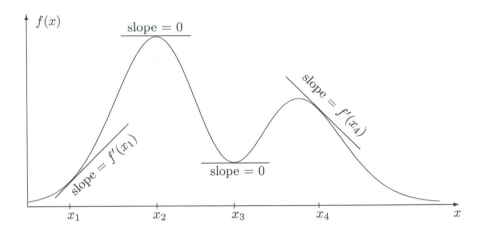

FIGURE A.1: *Derivative is the slope of a tangent line.*

and thus,

$$f'(x) = 0$$

at these points.

To find out where a function is maximized or minimized, we consider

– solutions of the equation $f'(x) = 0$,

– points x where $f'(x)$ fails to exist,

– endpoints.

The highest and the lowest values of the function can only be attained at these points.

A.4.5   Integrals

Integration is an action opposite to differentiation.

A function $F(x)$ is an **antiderivative (indefinite integral)** of a function $f(x)$ if

$$F'(x) = f(x).$$

Indefinite integrals are defined up to a constant C because when we take derivatives, $C' = 0$.

$$\underline{\text{NOTATION}} \quad \Big\| \quad F(x) = \int f(x)\,dx \quad \Big\|$$

An **integral (definite integral)** of a function $f(x)$ from point a to point b is the difference of antiderivatives,

$$\int_a^b f(x)\,dx = F(b) - F(a).$$

Improper integrals

$$\int_a^\infty f(x)\,dx, \quad \int_{-\infty}^b f(x)\,dx, \quad \int_{-\infty}^\infty f(x)\,dx$$

are defined as limits. For example,

$$\int_a^\infty f(x)\,dx = \lim_{b\to\infty} \int_a^b f(x)\,dx.$$

The most important integrals are:

Indefinite integrals

$$\int x^m\,dx \quad = \quad \frac{x^{m+1}}{m+1} \quad \text{for } m \neq -1$$

$$\int x^{-1}\,dx \quad = \quad \ln(x)$$

$$\int e^x\,dx \quad = \quad e^x$$

$$\int (f(x) + g(x))\,dx \quad = \quad \int f(x)\,dx + \int g(x)\,dx$$

$$\int Cf(x)\,dx \quad = \quad C\int f(x)\,dx$$

for any functions f and g and any number C

For example, to evaluate a definite integral $\int_0^2 x^3\,dx$, we find an antiderivative $F(x) = x^4/4$ and compute $F(2) - F(0) = 4 - 0 = 4$. A standard way to write this solution is

$$\int_0^2 x^3\,dx = \frac{x^4}{4}\Big|_{x=0}^{x=2} = \frac{2^4}{4} - \frac{0^4}{4} = 4.$$

Two important integration skills are *integration by substitution* and *integration by parts*.

Integration by substitution

An integral often simplifies when we can denote a part of the function as a new variable (y). The limits of integration a and b are then recomputed in terms of y, and dx is replaced by

$$dx = \frac{dy}{dy/dx} \quad \text{or} \quad dx = \frac{dx}{dy}\,dy,$$

whichever is easier to find. Notice that dx/dy is the derivative of the *inverse function* $x(y)$.

Integration by substitution

$$\int f(x)\,dx \quad = \quad \int f(x(y))\frac{dx}{dy}\,dy$$

For example,

$$\int_{-1}^2 e^{3x}\,dx = \int_{-3}^6 e^y \left(\frac{1}{3}\right)\,dy = \frac{1}{3}e^y\Big|_{y=-3}^{y=6} = \frac{e^6 - e^{-3}}{3} = 134.5.$$

Here we substituted $y = 3x$, recomputed the limits of integration, found the inverse function $x = y/3$ and its derivative $dx/dy = 1/3$.

In the next example, we substitute $y = x^2$. Derivative of this substitution is $dy/dx = 2x$:

$$\int_0^2 x\, e^{x^2}\, dx = \int_0^4 x\, e^y \frac{dy}{2x} = \frac{1}{2} \int_0^4 e^y\, dy = \frac{1}{2} e^y \bigg|_{y=0}^{y=4} = \frac{e^4 - 1}{2} = 26.8.$$

Integration by parts

This technique often helps to integrate a *product* of two functions. One of the parts is integrated, the other is differentiated.

> **Integration by parts**
>
> $$\int f'(x)g(x)dx = f(x)g(x) - \int f(x)g'(x)dx$$

Applying this method is reasonable only when function (fg') is simpler for integration than the initial function $(f'g)$.

In the following example, we let $f'(x) = e^x$ be one part and $g(x) = x$ be the other. Then $f'(x)$ is integrated, and its antiderivative is $f(x) = e^x$. The other part $g(x)$ is differentiated, and $g'(x) = x' = 1$. The integral simplifies, and we can evaluate it,

$$\int x\, e^x dx = x\, e^x - \int (1)(e^x)dx = x\, e^x - e^x.$$

Computing areas

Area under the graph of a positive function $f(x)$ and above the interval $[a, b]$ equals the integral,

$$(\text{area from } a \text{ to } b) = \int_a^b f(x)dx.$$

Here a and b may be finite or infinite; see Figure A.2.

Gamma function and factorial

Gamma function is defined as

$$\Gamma(t) = \int_0^\infty x^{t-1} e^{-x} dx \quad \text{for} \quad t > 0.$$

Taking this integral by parts, we obtain two important properties of a Gamma function,

$$\begin{aligned} \Gamma(t+1) &= t\Gamma(t) \quad \text{for any } t > 0, \\ \Gamma(t+1) &= t! = 1 \cdot 2 \cdot \ldots \cdot t \quad \text{for integer } t. \end{aligned}$$

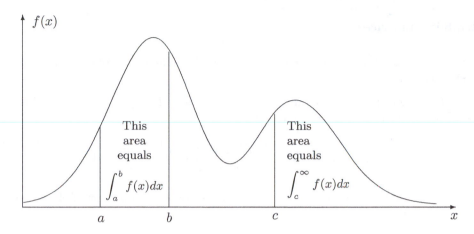

FIGURE A.2: *Integrals are areas under the graph of $f(x)$.*

A.5 Matrices and linear systems

A **matrix** is a rectangular chart with numbers written in rows and columns,

$$A = \begin{pmatrix} A_{11} & A_{12} & \cdots & A_{1c} \\ A_{21} & A_{22} & \cdots & A_{2c} \\ \cdots & \cdots & \cdots & \cdots \\ A_{r1} & A_{r2} & \cdots & A_{rc} \end{pmatrix}$$

where r is the number of rows and c is the number of columns. Every element of matrix A is denoted by A_{ij}, where $i \in [1, r]$ is the row number and $j \in [1, c]$ is the column number. It is referred to as an "$r \times c$ matrix."

Multiplying a row by a column

A row can only be multiplied by a column of the same length. The product of a row A and a column B is a number computed as

$$(A_1, \ldots, A_n) \begin{pmatrix} B_1 \\ \vdots \\ B_n \end{pmatrix} = \sum_{i=1}^{n} A_i B_i.$$

Example A.2 (MEASUREMENT CONVERSION). To convert, say, 3 hours 25 minutes 45 seconds into seconds, one may use a formula

$$(3 \ 25 \ 45) \begin{pmatrix} 3600 \\ 60 \\ 1 \end{pmatrix} = 12345 \ (\text{sec}).$$

$\Diamond$

Multiplying matrices

Matrix A may be multiplied by matrix B only if the number of columns in A equals the number of rows in B.

If A is a $k \times m$ matrix and B is an $m \times n$ matrix, then their product $AB = C$ is a $k \times n$ matrix. Each element of C is computed as

$$C_{ij} = \sum_{s=1}^{m} A_{is} B_{sj} = \left(\begin{array}{c} i^{\text{th}} \text{ row} \\ \text{of } A \end{array} \right) \left(\begin{array}{c} j^{\text{th}} \text{ column} \\ \text{of } B \end{array} \right).$$

Each element of AB is obtained as a product of the corresponding row of A and column of B.

Example A.3. The following product of two matrices is computed as

$$\left(\begin{array}{cc} 2 & 6 \\ 1 & 3 \end{array} \right) \left(\begin{array}{cc} 9 & -3 \\ -3 & 1 \end{array} \right) = \left(\begin{array}{cc} (2)(9) + (6)(-3), & (2)(-3) + (6)(1) \\ (1)(9) + (3)(-3), & (1)(-3) + (3)(1) \end{array} \right) = \left(\begin{array}{cc} 0 & 0 \\ 0 & 0 \end{array} \right).$$

$\Diamond$

In the last example, the result was a zero matrix "accidentally." This is not always the case. However, we can notice that matrices do not always obey the usual rules of arithmetics. In particular, a product of two non-zero matrices may equal a 0 matrix.

Also, in this regard, matrices *do not commute*, that is, $AB \neq BA$, in general.

Transposition

Transposition is reflecting the entire matrix about its main diagonal.

$$\underline{\text{NOTATION}} \; \| \, A^T \; = \; \text{transposed matrix } A \, \|$$

Rows become columns, and columns become rows. That is,

$$A_{ij}^T = A_{ji}.$$

For example,

$$\left(\begin{array}{ccc} 1 & 2 & 3 \\ 7 & 8 & 9 \end{array} \right)^T = \left(\begin{array}{cc} 1 & 7 \\ 2 & 8 \\ 3 & 9 \end{array} \right).$$

The transposed product of matrices is

$$\boxed{(AB)^T = B^T A^T}$$

Solving systems of equations

In Chapters 6 and 7, we often solve systems of n linear equations with n unknowns and find a *steady-state distribution*. There are several ways of doing so.

One method to solve such a system is by **variable elimination**. Express one variable in terms of the others from one equation, then substitute it into the unused equations. You will get a system of $(n-1)$ equations with $(n-1)$ unknowns. Proceeding in the same way, we reduce the number of unknowns until we end up with 1 equation and 1 unknown. We find this unknown, then go back and find all the other unknowns.

Example A.4 (Linear system). Solve the system

$$\begin{cases} 2x & + & 2y & + & 5z & = & 12 \\ & & 3y & - & z & = & 0 \\ 4x & - & 7y & - & z & = & 2 \end{cases}$$

We don't have to start solving from the first equation. Start with the one that seems simple. From the second equation, we see that

$$z = 3y.$$

Substituting $(3y)$ for z in the other equations, we obtain

$$\begin{cases} 2x & + & 17y & = & 12 \\ 4x & - & 10y & = & 2 \end{cases}$$

We are down by one equation and one unknown. Next, express x from the first equation,

$$x = \frac{12 - 17y}{2} = 6 - 8.5y$$

and substitute into the last equation,

$$4(6 - 8.5y) - 10y = 2.$$

Simplifying, we get $44y = 22$, hence $y = 0.5$. Now, go back and recover the other variables,

$$x = 6 - 8.5y = 6 - (8.5)(0.5) = 1.75; \quad z = 3y = 1.5.$$

The answer is $x = 1.75$, $y = 0.5$, $z = 1.5$.

We can check the answer by substituting the result into the initial system,

$$\begin{cases} 2(1.75) & + & 2(0.5) & + & 5(1.5) & = & 12 \\ & & 3(0.5) & - & 1.5 & = & 0 \\ 4(1.75) & - & 7(0.5) & - & 1.5 & = & 2 \end{cases}$$

$\Diamond$

We can also eliminate variables by multiplying entire equations by suitable coefficients, adding and subtracting them. Here is an illustration of that.

Example A.5 (ANOTHER METHOD). Here is a shorter solution of Example A.4. Double the first equation,

$$4x + 4y + 10z = 24,$$

and subtract the third equation from it,

$$11y + 11z = 22, \quad \text{or} \quad y + z = 2.$$

This way, we eliminated x. Then, adding $(y + z = 2)$ and $(3y - z = 0)$, we get $4y = 2$, and again, $y = 0.5$. Other variables, x and z, can now be obtained from y, as in Example A.4. $\Diamond$

The system of equations in this example can be written in a matrix form as

$$\begin{pmatrix} x & y & z \end{pmatrix} \begin{pmatrix} 2 & 0 & 4 \\ 2 & 3 & -7 \\ 5 & -1 & -1 \end{pmatrix} = \begin{pmatrix} 12 & 0 & 2 \end{pmatrix},$$

or, equivalently,

$$\begin{pmatrix} 2 & 2 & 5 \\ 0 & 3 & -1 \\ 4 & -7 & -1 \end{pmatrix} \begin{pmatrix} x \\ y \\ z \end{pmatrix} = \begin{pmatrix} 12 & 0 & 2 \end{pmatrix}.$$

Inverse matrix

Matrix B is the **inverse matrix** of A if

$$AB = BA = I = \begin{pmatrix} 1 & 0 & 0 & \cdots & 0 \\ 0 & 1 & 0 & \cdots & 0 \\ 0 & 0 & 1 & \cdots & 0 \\ \cdots & \cdots & \cdots & \cdots & \cdots \\ 0 & 0 & 0 & \cdots & 1 \end{pmatrix},$$

where I is the *identity matrix*. It has 1s on the diagonal and 0s elsewhere. Matrices A and B have to have the same number of rows and columns.

$$\underline{\text{NOTATION}} \ \| \ A^{-1} \ = \ \text{inverse of matrix } A \ \|$$

Inverse of a product can be computed as

$$\boxed{(AB)^{-1} = B^{-1}A^{-1}}$$

To find the inverse matrix A^{-1} by hand, write matrices A and I next to each other. Multiplying rows of A by constant coefficients, adding and interchanging them, convert matrix A to the identity matrix I. The same operations convert matrix I to A^{-1},

$$(\ A \ | \ I \ ) \longrightarrow (\ I \ | \ A^{-1} \ ).$$

Example A.6. Linear system in Example A.4 is given by matrix

$$A = \begin{pmatrix} 2 & 2 & 5 \\ 0 & 3 & -1 \\ 4 & -7 & -1 \end{pmatrix}.$$

Repeating the row operations from this example, we can find the inverse matrix A^{-1},

$$\left(\begin{array}{ccc|ccc} 2 & 2 & 5 & 1 & 0 & 0 \\ 0 & 3 & -1 & 0 & 1 & 0 \\ 4 & -7 & -1 & 0 & 0 & 1 \end{array}\right) \longrightarrow \left(\begin{array}{ccc|ccc} 4 & 4 & 10 & 2 & 0 & 0 \\ 0 & 3 & -1 & 0 & 1 & 0 \\ 4 & -7 & -1 & 0 & 0 & 1 \end{array}\right)$$

$$\longrightarrow \left(\begin{array}{ccc|ccc} 0 & 11 & 11 & 2 & 0 & -1 \\ 0 & 3 & -1 & 0 & 1 & 0 \\ 4 & -7 & -1 & 0 & 0 & 1 \end{array}\right) \longrightarrow \left(\begin{array}{ccc|ccc} 0 & 1 & 1 & 2/11 & 0 & -1/11 \\ 0 & 3 & -1 & 0 & 1 & 0 \\ 4 & -7 & -1 & 0 & 0 & 1 \end{array}\right)$$

$$\longrightarrow \left(\begin{array}{ccc|ccc} 0 & 4 & 0 & 2/11 & 1 & -1/11 \\ 0 & 3 & -1 & 0 & 1 & 0 \\ 4 & -7 & -1 & 0 & 0 & 1 \end{array}\right) \longrightarrow \left(\begin{array}{ccc|ccc} 0 & 1 & 0 & 1/22 & 1/4 & -1/44 \\ 0 & 3 & -1 & 0 & 1 & 0 \\ 4 & -10 & 0 & 0 & -1 & 1 \end{array}\right)$$

$$\longrightarrow \left(\begin{array}{ccc|ccc} 0 & 1 & 0 & 1/22 & 1/4 & -1/44 \\ 0 & 0 & -1 & -3/22 & 1/4 & 3/44 \\ 4 & 0 & 0 & 10/22 & 3/2 & 34/44 \end{array}\right) \longrightarrow \left(\begin{array}{ccc|ccc} 1 & 0 & 0 & 5/44 & 3/8 & 17/88 \\ 0 & 1 & 0 & 1/22 & 1/4 & -1/44 \\ 0 & 0 & 1 & 3/22 & -1/4 & -3/44 \end{array}\right)$$

The inverse matrix is found,

$$A^{-1} = \begin{pmatrix} 5/44 & 3/8 & 17/88 \\ 1/22 & 1/4 & -1/44 \\ 3/22 & -1/4 & -3/44 \end{pmatrix}.$$

You can verify the result by multiplying $A^{-1}A$ or AA^{-1}. ◇

For a 2×2 matrix, the formula for the inverse is

$$\begin{pmatrix} a & b \\ c & d \end{pmatrix}^{-1} = \frac{1}{ad - bc} \begin{pmatrix} d & -b \\ -c & a \end{pmatrix}.$$

Matrix operations in R

```
x <- c(1,8,0,3,3,-3,5,0,-1)  # Define a 1 × 9 vector and converting it...
A <- matrix(x,3,3)           # ... into a 3 × 3 matrix column by column
t(A)                         # Transposed matrix
B <- solve(A)                # Inverse matrix
A + B                        # Addition
A %*% B                      # Matrix multiplication
C <- A * B                   # Multiplying element by element, C_ij = A_ij B_ij
diag(n)                      # n × n identity matrix
matrix(rep(0,m*n),m,n)       # m × n matrix of 0s
cbind(A,B)                   # Joining matrices side by side (as columns)
rbind(A,B)                   # Joining matrices below each other (as rows)
A[2:3,]                      # Sub-matrix: rows 2-3 and all columns of A

# Calculation of a power of a matrix is a part of R package 'expm'
install.packages("expm")
library(expm)
A %^% 3                      # This calculates A^3 = A · A · A
solve(A %^% 3)               # The result is A^-3 = (A^3)^-1
```

Matrix operations in MATLAB

```
A = [1 3 5; 8 3 0; 0 -3 -1];   % Entering a matrix
B = [ 3 9 8
      0 0 2                     % Another way to define a matrix
      9 2 1 ];
A+B                            % Addition
A*B                            % Matrix multiplication
C=A.*B                         % Multiplying element by element, $C_{ij} = A_{ij}B_{ij}$
A^n                            % Power of a matrix, $A^n = \underbrace{A \cdot \ldots \cdot A}_{n \text{ times}}$

A'                             % transposed matrix
A(2:3,:)                       # Sub-matrix: rows 2-3 and all columns of A
inv(A) ⎫
A^(-1) ⎭                       % inverse matrix
eye(n)                         % $n \times n$ identity matrix
zeros(m,n)                     % $m \times n$ matrix of 0s
[ A B ]                        % Joining matrices side by side (as columns)
[ A
  B ]                          % Joining matrices below each other (as rows)
rand(m,n)                      % matrix of Uniform(0,1) random numbers
randn(m,n)                     % matrix of Normal(0,1) random numbers
```

A.6   Answers to selected exercises

Chapter 2

2.1. 1/15. **2.3.** 0.45. **2.5.** 0.72. **2.7.** 0.66. **2.8.** 0.9508. **2.9.** 0.1792. **2.12.** 0.9744. **2.13.** 0.992. **2.15.** 0.1364. **2.16.** (a) 0.049. (b) 0.510. **2.18.** 0.0847. **2.20.** 0.00534. **2.21.** 0.8854. **2.24.** (a) 5/21 or 0.238. (b) 10/41 or 0.244. **2.25.** 0.2. **2.29.** 0.1694.

Chapter 3

3.1. (a) P(0)=0.42, P(1)=0.46, P(2)=0.12. (b) Figure A.3. **3.2.** $E(Y) = 200$ dollars, $Var(Y) = 110,000$ squared dollars. **3.3.** $E(X) = 0.6$, $Var(X) = 0.24$. **3.4.** $E(X) = 3.5$, $Var(X) = 2.9167$. **3.7.** $E(Y) = 1.6$, $Var(Y) = 1.12$. **3.9.** This probability does not exceed 1/16. **3.10.** 0.28. **3.11.** (a) The joint pmf is in Table A.2. (b) They are dependent. (c) $P_X(1) = 11/36$, $P_X(2) = 9/36$, $P_X(3) = 7/36$, $P_X(4) = 5/36$, $P_X(5) = 3/36$, $P_X(6) = 1/36$. (d) $P\{Y = 5 \mid X = 2\} = 2/9$. **3.12.** (a) Dependent. (b) Dependent. **3.15.** (a) 0.48. (b) Dependent. **3.17.** Third portfolio. **3.18.** (a) E(Profit)=6, Var(Profit)=684. (b) E(Profit)=6, Var(Profit)=387. (c) E(Profit)=6, Var(Profit)=864. The least risky portfolio is (b); the most risky portfolio is (c). **3.20.** (a) 0.0596. (b) 0.9860. **3.21.** 0.2447. **3.22.** 0.0070. **3.23.** (a) 0.0055. (b) 0.00314. **3.24.** (a) 0.0328, (b) 0.4096. **3.26.** (a) 0.3704. (b) 0.0579. **3.28.** (a) 0.945. (b) 0.061. **3.30.** 0.0923. **3.31.** 0.0166. **3.33.** (a) 0.968. (b) 0.018. **3.38.** 0.827.

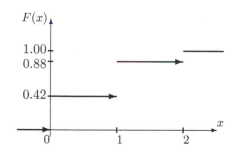

FIGURE A.3 *Cdf for Exercise 3.1.*

	$P(x,y)$	1	2	3	4	5	6
				y			
	1	1/36	1/18	1/18	1/18	1/18	1/18
	2	0	1/36	1/18	1/18	1/18	1/18
x	3	0	0	1/36	1/18	1/18	1/18
	4	0	0	0	1/36	1/18	1/18
	5	0	0	0	0	1/36	1/18
	6	0	0	0	0	0	1/36

TABLE A.2 *The joint pmf for Exercise 3.11.*

Chapter 4

4.1. 3; $1 - 1/x^3$; 0.125. **4.3.** 4/3; 31/48 or 0.6458. **4.4.** (a) 0.2. (b) 0.75. (c) $3\frac{1}{3}$ or 3.3333. **4.7.** 0.875. **4.8.** 0.264. **4.10.** 0.4764. **4.14.** (a) 4 and 0.2. (b) 0.353. **4.15.** (a) 0.062. (b) 0.655. **4.16.** (a) 0.8944. (b) 0.8944. (c) 0.1056. (d) 0.7888. (e) 1.0. (f) 0.0. (g) 0.84. **4.18.** (a) 0.9772. (c) 0.6138. (e) 0.9985. (g) -1.19. **4.20.** (a) 0.1336. (b) 0.1340. **4.22.** (a) 0.0968. (b) 571 coins. **4.24.** 0.2033. **4.30.** 0.1151. **4.31.** 0.567. **4.33.** (a) 0.00 (practically 0). (b) 38. **4.35.** (a) 0.75. (b) 0.4.

Chapter 5

Some answers in this chapter may differ from yours because of the use of Monte Carlo simulations. Answers marked with (*) require a large number of simulations to be accurate. **5.1.** $X = U^{2/3} = 0.01$. **5.3.** $X = 2\sqrt{U} - 1$. **5.5.** $X = (9U - 1)^{1/3} = 1.8371$. **5.7.** (d) $n = 38,416$ is sufficient. (e) 0.1847; 0.00000154*; 0.0108. **5.10.** (a) 73 days*. (b) 0.0012. (c) 2.98 computers. **5.11.** (a) 0.24. (b) 142 trees*. (c) 185*. (d) 0.98. (e) 0.13*. Yes. **5.17.** $p = \frac{1}{(b-a)c}$.

Chapter 6

6.1. (a) $P = \begin{pmatrix} 1/2 & 1/2 & 0 \\ 0 & 1/2 & 1/2 \\ 0 & 1/2 & 1/2 \end{pmatrix}$ (b) Irregular. (c) 0.125. **6.3.** (a) $P = \begin{pmatrix} 0.6 & 0.4 \\ 0.2 & 0.8 \end{pmatrix}$.

(b) 0.28. **6.4.** (a) $P = \begin{pmatrix} 0.6 & 0.4 \\ 0.3 & 0.7 \end{pmatrix}$. (b) 0.52. (c) 4/7 or 0.5714. **6.5.** 0.333. **6.9.** (a)

0.0579. (b) practically 0. **6.11.** (a) 0.05 min or 3 sec. (b) $\mathbf{E}(X) = 120$ jobs; $\text{Std}(X) = \sqrt{108}$ or 10.39 jobs. **6.13.** $\mathbf{E}(T) = 12$ sec; $\text{Var}(T) = 120$ sec^2. **6.15.** (a) 0.75 seconds. (b) $\mathbf{E}(T) = 5$ seconds; $\text{Std}(T) = 4.6$ seconds. **6.18.** 0.0162. **6.19.** (a) 1/16 min or 3.75 sec. (b) There is no indication that the arrival rate is on the increase. **6.20.** 0.8843. **6.22.** (a) 0.735. (b) 7500 dollars; 3354 dollars 10 cents. **6.23.** (a) 0.0028. (b) $\mathbf{E}(W_3) = 3/5$, $\text{Var}(W_3) = 3/25$. *The remaining answers in this chapter may differ slightly because of the use of Monte Carlo simulations.* **6.26.** 5102, 4082, 816. Yes. **6.28.** A sample path is on Figure A.4.

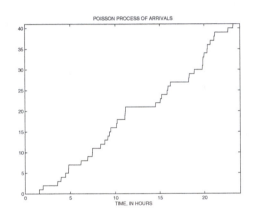

FIGURE A.4: *Trajectory of a Poisson process (Exercise 6.28).*

Chapter 7

7.1.
$$P = \begin{pmatrix} 35/36 & 1/36 & 0 & 0 & \cdots \\ 35/432 & 386/432 & 11/432 & 0 & \cdots \\ 0 & 35/432 & 386/432 & 11/432 & \cdots \\ 0 & 0 & 35/432 & 386/432 & \cdots \\ \cdots & \cdots & \cdots & \cdots & \cdots \end{pmatrix}$$

7.2. $\pi_0 = 0.5070$, $\pi_1 = 0.4225$, $\pi_2 = 0.0704$. **7.3.** 2/75 or 0.02667. **7.5.** $\pi_0 = 49/334 = 0.1467$, $\pi_1 = 105/334 = 0.3144$, $\pi_2 = 180/334 = 0.5389$. **7.8.** (a) 2/27 or 0.074. (b) 6 sec. **7.9.** (a) The job is expected to be printed at 12:03:20. (b) 0.064 or 6.4% of the time. **7.10.** (a) 4 min. (b) 1/3. **7.11.** (a) 4 and 3.2. (b) 0.8. (c) 0.262. **7.12.** (a) 7.5 min. (b) 0.64. (c) 0.6. **7.14.** (a) 3/7 or 0.4286. (b) 0.7 or 70% of time. (c) 30/7 or 4.2857 min. **7.15.** (a) $\pi_0 = 29/89$, $\pi_1 = 60/89$. (b) 0.0026. **7.18.** (a) $P = \begin{pmatrix} 0.8750 & 0.1250 & 0 \\ 0.0875 & 0.8000 & 0.1125 \\ 0.0088 & 0.1588 & 0.8325 \end{pmatrix}$.

(b) 0.3089, 0.4135, 0.2777. (c) 27.77%. (d) 27.77%. (e) $174.38. **7.20.** (a) 0.4286, 0.3429,

FIGURE A.5: *Stem-and-leaf plots and parallel boxplots for Exercise 8.1.*

0.1371, 0.0549, 0.0219, 0.0088, 0.0035, 0.0014, 0.0006, 0.0002, 0.0001. $\mathbf{E}(X) = 0.9517$. (b) 2.3792 min. (c) 0.3792 min, 0.1517 customers. (d) 0.9525 customers; 2.3813 min, 0.3813 min, 0.1525 customers. **7.22**. (a) M/M/∞. (b) 30. (c) 0.046. *The remaining answers in this chapter may differ slightly because of the use of Monte Carlo simulations.* **7.24**. 8.85 sec. **7.26**. (a) 0.085 min. (b) 3.42 min. (c) 0.04 jobs. (d) 2.64 min. (e) 2.15 jobs. (f) 0.91. (g) 0.76. (h) 89.1, 86.2, 74.2, and 51.4 jobs. (i) 392 min, 385 min, 352 min, 266 min. (j) 1.31 jobs. (k) 0.002%.

Chapter 8

8.1. (a) Figure A.5. (b) The five-point summaries are (37, 43, 50, 56, 60) and (21, 35, 39, 46, 53). Boxplots in Figure A.5. **8.2**. (a) 17.95, 9.97, 3.16. (b) 0.447. (c) (11.9, 15.8, 17.55, 19.9, 24.1). The boxplot is on Figure A.6. (d) $\widehat{IQR} = 4.1$. No outliers. (e) No. The histogram is on Figure A.6. **8.4**. 0.993. **8.6**. (a) 13.85 mln, 13.00 mln, 87.60 mln^2. (b) Figure A.6. **8.8**. (a) Left-skewed, symmetric, right-skewed (Figure A.7). (b) Set 1: 14.97, 15.5. Set 2: 20.83, 21.0. Set 3: 41.3, 39.5. Yes.

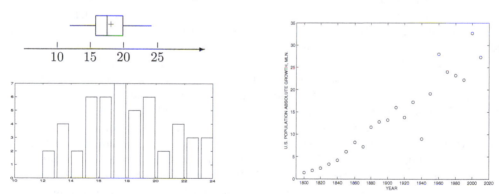

FIGURE A.6: *A boxplot and a histogram for Exercise 8.2. A time plot for Exercise 8.6.*

Chapter 9

9.1. (a) 0.625. (b) 0.625. (c) 0.171. **9.4**. The method of moments estimator is $\widehat{\theta} = 2$. The maximum likelihood estimator is $\widehat{\theta} = 2.1766$. **9.7**. (a) [36.19, 39.21]. (b) Yes, there is sufficient evidence. **9.9**. (a) 50 ± 33.7 or [16.3; 83.7]. (b) At this level of significance, the data does not provide significant evidence against the hypothesis that the average salary of all entry-level computer engineers equals \$80,000. (c) [11.6; 89.4]. **9.10**. (a) [0.073; 0.167].

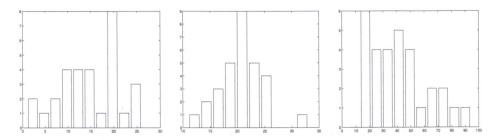

FIGURE A.7: *Histograms of three data sets in Exercise 8.8.*

(b) No. No. **9.11**. There is no significant evidence. P-value is 0.1587. **9.17**. 3.25%, 3.12%, and 4.50%. **9.18**. (a) [4.25, 15.35]. (b) Each test has a P-value between 0.0005 and 0.001. There is a significant reduction. **9.20**. P-value is between 0.02 and 0.05. At any level $\alpha \geq 0.05$, there is significant evidence that $\sigma \neq 5$. **9.22**. (a) There is significant evidence that $\sigma_X^2 \neq \sigma_Y^2$ ($P = 0.0026$), hence the method in Example 9.21 is chosen correctly. (b) [0.11, 0.60].

Chapter 10

The P-values in Exercises 10.1 and 10.4 may vary slightly depending on the choice of bins.
10.1. No, there is significant evidence that the distribution is not Poisson. $P < 0.001$.
10.4. Confirms. $P > 0.2$. **10.6**. No significant difference. $P > 0.2$. **10.7**. (a) Yes, $P \in (0.001, 0.005)$. (b) Yes, $P \in (0.005, 0.01)$. (c) Yes, $P < 0.001$. **10.9**. No, $P \in (0.1, 0.2)$.
10.11. Rejection region is $[0, 5] \cup [15, 20]$. Fifteen or more measurements on the same side of m justify recalibration. **10.13**. The null distribution is Binomial(60,0.75); $P = 0.1492$; there is no significant evidence that the first quartile exceeds 43 nm. **10.16**. Yes, there is significant evidence at the 5% level; $P = 0.01$. **10.18**. Yes, at the 5% level, there is significant evidence that the median service time is less than 5 min 40 sec. Assumption of a symmetric distribution may not be fully satisfied. **10.21**. (a) For the 1st provider, there is significant evidence for all $\alpha > 0.011$; $P = 0.011$. No significant evidence for the 2nd provider; $P = 0.172$. (b) For the 1st provider, there is significant evidence; $P \in (0.001, 0.005)$. For the 2nd provider, there is significant evidence for all $\alpha \geq 0.05$; $P \in (0.025, 0.05)$. Yes, the Wilcoxon test appears more sensitive. (c) No; $P \in (0.01, 0.025)$.
10.23. At any $\alpha \geq 0.05$, there is significant evidence that the median cost of textbooks is rising; $P \in (0.025, 0.05)$. **10.25**. Yes, there is a significant reduction of the median service time; $P = 0.0869$. **10.27**. (a) 352,716 ordered and 285,311,670,611 unordered bootstrap samples. (b) The bootstrap distribution of $\widehat{M^*}$ is given by the pmf, $P(44) = 0.0002$, $P(47) = 0.0070$, $P(52) = 0.0440$, $P(53) = 0.1215$, $P(54) = 0.2059$, $P(58) = 0.2428$, $P(59) = 0.2059$, $P(63) = 0.1215$, $P(68) = 0.0440$, $P(72) = 0.0070$, $P(77) = 0.0002$. (c) 4.1885. (d) [53, 63]. (e) 0.9928. *Answers to Exercises 10.29 and 10.30 may differ slightly due to the generated bootstrap samples.* **10.29**. 1.070, 0.107. **10.30**. [−0.9900 − 0.9564].
10.31. 4.5, 2.25. **10.33**. (b) 0.4, 0.02. (c) 0.22. (d) Reject H_0. **10.34**. (a) 17.661 (thousand of concurrent users). (b) [16, 765.7, 18, 556.5]. Given the observed data and the assumed prior distribution, there is a 90% posterior probability that the average number of concurrent users is between 16,765.7 and 18,556.5. (c) Yes. *The answers to Exercise 10.36 may differ slightly due to different choices available in question (a).* **10.36**. (a) For example, Normal(5.5, 0.255). (b) Normal(5.575, 0.245); the Bayes estimator is 5.575; its posterior risk is 0.060. (c) [5.095, 6.055]. The HPD credible set is much more narrow because besides the

data, it uses a rather informative (low-variance) prior distribution. **10.38**. (a) Pareto(α, σ).
(b) $\dfrac{(n+\alpha)\sigma}{n+\alpha-1}$; $\dfrac{(n+\alpha)\sigma^2}{(n+\alpha-1)^2(n+\alpha-2)}$. **10.40**. (a) 0.9, 0.0082. (b) [0.7743, 1]. (c) Yes.

Chapter 11

11.2. (a) $y = -9.71 + 0.56x$. (b) $SS_{REG} = 149.85$ with 1 d.f., $SS_{ERR} = 57.35$ with 73 d.f., $SS_{TOT} = 207.2$ with 74 d.f.; 72.32% is explained. (c) [0.455, 0.670]. Based on this interval, the slope is significant with a P-value less than 0.01. **11.4**. (a) 7.39. (b) There is significant evidence that the investment increases by more than \$1,800 every year, on the average; $P < 0.005$. (c) [50.40; 67.02]. **11.5**. (a) $b_0 = 13.36$ (if x =year-2000), $b_1 = 2.36$, $b_2 = 4.09$. (b) \$52,804.7. (c) Reduces by \$4,092.6. (d) $SS_{TOT} = 841.6$ with 10 d.f., $SS_{REG} = 808.0$ with 2 d.f., $SS_{ERR} = 33.6$ with 8 d.f.; $R^2 = 0.96$. Significance of the model: $F = 96.14$, significant at the 0.001 level. (e) The new variable explains additional 3.9% of the total variation. It is significant at the 0.025 level, but not at the 0.01 level. **11.10**. (a) Population $= -32.1 + 1.36$(year $- 1800$). (b) $SS_{TOT} = 203,711$ with 22 d.f., $SS_{REG} = 187,277$ with 1 d.f., $SS_{ERR} = 16,434$ with 21 d.f., $R^2 = 0.919$; $MS_{REG} = 187,277$, $MS_{ERR} = 783$, $F = 239$. (c) 260.4 mln., 267.2 mln. This is an obvious underestimation because the U.S. population has already exceeded 300 mln. The linear regression model does not yield accurate predictions because the population does not grow linearly. **11.11**.

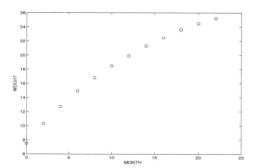

FIGURE A.8: *Time plot for Exercise 11.12.*

(a) $b_0 = 5.87$, $b_1 = 0.0057$, $b_2 = 0.0068$. The estimated regression equation is Population $= 5.87+0.0057$(year-1800)$+0.0068$(year-1800)^2. (b) 320.2 mln. in 2015 and 335.0 mln in 2020. (c) $SS_{TOT} = 203,711$ with 22 d.f., $SS_{REG} = 203,528$ with 2 d.f., $SS_{ERR} = 183$ with 20 d.f.; $MS_{REG} = 101,764$, $MS_{ERR} = 9.14$; $F = 11,135$. $R^2 = 0.999$. The quadratic term explains additional 7.9% of the total variation. (d) Adjusted R-square is 0.9154 for the linear model, 0.9990 for the full model, and 0.9991 for the reduced model with only the quadratic term. This last model is the best, according to the adjusted R-square criterion. (e) Yes, Figure 11.6 shows that the quadratic model fits the U.S. population data rather accurately. **11.12**. (a) The time plot on Figure A.8 shows a non-linear, slightly concave curve. A quadratic model seems appropriate. (b) $y = 7.716 + 1.312x - 0.024x^2$. The negative slope b_2 indicates that the growth is slowing with age. (c) 3.25%. Yes. (d) 25.53 lbs, 28.41 lbs. Quadratic prediction is more reasonable. (e) No. Assumption of independence is violated. **11.14**. (a) $\hat{y} = 0.376 - 0.00145x$, where $x =$ year-1800. $b_0 = 0.376$, $b_1 = -0.00145$. (b) $SS_{TOT} = 0.215$, 21 d.f.; $SS_{REG} = 0.185$, 1 d.f.; $SS_{ERR} = 0.030$, 20 d.f. $MS_{REG} = 0.185$; $MS_{ERR} = 0.0015$; $F = 123.9$; $R^2 = 0.86$. The linear model explains 86% of the total variation. (c) $F = 123.9$, P-value < 0.001. The slope is significant at the 0.1% level. The relative change in the U.S. population varies significantly in time. (d) $[-0.00172, -0.00118]$. (e) $[-0.031, 0.146]$, $[-0.046, 0.132]$. (f) See the histogram on Figure A.9. Its skewed shape does not quite support the assumption of a normal distribution.

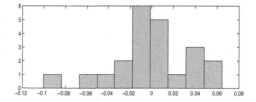

FIGURE A.9: *Histogram of regression residuals for Exercise 11.14.*

Index